Probabilistic Power System Expansion Planning with Renewable Energy Resources and Energy Storage Systems

Probabilistic Power System Expansion Planning with Renewable Energy Resources and Energy Storage Systems

Jaeseok Choi
Gyeongsang National University

Kwang Y. Lee
Baylor University

IEEE PRESS SERIES ON POWER ENGINEERING

IEEE PRESS

WILEY

Published by John Wiley & Sons, Inc., Hoboken, New Jersey.

Published simultaneously in Canada.

Library of Congress Cataloging-in-Publication Data Applied for:

ISBN: 9781119684138

Cover Design: Wiley
Cover Image: © zhengzaishuru/Shutterstock

Set in 9.5/12.5pt STIXTwoText by Straive, Pondicherry, India

10 9 8 7 6 5 4 3 2 1

MIX
Paper from responsible sources
FSC www.fsc.org FSC® C013604

"There is one God, the Father, ever-living, omnipresent, omniscient, almighty, the Maker of heaven and earth, and one Mediator between God and Man, the man Christ Jesus..."

Sir Isaac Newton (1642–1727)

"There is a God-shaped vacuum in the heart of every man, which only God can fill through His Son Jesus Christ."

Plaise Pascal (1623–1662)

"I shall be with Christ, and that is enough."

Michael Faraday (1791–1867)

"Almighty God, who has created man in Thine own image, ... and have dominion over Thy creatures, teach us to study the works of Thy hands, ... and strengthen the reason for Thy service; so to receive Thy blessed Word [Jesus], ... All of which we ask in the name of the same Jesus Christ, our Lord."

James Clerk Maxwell (1831–1879)

"For the invisible things of Him from the creation of the world are clearly seen, being understood by the things that are made, even His eternal power and Godhead, so that they are without excuse."

Romans 1:20

Dedicated to the Christ Jesus "through Him all things are made."

John 1:3

Contents

Author Biographies

Jaeseok Choi, IEEE Senior Member, received BS, MS, and PhD degrees from Korea University, Seoul, Korea, in 1981, 1984, and 1990, respectively. His research interest includes Power System Reliability Evaluation, Power System Planning, Outage Cost Assessment, Nodal Probabilistic Production Cost Simulation, Fuzzy Set Theory Application into Power Systems, Development of Flexible Method for Future Power Systems, Web-based Online Real-time Reliability-Integrated Information System considering Renewables, Energy Generation for Smart Grid, Reliability Analysis of Renewable Energy Resources and Distributed Generation, and Expansion Planning of Power Systems. In particular, his researches are focused on development and extension of model and methodology of reliability and economics evaluation of power systems considering renewable energy generation.

Since 1991, Dr. Choi has been a faculty of Gyeongsang National University, where he is currently a full professor. He has been a postdoctoral fellow at University of Saskatchewan in Canada in 1996. He was also a visiting professor at Cornell University, USA from 2004 to 2006. He has been also an adjunct professor of Illinois Institute of Technology, Chicago, USA from 2007 to 2010. Dr. Choi has over 480 journal and conference papers and 7 technical books. He has been involved in 48 R&D projects.

Dr. Choi is a Fellow of the Korean Institute of Electrical Engineers (KIEE) since 2018. He was Editor-in-Chief of Journal of Electrical Engineering and Technology (JEET) of KIEE for 2012–2014. He was the president of KIEE PES in 2016. He was the president of KIEE in 2020 and has been a Member of the Board of KEPCO-E&C since 2019. He has been organizing the International Symposium on Green Energy Policy Innovation (ISGEPI), an international conference, which was held annually in Korea since 2012.

Internationally, Dr. Choi is a Senior Member of IEEE since 2005. In addition, he is an active member of IEEE Power Engineering Society, participating in Reliability, Risk, and Probability Applications (RRPA) Subcommittee. He is an ITC member of Probabilistic Methods Applied to Power Systems (PMAPS) Conference since 2006. During this time, he has participated in international conferences such as the Program Committee, the Program Chair, the Reviewer, and the Special and Panel Sessions. He also gave keynote speeches as an editor of various conferences and international journals.

Kwang Y. Lee received the BS degree in Electrical Engineering in 1964 from Seoul National University, MS degree in Electrical Engineering in 1968 from North Dakota State University, and PhD degree in Systems Science in 1971 from Michigan State University. He is currently Professor and Chair of the Electrical and Computer Engineering Department at Baylor University, Waco, Texas. Before receiving his current calling, he worked as faculty at Michigan State University, Oregon State University, University of Houston, and Pennsylvania State University from 1972 to 2007.

His research interests include control, operation, and planning of power and energy systems; intelligent control and their applications to power and energy systems; and modeling, simulation, and control of micro-grids with renewable and distributed energy sources. He has published over 700 papers in archived journals and refereed conference proceedings and has authored chapters in several books. He has co-authored *Intelligent Network Integration of Distributed Renewable Generation* (Springer 2017), and has coedited *Modern Heuristic Optimization Techniques: Theory and Applications to Power Systems* (Wiley-IEEE Press 2008) and *Applications of Modern Heuristic Optimization Methods in Power and Energy Systems* (Wiley-IEEE Press 2020).

He was elected as a Fellow of IEEE in January 2001 for his contributions to the development and implementation of intelligent system techniques for power plants and power systems control. Dr. Lee has served as Editor of IEEE Transactions on Energy Conversion, Associate Editor of IEEE Transactions on Neural Networks, Associate Editor of IFAC Journal on Control Engineering Practice, Co-Editor-in-Chief of Journal of Electrical Engineering and Technology, and Editorial Board of Energies. He has served as the Chair of the IFAC Power & Energy Technical Committee (TC 6.3) and IEEE PES Working Group on Modern Heuristic Optimization Techniques.

Preface

This book, *Probabilistic Power System Expansion Planning with Renewable Energy Resources and Energy Storage Systems*, is written for power system expansion planning concept under the new environment. The objective of this book is to review the state-of-the-art technologies in power system expansion planning and present case studies on how these techniques have been applied in complex power system expansion planning problems. Emphasis is given to applications rather than theory, and the organization of the book is on an application basis rather than tools.

The book can serve as a resource book for power system engineers in utilities, government, and industry, who are interested in electric power system generation and transmission expansion planning based on deterministic as well as probabilistic approaches. The book can also be used as a reference for graduate students in advanced standing in power and energy systems. Power systems expansion planning tools in the book are key subjects for continuing education for engineers and managers in many utilities.

The book is organized into two parts: Part I is on Generation Expansion Planning (GEP). In addition to the conventional thermal generating power plants, this part includes expansion planning with renewable energy sources in the viewpoint of Best Generation Mix (BGM) and energy storage systems (ESS). Part II is on Transmission Expansion Planning (TEP). In addition to the conventional AC transmission network expansion, this part also includes high-voltage DC (HVDC) transmission expansion planning.

In Part I, background, cost assessment, and model and optimization methodologies are first introduced for GEP. Specially, effective load models for probabilistic production cost simulation are developed and long-range generation planning problem is formulated as a dynamic optimization problem and the maximum principle is used as a solution technique. Furthermore, this book is discussing a fuzzy set theory application to power system planning and applied to the BGM problem considering environmental constraints. The GEP problem is expanded to include

RES, and Reliability is evaluated considering Energy Storage Systems (ESSs). Finally, the GEP is improved with the Genetic Algorithm and augmented with the Reactive Power Planning (RPP) problem. Part I is organized in 10 chapters and their contents are highlighted in the following:

- Chapter 1 introduces the importance of electricity as energy in the modernization of the society and its expansion plans considering Sustainable Development Goal (SDG). The assumption of worldwide electricity production needed until 2040 is also briefly described with figures. The importance of renewables (solar, hydropower, or wind energy) as per environmental and economic issues concerned is also introduced. The basic and typical Generation Expansion Planning (GEP) criteria are derived in steps.
- GEP is one of the most important decision-making activities in electric utilities. Chapter 2 discusses the history of the methodologies and issues used in GEP and the least-cost generation expansion planning formulated which is to find a set of optimal decision vectors over a planning horizon that minimizes an objective function under several constraints.
- Chapter 3 defines and briefly discusses the types of costs associated with electric power-generating plants and systems in GEP and the methodologies used for solving the GEP problems. Therefore, while the terminology and conventions presented in this chapter are typical, they are not universal and are intended only to illustrate the basic concepts and categories of costs for power plants and electric generating systems.
- Chapter 4 presents a novel analytical approach to long-term GEP by representing random load fluctuations with Gaussian distribution functions, which probabilistically provides expected plant outputs, expected annual energy, expected reliability measures, expected annual cost including salvage cost, and expected marginal values. The optimal long-term GEP problem with a sufficiently accurate analytical cost formula is formulated as an investment optimization problem based on the discrete maximum principle that allows for the incorporation of any kind of constraints that are found necessary. The formulation enables to convert the multi-year high-dimensional optimization problem into a yearly low-dimensional Hamiltonian minimization problem with dynamic constraints. An advantage of the analytical approach is demonstrated when hydro resources are integrated for peak-shaving operation.
- Supply Reserve Rate (SRR) analysis becomes more important along with the huge expansion of power systems. The deterministic SRR and probabilistic Loss of Load Expectation (LOLE) relationship can be categorized by the exact method (Booth–Baleriaux method) and approximation methods (MONA). In Chapter 5, the probabilistic reliability evaluation and production cost simulation program, PRASim, was used to obtain these relationships.

- Chapter 6 proposes a fuzzy dynamic programming-based solution approach for the long-term generation mix with multi-stages (years) considering air pollution constraints on CO_2 emissions under uncertain circumstances and demonstrated on a KEPCO-system. This proposed approach can accommodate pumped storage hydro and nuclear plants with strict load following considering SO*x* and NO*x* emissions.
- In Chapter 7, an alternative approach for the long-term generation mix with multi-criteria considering air pollution constraints is proposed using linear programming (LP), which not only considers SO*x* and NO*x* but also CO_2 emission limitations. This flexible generation mix problem is proposed for the Korea Vision 2030. The approach provides a flexible solution to the generation mix plan.
- Chapter 8 has proposed a fuzzy LP-based approach for the long-term best generation mix with multi-criteria considering renewable energy generators such as wind turbine generators (WTGs) and solar cell generators (SCGs) and air pollution constraints. The proposed method can accommodate sensitivity analysis of capacity factor (CF) and apparent escalation rate (CER) of WTG and SCG including uncertainties. The effectiveness is demonstrated for the best generation mix in the Korea Vision 2030 which also includes other types of generation.
- Chapter 9 presents a new methodology for probabilistic power system reliability evaluation using a Monte Carlo simulation (MCS) when a multi-energy storage system (ESS) is integrated with wind farms in GEP. A large-scale WTG creates significant power fluctuations and affects the stability, frequency control, and reliability of the power system. A case study is demonstrated for the proposed model and methods in a power system in Jeju Island, Korea.
- Chapter 10 introduces an improved genetic algorithm (IGA) for a long-term least-cost generation expansion planning problem. It provides a better solution by incorporating improvements as compared to tunnel-constrained dynamic programming (TCDP) employed in the Wien Automatic System Planning (WASP) Package. A synthetic method of reactive power planning, modified simple genetic method (MSGA) approach, is introduced which uses not only objective function like SGA but also dual variable information.

In Part II, the TEP problem is decomposed into yearly and long-term transmission planning problems, and the Long-Term TEP is formulated as a dynamic optimization problem with the maximum principle as a solution technique. Different models of cost evaluation and methodologies for TEP are described with a special focus on Branch and Bound Method to solve the integer programming problem for the TEP. Probabilistic production cost models considering forced outage and capacity of transmission lines are extended and developed using the extended effective load and fictitious equivalent generator which is modeled as multi-states. Deterministic and probabilistic reliability constraints are presented for TEP. Deterministic reliability constraints are developed using the contingency analysis for security constraints and expanded to include bus marginal rate (BMR). As in GEP, the fuzzy set theory is also applied to TEP for decision making. The optimal deterministic and probabilistic reliability criteria are determined for composite power system expansion planning with an outage cost assessment. Finally, the TEP is considered for the composite power system expansion planning with wind turbine generators and HVDC. Part II is organized into eight chapters and their contents are highlighted in the following:

- Chapter 11 addresses the long-term transmission expansion planning (TEP) problem. The transmission planning algorithm is based on the linear power flow and failure probabilities for the operation cost model and the maximum principle for the long-term investment model. The optimization problem consists of the long-term problem to determine the annual investment and the yearly problem to determine the optimal operation. The high-dimensional long-term optimization problem is formulated as a low-dimensional yearly Hamiltonian minimization problem and the yearly problem is formulated as a linear programming problem.
- The premises in TEP are discussed in Chapter 12, which are demand forecasting and generation expansion schedule. The TEP problem formulation and cost evaluation are discussed where methodologies such as linear programming, dynamic programming, integer programming, a genetic algorithm with mixed-integer programming, and branch and bound algorithm were used to solve the TEP problem.
- Chapter 13 illustrates the concept and formulation of an extended effective load model for probabilistic production cost simulation and reliability evaluation at a load point in a composite power system. The proposed synthesized fictitious equivalent generator model is used for reliability evaluation of multi-area interconnected systems because the forced outage rates of transmission lines of an assisting system have been considered in the model.
- Chapter 14 presents a methodology for choosing the best transmission expansion plan using alternative security criteria based on different specifications of N-α contingencies. The chapter also addresses the TEP problem considering deterministic nodal delivery marginal rate criterion associated with construction cost using probabilistic reliability criterion, annual outage cost rate, and deterministic–probabilistic (D–P) criteria.
- Chapter 15 proposes an approach that is designed to facilitate sensitivity analysis and trend evaluation in view of a macro approach prior to the micro-level checking (stability and dynamics, etc.) in transmission expansion planning. A new

fuzzy branch and bound algorithm, which includes the use of network flow theory and the maximum flow-minimum cut-set theorem, is proposed.

- A methodology for selecting an optimal probabilistic reliability criterion in TEP is introduced in Chapter 16 in the viewpoint of the adequacy of power system energy delivery that uses probabilistic reliability index LOLE. Another deterministic reliability index, delivery marginal rate (DMR)-based methodology, is proposed for deciding the optimal reliability criterion for the composite power system expansion planning considering generation system as well as transmission system.
- Chapter 17 presents a method for grid expansion planning considering the probabilistic reliability of a composite power system with WTG. This utilizes a multi-state operation model obtained by combining the wind speed model and power output model. Useful information is provided according to the location and composite reliability of WTG's capacity by this method.
- Chapter 18 describes the basic model for probabilistic reliability-based grid expansion planning with high voltage direct current (HVDC) transmission lines in a power system integrated with WTG. A new HVDC contact point can be decided at the best reliability level. The chapter introduces two case studies – one is the NORDIC system, and another is Jeju Island power system in Korea.

There are the following books available under power system planning:

1) X. Wang and J. R. McDonald, Modern Power System Planning, McGraw-Hill 1944.
2) Robert L. Sullivan, Power System Planning, McGraw-Hill 1977.
3) Juergen Schlabbach and Karl-Heinz RofalskiJuergen Schlabbach, Power System Engineering: Planning, Design, and Operation of Power Systems and Equipment, Wiley 2008 [structures, substations, equipment design].
4) Fawwaz Elkarmi and Nazih Abu Shikhah Fawwaz Elkarmi, Power System Planning Technologies and Applications, IGI Global 2012 [technical planning at a micro level].
5) Seifi, Hossein, Sepasian and Mohammad Sadegh, Electric Power System Planning, Springer 2011.
6) Ankita, Sharma, Power System Planning, Genius Publications 2015 [finite element methods for structures].

The first two books [1, 2] are classic but outdated, and except book [5], all other books [3, 4, 6] are in micro-level technical aspects such as structure, substation, or equipment design. Book [5] is the only comprehensive book on power system planning, like this book. It covers the basic principles in power system planning, including optimization techniques, economics, generation expansion planning, network expansion planning, and planning under uncertainties. However, book [5] is rather narrow in techniques and scope in applicability.

The resourcefulness of this book is demonstrated through accurate mathematical and probabilistic formulations, abundant optimization techniques, consideration of several different kinds of reliability constraints, modern heuristic approaches in decision-making, consideration of environmental pollution, renewable energy generation, HVDC expansion planning, and Monte Carlo simulation based on the energy storage system (ESS) and reliability evaluation for power system expansion planning.

Acknowledgments

We wish to acknowledge many colleagues and students at Gyeongsang National University, Korea University, Konkuk University, Choongnam National University, Seoul National University, University of Saskatchewan, Cornell University, Illinois Institute of Technology, Waseda University, University of Houston, Pennsylvania State University, and Baylor University, who have contributed to the development of this book. In particular, the first author is indebted to his advisor, Professor Gil-young Song of Korea University, who introduced his dissertation topic on generation expansion and transmission expansion planning with branch and bound method and Professor Roy Billinton of University of Saskatchewan who invited him as a Post-Doctoral Fellow to work on probabilistic power system reliability. He also wishes to express thanks to Professors Robert Thomas, Timothy Mount, and Ray Zimmerman at Cornell University, Professor Mohammad Shahidehpour at IIT, and Professors Ryuichi Yokoyama and Junzo Watada at Waseda University for many stimulating discussions on the topics in this book. He appreciates the members of IEEE PES for their encouragement during the course of development of several topics in this book: J. Endrenyi, M.P. Bhavaraju, Channon Singh, Alex Schneider, Wenyuan Li, Papic Milorad, M. Fotuhi-Firuzabad, Rajesh Karki, and Lina Bertling of the Risk, Reliability and Probability Applications subcommittee, and M.L. Chan, Chair of Power System Planning & Implementation subcommittee. He also expresses thanks to Professors Jonggeun Park, Yongtae Yoon, and Junmin Cha, of Advanced Power System Network Research Center at Seoul National University, and power system reliability members of KEPCO, KPX, KERI, and KEEI in Korea for providing valuable data for simulation studies. The book is not possible without the dedicated work of his former students whom he expresses deep appreciation: Tran Trung Tinh, Vise Rector of CanTho University, Vietnam, Jintaek Lim, Ungjin Oh, Yeonchan Lee, and Nguyen Duy Phung.

The second author is indebted to Professor Young Moon Park of Seoul National University who suggested the innovative approach to the long-term generation expansion planning with Pontryagin's maximum principle and to the load model with Gaussian distribution function while he was visiting the University of Houston. He expresses thanks to his former student at the University of Houston, Professor Luis T.O. Yoon of Southern Illinois University – Edwardsville, who developed the new analytic approach for long-term generation expansion planning based on the maximum principle and Gaussian distribution function. He also expresses his gratitude to Professor Kern Joong Kim of Choongnam National University, who led the development of long-term transmission expansion planning with the maximum principle, and Professor Jongbae Park of Konkuk University, who led the development of the improved genetic algorithm for generation expansion planning. He continues to express his thanks to his former student at University of Houston, Professor Jorge T. Ortiz Alvarez of University of Puerto Rico – Mayaguez, who developed the optimal reactive power planning and former students at Penn State, Professor Maurice K. Mangoli of University of Nairobi, who developed the optimal long-term reactive power planning using decomposition technique, and Fred F. Yang from Tsinghua University, for the development of optimal reactive power planning using evolutionary algorithms. Finally, the second author expresses his thanks to Xiaomin Bai of China EPRI for the development of an optimization method for reactive power planning using a genetic algorithm while he was visiting Penn State.

We like to express thanks to Editor Mary Hatcher of Wiley-IEEE Press for her encouragement to submit the book proposal and Teresa Netzler, the Senior Managing Editor of Wiley, for working patiently with the authors. We also thank graduate students Semin Jeong, and Joochang Park of Power System Lab at GNU for their endless effort in typing and editing the manuscript.

Finally, we thank God for our family Jongin Cheng, Seonghwa, and Yeongsoo for the first author and Sangwol Lee, Edward, and Daniel for the second author; without their love, prayer, and sacrifice, this book would not have been possible.

Jaeseok Choi, Jinju, Korea
Kwang Y Lee, Waco, Texas, USA

"People may plan all kinds of things, but the LORD's will is going to be done."
Proverbs 19:21

Part I

Generation Expansion Planning

1

Introduction

1.1 Electricity Outlook

Electricity is at the heart of modern economies and it is providing a rising share of energy services. Demand for electricity is set to increase further as a result of the rising household incomes, with the electrification of transport and heat, and growing demand for digitally connected devices and air conditioning.

Rising electricity demand was one of the key reasons why global emissions from the power sector reached a record high in 2018, yet the commercial availability of a diverse suite of low emissions generation technologies also puts electricity at the vanguard of efforts to combat pollution. Renewable energy also has a major role to play in providing access to electricity for all.

Figure 1.1 shows the electricity generation by fuel and scenario, 2018–2040 [99].

The Stated Policies Scenario reflects the impact of the existing policy frameworks and today's announced policy intentions. The aim is to hold up a mirror to the plans of today's policy-makers and illustrate their consequences for energy use, emissions, and energy security. The world is not on track to meet the energy-related components of the Sustainable Development Goals (SDGs). The IEA's Sustainable Development Scenario (SDS) outlines a major transformation of the global energy system, showing how the world can change course to deliver on the main energy-related SDGs.

In the Stated Policies Scenario, global electricity demand grows at 2.1% per year to 2040, twice the rate of the primary energy demand. This raises electricity's share in total final energy consumption from 19% in 2018 to 24% in 2040. Electricity demand growth is set to be particularly strong in developing economies.

Government policies, market conditions, and the available technologies collectively set a course for electricity supply to shift toward low-carbon sources, with their share increasing from 36% today to 52% in 2040 in the Stated Policies Scenario.

Probabilistic Power System Expansion Planning with Renewable Energy Resources and Energy Storage Systems, First Edition. Jaeseok Choi and Kwang Y. Lee.

Published 2022 by John Wiley & Sons, Inc.

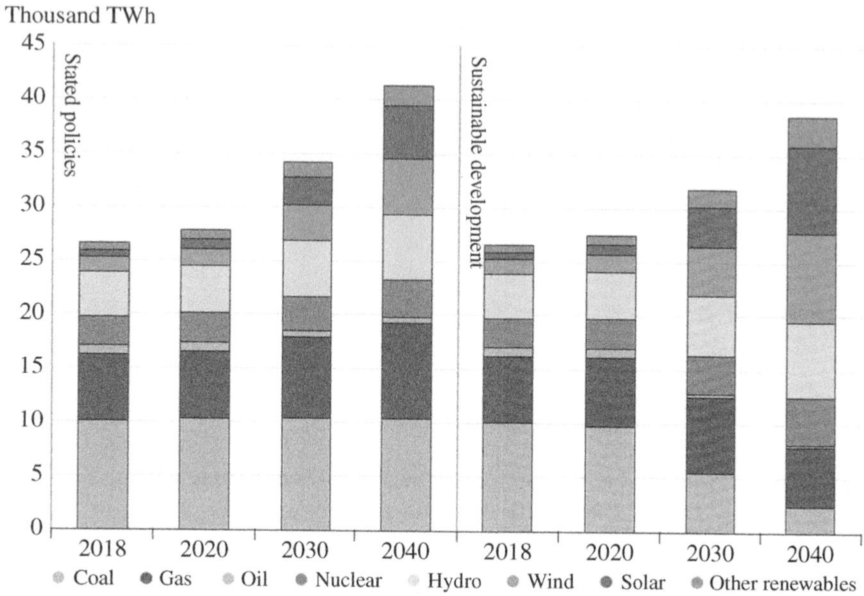

Figure 1.1 Electricity generation by fuel and scenario, 2018–2040. *Source:* Excecutive Summary, International Energy Agency, IEA [99].

In the Sustainable Development Scenario, electricity plays an even larger role, reaching 31% of the final energy consumption. In the Sustainable Development Scenario, electricity is one of the few energy sources that sees growing consumption in 2040 – mainly due to electric vehicles – alongside the direct use of renewables and hydrogen. The share of electricity in final consumption, less than half that of oil today, overtakes oil by 2040. Accelerated efforts on renewables, nuclear power, and carbon capture technologies rapidly decarbonize electricity supply, compensating for the sharp decline of coal-fired power generation and reducing power sector CO_2 emissions by three-quarters by 2040.

Electricity demand follows two distinct regional paths. In advanced economies, future growth linked to increasing digitalization and electrification is largely offset by energy efficiency improvements. Figure 1.2 shows the global electricity demand by region in the Stated Policies Scenario, 2000–2040 [99]. In developing economies, rising incomes, expanding industrial output, and a growing services sector push demand firmly up. Developing economies contribute nearly 90% of global electricity demand growth to 2040 in the Stated Policies Scenario, but demand per person in these economies remains 60% lower than in advanced economies.

In the Stated Policies Scenario, electricity generation from renewables increases rapidly, surpassing coal by 2026. Renewables contribute three-quarters of electricity

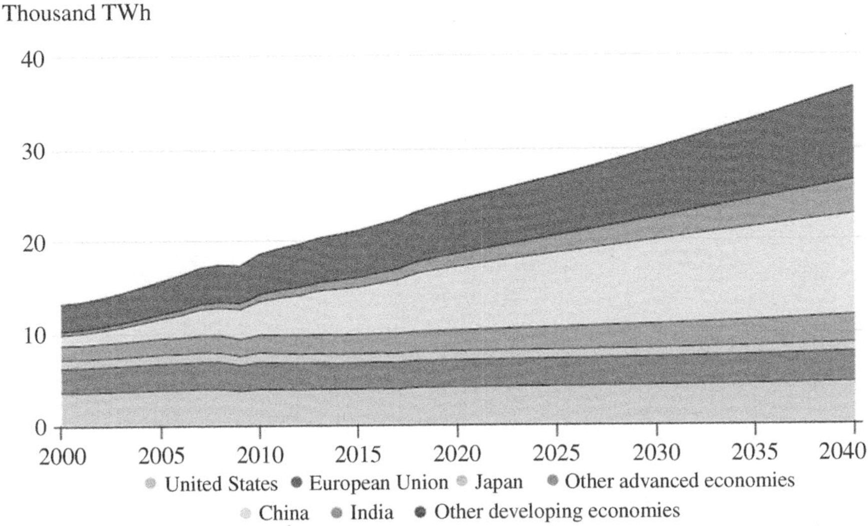

Figure 1.2 Global electricity demand by region in the Stated Policies Scenario, 2000–2040. *Source:* Excecutive Summary, International Energy Agency, IEA [99].

supply growth to 2040, underpinned by policy support in nearly 170 countries and falling costs. Coal-fired output remains broadly flat, though its share declines significantly, while natural gas and nuclear power maintain their share.

In the Sustainable Development Scenario, renewables provide two-thirds of electricity supply worldwide by 2040: solar PV and wind together provide 40%, with a further 25% from dispatchable renewables, including hydro and bioenergy. Nuclear power expands and close to 320 GW of coal and gas-fired capacity is equipped with carbon capture, utilization and storage (CCUS). Unabated coal-fired power is almost completely phased out by 2040, while gas-fired power remains an important source of flexibility (Figure 1.3).

The need for flexibility in power systems grows even faster than electricity demand, due to rising share of variable renewables and growing demand for cooling and electric vehicles. Figure 1.4 shows growth in electricity demand and flexibility needs by the selected region and scenario, 2018–2040 [99]. Power plants and networks remain the bedrock of power system flexibility, and demand-side response has a huge potential. As the fastest growing flexibility option, battery storage capacity rises 40-fold by 2040, due to its falling costs, short construction periods, widespread availability and scalability.

Figure 1.5 shows the installed power generation capacity by source in the New Policies Scenario, 2000–2040 [99]. In the Stated Policies Scenario, solar power becomes the largest source of installed capacity around 2035, surpassing coal

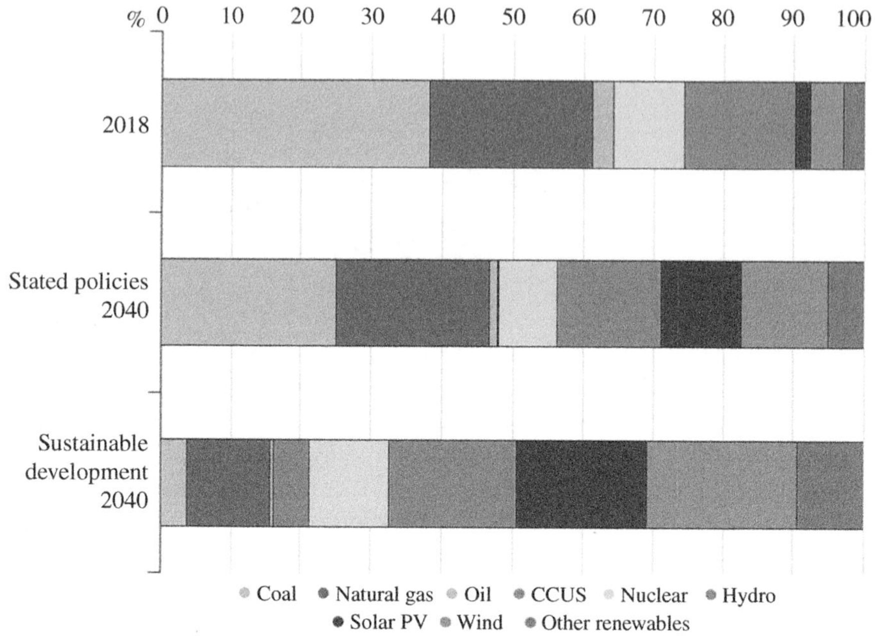

Figure 1.3 Global electricity generation mix by scenario, 2018, Stated Policies and Sustainable Development Scenarios, 2040. *Source:* Excecutive Summary, International Energy Agency, IEA [99].

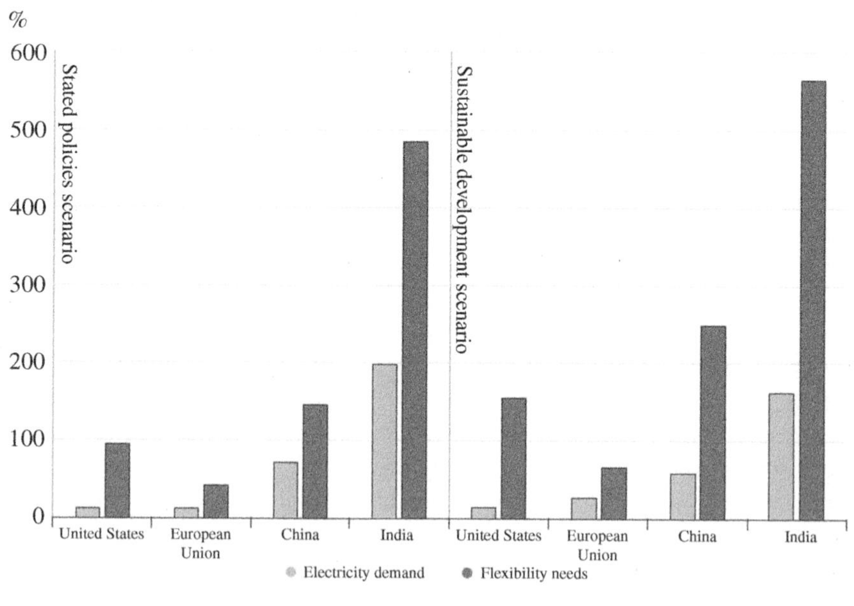

Figure 1.4 Growth in electricity demand and flexibility needs by the selected region and scenario, 2018–2040. *Source:* Excecutive Summary, International Energy Agency, IEA [99].

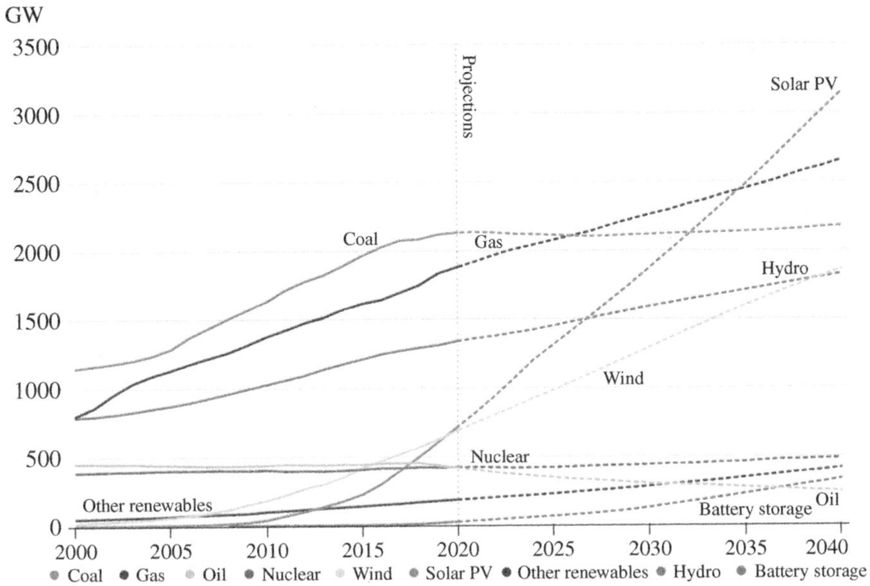

Figure 1.5 Installed power-generation capacity by source in the New Policies Scenario, 2000–2040. *Source:* Excecutive Summary, International Energy Agency, IEA [99].

and gas. Global coal-fired capacity plateaus, with the project pipeline of 710 GW, mainly in Asia, just exceed coal plant retirements, mainly in advanced economies. Renewables make up for two-thirds of all capacity additions to 2040 globally. Wind power capacity triples, with offshore wind taking off in Europe, China, and the United States. Gas-fired capacity grows in most markets for reliability purposes and battery storage skyrockets.

Figure 1.6 shows global annual average power sector investment, historical, and by scenario, 2019–2040 [99]. In the Stated Policies Scenario, global power sector investment totals $20 trillion over the period to 2040, on average, this is 20% higher than annual spending from 2010 to 2018. The power sector represents 50% of total energy supply investment worldwide and includes spending on new power plants, transmission, and distribution lines, as well as refurbishments and upgrades.

In the Sustainable Development Scenario, power sector investment is nearly $1.2 trillion per year on average to 2040, some 60% higher than recent spending levels. In this scenario, power sector investment accounts for two-thirds of total energy supply investment, compared with almost half today. Annual spending on renewables doubles.

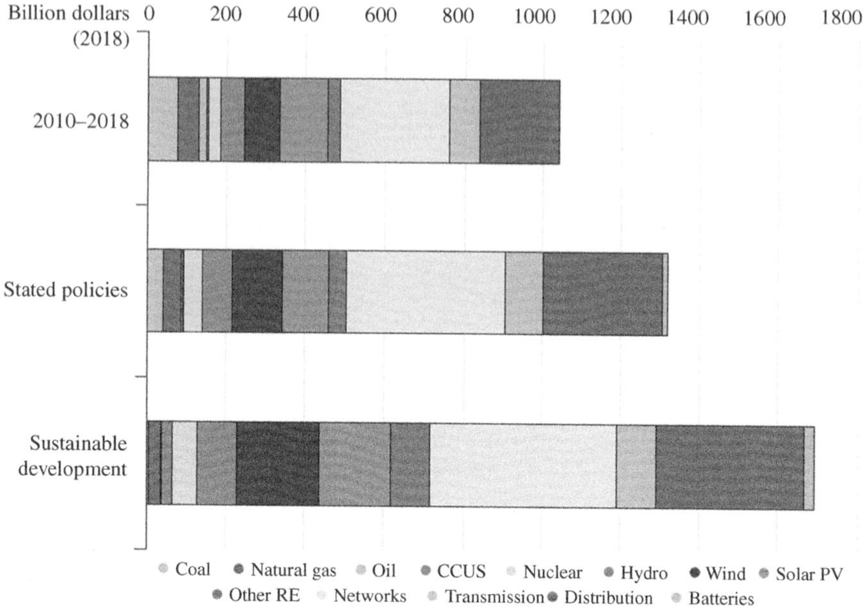

Figure 1.6 Global annual average power sector investment, historical and by scenario, 2019–2040. *Source:* Excecutive Summary, International Energy Agency, IEA [99].

1.2 Renewables

Electricity moves to the heart of modern energy security. Cost reductions in renewables and advances in digital technologies are opening huge opportunities for energy transitions, while creating some new energy security dilemmas. Wind and solar PV provide more than half of the additional electricity generation to 2040 in the Stated Policies Scenario and almost all the growth in the Sustainable Development Scenario. Policy-makers and regulators will have to move fast to keep up with the pace of technological change and the rising need for flexible operation of power systems. Issues such as the market design for storage, the interface between electric vehicles and the grid, and data privacy all have the potential to expose consumers to new risks.

Solar PV becomes the largest component of global installed capacity in the Stated Policies Scenario (Figure 1.5). The expansion of generation from wind and solar PV helps renewables overtake coal in the power generation mix in the mid-2020s. By 2040, low-carbon sources provide more than half of the total electricity generation. Wind and solar PV are the star performers, but hydropower (15% of total generation in 2040) and nuclear (8%) retain major shares.

The speed at which battery costs decline is a critical variable for power markets as well as for electric cars. India is the largest overall source of energy demand growth in this year's Outlook, and it is examined how a cost-effective combination of cheaper battery storage and solar PV could reshape the evolution of India's power mix in the coming decades. Battery storage is well suited to provide the short-term flexibility that India needs, allowing a lunch-time peak in solar PV supply to meet an early evening peak in demand. In the Stated Policies Scenario, a major reduction in battery costs means that some 120 GW of storage are installed by 2040.

The possibility is also examined that battery costs could decline even faster – an extra 40% by 2040 – because of the greater industrial economies of scale or a breakthrough in battery chemistry, for example. In this case, combined solar and battery storage plants would be a very compelling economic and environmental proposition, reducing sharply India's projected investment in new coal-fired power plants.

The past decade has seen strong growth in the deployment of renewable energy technologies, with the power sector leading the way thanks to sharp cost reductions for solar photovoltaic (PV) and wind power. However, the uptake of renewables has been slower in end-use sectors such as industry and buildings.

Figure 1.7 shows the renewable electricity generation by region and scenario, 2018–2040 [99]. Electricity generation from renewables continued to grow in

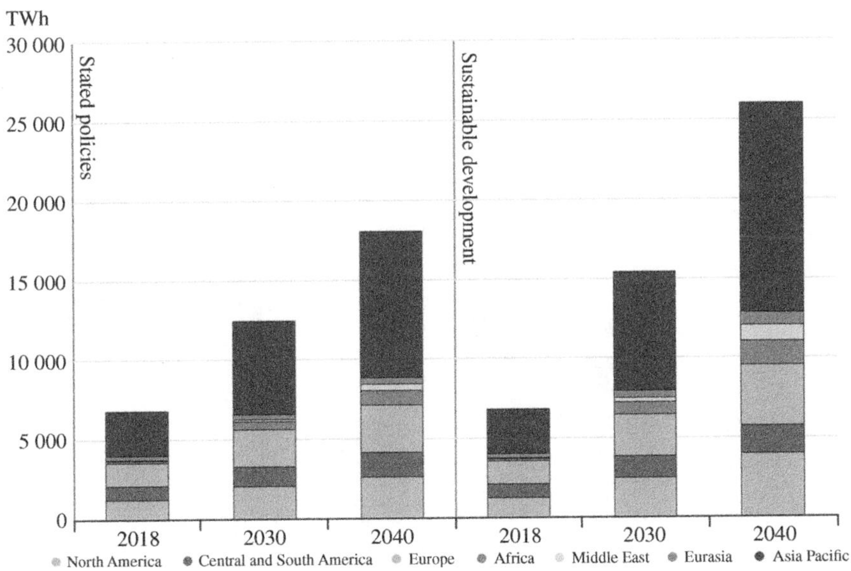

Figure 1.7 Renewable electricity generation by region and scenario, 2018–2040. *Source:* Excecutive Summary, International Energy Agency, IEA [99].

2018 with output up by 450 terawatt-hours (TWh) (or 7%) compared to the previous year, accounting for more than a quarter of total power generation. Growth in output from solar PV, wind and hydro accounted for 90% of the increase. Figure 1.8 shows the renewable electricity capacity by region and scenario, 2018–2040 [99]. Around 180 gigawatts (GW) of new renewable power capacity were added in 2018, which is the same level as the previous year, although the IEA's estimate for 2019 suggests a resumption of robust growth in annual renewable additions.

Cost reductions in renewables and advances in digital technologies are opening huge opportunities for energy transitions. Wind and solar PV provide more than half of the additional electricity generation to 2040 in the Stated Policies Scenario and almost all the growth in the Sustainable Development Scenario.

In the Stated Policies Scenario, the amount of renewables (excluding traditional use of biomass) in final energy consumption increases from more than 990 Mtoe today to almost 2260 Mtoe in 2040. The share of renewables in global heat increases by 60% and reaches nearly 940 Mtoe in 2040 thanks to substantial growth in the modern use of bioenergy (pellets in boilers and stoves, biogas and biomethane, and biofuels), renewable electricity and also of solar thermal. Meanwhile, the contribution of renewables to the transport sector triples to around 300 Mtoe, three-quarters of which comes from biofuels.

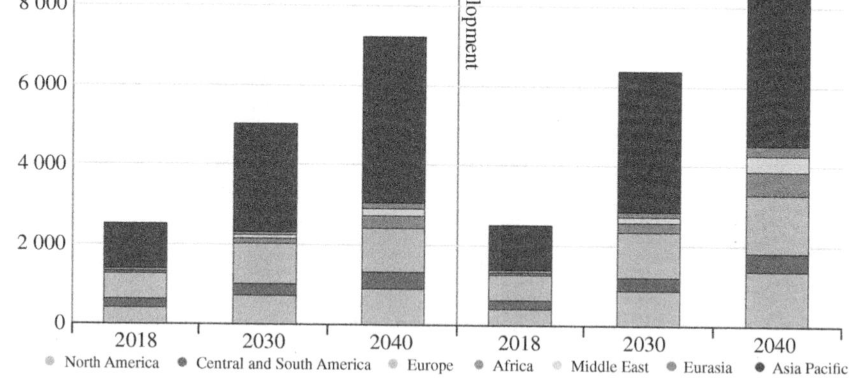

Figure 1.8 Renewable electricity capacity by region and scenario, 2018–2040. *Source:* Excecutive Summary, International Energy Agency, IEA [99].

In the Sustainable Development Scenario, additional measures to incentivize investment in renewables-based electricity, bioenergy, solar heat, geothermal heat, and electrification push the share of renewables to two-thirds of electricity generation output and 37% of final energy consumption. By 2040, expected output from wind (8300 TWh) and solar PV (7200 TWh) are expected to exceed hydropower (6 950 TWh), while the share of heat coming from renewables in 2040 increases to 30% of the total or to 1200 Mtoe. In the transport sector, consumption of energy from renewable sources is projected to increase to 600 Mtoe, with biofuels accounting for around 60% and electricity from renewable sources consumed by electric vehicles and rail accounting for the remainder.

Figure 1.9 shows the renewables share in capacity additions by region in the Stated Policies and Sustainable Development scenarios, 2019–2040 [99]. In the Stated Policies Scenario, nearly 8500 GW of new power capacity is added globally by 2040, of which two-thirds are renewables. Renewables account for the majority of capacity additions in most regions.

This includes about 80% of additions in the European Union and China, but they provide less than half of additions in Southeast Asia and the Middle East. Solar PV provides the largest share of renewable capacity additions in most regions, including China and India.

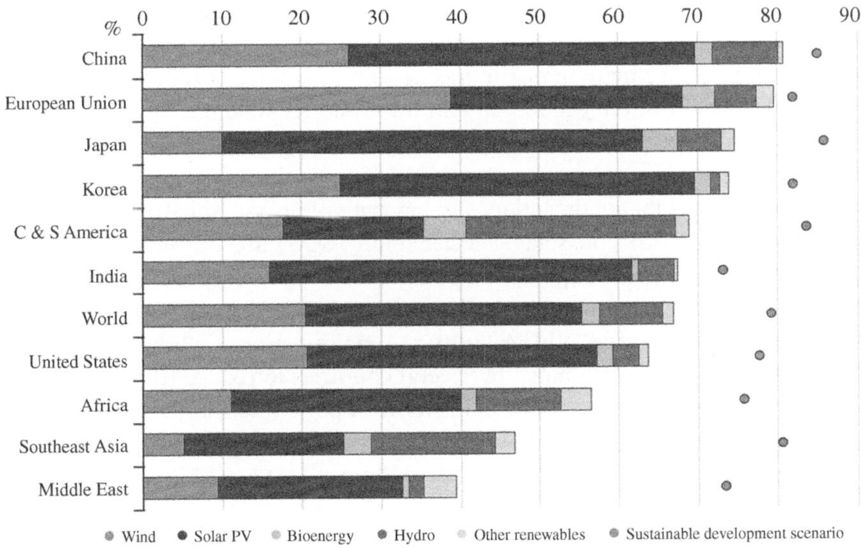

Figure 1.9 Renewables' share in capacity additions by region in the Stated Policies and Sustainable Development scenarios, 2019–2040. *Source:* Excecutive Summary, International Energy Agency, IEA [99].

In the Sustainable Development Scenario, renewables account for around 80% of capacity additions in all regions, complemented mainly by nuclear power and carbon capture technologies.

Renewables-based power investment declined slightly in 2018 around $390 billion, yet a dollar of renewables spending continued to buy more generation capacity than in the past. Whichever scenario the world follows, investment in renewables technologies would need to accelerate.

In the Stated Policies Scenario, investment in renewables reaches a cumulative total between now and 2040 of around $10 trillion. In the Sustainable Development Scenario, investment grows at a much faster rate, reflecting a stronger policy support and the central role that these clean energy technologies play in reaching sustainable energy goals.

1.3 Power System Planning

Power system planning is part of a more general problem, that of energy and economic development planning. Its objective is therefore to determine a minimum cost strategy for long-range expansion of the generation, transmission, and distribution systems adequate to supply the load forecast within a set of technical, economic, and political constraints. Traditionally, power system planning has been mainly related to the generation expansion planning. This is because investment in transmission lines is a relatively small fraction of the investment in power plants, and that investment in distribution systems, although sizable, is to a large extent independent of the generation and transmission systems.

In recent years, however, the use of renewable energy sources as a generation resource in electric power systems is growing rapidly due to its advantages that are minimal pollution, nondepletion, and low operating cost. The use of wind turbine generator (WTG) and solar cell generator (SCG) are the fastest growing energy sources on a percentage basis. In the past 20 years, for example, the global wind energy capacity has increased by 25-fold – from 24 GW in 2001 to almost 200 GW by the end of 2010 and 591 GW by 2018. A major reason for this development is the incentive policies promoted by several countries to reduce the emission of pollutant gases. As a result, it is becoming more important to analyze the best generation mix including WTG and SCG. Also, as the number of renewable energy sources increases, it is expected that Energy Storage Systems (ESS) will increase as well.

The main steps in power system planning may be summarized as follows [100]:

a) Study of the electric load forecast 5–30 years into the future, based on the most reliable information;

b) Evaluation of the energy resources available in the future for electricity generation and the foreseeable trends in technical and economic developments;
c) Evaluation of the economic and technical characteristics of the existing system of generating units and of the plants that are considered as potential units for system expansion. These characteristics include capital investment cost, fuel cost, operation and maintenance costs, efficiencies, construction times, etc.;
d) Determination of technical and cost characteristics of the plants available for expansion.
e) Determination of the economic and technical parameters affecting decisions such as discount rate, level or reliability required from the generating system, etc.;
f) Choice of a procedure to determine the optimal expansion strategy within the imposed constraints; and
g) Qualitative review of the results to estimate the viability of the proposed solution.

Figure 1.10 shows the basic and typical concept of Power System Expansion Planning.

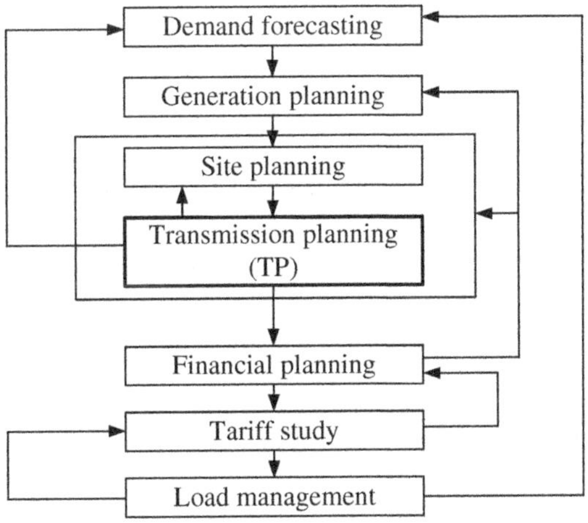

Figure 1.10 The basic and typical concept of Power System Expansion Planning.

2

Background on Generation Expansion Planning

2.1 Methodology and Issues

Generation expansion planning (GEP) is one of the most important decision-making activities in electric utilities. The electric power industry in the United States is being restructured with the passage of regulation by the Federal Energy Regulatory Commission (FERC). Several states have proposed plans to restructure the industry within their geographic regions. While the transmission system continues to be regulated, generation supply is being opened to competition. Competition in the generation supply industry will fundamentally alter the way GEP is conducted. Least-cost GEP is to determine the minimum-cost capacity addition plan (i.e., the type and number of candidate plants) that meets forecasted demand within a prespecified reliability criterion over a planning horizon [102].

A least-cost GEP problem is a highly constrained nonlinear discrete dynamic optimization problem that can only be fully solved by complete enumeration in its nature. Therefore, every possible combination of candidate options over a planning horizon must be examined to get the optimal plan, which leads to the computational explosion in a real-world GEP problem. To solve this complicated problem, a number of salient methods have been successfully applied during the past decades. Masse and Gilbrat [8] applied a linear programming approach that necessitates the linear approximation of an objective function and constraints. Bloom [9] applied a mathematical programming technique using a decomposition method and solved it in a continuous space. Park et al. [10] applied Pontryagin's maximum principle whose solution also lies in a continuous space. Although the above-mentioned mathematical programming methods have their own advantages, they possess one or both of the following drawbacks in solving a GEP problem. That is, they treat decision variables in a continuous space. And there is no guarantee to get the global optimum since the problem is not mathematically convex. Dynamic programming (DP)-based framework is one of the most widely used

Probabilistic Power System Expansion Planning with Renewable Energy Resources and Energy Storage Systems, First Edition. Jaeseok Choi and Kwang Y. Lee.

Published 2022 by John Wiley & Sons, Inc.

algorithms in GEP. However, the so-called "curse of dimensionality" has interrupted the direct application of the conventional full DP in practical GEP problems. For this reason, the Wien Automatic System Planning Package (WASP) [11] and the Electric Generation Expansion Analysis System (EGEAS) [7] use a heuristic tunneling technique in the DP optimization routine where users prespecify states and successively modify tunnels to arrive at a local optimum. David and Zhao [12] developed a heuristic-based DP and applied the fuzzy set theory to reduce the number of states. Recently, Fukuyama and Chiang [13] and Park et al. [14] applied a genetic algorithm (GA) to solve sample GEP problems, and showed promising results. However, an efficient method for a practical GEP problem that can overcome a local optimal trap and the dimensionality problem simultaneously has not been developed, yet. The GA is a search algorithm based on the hypothesis of natural selections and natural genetics. Recently, a global optimization technique using GA has been successfully applied to various areas of power system such as economic dispatch, unit commitment, reactive power planning, and power plant control. The GA-based approaches for least-cost GEP have several advantages. Naturally, they can not only treat the discrete variables but also overcome the dimensionality problem. In addition, they have the capability to search for the global optimum or quasi-optimums within a reasonable computation time. However, there exist some structural problems in the conventional GA, such as premature convergence and duplications among strings in a population as generation progresses [14].

Optimal GEP is traditionally focused on minimizing the total investment and satisfying operation constraints. During the last decades, generation industries have faced a significant transformation from a regional monopoly to a deregulated market. Planners must seriously evaluate what resources should be chosen to increase their competitions. In a competitive electricity market, all utilities including independent power producers (IPPs) need to maximize the profit when the power had been completely auctioned. In order to achieve the objective, they will seriously consider options of either constructing new generating units or purchasing electricity from other utilities or IPPs. Thus, GEP is an important decision-making activity in a competitive market. Generating technologies consist of nuclear, coal, gas, and oil-fired plants. There are three types of generators depending upon generation characteristics: the base-type, middle-type, and peak-type. The scheduling order for generations to satisfy the load profile is generally nuclear, coal, oil, and liquefied natural gas (LNG) or gas generations. The IPPs want to sell as much electricity as possible for various load profiles.

The utilities need to minimize the total cost under operational constraints for all types of generations. It is important to determine what type of generating units to be constructed and when the unit to be on line over a planning horizon to maximize profits. Besides minimizing the cost, environmental issues are also important. Fossil-fired plants such as coal, gas, and oil produce atmospheric emissions with various fuels at various cost bases. The power industry is certainly a major contributor (about one-third) to global CO_2 emissions. In recent years, rigid environmental regulations and CO_2 emission force utility planners to consider emission as a cost and an important constraint in GEP. Besides, the plan must satisfy the desired level of reliability generally defined by two indices: the loss-of-load probability (LOLP) and the expected energy demand not served (EENS) [15, 16].

The GEP is a large-scale stochastic nonlinear optimization problem. To handle the problem complexity, decomposition schemes have been used. Usually, such schemes divide the expansion problem into two subproblems: one related to the construction of new plants (investment subproblem) and another dealing with the task of operating the system (operation subproblem). Table 2.1 shows Typical Energy Supply Technical Data Requirements for GEP [102].

Table 2.1 Typical energy supply technical data requirements.

Parameters	Examples
Engineering performance data:	
Energy output	Type of products' output
	Range of output
Energy input	Input materials
	Restrictions on inputs
Thermodynamic efficiency	Current and future improvements
Performance limits	Capacity-design, maximum, minimum Operational limitations
	Reliability
Technology status	Commercially available
	Research
	Pilot plant
Economic data:	
Capital cost	Labour, materials
	Interest during construction
	Foreign and domestic component
	Per-unit of output
	Current, future

(*Continued*)

Table 2.1 (Continued)

Parameters	Examples
Non-fuel-operating cost	Labor, materials
	Taxes
	Per-unit of output
Output energy cost	Including capital charges, fuel costs, non-fuel operating costs
	Taxes
	Per unit of output
Financial data	Interest rates[a]
	Return on investment[a]
	Discount rate[a]
	Foreign exchange implications
Ancillary data:	
Environmental burdens	Air pollution
	Water pollution
	Solid waste
	Noise
Labor requirements for construction and operation	Quantity
	Skilled, unskilled
	Foreign, domestic
Barriers to implementation	Social acceptance
	Policy issues

[a]Generally not technology-specific.
Source: International Atomic Energy Agency [102].

2.2 Formulation of the Least-Cost Generation Expansion Planning Problem

Mathematically, solving a least-cost GEP problem is equivalent to finding a set of optimal decision vectors over a planning horizon that minimizes an objective function under several constraints. The GEP problem to be considered is formulated as follows [14]:

$$\underset{U_1,\cdots,U_T}{Min} \sum_{t=1}^{T}\{ f_t^1(U_t) + (f_t^2)(X_t) - f_T^3(U_T)\} \tag{2.1}$$

$$s.t.\, X_t = X_{t-1} + U_t \quad (t = 1, \ldots, T) \tag{2.2}$$

$$LOLP(X_t) < \varepsilon \quad (t = 1, \ldots, T) \tag{2.3}$$

$$\underline{R} \leq R(X_t) \leq \overline{R} \quad (t = 1, \ldots, T) \tag{2.4}$$

$$\underline{M}_t^j \leq \sum_{i\in\Omega_j} x_t^i \leq \overline{M}_t^j \quad (t = 1, \ldots, T \text{ and } j = 1, \ldots, J) \tag{2.5}$$

$$0 \leq U_t \leq \overline{U}_t \quad (t = 1, \ldots, T) \tag{2.6}$$

where
T: number of periods (years) in a planning horizon,
J: number of fuel types,
Ω_j: index set for jth fuel-type plant,
X_t: cumulative capacity (MW) vector of plant types in year t,
x_t^i: cumulative capacity (MW) of ith plant type in year t,
U_t: capacity addition (MW) vector by plant types in year t,
$\overline{U}_t$: maximum construction capacity (MW) vector by plant types in year t,
u_t^j: capacity addition (MW) of ith plant in year t,
$LOLP(X_t)$: loss of load probability (LOLP) with X_t, in year t,
$R(X_t)$: reserve margin with X_t, in year t,
ε: reliability criterion expressed in LOLP,
$\overline{R}, \underline{R}$: upper and lower bounds of reserve margin,
$\overline{M}_t^j, \underline{M}_t^j$: upper and lower bounds of jth fuel type in year t,
$f_t^1(U_t)$: discounted construction costs ($) associated with capacity addition U_t in year t,
$f_t^2(X_t)$: discounted fuel and O&M costs ($) associated with capacity X_t, in year t,
$f_t^3(U_t)$: discounted salvage value ($) associated with capacity addition U_t in year t.

The objective function is the sum of tripartite discounted costs over a planning horizon. It is composed of discounted investment costs, expected fuel and operation and maintenance (O&M) costs, and salvage value. To consider investments with longer lifetimes than a planning horizon, the linear depreciation option is utilized. Equation (2.2) implies a state equation for a dynamic planning problem. Equations (2.3) and (2.4) are related to the LOLP reliability criteria and the reserve margin bands, respectively. The capacity mixes by fuel types are considered in (2.5). Plant types give another physical constraint in (2.6), which reflects the yearly construction capabilities.

3

Cost Assessment and Methodologies in Generation Expansion Planning

This section defines and briefly discusses the types of costs associated with electric power-generating plants and systems in generation expansion planning (GEP) and methodologies used for solving the GEP problems. Cost accounting practices and terminology vary from country to country and, in many instances, within countries. Furthermore, special terminology and conventions are often used in conjunction with specific types of generating units such as coal and nuclear power plants. Therefore, while the terminology and conventions presented in this section are typical, they are not universal and are intended only to illustrate the basic concepts and categories of costs for power plants and electric generating systems. Some of the terms defined here may be defined differently elsewhere depending on the cost accounting system used or the type of analysis being performed [102].

3.1 Basic Cost Concepts

From an economic point of view, it is desirable (but seldom possible) to expand a power-generating system by adding plants that are both cheap to build and that produce electrical power at the lowest possible cost. Two distinct figures of merit are therefore important when discussing or comparing the economics of power-generating technologies: (i) *capital investment costs*, expressed in ($/kW) of installed capacity, that denote the capital outlay necessary to build a power plant; and (ii) *power generation costs*, expressed in ($/kW)-h of generation, that represent the total cost of generating electricity. Power-generation costs consist of the costs associated with the initial capital investment in a power plant (*fixed investment charges*), fuel costs, and operation and maintenance (O&M) costs. For discussion, these costs can be divided into two broad categories: *fixed costs* and *variable costs*. A breakdown of the general categories of costs for power-generating technologies is presented in Figure 3.1. As illustrated, fuel and O&M costs have both fixed cost

Probabilistic Power System Expansion Planning with Renewable Energy Resources and Energy Storage Systems, First Edition. Jaeseok Choi and Kwang Y. Lee.

Published 2022 by John Wiley & Sons, Inc.

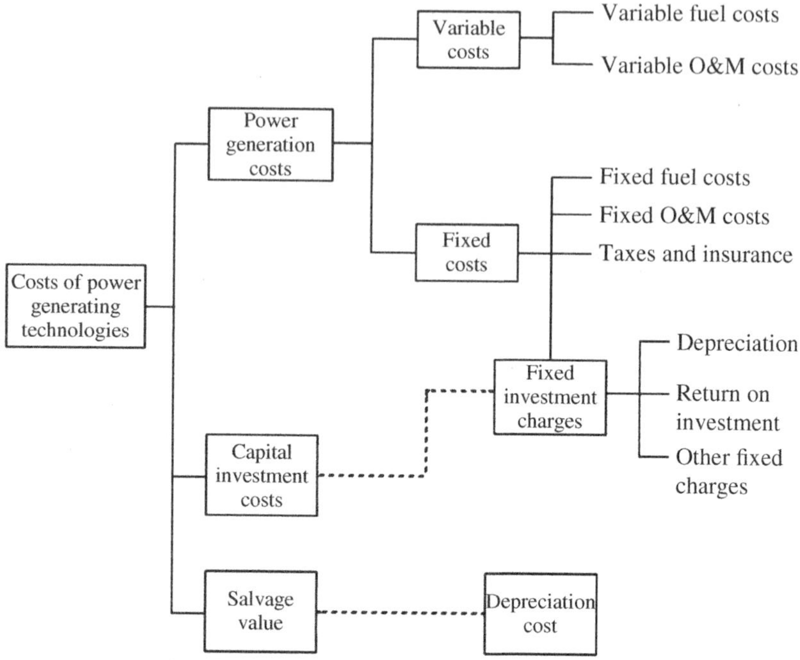

Figure 3.1 Categories of costs for power-generating technologies. *Source:* International Atomic Energy Agency [102].

and variable cost components. The dashed line indicates that the fixed investment charges are a function of the capital investment costs. The levels of costs for the cost categories identified in Figure 3.1 will vary considerably depending on the technology examined. For example, nuclear power plants are characterized by high capital investment costs and low fuel costs, while no fuel costs are usually associated with a hydroelectric power plant [102].

3.1.1 Annual Effective Discount Rate

The annual effective discount rate is the amount of interest paid or earned as a percentage of the balance at the end of the annual period. It is the annual discount factor to be applied to the future cash flow; to find the discount, it is subtracted from a future value to find the value one year earlier. This is in contrast to the effective rate of interest, which expresses the amount of interest as a percentage of the balance at the start of the period. For example, suppose there is a government bond that sells for $95 and pays $100 in a year's time. The discount rate according to the given definition is

$$\frac{100-95}{100} = 5.00\%$$

The interest rate is calculated using 95 as its base:

$$\frac{100-95}{95} = 5.26\%$$

For every annual effective interest rate, there is a corresponding annual effective discount rate, given by the following formula

$$d = \frac{i}{1+i} \approx i - i^2 \quad (3.1)$$

or inversely,

$$i = \frac{d}{1-d} \approx d - d^2 \quad (3.2)$$

where the approximations apply for small i and d; in fact, $i - d = id$.

3.1.2 Present Value

Present value is the value on a given date of a future payment or series of future payments, discounted to reflect the time value of money and other factors such as investment risk. Present value calculations are widely used in business and economics to provide a means to compare cash flows at different times on a meaningful "like-to-like" basis. The most commonly applied model of the time value of money is compound interest. To someone who can lend or borrow for years at an interest rate per year (where interest of "5%" is expressed fully as 0.05), the present value of the receiving monetary units' years in the future is:

$$C = C(1+i)^{-t} = \frac{C}{(1+i)^t} \quad (3.3)$$

This is also found from the formula for the future value with negative time. The purchasing power in today's money of an amount C of money, t years into the future, can be computed with the same formula, where, in this case, i is an assumed future inflation rate. The expression $(1+i)^{-t}$ enters almost all calculations of present value. Where the interest rate is expected to be different over the term of the investment, different values may be included; an investment over a two years' period would then have PV of:

$$PV = \frac{C}{(1+i_1)(1+i_2)} \quad (3.4)$$

Net present value (NPV) or net present worth (NPW) is defined as the total present value (PV) of a time series of cash flows (Figure 3.2). It is a standard method for using the time value of money to appraise long-term projects. Used for capital

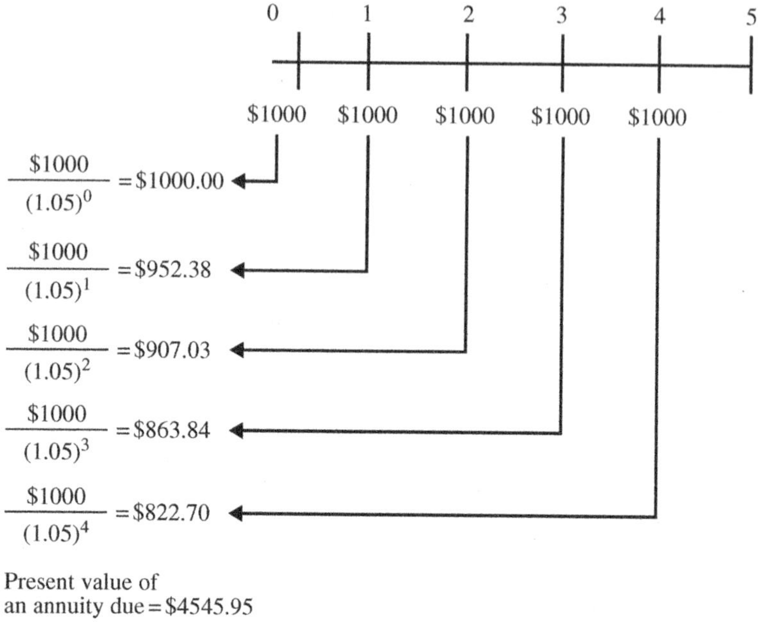

Figure 3.2 Present value.

budgeting, and widely throughout economics, it measures the excess or shortfall of cash flows, in present value terms, once financing charges are met.

3.1.3 Relationship Between Salvage Value and Depreciation Cost

Graphically, the relationship between salvage value (SV) and depreciation cost (*D*) can be represented as in Figure 3.3.

Basic Methods for Calculating Depreciation Cost are as following:

1) *Straight-line Depreciation*

$$D = \frac{OC - SV}{T} \tag{3.5}$$

where
D: depreciation cost,
OC: original cost,
SV: salvage value,
T: useful life of asset (years).

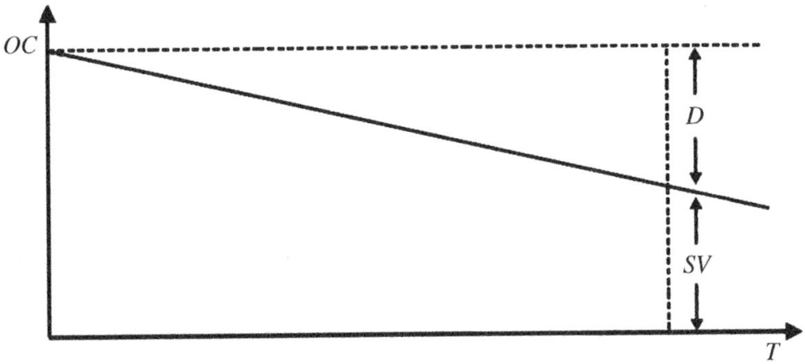

Figure 3.3 Relationship between salvage value and depreciation cost.

2) *Declining-Balance Method*

$$D = (OC - BVB) \times DR \tag{3.6}$$

where
D: depreciation cost,
OC: original cost,
BVB: book value at the beginning of the year,
DR: depreciation rate.

3) *Units-of-Production Depreciation Method*

$$D = (OC - SV) \times \frac{UP}{TUP} \tag{3.7}$$

where
D: depreciation cost,
OC: original cost,
SV: salvage value,
UP: units produced during the year,
TUP: total units of production.

4) *Sum-of-Years' Digits Method*

$$D = (OC - SV) \times \frac{N}{1 + 2 + 3 + \cdots + n} \tag{3.8}$$

where
D: depreciation cost,
OC: original cost,
SV: salvage value,

n: years from original year to salvage year,
N: years from $n + 1$ years to life-time year.

3.2 Methodologies

In traditional generation planning, mathematical methodologies used are linear programming, non-linear programming, dynamic programming, mix-integer programming, and the Pontryagin's maximum principle [10]. With the non-linearity and discrete nature considered in GEP, the problem becomes more difficult to solve. New algorithms based on artificial intelligence (AI) have been developed, such as simulated annealing (SA), genetic algorithm (GA), and expert systems and particle swarm optimization (PSO). Recently, the concept of integrated resource planning (IRP) is applied in the deregulated market to reach an efficient least-cost plan. Solution strategies proposed by most AI algorithms need to consider a large solution space. Conventional methods may be faster; they are very often limited by the problem structure and may diverge or could lead to a local minimum. Immune algorithm (IA) is based on the operation of human immune system. The IA takes advantage of the conventional GA and Tabu search, but the choice of a proper probability for crossover and mutation is still a dilemma. It has advantages and some characteristics to show a better performance than many other algorithms [103].

3.2.1 Dynamic Programming

Dynamic programming is both a mathematical optimization method and a computer programming method. In both contexts, it refers to simplifying a complicated problem by breaking it down into simpler sub-problems in a recursive manner. In 1957, Richard Bellman's "Principle of Optimality" describes how to do this [60]: Principle of Optimality: *An optimal policy has the property that whatever the initial state and initial decision are, the remaining decisions must constitute an optimal policy with regard to the state resulting from the first decision.*

Applying the principle of optimality, the maximization problem can be rewritten as a recursive relation of the value function V:

$$V(x_0) = \max_{a_0} \{F(x_0, a_0) + \beta V(x_1)\} \tag{3.9}$$

Therefore,

$$V(x) = \max_{a\in\Gamma(x)} \{F(x,a) + \beta V(T(x,a))\} \tag{3.10}$$

Where, V is the *value function* and T is the transition function. Recall that the value function describes the best possible value of the objective, as a function of the state x. Optimal state x can be obtained using the Bellman's optimal policy satisfying the recursive equation.

3.2.2 Linear Programming

3.2.2.1 Investment Cost (Capital Cost)

$$C'_{fi}(X_j) = \left[\frac{(1+r)^{(T_j+1)} - 1}{(1+r)^{(T_j+1)}}\right] C_{oj} a_{oj} G_{xj} \tag{3.11}$$

The present worth of the above cost is represented as [104]

$$C_{fi}(X_j) = \left[\frac{1+i}{1+r}\right]^t C'_{fi}(X_j) \tag{3.12}$$

where "t" is the time in years at which the investment is made, "i" the rate of inflation, "r" the rate of interest, T_j the life span in the year of jth unit, a_{oj} the cost in money units per MW capacity of jth unit, G_{xy} the unit capacity, and C_{oj} are the carrying charges of the jth unit.

The total cost is represented as

$$C_f(X) = \sum_{J=1}^{N} C_{fi}(X_j) \tag{3.13}$$

3.2.2.2 Operating Cost

The total operating cost is given by

$$C_o(P_o) = \sum_{j=1}^{NG} C'_o(P_{oj}) \tag{3.14}$$

The objective function to be minimized is the sum of the operating cost and the capital cost and is given by

$$CT = C_f(X) + C_o(P_o) \tag{3.15}$$

3.2.2.3 LP Formula

$$\begin{aligned}&\text{Minimize}: C_f(X) + C_o(P_o)\\&\text{subject to}\\&(\text{a})\ P_{oj} \geq P_j(\min)\\&(\text{b})\ P_{oj} \leq P_j(\max)\ \text{and}\\&(\text{c}) \sum P_j\text{old} + \sum P_j\text{new} \geq 1.2\text{D}\end{aligned} \tag{3.16}$$

where D: peak demand (kW) with reserve rate 20%.

3.2.3 Integer Programming

Power GEP inherently involves multiple, conflicting, and incommensurate objectives. Therefore, mathematical models become more realistic if distinct evaluation aspects, such as cost and environmental concerns, are explicitly considered as objective functions rather than being encompassed by a single economic indicator. With the aid of multiple objective models, decision makers may grasp the conflicting nature and the trade-offs among the different objectives in order to select satisfactory compromising solutions. A multiple objective mixed-integer linear programming model for power GEP that allows the consideration of modular expansion capacity values of supply-side options may be formulated. This characteristic of the model avoids the well-known problem associated with continuous capacity values that usually have to be discreetized in a post-processing phase without feedback on the nature and importance of the changes in the attributes of the obtained solutions. Demand-side management (DSM) is also considered as an option in the planning process, assuming there is a sufficiently large portion of the market under franchise conditions. As DSM full costs are accounted for in the model, including lost revenues, it is possible to perform an evaluation of the rate impact in order to further inform the decision process.

3.2.4 Multi-objective Linear Programming

The planning of new units for electrical power generation is a problem that involves different, conflicting, and incommensurate aspects, some of which are not directly quantifiable by an economic indicator. The models which express the qualitatively different aspects of these complex problems in currency units in order to encompass them in an economic objective function fail to capture explicitly the distinct aspects arising in the evaluation of power expansion policies. The problem is multiple objective in nature: economic, technological, environmental, and reliability aspects, among others, must be explicitly taken into account. With the aid of multiple objective models, decision makers may grasp the conflicting nature of the objectives and the trade-offs to be made in order to obtain satisfactory compromising solutions from the set of nondominated solutions. In a multiple objective context, the concept of optimal solution gives place to one of nondominated solutions (feasible solutions for which no improvement in any objective function is possible without sacrificing at least one of the other objective functions) [105].

Although power generation expansion problems have been modelled as single objective programming problems by considering the total cost only, some attention has been paid to the development of multiple objective models [105], where a multiple objective linear programming (MOLP) model was used to characterize the whole nondominated surface by computing all the nondominated extreme points. Further experiments with more realistic formulations, thus augmenting the dimension of the MOLP model, showed the disadvantages of using methods to compute all nondominated extreme points.

The MOLP problem considers the following:

$$\begin{aligned} &\text{Maximize}: && f(x) = C_x \\ &\text{subject to} && x \in X = \{x \in \mathbb{R}^n : x \geq 0, Ax = b\} \end{aligned} \tag{3.17}$$

Given a vector-valued function $f(x) = C_x : \mathbb{R}^n \rightarrow \mathbb{R}^p$, the set of efficient (or Pareto optimal) solutions is defined by

$$X_E = \{x \in X | x' \in X : f(x') \geq f(x)\} \tag{3.18}$$

3.2.5 Genetic Algorithm

Genetic algorithms (GAs) are different from normal optimization and search procedures. First, GAs work with the coding of the parameter set, not the parameters themselves, which require to discretize the parameters of the optimization problem. Secondly, GAs use payoff (objective function) information directly, not derivatives of other auxiliary knowledge. Furthermore, GAs must have fitness functions (mapped from objective function) that reflect information about both the quality and feasibility of solutions, which requires a special method to adjoin inequality constraints to the objective function of the optimization problem. A similar alternative method proposes an improved genetic algorithm (IGA) to solve the investment subproblem. The basic idea is to use a special type of chromosome, christened pointer-based chromosome (PBC), and the particular structure of that subproblem, to transform an integer-constrained problem into an unconstrained one. In [18] study, IGA's results were compared to those of a branch and bound (B&B) algorithm – provided by a commercial package – in three different case studies of growing complexity, respectively, containing 144, 462 and 1845 decision variables. These results indicate that the IGA is an effective alternative to the solution of the investment subproblem [14, 18]:

Minimize:

$$f(i) = \sum_{i \in c_i} C_i[I'(i) + I''(i)]X_i + \sum_{i=1}^{n} [B_i Q_i + B_{il} Q_{il}] \tag{3.19}$$

Subject to

$$\sum_{i \in c_i} I'(i) \times X_i \times AV_i + \sum_{i=1}^{n} I''(i) \times X_h(i) \times AV_i \geq C_{tf} \tag{3.20}$$

$$\sum_{i-1} I'(i) \times X_i \times AV_i + \sum_{i \in m_i}^{n} I''(i) \times [X_l(i) + X_h(i)] \times AV_i \geq L^m \tag{3.21}$$

$$0 \leq I'(i) + I''(i) \leq K_i \tag{3.22}$$

$$I''(i) = 0 \quad \text{for} \quad i \in b_i \cup c_i \tag{3.23}$$

$$0 \leq X(i) \leq X_i \quad \text{for} \quad i \in I' \tag{3.24}$$

$$X(i) = X_i \times PP_i \quad \text{for} \quad i \in I'' \tag{3.25}$$

where Q_i and Q_{il} are respectively the output energy corresponding to monoblock state and base load state block of generator unit i, resulting from the load points of these blocks and the load duration curve $L(x)$.

3.2.6 Game Theory

This method is that each firm chooses an output quantity to maximize profit using the Cournot model instead of minimizing the total cost of the traditional model. The model is summarized in [101].

The following variables are defined:

MCP_t: fuel cost of the marginal unit in the energy market at time t,
RTP_t: fuel cost of the marginal unit in real-time market at t,
L_t: load at time t,
e_{ti}: energy production commitment of unit i at time t,
r_{ti}: real-time production of unit i at time t,
x_{ci}: generation capacity expansion decision of unit i,
x_{oi}: capacity reservation decision of unit i,
X_o: industry capacity reserve offering,
p_i: forced outage rate of unit i,
C_0: total capacity of the existing system,
$ELDC_0$: equivalent load duration curve of the initial system,
a: system-wide average customer outage cost,
T: time duration of the planning period.

Profit is the difference between revenues earned and costs incurred from providing electric service. Revenues consist of energy and real-time auction market payments and contracted reserve payments. Expenses include operating costs, capital investment costs, and outage costs. Let $i(x_{ci}, x_{oi})$ represent the profit earned by a generating unit indexed by i, of size x_{ci} (MW), that devotes x_{oi} (MW) of its capacity to supplying reserves. Unit i's profit function is then

$$\pi_i(x_{ci}, x_{oi}) = R_i(x_c, x_o) + P(X_o)\cdot x_{oi} - F_i(x_c, x_o) - C_i(x_{ci}) - O_i(x_c, x_o) \quad (3.26)$$

where x_c is the capacity expansion vector comprised of all players' decisions; x_o is the capacity reserve vector; and $P(X_o)$ is the price of contracted reserves as a function of industry reserve offering. $R_i(.)$ represents unit i's revenue from energy and real-time market production; $F_i(.)$ is unit i's production cost; $C_i(.)$ is i's annualized capital investment cost; and $O_i(.)$ is i's outage cost. Whereas the production and capital investment costs are traditional components in generation planning, the other components are additional considerations. Consequently, they are detailed in the following equations [101]:

$$R_i(x_c, x_o) = \sum_{t=1}^{T} MCP_t\cdot e_{ti} + \sum_{t=1}^{T} RTP_t\cdot r_{ti} \quad (3.27)$$

$$P(X_o)\cdot x_{oi} = T\cdot(1-p)\cdot ELDC_o(C_o + X_o)\cdot a\cdot x_{oi} \quad (3.28)$$

$$O_i(x_c, x_o) = \sum_{t=1}^{T} RTP_t\cdot p_i\cdot(e_{ti} + r_{ti}) \quad (3.29)$$

Based on the definitions and profit function previously described, a player's expansion planning problem can be formulated as the following mathematical program:

$$\sum_{i=1}^{NG} NO_{Xin}\xi_i y_n \leq NO_{XMAX_n} \quad (3.30)$$

$$\text{s.t.} \quad 0 \leq x_{oi} \leq x_{ci} \quad (3.31)$$

$$e_{ti} \leq x_{ci} - x_{oi} \forall t \quad (3.32)$$

$$\sum_i e_{ti} \leq L_t \forall t \quad (3.33)$$

$$r_{ti} \leq x_{ci} - e_{ti} \forall t \quad (3.34)$$

$$\sum_i r_{ti} \leq \sum_i p_i\cdot e_{ti} \forall t \quad (3.35)$$

Each player seeks to maximize its profit subject to operational and physical constraints. Constraint (3.31) prevents a unit's capacity reserve commitment from exceeding its capacity rating. Constraint sets (3.32) and (3.33) are production constraints limiting participation in the energy and real-time markets according to the resource availability of a unit at any given time. The former constraint set limits the amount of capacity a unit can supply in the energy market to its plant capacity minus its contracted reserve commitment. Similarly, the latter constraint set prevents the amount of capacity a unit supplies in the real-time market from exceeding the amount available after discounting energy market commitments. Constraint sets (3.34) and (3.35) are analogous to energy balance constraints in the energy and real-time markets, respectively. In the program formulation above, demand in the energy market and demand in the real-time market are assumed to be inelastic. So, constraints (3.33) and (3.35) take the form of inequality constraints [101].

3.2.7 Reliability Worth

This model uses objective function considering outage cost (reliability worth) conversed from expected energy not served. Table 3.1 shows a comparison between the conventional and the outage cost-considered model.

3.2.8 Maximum Principle

The long-range problem for a system of generating plants is to minimize the present value of total costs subject to reliability constraints, and is now stated formally as follows:

$$\underset{u_i^j}{\text{minimize}} \quad G = \sum_{i=1}^{I} G_i\left(x_i^j, u_i^j\right) \tag{3.36}$$

Table 3.1 Two main kinds of optimal model for power system expansion planning problem.

	Model I	**Model II**
Objective function	Minimize $z =$ Investment cost + Operating cost	Minimize $z =$ Investment cost + Operating cost + Outage cost
Main constraint	Reliability criterion (*LOLE*)	—
method for assessing outage cost	—	$IEAR_{sys} \times EENS_{sys}$
Remarks	Optimal reliability criterion (*LOLE*) is necessary	Program assessing outage cost is necessary

such that

$$x_{i+1}^j = x_i^j + u_i^j \tag{3.37}$$

and

$$R_i\left(x_i^j, u_i^j\right) \leq \varepsilon_i \tag{3.38}$$

$$u_i^j \geq 0 \tag{3.39}$$

where
$i = 1, 2, \ldots, I$: index of years in the planning horizon,
$j = 1, 2, \ldots, J$: index of plant types,
x_i^j: previously installed capacity of plant type j in year i (MW) (state variables),
u_i^j: newly added capacity of plant type j in year i (MW) (decision variables),
G_i: present-value expected annual cost ($),
R_i: expected annual unserved energy (MWh),
ε_i: desired reliability level of annual unserved energy (MWh).

The (3.37) is the dynamic state equation describing the state of installed capacities each year, that is, the sum of previously installed capacity and the newly added capacity is the installed capacity for the next year. Considering the newly added capacities u_i^j as decision or control variables, the master problem (3.36)–(3.39) is now formulated as an optimal control problem.

According to the Pontryagin's minimum principle of modern optimal control theory [10, 19], the optimal control must satisfy the necessary conditions:

$$x_{i+1}^j = x_i^j + u_i^j, x_i^j = \text{specified (state equation)} \tag{3.40}$$

$$\lambda_i^j = \frac{\partial G_i}{\partial x_i^j} + \lambda_{i+1}^j; \lambda_{i+1} = 0\,\text{specified (costate equation)} \tag{3.41}$$

and u_i^j minimizes the Hamiltonian

$$H_i\left(x_i^j, u_i^j, \lambda_{i+1}^j\right) \triangleq G_i\left(x_i^j, u_i^j\right) + \sum_{J=1}^{J} \lambda_{i+1}^j\left(x_i^j + u_i^j\right) \tag{3.42}$$

subject to

$$R_i\left(x_i^j, u_i^j\right) \leq \varepsilon_i \text{ (reliability constraint)} \tag{3.43}$$

$$u_i^j \geq 0, \text{(control constraint)} \tag{3.44}$$

where λ_i^j is the costate (adjoint) of plant type j in year i.

Note that the minimization of Hamiltonian H_i is to be done for one year at a time while the minimization of the total cost G in (3.36) needs to be performed for all years in the planning horizon simultaneously. This reduces the number of decision variables u_i^j to J from $J \times I$ in the minimization algorithm.

The minimization of Hamiltonian (3.42) subject to inequality constraints (3.43)–(3.44) is a typical nonlinear programming problem. It can be solved by using a version of the Gradient Projection Method (GPM) [20], which has been developed specially for this purpose. The results from several sample studies show remarkable advantages of this method in computational efficiency and reliability. The reliability constraint (3.43) was linearized using the gradients or marginal values $\partial R_i/\partial u_{i\text{-}k}^k$. The marginal values of expected annual costs $\partial G_i/\partial x_i$ and $\partial G_i/\partial u_i^k$ with respect to marginal changes in the previously installed capacity and the newly added capacity, respectively, are also required in the GPM. Since these marginal values are not only required to find the optimal solution, but also provide a planner useful technical and managerial information, they are computed in the basic subproblems. Complete analytic formulas are developed in Section 4.2.

3.3 Conventional Approach for Load Modeling

3.3.1 Load Duration Curve

In the conventional method, the production costing model is based on the load duration curve. The time-dependent load curve is converted into the inverted load duration curve that gives just the percentage of time that each demand level occurs, as shown in Figure 3.4.

The generating units are loaded according to a merit order that reflects increasing operating costs per (MWh) produced by using the convolution processes. Figure 3.5 illustrates how the generating unit may be convolved and how the equivalent load duration curve on the second unit in the loading order may be found for a two-state forced outage probability distribution. This curve is used to find the expected energy generated by a particular unit by taking the product of the unit's availability and the area under the curve between the unit's loading point and the loading point of the next unit.

A new load curve is computed each time a unit is brought online. Finally, after the outage of all available units have been added to customer demand, the expected unserved energy may be obtained as an area under the curve, thus providing a measure of system reliability. The height of that same curve at the capacity point of the last unit in the loading order is the loss-of-load probability (LOLP), that is, the expected proportion of time that customer demand may exceed

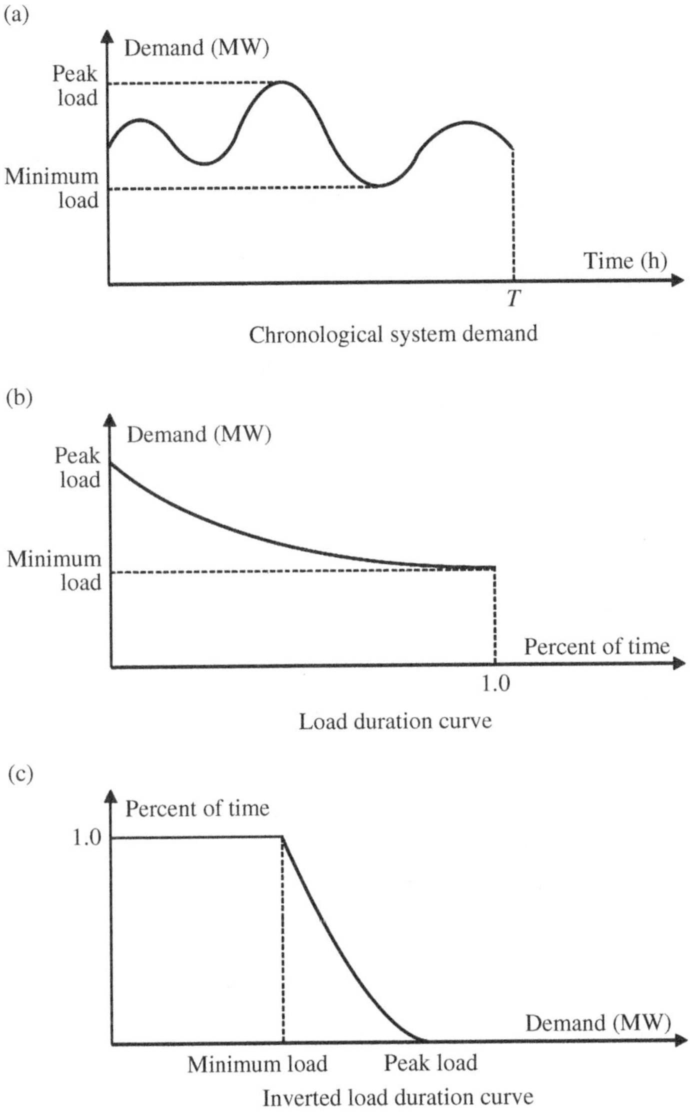

Figure 3.4 Different representations of the load curve.

available generating capacity. This is shown in Figure 3.5e, where *IC* is the total installed capacity (MW). Power system expansion planning needs huge numerical convolution integral computation for probabilistic production cost.

(a)

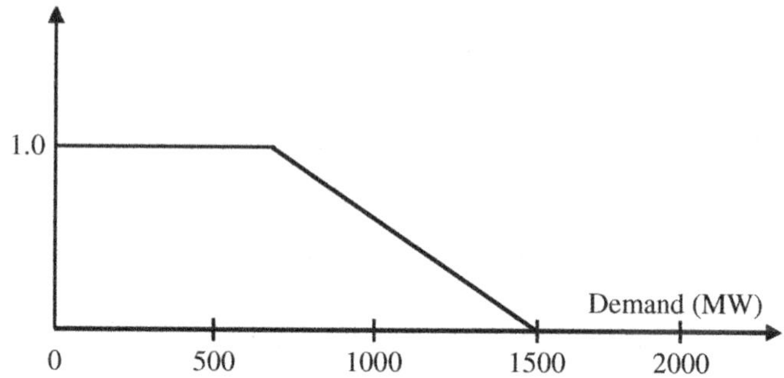

Customer demand represented by a load duration curve.

(b)

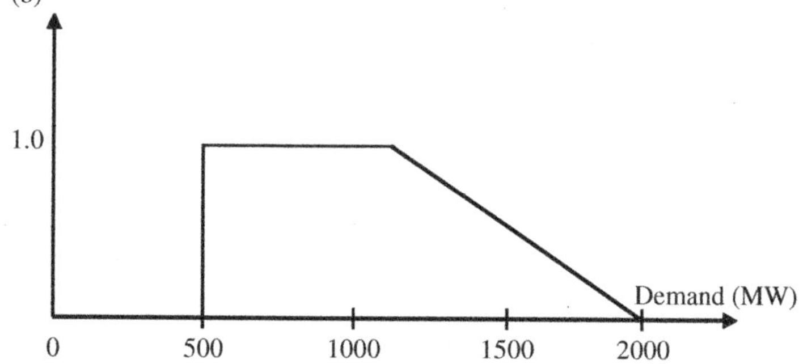

If the first plant fails, the second plant sees the original customer curve. This event has a probability of 0.10.

(c)

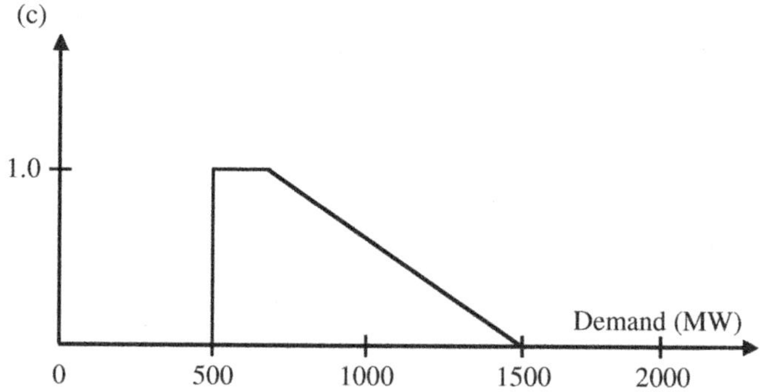

If the first plant operates, the second plant does not see the first 500 (MW) of customer demand. This event has a probability of 0.90.

Figure 3.5 Graphic illustration of convolution: loading of a 500 MW plant with 90% availability.

(d)

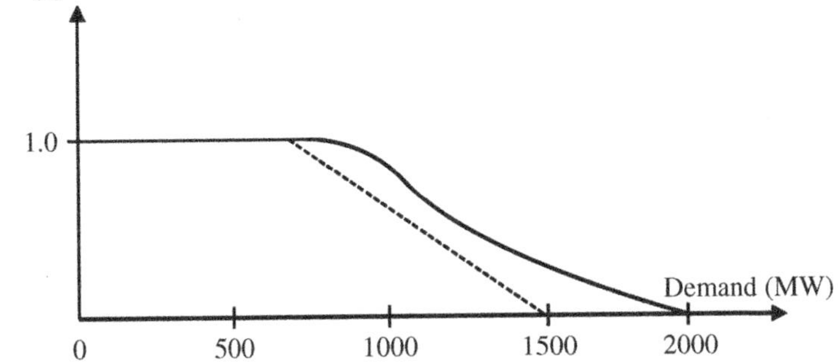

The equivalent load curve for the second plant is the sum of the two curves weighted by their respective probabilities.

(e)

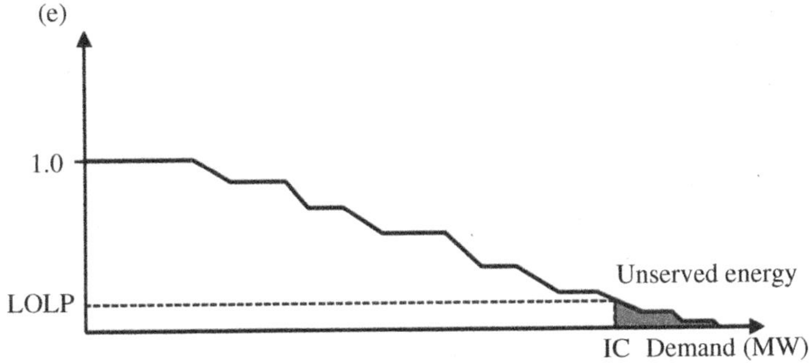

The final equivalent load curve after all plants are convolved.

Figure 3.5 (Continued)

It is well known that the computational effort and burden of most planning packages is primarily due to the numerical convolution processes involving the inverted load duration curve as described above. The numerical error accumulates as the convolution is repeated each time and the convolution increases if the number of points to represent the load duration curve or the increment for a piecewise linear representation of each equivalent inverted load duration curve is increased. Also, when the hydro and other limited energy plants are part of the generating system and must be optimally dispatched taking into account thermal units' forced outages requiring repeated loading and unloading of thermal units (convolutions and deconvolutions), the efficiency-accuracy trade-offs become particularly pronounced. In the conventional method, it is common to use only one load curve to represent a yearly load demand. If the different load curve for each season of a year is used, the number of convolution processes would increase as well, thus resulting in a big increase in computational effort.

4

Load Model and Generation Expansion Planning

4.1 Introduction

The problem of long-range generation expansion planning has received considerable attention and the motivation for more efficient and more sophisticated techniques for evaluating utility expansion policies has been increased during the past decades. In these studies, there has always been an issue of trade-off between computational effort and accuracy of the desired solution.

The basic idea of new approach is similar to the probabilistic production costing of Booth [21] and Baleriaux [22], where the load is modeled as a random variable and generating units are modeled as power sources with the available capacity represented also by a random variable. However, Park et al. [10] presented a new approach for the long-range generation expansion planning, which provides a comparable accuracy with much less computational effort than the conventional one. A completely new analytical approach for the production costing model was developed under the assumption of Gaussian probabilistic distribution of random load fluctuations and plant outages. Analytic formulas were derived not only for operating costs and reliability measures (unserved energy, loss-of-load probability, etc.), but also for their marginal values.

The problem considered is to determine the most economical and reliable generation expansion plan in order to meet forecasted loads over a long-range horizon (usually 5–30 years), subject to a multitude of technical, economical, and social constraints. The method presented in [10] uses the discrete optimal control theory to determine optimally the annual investment in new generating capacity, while the Hamiltonian minimization problem within the dynamic optimization process is solved by using a version of the gradient projection method which has been developed specially for this purpose. In the new algorithm, system reliability measures such as loss-of-load probability (LOLP) or expected unserved energy can be

Probabilistic Power System Expansion Planning with Renewable Energy Resources and Energy Storage Systems, First Edition. Jaeseok Choi and Kwang Y. Lee.

Published 2022 by John Wiley & Sons, Inc.

considered as operating constraints rather than supply-shortage cost. Also, a variety of constraints (financial, social, etc.) can be introduced similarly as operating constraints.

4.2 Analytical Approach for Long-Term Generation Expansion Planning

This section develops analytic formulas required to compute the expected annual cost and reliability measures. The basic idea of the analytical approach is similar to the probabilistic production costing model of Booth and Baleriaux [21, 22], where the load is modeled as a random variable and generating units are modeled as power sources with the available capacity represented also by a random variable. However, the analytical production-costing model for the long-range generation expansion planning problem is developed under the assumption of Gaussian probability distribution of random load fluctuations and plant outages.

One of the primary features of the presented approach is the direct use of load curves rather than the equivalent inverted load duration curves (LDCs). It is well known that the computational burden of most planning packages is primarily due to the numerical convolution processes involving LDCs. The numerical error accumulates as the convolution is repeated each year. By avoiding the use of LDCs, not only is the computational requirement reduced in the order of magnitudes, but also is the great accuracy achieved as will be demonstrated in later sections. In order to reduce the computational requirement, an effort is made to develop analytical formulas rather than numerical (convolution) approaches. This is achieved by assuming Gaussian probability distribution on the random load fluctuations and plant outages.

As another feature, several load curves can be used to represent an annual load demand without increasing the computational effort too much as in the conventional method. Thus, a representation of annual demand becomes more realistic and accurate.

Finally, the analytical approach enables the use of optimal control theory for the long-range generation expansion planning problem to determine optimally the annual investment in new generating capacity, where the Hamiltonian minimization problem within the dynamic optimization process reduces the computational burden in the order of magnitude of the number of investment options. In this approach, the system reliability measures such as loss-of-load probability or expected unserved energy can be considered as operating constraints rather than supply-shortage cost. Also, a variety of constraints (financial, social, etc.) can be introduced similarly as operating constraints.

4.2.1 Representation of Random Load Fluctuations

The load at a particular time of the day of a week in a given season fluctuates randomly and it is reasonable to assume that it behaves with Gaussian distribution since its value is obtained from a large number of previous historical data. Figure 4.1 shows the Gaussian random load fluctuations.

At time t_k, $k = 1, 2, 3$, the random load is represented by a normal distribution with the statistical mean $\overline{L}_k$ and variance σ_k. Any kind or length of the load curve can be used for the generation expansion purpose.

It is, however, very common to use a weekly load curve because it represents more accurately the behavior of load demand for a season. When a weekly load curve is used, there are several options to represent it as follows:

1) use seven daily load curves, each with T time-bands, or
2) use one equivalent weekday load curve with a duration of five days and T time-bands, and one equivalent weekend load curve with a duration of two days and T time-bands, or
3) use one equivalent load curve with a duration of seven days and T time-bands.

It is important to remember that in a planning problem, although the overall pattern of the load on an electric utility is predictable, it is difficult to make hourly

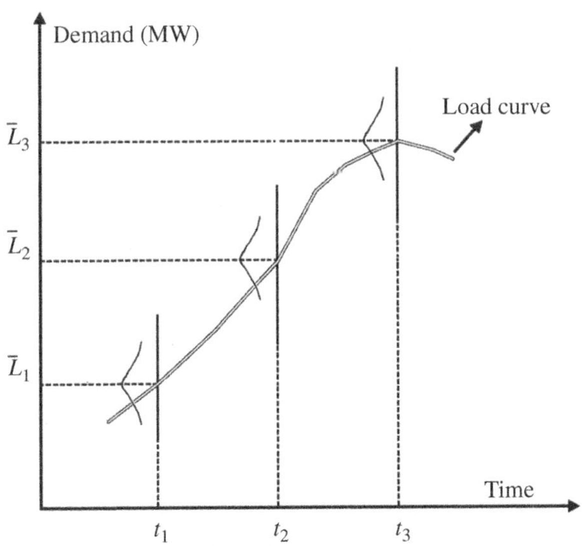

Figure 4.1 Gaussian random load fluctuations.

prediction due to its randomness. For this reason, in the conventional method, the time-independent load curve is converted into an LDC. Even in generation planning with limited energy and storage plants, the model of storage is an approximation using LDCs, which ignores the inherent chronological nature of storage operations since a chronological model would require too much detail and computation to be suitable for use in a long-range generation planning model.

In the analytical method, the load curve can be either time-dependent or time-independent. For example, option (1) above mentioned is time-dependent and options (2) and (3) are time-independent. Options (2) and (3) would be more suitable for long-range generation planning problem. The study shows that the difference among the results obtained by using different options of load curve is insignificant.

The statistical means and variances for past years can be easily computed following standard statistical analysis. However, they cannot be computed for future years and must be extrapolated on the basis of past values as well as other economic and social considerations. This is normally done in load forecasting and the long-range planning problem is to determine the most economical and reliable generation expansion plan to meet these forecasted loads.

The random load $L_{i,s,t}$ (MW) in each time-band t can be represented by a normal distribution with its statistical mean $\overline{L}_{i,s,t}$ (MW) and variance $\sigma^{L^2}_{i,s,t}$ (MW^2). Given the statistical values of future loads on the basis of one equivalent load curve and load forecasting, the expected annual energy demand for year i, $\overline{D_i}$, can be accounted as

$$\overline{D_i} = \sum_{s=1}^{S}\sum_{t=1}^{T} n_s \tau_t \overline{L}_{i,s,t} = \sum_{s=1}^{S}\sum_{t=1}^{T} n_s \tau_t \overline{L}_{\Delta} \text{ (MWh)} \tag{4.1}$$

where

$\Delta = i, s, t$: index of year, season, and time-band, respectively,

$\overline{L}_{\Delta}$: the statistical mean of the random load L_{Δ} for year i, season s, and time-band t(MW),

τ_t: length of time-band t(h),

T: number of time-bands in a load cycle,

n: number of load cycles in season s,

S: number of seasons in a year.

Here, the number of time-bands, T, represents the number of intervals that a load cycle can be broken into, and τ_t is the length of each time-band, not necessarily of the same length, resulting from the division. Similarly, the number of seasons, s, can be chosen to reflect how many distinct seasons one wishes to use to represent a year.

4.2.2 Available Generation Capacities

The generating plants are grouped into several types, such as nuclear, base oil, intermediate oil, base coal, intermediate coal, peaking gas turbine, etc. All units of the same plant type are assumed to have the same economic and technical characteristics. This makes it possible to represent each type with a number of identical single units of the same size. The unit capacity, of course, will differ from type to type.

We will first consider the installed capacity for each type and then represent the available generation capacity by incorporating random plant outages, maintenance requirements, energy resource availabilities, and the teething factors. The teething factor is defined to be a discount factor to reflect the nature of a newly installed unit which cannot operate at its full capacity during the first year of operation.

The installed capacity of plant type j in year i, g_i^j, is the sum of the previously installed capacity and newly added capacity, i.e.

$$g_i^j = x_i^j + u_i^j \tag{4.2}$$

where

x_i^j: previously installed capacity of plant type j in year i (MW) (state variables),
u_i^j: newly added capacity of plant type j in year i (MW) (decision variables).

The probabilistic nature of plant outages of units in each plant type j makes the available generation capacity y_Δ^j (in time-band t of seasons of year i) as a random variable. Each unit is considered to have two stages of operation: full-capacity operation with availability p, and full outage with failure rate $1 - p$. Each plant type has many such single units, which are assumed to be mutually independent. Therefore, according to the law of large numbers, the total available generation capacity y_Δ^j for type j can be represented by a normal distribution with its statistical mean $\bar{y}_\Delta^j$ and variance $\sigma_\Delta^{j^2}$, which are derived as

$$\bar{y}_\Delta^j = p^j\left(1 - v_{i,s}^j\right)\beta_\Delta^j\left(x_i^j + \alpha^j u_i^j\right) \quad \text{(MW)} \tag{4.3}$$

$$\sigma_\Delta^{j^2} = p^j(1 - p^j)\left(1 - v_{i,s}^j\right)\beta_\Delta^j a^j\left(x_i^j + \alpha^j u_i^j\right) \quad \text{(MW)} \tag{4.4}$$

where

$\Delta = i, s, t$: index of year, season, and time-band, respectively,
x_i^j: previously installed capacity of plant type j in year i(MW),
u_i^j: newly added capacity of plant type j in year i(MW),

α^j: teething factor for units in type j(p.u.),

β^j_Δ: energy resource distribution factor of plant type j in time-band t of season s of year i (= 1 for non-hydro plant type) (p.u.),

$v^j_{i,s}$: maintenance rate of plant type j in seasons s of year i(HW),

p^j: availability of units in plant type j(p.u.),

a^j: unit capacity of plant type j(MW).

The energy resource distribution factors β's are fixed to one for non-hydro plants. But they can be determined optimally for hydro plants to represent the optimal use of hydro resources. The maintenance rates v's are also fixed in our problem but can be determined optimally from the maintenance scheduling subproblem.

4.2.3 Expected Plant Outputs

Comparing forecasted loads and available generation capacities, the expected plant outputs can be determined optimally in the production scheduling. If the generating plant types are confined to the non-hydros, the optimal solution to this operating problem can be based upon the conventional loading-order concept. The 1oading-order is defined as the economic merit order in which plant type j is loaded in the order of increasing operating costs.

Figure 4.2 shows an example of the loading of generating plants in order to meet the demand at time-band k. Although the generating units use the same fuel type, they can be grouped into different plant types according to their characteristics.

In the conventional method, when the hydro plants are part of the generating systems, the peak-shaving operation is equivalent to finding the position for the hydro unit as high as possible in the order which will fully utilize the assigned energy from the inverted LDC. This process is complex and involves the loading and unloading of generating units, i.e., convolutions and deconvolutions.

The use of the load curve and the analytical production costing model, however, eliminates the complicated procedure for finding the loading position of the hydro unit.

Let $j = 1, 2, \ldots, J$ be the indices of plant types already ordered in the order of increasing operating costs. The total power output from plant type 1 up to type j, ${}^jP_\Delta$, does not exceed the loads, and can be expressed as

$${}^jP_\Delta = \min\left(L_\Delta, \sum_{k=1}^{j} y^k_\Delta\right) = \sum_{k=1}^{j} P^k_\Delta \tag{4.5}$$

where L_Δ is the random load, y^k_Δ is the randomly available generation capacity of plant type k, and P^k_Δ is the power output of plant type k alone. Since the load and the available generation capacities have the Gaussian probability distributions, the

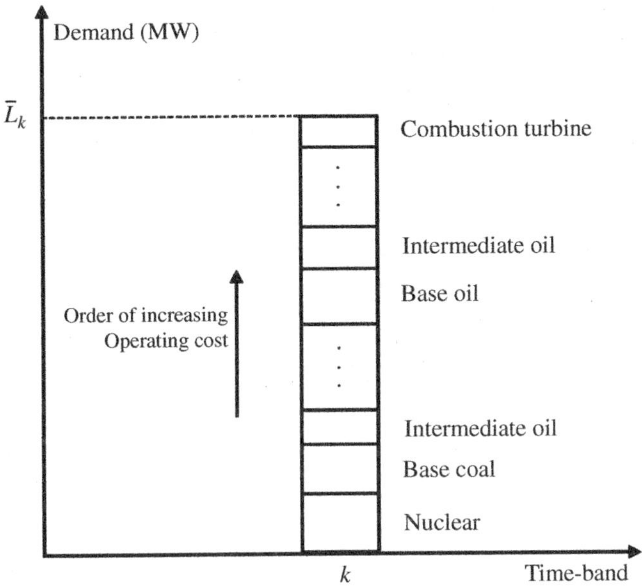

Figure 4.2 Example of loading of generating plants at time-band k to meet the demand $\overline{L}_k$.

total power output ${}^jP_\Delta$ is also a random variable with the Gaussian distributions, whose statistical mean (expected value) and variance are computed as follows:

The probability density function of the total power output (4.5) is [90]

$$f_P(z) = [1 - F_L(z)]f_y(z) + \left[1 - F_y(z)\right]f_L(z) \tag{4.6}$$

where f_P, f_y, and f_L, respectively, are the probability density functions of ${}^jP_\Delta$, $\sum_{k=1}^{j} y_\Delta^k$, and L_Δ, respectively, and F_L and F_y are corresponding distribution functions. Accordingly, the expectation of (4.5) is expressed by

$${}^j\overline{P}_\Delta = \int_{-\infty}^{\infty} z \cdot f_P(z)\ dz \tag{4.7}$$

On the other hand, the assumed Gaussian distributions of the loads and the available capacities and mutual independence among them give the following probability density and distribution functions:

$$f_L(z) = \left[1/\sqrt{2\pi}\sigma_\Delta^L\right] \exp\left[-0.5\left(z - \overline{L}_\Delta\right)^2/\sigma_\Delta^{L^2}\right] \tag{4.8}$$

$$f_y(z) = \left[1/\left(\sqrt{2\pi}\sqrt{\sum_{k=1}^{j} \sigma_{\Delta}^{k^2}}\right)\right] \exp\left[-0.5\left(z - \sum_{k=1}^{j} \overline{y}_{\Delta}^{k}\right)^2 / \sum_{k=1}^{j} \sigma_{\Delta}^{k^2}\right] \tag{4.9}$$

$$F_L(z) = 0.5 + erf\left[(z - \overline{L}_{\Delta})/\sigma_{\Delta}^{L}\right] \tag{4.10}$$

$$F_y(z) = 0.5 + erf\left[\left(z - \sum_{k=1}^{j} \overline{y}_{\Delta}^{k}\right) / \sqrt{\sum_{k=1}^{j} \sigma_{\Delta}^{k^2}}\right] \tag{4.11}$$

where $\overline{L}_{\Delta}$ and σ_{Δ}^{2} are mean and variance of L_{Δ}, $\overline{y}_{\Delta}^{k}$ and $\sigma_{\Delta}^{k^2}$ are mean and variance of y_{Δ}^{k}, and $erf(\bullet)$ and $\exp(\bullet)$ are error function and exponential function, respectively.

Substituting (4.8) through (4.11) in (4.6) and then using (4.7), the mean and variance of ${}^{j}P_{\Delta}$ in (4.5) is derived as [10]:

$${}^{j}\overline{P}_{\Delta} = \overline{L}_{\Delta} - {}^{j}\overline{Z}_{\Delta}\left[0.5 + erf\left({}^{j}\overline{Z}_{\Delta}/{}^{j}\sigma_{\Delta}\right)\right] - \left({}^{j}\sigma_{\Delta}/\sqrt{2\pi}\right) \exp\left(-0.5\,{}^{j}\overline{Z}_{\Delta}^{2}/{}^{j}\sigma_{\Delta}^{2}\right) \text{ (MW)} \tag{4.12}$$

$${}^{j}\sigma_{\Delta}^{2} = \sigma_{\Delta}^{L^2} + \sum_{k=1}^{j} \sigma_{\Delta}^{k^2} \quad (\text{MW}^2) \tag{4.13}$$

where

$${}^{j}\overline{Z}_{\Delta} \triangleq \overline{L}_{\Delta} - {}^{j}\overline{y}_{\Delta} = \overline{L}_{\Delta} - \sum_{k=1}^{j} \overline{y}_{\Delta}^{k} \quad (\text{MW}) \tag{4.14}$$

Here ${}^{j}\overline{Z}_{\Delta}$ represents the difference between the expected load and the sum of the available generating capacities from type 1 to j, and the negative value implies the shortage of generation.

Thus, the expected power output of each plant type, $\overline{P}_{\Delta}^{j}$, can be simply computed as

$$\overline{P}_{\Delta}^{j} = \begin{cases} {}^{j}\overline{P}_{\Delta} - {}^{j-1}\overline{P}_{\Delta} & j = 2, 3, \ldots, J \\ {}^{1}\overline{P}_{\Delta} & j = 1 \end{cases} \quad (\text{MW}) \tag{4.15}$$

where

$\Delta = i, s, t$: index of year, season, and time-band, respectively.

${}^{j}\overline{P}_{\Delta}$: the expected total power output of all plants from type 1 up to type j,

$\overline{P}_{\Delta}^{j}$: the expected power output of plant type j alone.

4.2.4 Expected Annual Energy

The expected annual energy generated by plant type j, $\overline{E}_i^j$, and the expected total energy in year i, $\overline{E}_i$, are

$$\overline{E}_i^j = \sum_{s=1}^{S} \sum_{t=1}^{T} n_s \tau_t \overline{P}_{\Delta}^j \quad \text{(MWh)} \tag{4.16}$$

$$\overline{E}_i = \sum_{j=1}^{J} \overline{E}_i^j \quad \text{(MWh)} \tag{4.17}$$

where τ_t and n_s are the length of a time-band t and the number of load cycles in a season s, respectively, as defined in (4.1).

4.2.5 Reliability Measures

4.2.5.1 Expected Annual Unserved Energy

The expected unserved power $\overline{N}_{\Delta}$ is the excess of the expected load compared to the expected total power output from all J types, i.e., by setting $j = J$ in (4.6),

$$\overline{N}_{\Delta} = \overline{L}_{\Delta} - {}^{J}\overline{P}_{\Delta} \quad \text{(MWh)} \tag{4.18}$$

Consequently, the expected annual unserved energy over a year is

$$R_i = \sum_{s=1}^{S} \sum_{t=1}^{T} n_s \tau_t \overline{N}_{\Delta} \quad \text{(MWh)} \tag{4.19}$$

where

$\Delta = i, s, t$: index of year, season, and time-band, respectively,
$\overline{L}_{\Delta}$: the expected load (MW),
${}^{J}\overline{P}_{\Delta}$: the expected plant total power output from all plant types, i.e., from type 1 to type J (MW).

4.2.5.2 Annual Loss-of-Load Probability

The positive value of the random variable ${}^{j}\overline{Z}_{\Delta}$ defined in (4.8) is a shortage of generation. The probability density function of ${}^{j}\overline{Z}_{\Delta}$, f_Z, is Gaussian since f_y and f_L are Gaussian. Thus the loss-of-load probability (LOLP) in each time-band is

$$\begin{aligned} LOLP_\Delta &= \int_0^\infty f_Z(z)\ dz \\ &= 1/\left(\sqrt{2\pi}\,{}^J\sigma_\Delta\right)\int_0^\infty \exp\left[-\left(z-{}^JZ_\Delta\right)^2/\left({}^J\sigma_\Delta\right)^2\right]\ dz \quad \text{(p.u.)} \\ &= 0.5 + erf\left({}^J\overline{Z}_\Delta/{}^J\sigma_\Delta\right) \end{aligned} \tag{4.20}$$

where ${}^J\overline{Z}_\Delta$ and ${}^J\sigma_\Delta$ are defined in (4.8) and (4.7), respectively.

Consequently, the annual LOLP is integrated over a year is simply obtained by integrating $LOLP_\Delta$ over a year as

$$LOLP_i = \left(\sum_{s=1}^{S}\sum_{t=1}^{T} n_s\tau_t \cdot LOLP_\Delta\right)/w \quad \text{(p.u.)} \tag{4.21}$$

where w is the total hour in a year.

4.2.6 Expected Annual Cost

Having computed the generation capacities and the expected total energy generated, the economic annual cost can be computed in a straightforward manner.

The annual cost G_i is made of three cost terms: the capital cost C_i, fuel cost F_i, and the other non-fuel maintenance cost term M_i. All costs are, of course, present-values, and are escalated and discounted during the projected period. Moreover, in the case of capital cost, a credit is given as a salvage value for the unused portion of the plant life. In summary, the expected annual cost is

$$\begin{aligned} G_i\left(x_i^j, u_i^j\right) &= C_i + F_i + M_i \\ &= \sum_{j=1}^{J}\left\{pc_i^j c_i^j r_i^j u_i^j + pf_i^j f_i^j \overline{E}_i^j + pm_i^j m_i^j\left(x_i^j + u_i^j\right)\right\} \end{aligned} \tag{4.22}$$

where

pc_i^j, pf_i^j, pm_i^j: factors to levelize the resultant present-value unit capacity cost, unit fuel price and unit non-fuel maintenance cost, respectively, of plant type j in year i(p.u.),

c_i^j, f_i^j, m_i^j: unit capacity cost ($/MW), unit fuel price ($/MWH), and unit non-fuel maintenance cost ($/MW), respectively, of plant type j in year i(p.u.),

r_i^j: factor for the salvage value of capital cost for type j installed in year i(p.u.),

x_i^j: previously installed capacity of plant type j in year i (MW) (state variables),

u_i^j: newly added capacity of plant type j in year i (MW) (decision variables),

$\overline{E}_i^j$: the expected total energy generation of plant type j in year i (MWh) defined in (4.17).

The decision variables are the newly added capacities u_i^j for each year and are determined optimally in the optimization problem.

The factors to levelized the resultant present values and the factor for the salvage value of capital cost are obtained by using the interest rate and/or discount rate as follows:

$$pc_i^j = \left(1 + e_c^j\right)^{i+b} / (1 + d)^{i+b} \tag{4.23}$$

$$pf_i^j = \left(1 + e_f^j\right)^{i+b} / (1 + d)^{i+b+0.5} \tag{4.24}$$

$$pm_i^j = \left(1 + e_m^j\right)^{i+b} / (1 + d)^{i+b+0.5} \tag{4.25}$$

$$r_i^j = \begin{cases} (1 + I - i)/L^j, & \text{for straight-line depreciation} \\ \left[1 - (1 + d)^{I+1-i-L^j}\right] / \left[1 - (1 + d)^{-L^j}\right], & \text{for sinking-fund depreciation} \end{cases} \tag{4.26}$$

where

e_c^j, e_f^j, e_m^j: escalation for capital, fuel, and non-fuel maintenance costs, repectively (p.u.)
d: discount rate for costs (p.u.)
b: number of years from the escalation base year or from the present-value base year to the first year of study,
L^j: economic life of planning years.

4.2.7 Expected Marginal Values

The expected marginal values of annual costs and reliability measure are used to find the optimal planning solution. The expected marginal values are computed analytically by taking the partial derivatives of the reliability measure R_i in (4.19) and the expected annual cost G_i in (4.22) with respect to the previously installed capacity x_i^k and the newly installed capacity u_i^k. They are summarized as follows:

$$\begin{aligned} \frac{\partial R_i}{\partial u_i^k} = \alpha^k \sum_{s=1}^{S} \sum_{t=1}^{T} n_s \tau_t \Big[& - P^k (1 - v_{i,s}^k) \beta_\Delta^I \{0.5 + erf\{(J_{Z_\Delta}/J_{\sigma_\Delta})\} \\ & + \left\{ P^k (1 - P^k)(1 - v_{i,s}^k) \beta_\Delta^k a^k / \left(2\sqrt{2\pi} J_{\sigma_\Delta}\right) \right\} \exp\left(-0.5 J_{Z_\Delta}^2 / J_{\sigma_\Delta}^2 \right) \Big] \end{aligned} \tag{4.27}$$

$$\begin{aligned}\frac{\partial G_i}{\partial x_i^k} &= \frac{\partial F_i}{\partial x_i^k} + \frac{\partial M_i}{\partial x_i^k} \\ &= m_i^k + \sum_{j=1}^{J} f_i^j \sum_{s=1}^{S}\sum_{t=1}^{T} n_s\tau_t \left[P^k\left(1-v_{i,s}^k\right)\beta_\Delta^k(h(j-k)\{0.5\right. \\ &\quad + erf\left(j_{\overline{Z}_\Delta}/j_{\sigma_\Delta}\right)\} - h(j-1-k)\left\{0.5 + erf\left(j-1_{\overline{Z}_\Delta}/j-1_{\sigma_\Delta}\right)\right\}) \\ &\quad + P^k\left(1-P^k\right)\left(1-v^k\right)\beta_\Delta^k a^k/\left(2\sqrt{2\pi}\, j_{\sigma_\Delta}\right)\}(-h(j-k) \\ &\quad \left.\cdot \exp\left(-0.5 j_{\overline{Z}_\Delta}^2/j_{\sigma_\Delta}^2\right) + h(j-1-k)\exp\left(-0.5j-1_{\overline{Z}_\Delta}^2/j-1_{\sigma_\Delta}^2\right)\right] \ (\$/\text{MW})\end{aligned} \tag{4.28}$$

$$\begin{aligned}\frac{\partial G_i}{\partial u_i^k} &= \frac{\partial C_i}{\partial u_i^k} + \frac{\partial F_i}{\partial u_i^k} + \frac{\partial M_i}{\partial u_i^k} \\ &= c_i^k r_i^k + m_i^k + \alpha^k \frac{\partial F_i}{\partial x_i^k} \quad (\$/\text{MW})\end{aligned} \tag{4.29}$$

where

$$h(x) = \begin{bmatrix} 1, & x \geq 0 \\ 0, & x < 0 \end{bmatrix} \tag{4.30}$$

The expected marginal values of annual costs and reliability measures are computed in the basic annual operation problems and then transferred to the master investment problem to find the optimal long-term generation expansion planning solution.

4.3 Optimal Utilization of Hydro Resources

4.3.1 Introduction

In the previous section, a new analytical production costing model was developed for use in long-range generation expansion planning. In the model, all generating plants were considered to be thermal type, such as fossil-fueled or nuclear, which are assumed to produce unlimited energy.

In modeling a real utility system, it is often necessary to include hydro plants in generating system. The hydro plants are considered as limited energy plants since they, unlike thermal plants, are subject to reservoir constraints and seasonal variations in rainfall, which limit the maximum available energy for generation. Because of the low production cost associated with this type of generation, it is economical to fully utilize all available hydro energy replacing high-cost fossil peaking energy. This is defined as a peak-shaving operation. Often, a portion of the hydro capacity is unavailable for peak-shaving due to run-of-river hydro capacity requirements. Thus, two types of hydro capacity, continuous-on-stream and peak-shaving, are represented as a two-piece unit. The base piece capacity equaling the run-of-river capacity is considered the same as the thermal unit in the production costing model.

In this section, a new method of peak-shaving operation is presented to determine the optimal use of the available hydro resources, when the hydro plants are part of the generating system, based on the production costing model of the previous section.

4.3.2 Conventional Peak-Shaving Operation and its Problems

Using hydro units to peak shave is a sound operation strategy because of the low production cost associated with this type of generation, i.e., no fuel cost. As the thermal units with expensive operating cost occupy high positions in the loading order, it is highly desirable to replace the expensive fossil peaking energy by hydro energy. The peak-shaving operation is illustrated in Figure 4.3.

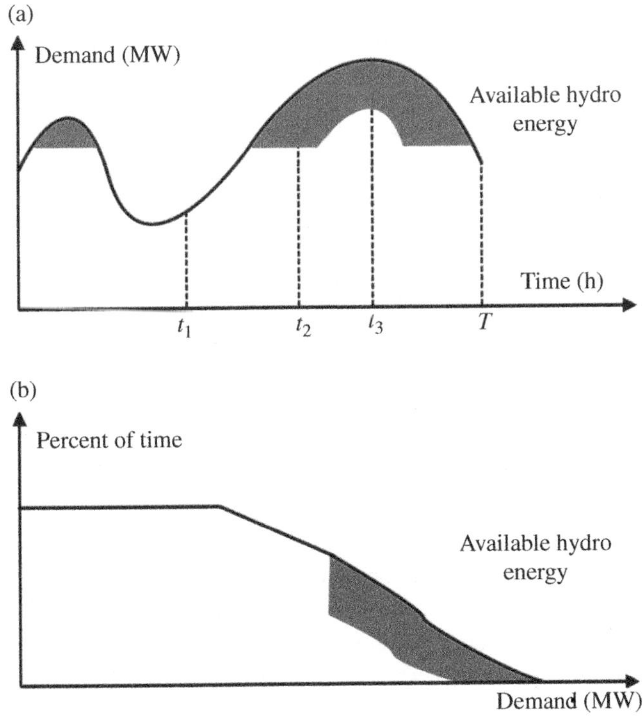

Figure 4.3 Peak-shaving operating from different type of load curves.

In the conventional method, the peak-shaving operation is equivalent to finding the optimal loading order which will fully utilize the assigned energy based on the inverted LDC. This is shown in Figure 4.3b.

There have been proposed several different algorithms to determine the optimal loading position of the hydro unit [17, 24–26]. These algorithms, however, are generally complex and require repeated loading and unloading of thermal units until the optimal loading position of the hydro unit is found. Thus, the efficiency-accuracy trade-offs become more pronounced when the hydro plants are included in the generating system.

4.3.3 Peak-Shaving Operation Based on Analytical Production Costing Model

4.3.3.1 Basic Concept

The reason why the conventional method involves convolution and deconvolution processes to load and unload the generating unit is that it is based on the LDC. The analytical production costing model based on the original load curve makes it possible to eliminate the convolution process to load the unit. Now, the optimal peak-shaving operation also can be determined without the use of convolution and deconvolution processes by applying the concept of peak-shaving operation from the original load curve, as shown in Figure 4.3a, and by using the analytical production costing model.

The hydro plant has a limit for its energy to be generated in each time-band. It is constrained not only by the installed capacity of the plant, but also by the available hydro resources. When (4.3) was derived to compute the available generating capacity of a plant in each time-band, the energy resource distribution factor β_{Δ}^{j} was introduced in order to take into account its limitation by the available hydro resources.

Suppose that the load curve of Figure 4.3a is the Gaussian load curve represented only by statistical means. As the most economic use of the available hydro energy is realized when it replaces high-cost fossil peaking energy, its distribution must start from the top portion of the load curve. Suppose the peak-shaving operation of one hydro plant is executed up to dotted line. Then, at time t_3, it is expected to generate the hydro energy as much as possible by using its total available generating capacity, i.e. $\beta_{t3} = 1$. At time t_2, it is necessary to limit its available capacity, i.e., $0 < \beta_{t2} < 1$. At time t_1, as no hydro energy is assigned to be generated, the available generating capacity must be zero. It implies that it is not economical to replace the expensive peaking energy by hydro energy at this time, i.e., $\beta_{t1} = 0$.

Therefore, the peaking operation can be accomplished if the energy distribution factors are determined optimally for all time. They can be determined optimally by minimizing total fuel cost subject to hydro energy constraint since the objective of the optimal peak-shaving operation is to minimize the operation cost, particularly the fuel cost.

4.3.3.2 Peak-Shaving Operation Problem

The peak-shaving operation problem is to determine the optimal use of the available hydro resources and can be stated as follows:

$$\text{Mimimize} \quad F\left(\beta_t^j\right) = \sum_{j=1}^{J} f^j \overline{E}^j \tag{4.31}$$

$$\text{Subject to} \quad \sum_{t=1}^{T} \overline{E}_t^j = \overline{W}^j \tag{4.32}$$

where

F: total fuel cost ($),
β_t^j: energy resource distribution factor of plant j in time-band t (p.u.)
f^j: unit fuel price of plant type j ($/MWh)
$\overline{E}^j$: total energy generated by plant type j (MWh)
$\overline{E}_t^j$: energy generated by plant type j in time-band t (MWh)
$\overline{W}^j$: available hydro energy of plant type j (MWh)

The (4.32) represents the hydro balance constraint, i.e., the sum of energy generated in each time-band t must be equal to its available hydro energy.

This is a typical nonlinear programming problem with nonlinear constraints. Although it can be solved by using the conventional optimization technique, such as the gradient projection method, there is a simple way of solving it by combining the concept of peak-shaving operation from the original load curve with the analytical production costing model developed in the previous section.

4.3.4 Optimization Procedure for Peak-Shaving Operation

Once again, consider a smooth and continuous Gaussian load curve for better illustration of the optimization procedure, although it is divided into T time-bands. Figure 4.4 shows the peak-shaving operation performed by two hydro plants.

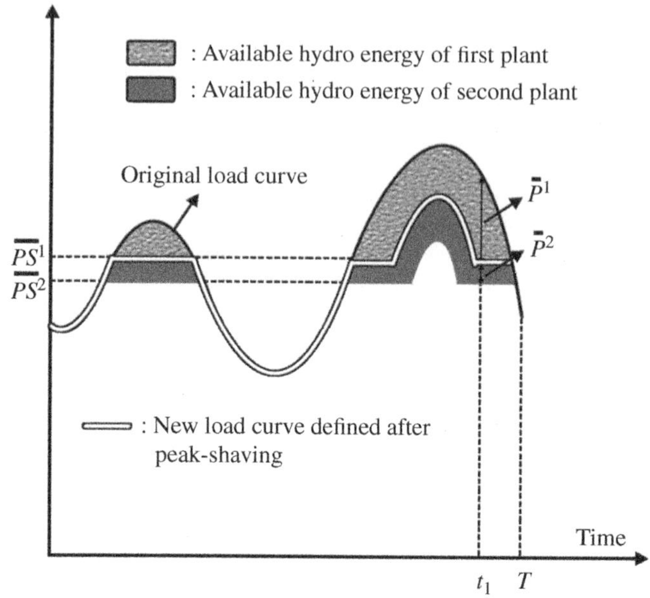

Figure 4.4 Peak-shaving operation performed by two hydro plants.

where

$\overline{P}^1$: expected power output of the first hydro plant
$\overline{P}^2$: expected power output of the second hydro plant

The power level $\overline{PS}^j, j = 1, 2$, indicates the optimum power level of plant j for peak-shaving, i.e., each of the two different shaded areas is equal to the available hydro energy of each plant.

In order to compute the optimal $\beta^1_{t_1}$ at time t_1 for the first hydro plant, it is necessary to find $\overline{PS}^1$ first. Once it is found through the optimization procedure, a new load curve represented by the dotted line in Figure 4.4 is defined. Then, the expected power output of the first plant at time t_1 can be calculated by computing a difference between loads of the original load curve and new load curve at time t_1.

Although the total output of the system is generally different from the total generating capacity at a particular time, the power output of each plant is loaded with its full capacity. Thus, for the total hydro capacity less than the minimum load, the optimal β^j_t can be determined accurately by (4.3). Even in the case that the total hydro capacity is greater than the minimum load, the values computed from (4.3) are near-optimal values. If these values do not satisfy the convergence requirement to be the optimal solution, the penalty factor can be introduced in (4.3) or the gradient projection method can be used adopting the solution found by this method as an initial trial solution.

The optimization procedure to find optimal $\overline{PS}^j$ is as follows:

1) $jh = 1$
2) Load non-hydro plants in each time-band under the original load curve and compute $\overline{P}^j$, $\overline{E}^j$, $\overline{E}$, and F by using (4.15)–(4.17). Here the system is considered to consist only of thermal plants and the indices of season and year are disregarded in the equations.
3) Set the initial optimal power level of hydro plant jh, $\overline{PS}^{jh}$. It can be set arbitrarily, for example, 15% below the value of peak demand in the original load curve.
4) Define the new load curve as shown in Figure 4.4. Consider it as an original load curve and repeat from Step (2).
5) Compute the difference in hydro energy balance constraint (4.32).

$$\overline{ED} = [\overline{E} \text{ in } 1)] - \left(\overline{W}^{jh} + [\overline{E} \text{ in } 4)]\right) \tag{4.33}$$

where
$\overline{E}$: total expected energy (MW)
$\overline{W}^{jh}$: available hydro energy of plant type jh (MWh)
$\overline{ED}$: difference in energies (MWh)

6) a. If $|\overline{ED}| < \varepsilon$, go to Step (9)
   b. Otherwise,

$$\overline{PS}^{jh} \rightarrow \overline{PS}^{jh} + \overline{PS} \quad \text{if } \overline{ED} > 0$$
$$\overline{PS}^{jh} \rightarrow \overline{PS}^{jh} - \overline{PS} \quad \text{if } \overline{ED} < 0$$

where
$\overline{PS}$: pre-fixed increment of power level (MW)
ε: convergence criteria (MWh)

7) Repeat Steps (3)–(5).
8) Compare the signs of $\overline{ED}$ calculated in Step (7) within Step (6).
   a. If both are the same, go to Step (6).
   b. If both are different, proceed.
9) Find the optimal power level $\overline{PS}^{jh}$ by using the bisection method. The change of sign in $\overline{ED}$ indicates that the optimum value of $\overline{PS}^{jh}$ lies between two power levels found in the last two iterations. The bisection method is simple and

powerful to find the optimal $\overline{PS}^{jh}$ because the convergence of the solution is guaranteed and the value of $\overline{ED}$ is provided as a byproduct. Steps (3)–(5) are repeated in this step until the optimum solution is found.

10) Load the hydro and non-hydro plants in each time-band under the original load curve and compute $\overline{E}^j$, $\overline{E}$, LOLP, unserved energy, etc.
11) Change the original load curve by the latest new load curve found in Step (9).
12) If $jh = JH$, where $JH =$ the total number of hydro plants, stop. Otherwise, $jh \rightarrow jh + 1$, and go back to Step (2).

4.4 Long-Range Generation Expansion Planning

4.4.1 Statement of Long-Range Generation Expansion Planning Problem

The long-range problem for a system of generating plants is to minimize the present value of total costs subject to reliability constraints and is now stated formally as follows:

$$\underset{u_i^j,\,\beta_\Delta^j}{\text{minimize}} \quad G = \sum_{i=1}^{I} G_i\left(x_i^j, u_i^j, \beta_\Delta^j\right) \tag{4.34}$$

such that

$$x_{i+1}^j = x_i^j + u_i^j \tag{4.35}$$

subject to

$$R_i\left(x_i^j, u_i^j\right) \le \varepsilon_i \tag{4.36}$$

$$u_i^j \ge 0 \tag{4.37}$$

(For hydro plants, only $j = 1,\ldots, JH$)

$$\sum_{s=1}^{S}\sum_{t=1}^{T} n_s \tau_t \overline{P}^j = \sum_{s=1}^{S} \overline{E}_{i,s}^j = \overline{W}_i^j \tag{4.38}$$

$$\overline{E}_{i,s}^j \le \overline{EM}_{i,s}^j \tag{4.39}$$

$$0 \le \beta_\Delta^j \le 1 \tag{4.40}$$

where

$i = 1, 2, \ldots, I$: index of years in the planning horizon,

$s = 1, 2, ..., S$: index of seasons in a year,
$j = 1, 2, ..., J$: index of plant types,
x_i^j: previously installed capacity of plant type j in year i(MW) (state variables),
u_i^j: newly added capacity of plan type j in year i(MW) (decision variables),
β_{Δ}^j: energy resource distribution factor of plant type j in time-band t of season s of the year i (p.u.),
G_i: present-value expected annual cost (\$),
R_i: expected annual unserved energy (MWh),
ε_i: desired reliability level of annual unserved energy (MWh),
$\overline{W}_i^j$: available hydro energy of plant type j in the year i (MWh)
$\overline{EM}_{i,s}^j$: maximum available hydro energy of plant type j in season s of the year i (MWh),
$\overline{P}_{\Delta}^j$: power output of plant type j in time-band t of season s of the year i (MW),
JH: total number of hydro units.

This problem can be decomposed into a master problem, basic subproblems, and hydro subproblem. The master problem is to determine optimally the yearly investment in new generating capacity. The basic subproblems are used to determine the minimum operating cost of generating systems and annual reliability measures. When hydro plants are part of the generating system, the hydro subproblem determines the optimal distribution of hydro resources by minimizing the operating cost.

4.4.1.1 Master Problem and Basic Subproblems

The master problem for the long-range generation expansion planning is stated as follows:

$$\underset{u_i^j}{\text{minimize}} \quad G = \sum_{i=1}^{I} G_i\left(x_i^j, u_i^j\right) \tag{4.41}$$

such that

$$x_{i+1}^j = x_i^j + u_i^j \tag{4.42}$$

subject to

$$R_i\left(x_i^j, u_i^j\right) \le \varepsilon_i \tag{4.43}$$

$$u_i^j \ge 0 \tag{4.44}$$

Equation (4.31) is the dynamic state equation describing the state of installed capacities each year, that is, the sum of previously installed capacity and the newly added capacity is the installed capacity for the next year. Considering the newly added capacities u_i^j as decision or control variables, the master problem (4.41)–(4.44) is now formulated as an optimal control problem.

According to the Pontryagin's minimum principle of modern optimal control theory [19], the optimal control must satisfy the necessary conditions:

$$x_{i+1}^j = x_i^j + u_i^j, \quad x_i^j = \text{specified (state equation)} \tag{4.45}$$

$$\lambda_i^j = \frac{\partial G_i}{\partial x_i^j} + \lambda_{i+1}^j; \lambda_{i+1} = 0 \quad \text{specified (costate equation)} \tag{4.46}$$

and u_i^j minimizes the Hamiltonian

$$H_i\left(x_i^j, u_i^j, \lambda_{i+1}^j\right) \triangleq G_i\left(x_i^j, u_i^j\right) + \sum_{J=1}^{J} \lambda_{i+1}^j\left(x_i^j + u_i^j\right) \tag{4.47}$$

subject to

$$R_i\left(x_i^j, u_i^j\right) \leq \varepsilon_i \quad \text{(reliability constraint)} \tag{4.48}$$

$$u_i^j \geq 0, \quad \text{(control constraint)} \tag{4.49}$$

where λ_i^j is the costate (adjoint) of plant type j in year i.

Note that the minimization of Hamiltonian H_i is to be done for one year at a time while the minimization of the total cost G in (4.41) needs to be performed for all years in the planning horizon simultaneously. This reduces the number of decision variables u_i^j to J from $J \times I$ in the minimization algorithm.

The minimization of Hamiltonian (4.47) subject to inequality constraints (4.48)–(4.49) is a typical nonlinear programming problem. It is solved by using a version of the Gradient Projection Method (GPM) [20], which has been developed specially for this purpose. The results from several sample studies show remarkable advantages of this method in computational efficiency and reliability.

The reliability constraint (4.48) was linearized using the gradients or marginal values $\partial R_i/\partial u_{i \cdot k}^k$ in (4.27). The marginal values of expected annual costs $\partial G_i/\partial x_i$ in (4.28) and $\partial G_i/\partial u_i^k$ in (4.29) with respect to marginal changes in the previously installed capacity and the newly added capacity, respectively, are also required in the GPM. Since these marginal values are not only required to find the optimal solution, but also provide a planner-useful technical and managerial information, they are computed in the basic subproblems.

4.4.1.2 Hydro Subproblem

The hydro subproblem is to determine the optimal distribution of hydro resources in each year and can be stated as follows:

$$\underset{\beta_\Delta^j}{\text{minimize}} \quad F_i = \sum_{j=1}^{J} pf_i^j f_i^j \overline{E}_i^j \left(\beta_\Delta^j\right) \tag{4.50}$$

subject to

$$\sum_{s=1}^{S} \sum_{t=1}^{T} n_s \tau_t \overline{P}^j = \sum_{s=1}^{S} \overline{E}_{i,s}^j = \overline{W}_i^j \tag{4.51}$$

$$\overline{E}_{i,s}^j \le \overline{EM}_{i,s}^j \tag{4.52}$$

$$0 \le \beta_\Delta^j \le 1 \tag{4.53}$$

(4.51) is the hydro energy balance equation describing that the sum of energies generated by the hydro plant type j in season s during the year i is equal to its annual available hydro energy. Equation (4.52) is the seasonal hydro energy constraint to prevent the case that the amount of distributed energy from the annual available hydro energy to each plant exceeds its available hydro resources.

The optimization method for hydro subproblem is similar to the optimization procedure for peak-shaving operation of the hydro plants, presented in Section 4.4.

4.4.2 Optimization Procedures

The computational steps of solving the optimal control problem (4.34)–(4.40) are summarized as follows:

1) Initially, guess $u_i^j, j = 1, 2, ..., J$, $i = 1, 2, ..., I$, and set $\beta_\Delta^j = 1$ for all time-bands and all plant types.
2) Solve the state (4.45) forward to determine x_i^j, where $j = 1, 2, ..., J$ and $i = 1, 2, ...I$.
3) Solve the costate (4.46) backward to determine λ_i^j, where $j = 1, 2, ..., J$ and $i = I, I-1, I-2, ..., 2, 1$.
4) Set $i = I$.
5) Solve the basic subproblems to determine the optimal annual operation and get other necessary information for year i.
6) Minimize the Hamiltonian H_j in (4.47) subject to the inequality constraints (4.48)–(4.49) to update u_i^j where $j = 1, 2, ..., J$.
7) Solve the hydro subproblem, if any, to update β_Δ^j, where $j = 1, 2, ..., J$, $t = 1, 2, ..., T$, and $s = 1, 2, ..., S$.
8) Solve x_{i+1}^j forward by (4.45).

9) Increase $i \rightarrow i+1$.
10) If i is greater than I, go to Step (11). Otherwise, go to Step (5).

If the cost function in (4.41) is sufficiently converged, terminate. Otherwise, go to Step (3).

4.5 Case Studies

The accuracy and efficiency of the presented method were tested in several different systems. Among the results, two of them are shown in this section: the one for the accuracy test of the analytic formulas developed in Section 4.2 for Optimal Operation and the other for the convergence test of the computational algorithm developed in Section 4.4 for Optimal Investments.

4.5.1 Test for Accuracy of Formulas

The load cycle of Scenario B and the full-scale system of Scenario D from EPRI Synthetic Utility Systems [27] are used as load and generation system data in order to compare the accuracy of solutions by the presented formulas and by the conventional method which is based on the use of equivalent LDCs and their recursive convolutions.

The result of using the conventional method was obtained from EPRI Report EA-1411 [24]. It is shown in Table 4.1 and compared with the result computed using the presented formulas.

Simulation period, peak load, and minimum load are 728 hours, 26 000 MW, and 13 842 MW, respectively. According to the given data, the total installed capacity up to the fifth plant type is 12 600 MW. It is less than the minimum load, thus the capacity factor of each plant type from 1 to 5 should be the same as its availability.

The result by the conventional method, however, as shown in Table 4.1, exhibits inconsistencies as the number of plant types increases. For example, for the fifth plant type, the capacity factor is 0.8399 instead of 0.870. This is due to the recursive convolutions which cause an accumulated error. It can be observed from Table 4.1 that up to the fifth plant type, each capacity factor obtained by using the presented analytic formulas is in perfect match with its corresponding availability. For example, for the fifth plant type, both capacity factor and availability are 0.870.

The overall accuracy was also evaluated by comparing the energy balance in both methods. The sum of total generated energy and unserved energy in each case is compared with the total load demand of 15 017.24 GWH found directly from the load curve used [27]. The result from the conventional method is 14 293.74 GWH

Table 4.1 Comparison between conventional and analytical methods for formula accuracy.

Plant type	Total capacity (MW)	Availability	Conventional method: Energy generated (GWH)	Conventional method: Capacity factor (%)	Analytical method: Energy generated (GWH)	Analytical method: Capacity factor (%)
1. Nuclear	7 200	0.850	4 455.36	85.00	4 455.340 0	85.00
2. Nuclear	800	0.850	495.04	85.00	495.038 8	85.00
3. Coal	800	0.760	442.16	75.92	442.623 2	76.00
4. Coal	1 800	0.790	1 028.11	78.46	1 035.212 9	79.00
5. Coal	2 000	0.870	1 222.92	83.99	1 266.650 0	87.00
6. Coal	6 600	0.920	3 492.61	72.69	4 127.1800	85.90
7. Oil	800	0.760	214.00	53.92	334.931 1	57.51
8. Oil	1 800	0.790	635.52	48.50	733.357 9	55.96
9. Oil	800	0.870	298.41	51.24	333.636 5	57.29
10. Oil	4 600	0.926	1 597.30	47.70	1 465.829 0	43.77
11. Combu. Turbine	4 600	0.760	298.19	8.53	309.3273	8.85
A. Total energy generated (GWH)			14 279.62		14 999.118 7	
B. Unserved energy (GWH)			14.11		18.083 5	
C. Total (A + B)			14 293.73		15 017.202 2	
D. Total load demand in both cases (GWH)				15 017.24		
E. (%) Error in energy balance			4.82		0.00	
F. LOLP (p.u.)			0.0241		0.0256	

or 4.82% error in energy balance, while the result from the analytic formulas is 15 017.20 GWH or 0.00% error.

The LOLP calculated by the conventional method is normally inaccurate since it is obtained from the final equivalent LDC after all units (great in number) are convolved. Thus, in view of the above discussions, the LOLP calculated by simple analytical expression, (4.21), is believed to be more accurate.

In conclusion, the results of the presented analytical formulas are shown to be more accurate in spite of the Gaussian probabilistic approximation of load fluctuations and plant outages in the analytical formulation.

The CPU time required for this example by the presented method is 1.30 seconds.

4.5.2 Test for Solution Convergence and Computing Efficiency

The generation expansion planning problem based on a full-scale system of Scenario D from EPRI Synthetic Utility Systems [27] is considered in order to test the convergence characteristics of the optimal solution and to evaluate the computing efficiency of the presented method. The system has 174 existing units and is described briefly in Table 4.2.

All three weekly load cycles of Scenario D corresponding to spring (or fall), summer, and winter, respectively, are used in order to represent more realistic energy demand in a year since the analytical nature of the presented method allows for the use of different load curves in each season.

The planning horizon (I) of 15 years beginning in 1988, 4 seasons (S) for each year, 4 time-bands (T) for each day, and 1980 as the base year for costs and loads are used in planning, and the growth in energy demand in subsequent years is set at an annual rate of 5.3%. Peak load and energy in the base year are taken to be 19 774 MW and 129 142 GWH, respectively. The reliability standard, expected unserved energy, is set at 0.5% of the energy demand. Future costs escalate at an annual rate of 10%. A discount rate of 15% is used in computing present values. The costs are based on projections for 1980 escalated to the beginning of the

Table 4.2 Data for the existing units.

Plant type	Capacity (MW)	Availability	Capital cost ($/KW)	Fuel cost ($/MWh)	Maintenance cost ($/MW)	Maintenance rate
Nuclear (6 units)	1 200 (7 200)	0.851		2.41	210	0.134 2
Coal (44 units)	600–200 (12 200)	0.845–0.917		4.21	450–516	0.115 1–0.057 5
Oil (28 units)	800–200 (7 800)	0.824–0.919		11.30	195–200	0.095 9–0.575
Combu. turbine (96 units)	50 (4 800)	0.895		12.16	235	0.038 4

Table 4.3 Data for candidate units, availability each year.

Plant type	Capacity (MW)	Availability	Capital cost ($/KW)	Fuel cost ($/MWh)	Maintenance cost ($/MW)	Maintenance rate
1. Nuclear	1200	0.851	1157	2.41	210	0.1342
2. Coal I	400	0.876	735	4.21	450	0.0959
3. Coal II	200	0.917	695	4.21	516	0.0575
4. Oil	400	0.876	341	11.30	195	0.0575
5. Combu. Turbine	50	0.895	152	12.16	235	0.0384

planning horizon. The energy resource distribution factor β_{Δ}^{j} is 1 since all units are non-hydro type. Other necessary data are based on EPRI Report EL2561 [7].

Five types of alternative units are considered to be available for installation in each year, as described in Table 4.3. The teething factor for each plant type is 0.8.

The computed optimal solution is shown in Table 4.4. The third plant type (Coal II) has more added capacity during the planning period than the second plant type (Coal I). This is because, though the Coal I units have slightly lower maintenance cost, the Coal II units have higher availability and lower maintenance rate but the same fuel cost. By varying the planning horizon, it was noticed that each run gives almost identical capacity additions compared to the 15-year run. Most of the combustion turbine units were installed during the last years of the planning period. This is a reasonable solution path since the fuel savings associated with the base nuclear units are less pronounced during the last years of the planning period. If a salvage value assigned to the capital stock along with associated costs is not considered, there will be a tendency of installing more capacities to the lower capital cost units as the year increases.

In this example, the variation of capital costs showed to give a sensitive change to a configuration of added capacities in Table 4.4. For example, an increase of 10% in capital cost of nuclear units caused 36% reduction in their total added capacity while most of this reduced capacity was dislocated to the Coal I and II units as expected. The grand total of added capacities was decreased by 3.2% since the units with higher availability and lower maintenance rate need less capacity to be installed in order to meet the reliability constraints and the load demands. The

Table 4.4 Computed result of 15-year generation expansion planning.

Plant type	**New added capacities (MW)**				
Year	**1**	**2**	**3**	**4**	**5**
1988	1 021		6 898		
1989	1 681				637
1990	1 559				98
1991	1 721	879			
1992	1 942		107		
1993	1 968		130		
1994	1 984				254
1995	2 085		312		
1996	2 633	144	207		
1997	2 820	203	275		
1998	3 037	172	233		
1999	3 118	219	290		
2000	3 156	215	414		
2001	3 329	200	454		
2002		137	1 664		1 824
Total (MW)	32 054	2 287	10 984	0	2 813
Grand total (MW)	48 138				
Total cost ($)	$3.882\,61 \times 10^{10}$				

Table 4.5 CPU time requirements.

No. of load curves	**CPU time (seconds)**
1	11.98
2	16.02
3	22.32

total cost, though the capital cost was increased for nuclear units, decreased by 0.7% due to the less amount of the total installed capacity.

The optimization algorithm of the master routine was iterated nine times by means of AS 9000/N computer system. The CPU time for this example is

22.32 seconds, which shows that the presented method is significantly much faster than the conventional one due to its analytical nature. Table 4.5 shows CPU time requirements with respect to the number of different load curves or seasons used in a year. The conventional method normally uses one shape of the load curve for the whole planning period.

4.6 Conclusion

This chapter presents a new analytical approach to optimal long-range generation planning with the following conclusions:

1) The analytic formulas are shown to be sufficiently accurate in spite of the assumed Gaussian distribution of loads and plant availability.
2) The investment optimization algorithm, based on the discrete maximum principle and the gradient projection method is shown to be reliable and much faster than the conventional one due to its analytical nature.
3) The new algorithm allows for the incorporation of any kind of constraints that are found necessary.
4) The extended studies on both the hydro and maintenance schedule subproblems will supplement and complete the research on the optimal long-range generation expansion planning.

5

Probabilistic Production Simulation Model

5.1 Introduction

The analysis of the supply reserve rate (SRR) is becoming more important because the power system is larger than in the past and it is expected that new renewable generation will be installed aggressively. Korea has been using the LOLE (Loss of Load Expectation) criterion to determine the adequacy of installed capacity requirement. The criterion is that load shall not exceed the available capacity more than 5 days in 10 years. It is difficult to find the adequacy of SRR that meets this criterion. One of the reasons is that the LOLE criterion is one of the probabilistic reliability criteria; however, SRR is one of the deterministic reliability criteria. In order to identify the relationship between the two, a fast and accurate algorithm, called PRASim, is developed, which is based on an exact method, Booth-Baleriaux method, and approximation methods, Cumulant method and Mixture of Normals Approximation method (MONA). In this chapter, the probabilistic reliability evaluation and the PRASim was used to obtain the relationship. Although there are references in detail, case studies about the relationship between LOLE and SRR in Korea power system are introduced in Section 5.5 (Appendix). [2, 15, 16, 28, 87, 106, 116, 117, 118].

5.2 Effective Load Distribution Curve

The actual system at HLI in Figure 5.1a can be represented by an equivalent system, as shown in Figure 5.1b. The capacity and forced outage rate (FOR) of the ith generator are C_i (MW) and q_i respectively in Figure 5.1a. It is equivalent to considering the capacity and FOR of the generator as C_i (MW) and 0 with the fictitious load added to the original load. The capacity and FOR of the fictitious load for the ith generator are C_i (MW) and q_i respectively as shown in Figure 5.1b. The effective

Probabilistic Power System Expansion Planning with Renewable Energy Resources and Energy Storage Systems, First Edition. Jaeseok Choi and Kwang Y. Lee.

Published 2022 by John Wiley & Sons, Inc.

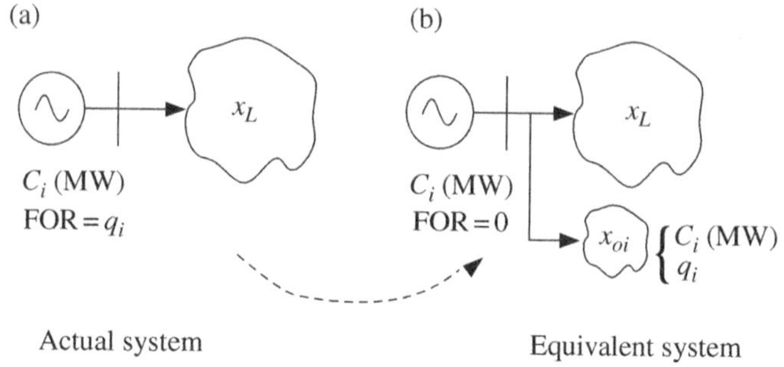

Figure 5.1 Actual and equivalent systems at HL1.

load is defined as the sum of the original load and the probabilistic additional load caused by the forced outage of the generators. The effective load is formulated as

$$x_e = x_L + \sum_{i=1}^{NG} x_{oi} \tag{5.1}$$

where

x_e: random variable of the effective load,
x_L: random variable of the original load,
x_{oi}: random variable of the probabilistic load caused by the forced outage of the ith generator,
NG: the total number of generators.

The ELDC at HLI can be calculated with

$$\begin{aligned}\Phi_i(x_e) &= \Phi_{i-1}(x_e) \otimes f_{oi}(x_{oi}) \\ &= \int \Phi_{i-1}(x_e - x_{oi}) f_{oi}(x_{oi}) dx_{oi}\end{aligned} \tag{5.2}$$

where

$\otimes$: convolution symbol,
$f_{oi}(x_{oi})$: probability distribution function (PDF) of the outage capacity of the ith generator,
$\Phi_i(x_e)$: the PDF of the effective load duration curve (ELDC) convolved with the PDFs of the outage capacities of the generators, 1st to ith,
$\Phi_{i-1}(x_e)$: the PDF of the ELDC convolved with the PDFs of the outage capacities of the generators, 1st to $(i-1)$th,
$\Phi_{i-1}(x_e - x_{oi})$: the PDF of the integral convolution-styled ELDC convolved with the PDFs of the outage capacities of all the generators, 1st to $NG-1$.

The expected energy E_i of the ith generator can be derived by

$$E_i = Tp_i \int_{U_{i-1}}^{U_i} \Phi_{i-1}(x)dx \tag{5.3}$$

where

$$U_i = \sum_{j=1}^{i} C_j = C_1 + C_2 + \cdots + C_i$$

P_i: availability of the ith generator,
T: period (hours).

When the expected value, "$E_k(L_d)$" means the expected energy not served (EENS), which is the energy of the load over installed capacity, it can be formulated as follows. Here, L_p is the peak load and IC is the total installed capacity, and $\Phi(.)$ is the probability density function of the outage load:

$$\begin{aligned} E_k(L_d)_{NG} &= \int_{-\infty}^{\infty} L_d\Phi(L_d)dL_d \\ &= \int_{0}^{IC+L_p} L_d\Phi(L_d)dL_d \\ &= \int_{0}^{IC} L_d\Phi(L_d)dL_d + \int_{IC}^{IC+L_p} L_d\Phi(L_d)dL_d \end{aligned} \tag{5.4}$$

where

L_d is $x - IC$, (when $IC \leq x \leq IC + L_p$),
L_d is zero, (when $x < IC$).

Because IC is constant, the following relationship can be established:

$$\Phi(L_d)_{NG} = \Phi(x = L_d + IC) \quad (IC \leq x \leq IC + L_P)$$

$$\int_{0}^{IC} L_d\Phi(L_d)dL_d = 0$$

With the relationship, (5.4) can be written as follows (5.5):

$$\begin{aligned} \in (L_d)_{NG} &= \int_{IC}^{IC+L_p} (x - IC)\ \Phi(x)dx \\ &= \int_{\Phi^*(x)}^{1.0} (x - IC)\ d\Phi^*(x) \\ &= \int_{\Phi(IC)}^{0} (x - IC)\ d\Phi(x) \\ &= \int_{0}^{0} d[(IC - x)\Phi(x)] + \int_{IC}^{IC+L_P} \Phi(x)dx \\ &= \int_{IC}^{IC+L_P} \Phi(x)dx \end{aligned} \tag{5.5}$$

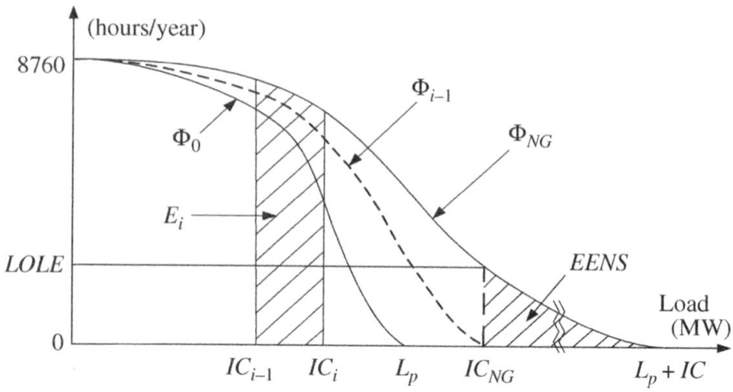

Figure 5.2 Reliability indices and ELDC (Effective Load Duration Curve).

Where $\Phi^*(x)$ is the inverse function of $\Phi(x)$ and $\Phi_{NG}(x)$ is the function of the ELDC convoluted from #1 to #NG generators using an economical priority list. Using the $\Phi_{NG}(x)$, *LOLE* and *EENS* which are the indices of probabilistic reliability can be computed as (5.6) and (5.7) each using ELDC (Figure 5.2), where ΔE_i is a probabilistic generated power; ΔPC_i is a probabilistic production cost; λ_a is an increment cost coefficient; λ_b is a cost constant; and λ_a and λ_b can be calculated by using the fuel cost curve.

$$LOLE = \Phi_{NG}(x)|_{x=IC} \text{ hours/year} \tag{5.6}$$

$$EENS = \int_{IC}^{IC+L_p} \Phi_{NG}(x)dx \text{ MWh/year} \tag{5.7}$$

$$\Delta E_i = EENS_{i-1} - EENS_i \tag{5.8}$$

$$\Delta PC_i = \lambda_a \times \Delta E_i + \lambda_b \times \left(1 - \sum_1^i \text{FOR}\right) \times LOLE_{i-1} \tag{5.9}$$

The conventional deterministic reliability index, the SRR can be formulated as follows:

$$SRR = \frac{SRP - L_p}{L_p} \times 100[\%] = g(C_i, L_p) \tag{5.10}$$

where

SRP : Supply Reserve Power (=IC-MC) [MW],
IC : Installed capacity [MW] ($= \sum C_i$),
C_i : #i generator capacity [MW],
L_p : Peak load [MW],
MC : Yearly average maintenance capacity [MW],

$$MC = \frac{\sum C_i \times MD_i \times 24}{8760} \text{ [MW]},$$

MD_i : $\#i$ generator maintenance dates [days].

Analytical relationship between LOLE and SRR is not available. However, simulation technique is used in order to obtain the relationship.

5.3 Case Studies

Using the effective load model concept in Figure 5.3, which is a well-known model in HLI and generation expansion planning problems, three sample case studies for the system in Figures 5.4 and 5.5 are demonstrated. The model has been used for calculating probabilistic production cost simulation and probabilistic reliability since long years ago (Table 5.1).

5.3.1 Case Study I: Sample System I With One 30 MW Generator Only

Figure 5.6 shows System I with one 30 MW generator for studying this case with load model shown in Figure 5.7. The equivalent system model and the original load function are shown in Figures 5.8 and 5.9, respectively. The outage capacity PDF and effective load PDF of the 30 MW generator are given in Table 5.2 and Figure 5.10, respectively. Equations (5.11) and (5.12) is used as shown in Table 5.3 and plotted in Figure 5.11. Finally, the ELDC and the EENS for the case study are illustrated in Figures 5.12 and 5.13, respectively.

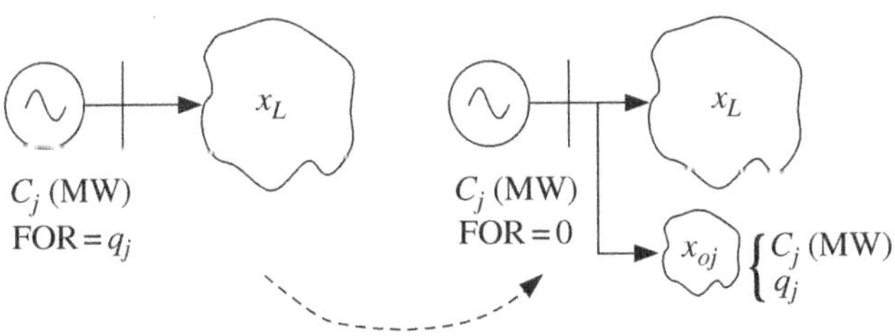

Figure 5.3 Effective load model concept of HLI used for sample case studies.

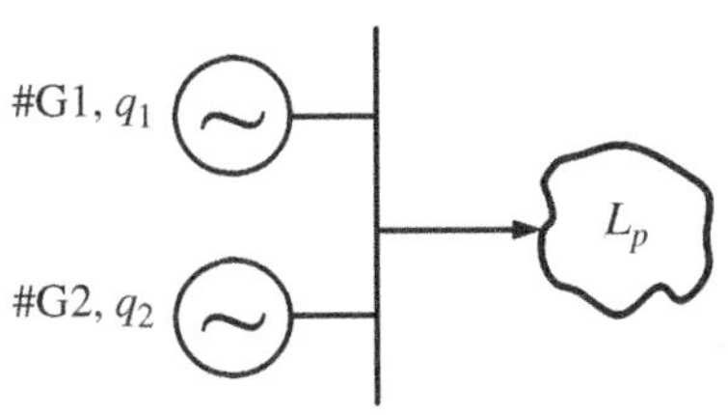

Figure 5.4 System for sample case studies.

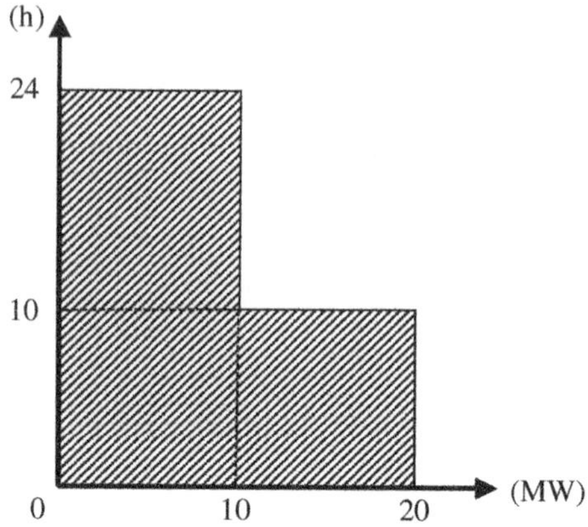

Figure 5.5 Load model (Φ0) .

Table 5.1 Generator data for sample case studies with two generators.

No.	Capacity (MW)	FOR (q)	No.
#1	30	0.1	#1
#2	10	0.1	#2

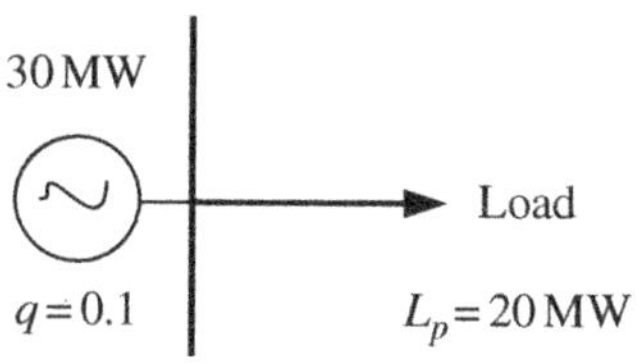

Figure 5.6 Sample Case Study I: System I with one 30 MW generator only.

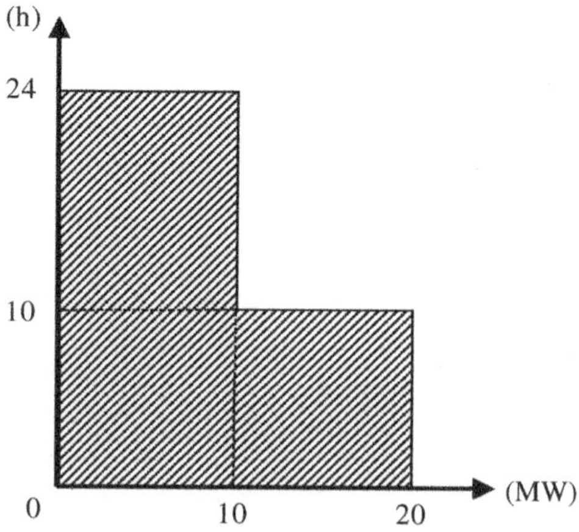

Figure 5.7 Load model in Figure 5.6 (Sample System I).

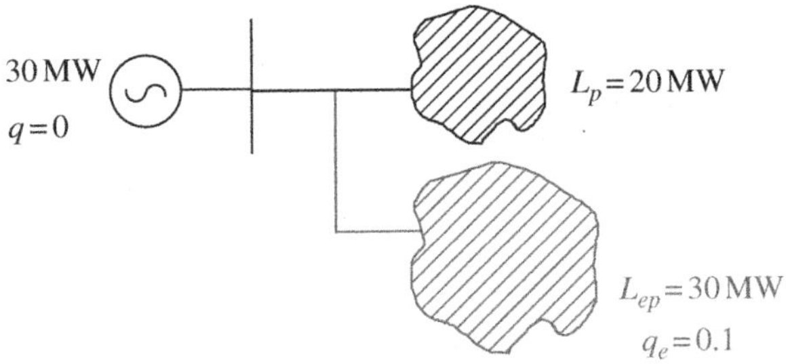

Figure 5.8 Equivalent model for Figure 5.6 (Sample System I).

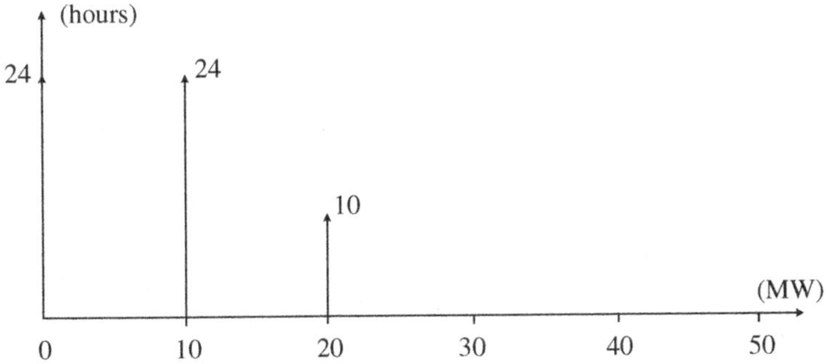

Figure 5.9 Original load function (Φ_0) (not PDF but actual value (time)) function.

Table 5.2 Outage capacity probability distribution function (PDF) of the 30 MW generator.

Outage capacity (MW)	PDF (*pdf* (f_{01}))
0	0.9
10	0.0
20	0.0
30	0.1

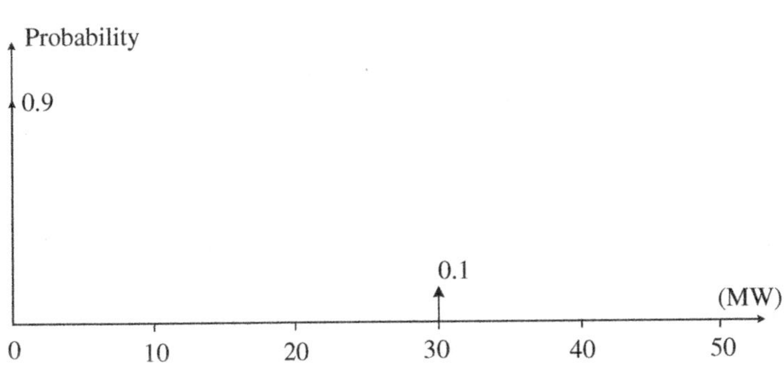

Figure 5.10 Effective load *pdf* (f_{0e}) of the 30 MW generator.

Table 5.3 ELDC using recursive equations of Sample Case Study I.

Outage capacity (MW)	P_n	Eq. (5.11) $P'_n(\Phi_1)$	Eq. (5.12) P'_n
0	24	$24 \times 0.9 + 24 \times 0.1 = 24$	$24 \times 0.1 + 0 \times 0.9 = 2.4$
10	24	$24 \times 0.9 + 24 \times 0.1 = 24$	$24 \times 0.1 + 0 \times 0.9 = 2.4$
20	10	$10 \times 0.9 + 24 \times 0.1 = 11.4$	$10 \times 0.1 + 0 \times 0.9 = 1.0$
30	0	$0 \times 0.9 + 24 \times 0.1 = 2.4$	
40	0	$0 \times 0.9 + 24 \times 0.1 = 2.4$	
50	0	$0 \times 0.9 + 10 \times 0.1 = 1.0$	

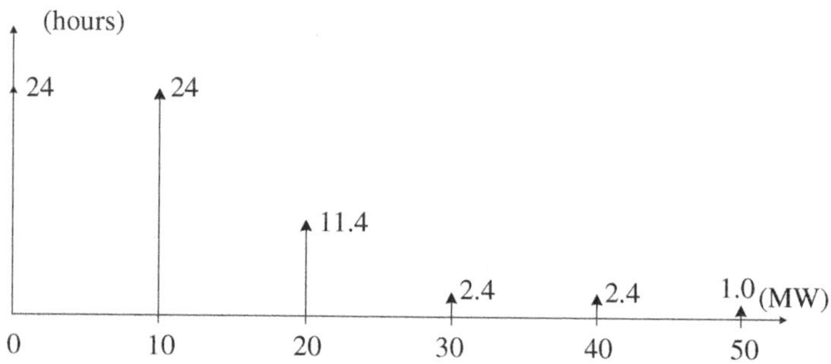

Figure 5.11 Effective load duration curve (ELDC) function (Φ') for Case Study I (Sample System I with one 30 MW generator only) (actual function).

The ELDC can be generated by convolving the load curve and the plant outage capacity as defined by (5.2). For Sample Case I, the convolution leads to the following ELDC:

$$P'_n(x) = (1-q_i)P_n(x) + q_iP_n(x-c) \tag{5.11}$$

where q_i is the outage rate of the ith generator and c is the discrete step size of the load.

When only ELDC for EENS area is considered as defined in Figure 5.2, it can be obtained through by

$$P'_n(x) = q_iP_n(x) + (1-q_i)P_n(x+c) \tag{5.12}$$

Equations (5.11) and (5.12) as shown in Table 5.3 and plotted in Figure 5.11. Finally, the ELDC and the EENS for the case study are illustrated in Figures 5.12 and 5.13, respectively.

The *pdf* (probabilistic density function) (f_{01}): same with PDF.

Therefore, reliability results for a Sample Case Study I follow:

LOLE = 2.4 hours/day
EENS = 2.4 hours × 10 MW + 1.0 hours × 10 MW = 34 MWh/day
$EES_{30\,MW} = EENS_0 - EENS_1 = 340 - 34 = 306$ MWh/day

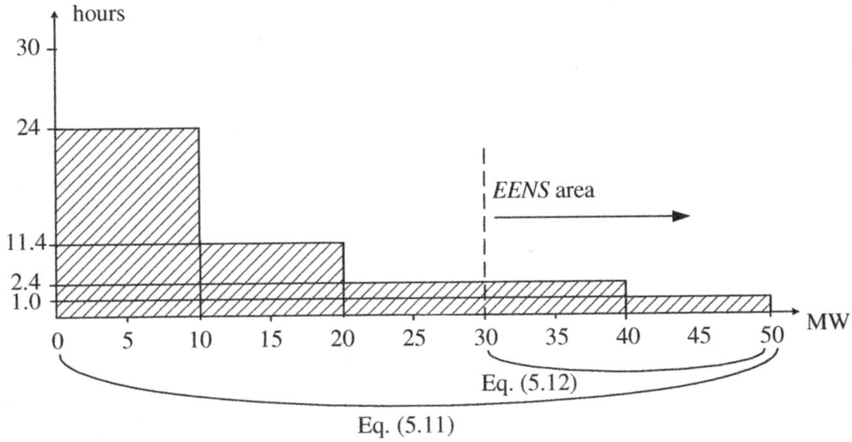

Figure 5.12 ELDC and EENS curve for the Sample Case Study I.

Figure 5.13 Remaining load (not supplied load) after loading 30 MW generator with zero FOR(q = 0).

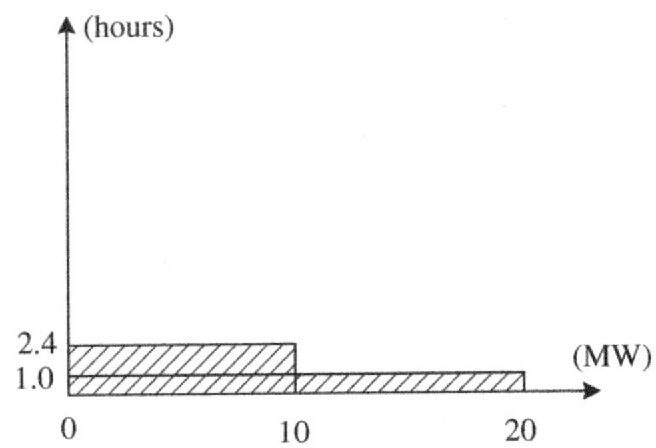

where

$EES_{30\,MW}$: Expected Energy Supplied (Probabilistic Production Energy) of 30 MW generator (MWh/day).

$EENS_0$: Original Load (Demand) Energy (MWh/day),

$EENS_1$: Expected Energy Not Served after loading 30 MW generator (MWh/day).

5.3.2 Case Study II: Sample System II With One 10 MW Generator Only

Figure 5.14 shows System I with one 10 MW generator only for studying this case with load model shown in Figure 5.15. The equivalent system model and the original load function are shown in Figures 5.16 and 5.17, respectively. The outage capacity PDF and effective load PDF of the 10 MW generator are given in Table 5.4 and Figure 5.18, respectively. The ELDC was calculated using the recursive equations (5.11) and (5.12) as shown in Table 5.5 and the ELDC, and the EENS for the case study are illustrated in Figures 5.19 and 5.20, respectively.

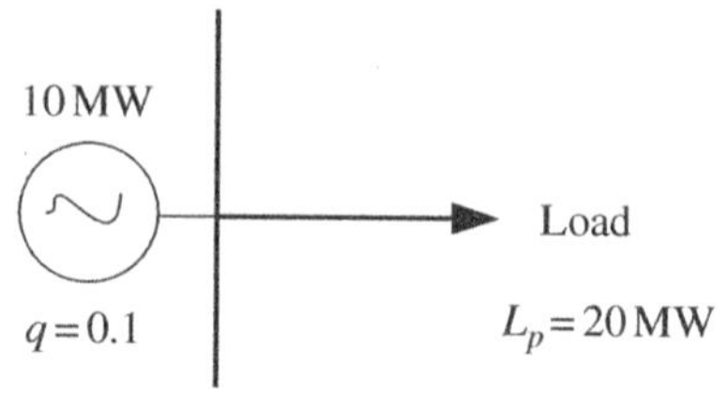

Figure 5.14 Sample Case Study II: System II with one 10 MW generator only.

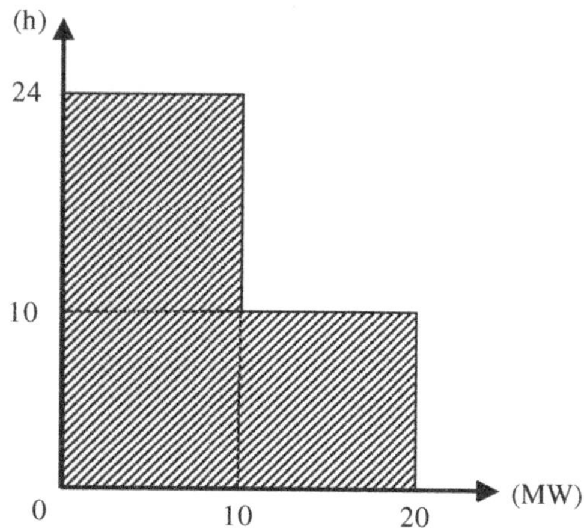

Figure 5.15 Load model in Figure 5.14 (Sample Case Study II).

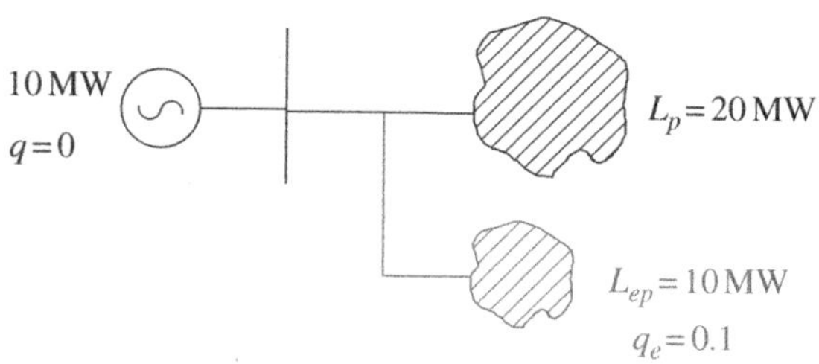

Figure 5.16 Equivalent model for Figure 5.14 (Sample System II).

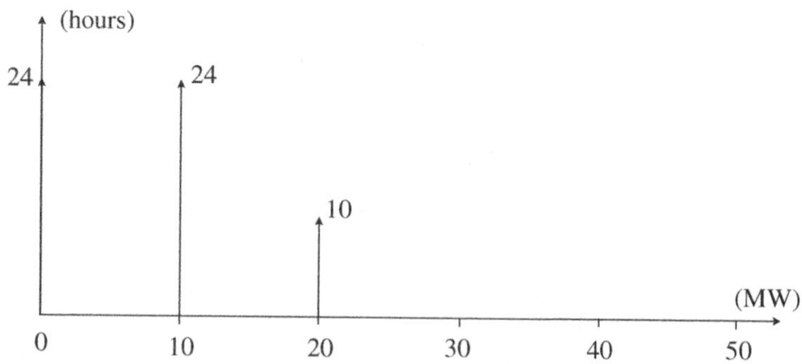

Figure 5.17 Original load function (Φ_0) (not PDF but actual value (time)) function.

Table 5.4 Outage capacity probability distribution function (PDF) of the 10 MW generator.

Outage capacity (MW)	PDF (pdf (f_{02}))
0	0.9
10	0.1
20	0.0
30	0.0

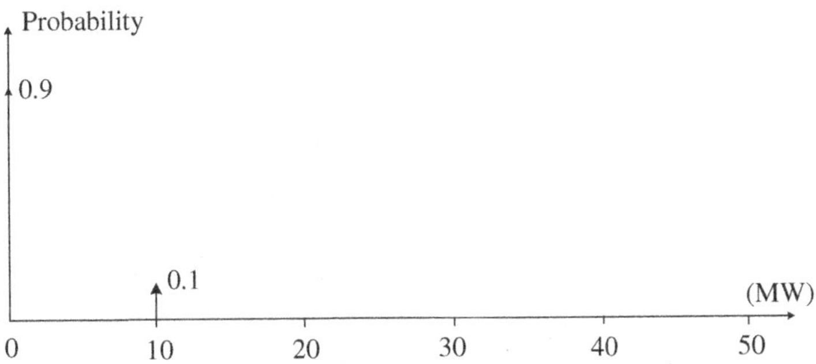

Figure 5.18 Effective load pdf (f_{0e}) for 10 MW generator with FOR = 0.1.

Table 5.5 ELDC using recursive equations of Sample Case Study II.

Outage capacity (MW)	$P_n(x)$	Eq. (5.11) $P'_n(\Phi_1)$	Eq. (5.12) P'_n
0	24	$24 \times 0.9 + 24 \times 0.1 = 24$	$24 \times 0.1 + 24 \times 0.9 = 24$
10	24	$24 \times 0.9 + 24 \times 0.1 = 24$	$24 \times 0.1 + 10 \times 0.9 = 11.4$
20	10	$10 \times 0.9 + 24 \times 0.1 = 11.4$	$10 \times 0.1 + 0 \times 0.9 = 1.0$
30	0	$0 \times 0.9 + 10 \times 0.1 = 1.0$	

The pdf (probabilistic density function) (f_{02}): same with PDF.

Therefore, reliability results for Sample Case Study II.

LOLE = 11.4 hours/day
EENS = 11.4 hours × 10 MW + 1.0 hours × 10 MW = 124 MWh/day
$EES_{10\,MW} = EENS_0 - EENS_2 = 340 - 124 = 216$ MWh/day

where

$EES_{10\,MW}$: Expected Energy Supplied (Probabilistic Production Energy) of 10 MW generator (MWh/day).

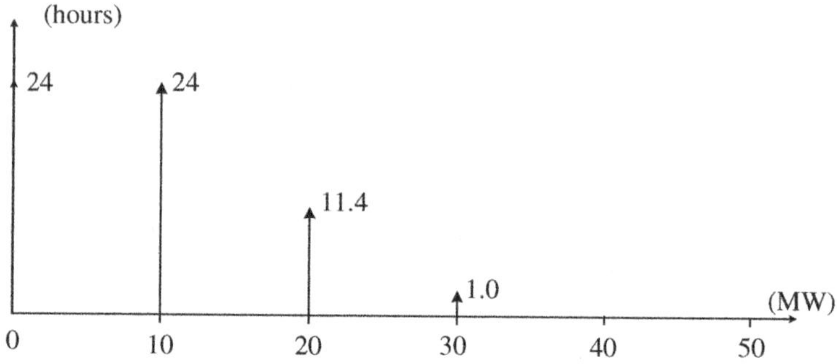

Figure 5.19 Effective load duration curve (ELDC) function (Φ') for Sample Case Study II (Sample System II with one 10 MW generator only) (actual function).

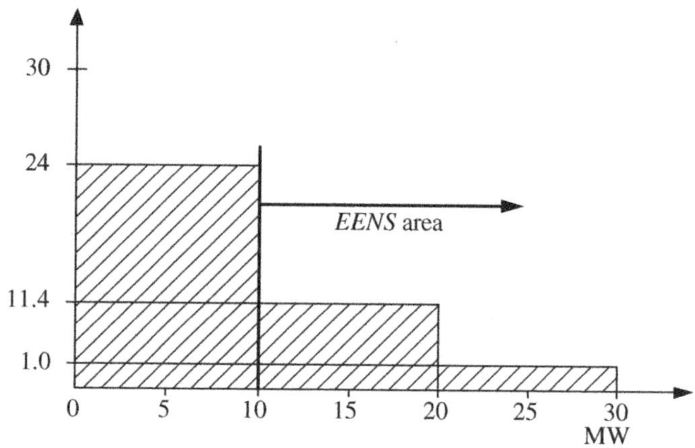

Figure 5.20 ELDC and EENS curve for Sample Case Study II.

$EENS_0$: Original Load (Demand) Energy (MWh/day),
$EENS_2$: Expected Energy Not Served after loading 10 MW generator (MWh/day).

Comments: Sample Case Study I is a higher reliable system than Sample Case Study II as results.

5.3.3 Case Study III: Sample System III With Two Generators – 30 and 10 MW

This simple system has two generators and a single load as shown in Figure 5.21.
The ELDCs after loading generator #1 and generator #2 are shown in Table 5.6 and Figures 5.22 and 5.23.

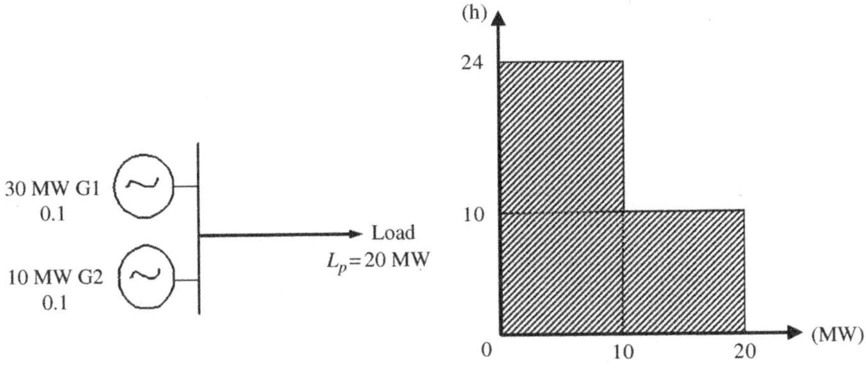

Figure 5.21 Sample Case Study III: two generators' sample system.

Table 5.6 Final ELDC (Φ_2) after loading G1 and G2.

Outage capacity (MW)	P_n	+30 MW(Φ_1)	+10 MW(Φ_2)
0	24	24	$24 \times 0.9 + 24 \times 0.1 = 24$
10	24	24	$24 \times 0.9 + 24 \times 0.1 = 24$
20	10	11.4	$11.4 \times 0.9 + 24 \times 0.1 = 12.66$
30	0	2.4	$2.4 \times 0.9 + 11.4 \times 0.1 = 3.3$
40	0	2.4	$2.4 \times 0.9 + 2.4 \times 0.1 = 2.4$
50	0	1.0	$1.0 \times 0.9 + 2.4 \times 0.1 = 1.14$
60	0	0	$0.0 \times 0.9 + 1.0 \times 0.1 = 0.1$
	Φ_0	Φ_1	Φ_2

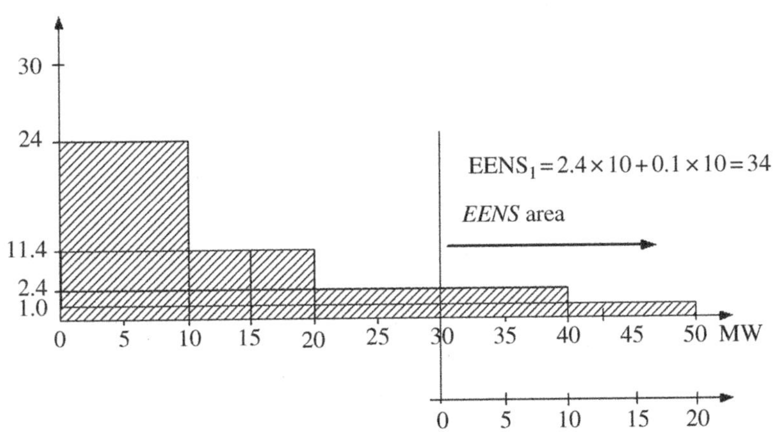

Figure 5.22 ELDC after loading Gen #1.

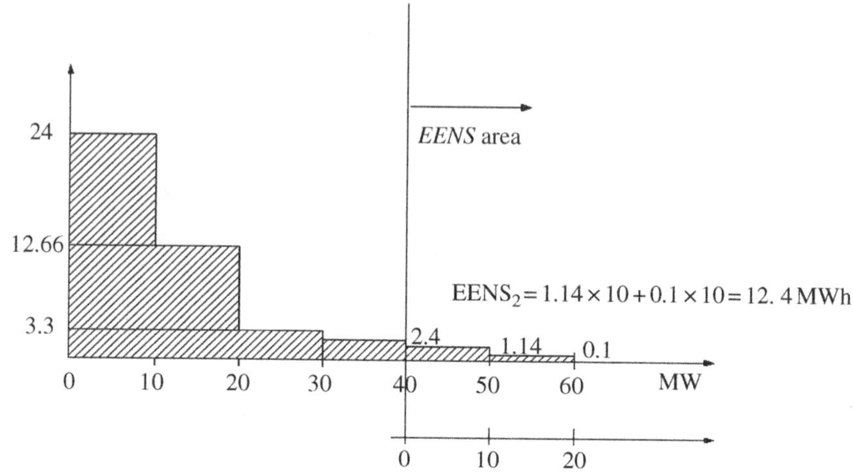

Figure 5.23 ELDC after loading Gen #1 and #2.

Expected Energy Supplied by Gen #1 = $EENS_0 - EENS_1$ = 340 − 34 = 306 MWh/day.

Where

$EENS_0$: Original Load (Demand) Energy (MWh/day),
$EENS_{1\&2}$: Expected Energy Not Served after loading 30 MW generators (MWh/day).

Therefore
Expected Energy Supplied by Gen #2 = $EENS_1 - EENS_{1\&2}$ = 34 − 12.4 = 21.6 MWh/day.

Where

$EENS_1$: Expected Energy Not Served after loading 30 MW generator (MWh/day)
$EENS_{1\&2}$: Expected Energy Not Served after loading 30 and 10 MW, two generators (MWh/day)

Considering both generators 1 and 2, the reliability indices can be calculated as follows:

Probabilistic Reliability and Production-Simulated Results with both generators 1 and 2 operating:

i) *Final Reliability Evaluation of the System*
LOLE = 1.14 hours/day
EENS = 340 − 327.6 = 12.4 MWh/day

EIR (Expected Index of Reliability) = 1 − 12.4/340 = 1 − 0.036 470 588 = 0.963 529 412 p.u.

ii) *Probabilistic Production Energy of the System*

Expected Energy Supplied by Gen #1, $EES_{G1} = EENS_0 - EENS_1 = 340 - 34 = 306$ MWh/day

Expected Energy Supplied by Gen #2, $EES_{G2} = EENS_1 - EENS_2 = 34 - 12.4 = 21.6$ MWh/day

Total probabilistic production energy = Total EES (Expected Energy Supplied) of Gen #1 and Gen #2 = 306 + 21.6 = 327.6 MWh/day

An alternative method is demonstrated for the same Sample Study III. Using $\Phi_2 = \Phi_o \otimes f_{os}$, we can calculate the final ELDC, Φ_2 also (Tables 5.7–5.9).

Table 5.7 Synthesized *pdf* (f_{os}) of effective loads of G1 and G2.

Outage capacity (MW)	30 MW (f_{o1})	30 MW + 10 MW ($f_{os} = f_{o1} \otimes f_{o2}$)	Cumulative
0	0.9	$0.9 \times 0.9 + 0 \times 0.1 = 0.81$	1.0
10	0	$0 \times 0.9 + 0.9 \times 0.1 = 0.09$	0.19
20	0	$0 \times 0.9 + 0 \times 0.1 = 0$	0.1
30	0.1	$0.1 \times 0.9 + 0 \times 0.1 = 0.09$	0.1
40	0	$0 \times 0.9 + 0.1 \times 0.1 = 0.01$	0.01

Table 5.8 Convolution integral starting step between synthesized *pdf* (f_{os}) of effective loads of G1 and G2 and original LDC.

40	0.01		
30	0.09		
20	0		
10	0.09		
0	0.81		
(MW)	$f_{o1} \otimes f_{o2} = f_{os}$	(MW)	LDC (Φ_o)
		0	24
		10	24
		20	10

Table 5.9 Convolution integral process between synthesized *pdf* (f_{os}) of effective loads of G1 and G2 of Sample Study III and LDC.

Gen (MW)								
Load (MW)		**0**	**−10**	**−20**	**−30**	**−40**	**Cum. Prob.** $\Phi_2 = \Phi_o \otimes f_{os}$	**Exact. Prob.**
0		24 × 0.81	24 × 0.09	24 × 0	24 × 0.09	24 × 0.01	24.0	0
10		24 × 0.81	24 × 0.09	24 × 0	24 × 0.09	24 × 0.01	24.0	11.34
20		10 × 0.81	24 × 0.09	24 × 0	24 × 0.09	24 × 0.01	12.66	9.36
30		0	10 × 0.09	24 × 0	24 × 0.09	24 × 0.01	3.3	0.9
40	0	0	0	10 × 0	24 × 0.09	24 × 0.01	2.4	1.26
50	10	0	0	0	10 × 0.09	24 × 0.01	1.14	1.04
60	20	0	0	0	0	10 × 0.01	0.1	0.1

The typical process for calculating ELDC using an effective load model is demonstrated in Figure 5.24. Using this recursive process, many generators can be considered on a probabilistic approach for the generation expansion planning problem.

5.4 Probabilistic Production Simulation Algorithm

5.4.1 Hartley Transform

A real integral transform was presented by R.V.L. Hartley [117, 118]. The Hartley transform of $f(x)$ is defined as

$$H(\omega) = (2\pi)^{-1/2} \int_{-\infty}^{\infty} f(x)(\cos \omega x + \sin \omega x) dx \tag{5.13}$$

where

ω: angular velocity.

While the Fourier transform defined in (5.14) uses ($\cos t\omega x - j\sin\omega x$), the Hartley transform uses ($\cos t\omega x + j\sin\omega x$) as shown in (5.13). Therefore, the convolution integral procedure in Hartley transform is as fast as or faster than that in Fourier transform.

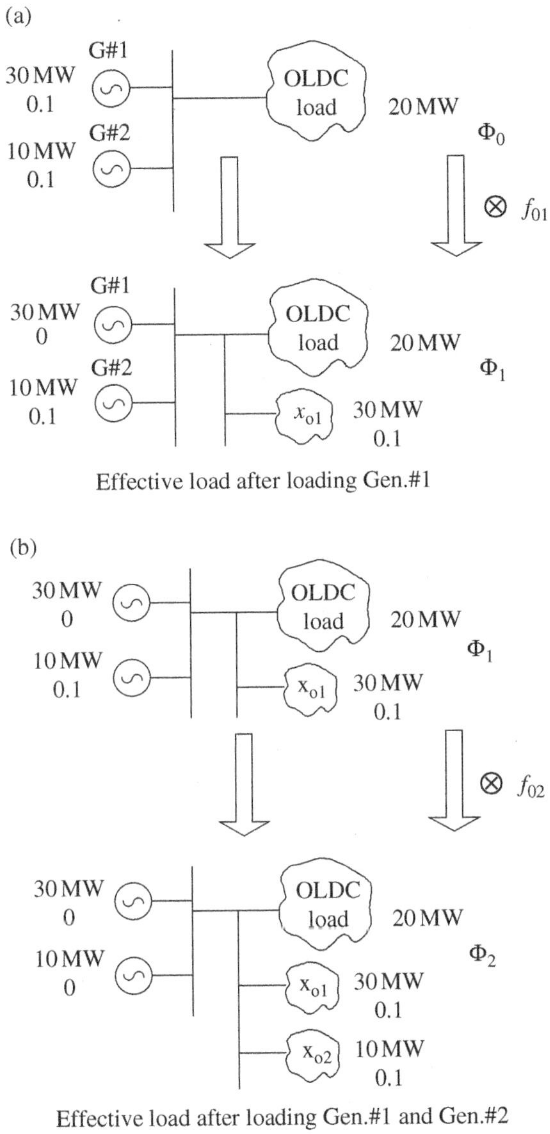

Figure 5.24 Typical recursive process for calculating ELDC using effective load model at HLI.

$$F(\omega) = (2\pi)^{-1/2} \int_{-\infty}^{\infty} f(x)(\cos \omega x - j \sin \omega x) dx \tag{5.14}$$

The Discrete Hartley Transform for a practical application can be expressed as follows:

$$H(\omega) = N^{-1/2} \sum_{x=0}^{N-1} f(x)(\cos \omega x + \sin \omega x) \tag{5.15}$$

$$= DHT \; f(x) \tag{5.16}$$

where N is the number of discrete data which means the period. DHT is the operator representing the Discrete Hartley Transform.

In Hartley Transform with a fast algorithm, the given function $f(x)$ is partitioned and substituted into (5.15). Then the decomposition formula based on the stretch theorem and the shift theorem is applied to $H(\omega)$. As the procedures which bisect the data sequences progressively go forward, only two final elements can be obtained. Therefore, Hartley Transformed function for a given function can be obtained faster. If the DHT is evaluated numerically from (5.15), the computer time for a data sequence of length N is proportional to N^2. However, with this fast algorithm, the computer time can be reduced to $N\log_2 N$.

1) *Thermal and nuclear plants* – Because thermal and nuclear generators have an identical characteristic of fuel expenses, they can be considered together. The equation for ELDC (5.2) by Fast Hartley Transform (FHT) can be formulated using

$$\Phi_i(x) = FHT^{-1}[FHT \;\; \Phi_{i-1}(x) \otimes FHT \;\; f_{oi}(x)] \tag{5.17}$$

where

$$\Phi_i(x) = \sum_{m=0}^{N-1} \alpha_m \delta(x - L_m), \quad i = 1, 2, 3, ..., NG$$

$$f_{oi}(x) = \sum_{m=0}^{N-1} \beta_m \delta(x - G_m), \quad i = 1, 2, 3, ..., NG$$

α_m: probability of the existing load $L_{m.}$,
β_m: probability of the existing generator output $G_{m.}$,
δ: delta function.The new algorithm for simulating thermal unit by the FHT method is summarized in the following steps.

Step 1. Evaluate the Hartley transform $Ho(v)$ for a given inverted cumulative load duration PDF $\Phi_o(x)$ using (5.18) by the fast algorithm.

$$H_o(v) = N^{-1/2} \sum_{x=1}^{N-1} \Phi_o(x)\{\cos(2vx/N) + \sin(2vx/N)\} \tag{5.18}$$

Step 2. Set $i = 1$

Step 3. Evaluate the ELDC $\Phi_{i-1}(x)$ which was convolved until the $(i-1)$th generator by substituting $H_{i-1}(v)$ to (5.19) which represents the inverted Hartley transform.

$$\Phi_{i-1}(x) = N^{-1/2} \sum_{x=1}^{N-1} H_{i-1}(v)\{\cos(2\pi vx/N) + \sin(2\pi vx/N)\} \tag{5.19}$$

And, calculate the expected energy E_i of the ith generator using (5.3)

Step 4. Evaluate the capacity outage PDF $f_{oi}(x)$ of the ith generator and its Hartley transform $T_i(v)$ as (5.20) by the fast algorithm.

$$T_i(v) = N^{-1/2} \sum_{x=1}^{N-1} f_{oi}(x)[\cos(2\pi vx/N) + \sin(2\pi x/N)\} \tag{5.20}$$

Step 5. Calculate the $H_i(v)$ which represents the convolution of $T_i(v)$ with $H_{i-1}(v)$ as

$$H_i(v) = H_{i-1}(v)T_{ie}(v) + H_{i-1}(N-v)T_{io}(v) \tag{5.21}$$

where

$$T_{ie}(v) = [T_i(v) + T_i(N-v)]/2$$
$$T_{io}(v) = [T_i(v) - T_i(N-v)]/2$$

Step 6. If i equals NG, go to step 7. Otherwise, set $i = i + 1$ and go to step 3.

Step 7. Calculate $LOLE$ by (5.6).

2) *Hydro plant* – For real utility systems, it is often necessary to include a hydro plant. The total energy production of a hydro unit is constrained by the reservoir size and the amount of rainfall. Its operation can be simulated by degenerating thermal power plants in reciprocal merit order until its energy limit is satisfied. So, the effective load duration PDF $\Phi_i(x_e)$ must be deconvolved. The deconvolution process in FHT can be formulated as

$$H_{i-1} = \frac{[H_{ie}(T_{ie} - T_{io}) + H_{io}(T_{ie} + T_{io})]}{(T_{ie} + T_{io})} \tag{5.22}$$

From $\Phi_i(x)$, $\Phi_{i-1}(x)$ can be obtained using (5.22). The generated energy, E_H of hydro power plant with capacity C_H is equal to the sum of the expected energy of degenerating thermal units

$$E_H = \sum_{\varepsilon_H} \Delta E_{Hi} \tag{5.23}$$

where

$$\begin{aligned}
\Delta E_{Hi} &= E_{i1} - \left({}_{PH}E_{i2} + {}_{qH}E_{i3}\right) \\
E_{i1} &= Tpi\int_{U_{i-1}}^{U_i} \Phi_{i-1}(x)dx \\
E_{i2} &= Tpi\int_{U_{i-1}+C_H}^{U_i+C_H} \Phi_{i-1}(x)dx \\
E_{i3} &= E_{i1}
\end{aligned} \tag{5.24}$$

P_H: the availability of hydro plant.
ε_H: a set of thermal units degenerated by hydro plant.

The algorithm for simulating the operation of hydro power plant is summarized in the following steps:

Step 1. Set $i = NG$ and calculate the degenerating energy, ΔE_{Hi} of the ith thermal unit as in (5.24)

Step 2. Check whether (5.25) is true or not.

$$\sum_{j=1}^{i} \Delta E_{hj} \geq C_H T C_F \tag{5.25}$$

where
C_F: capacity factor of hydro plant.

Step 3. If it is true, go to step 4. Otherwise, set $i = i - 1$ and go to step 2.

Step 4. This hydro plant can be set at the optimal loading point. Hydro-generated energy must be equal to the required energy. After setting $k = i$, adjust thermal and hydro-generated energy as (5.26).

$$\begin{aligned}
\Delta E'_{Hk} &= \Delta E_{Hk} - \Delta E_x \\
E'_{k2} &= E'_{k2} + \Delta E_x
\end{aligned} \tag{5.26}$$

where

$$\Delta E_x = \sum_{j=NG}^{k} \Delta E_{Hj} - C_H T C_F$$

3) *Pumped-storage plant* – The stored energy is limited depending on the reservoir size. And the stored energy should not be discharged before it is charged. It would be necessary to do a chronological simulation of the charging/discharging cycles over a long period of time, But, such an approach requires excessive computer time in the case of a long-range planning model.In this chapter, we only considered the total stored and discharged energy over a considerable

period of time. The total installed capacity excluding the pumped-storage unit is assumed to be larger than the system peak load. Two kinds of operations are simulated. They are compulsory pumping and economic operations.

The stored energy is limited depending on the reservoir size. And the stored energy should not be discharged before it is charged. It would be necessary to do a chronological simulation of the charging/discharging cycles over a long period of time, But, such an approach requires excessive computer time in the case of a long-range planning model.

In this chapter, we only considered the total stored and discharged energy over a considerable period of time. The total installed capacity excluding a pumped-storage unit is assumed to be larger than the system peak load. Two kinds of operations are simulated. They are compulsory pumping and economic operations.

4) *Compulsory pumping operation* – Because nuclear power generators' lack load-following ability, we usually encounter the problem of excessive generation during midnight. The energy, TE_{cp}, by compulsory pumping operation is obtained from

$$TE_{CP} = T\sum_{i\in CG}\left\{\int_{U_{i-1}-pgcap}^{U_i-pgcap}\Phi_{i-1}(x)dx - \int_{U_{i-1}}^{U_i}\Phi_{i-1}(x)dx\right\} \tag{5.27}$$

where
pgcap: the total capacity of pumped-storage plant,
CG: a set of must-run generators.

5) *Economic operation* – The pumped-storage plant operation is justified when the economy between pumping and generating is expected. Therefore, its operation is formulated as follows:

$$\min\ Z = \left|\sum_{i\in\varepsilon_p}\Delta E_{pi} - \sum_{i\in\varepsilon_p}\Delta E_{gi}\right| \tag{5.28}$$

$$\begin{aligned} \text{sub}\ \ \Delta E_{pi} + TE_{CP} &\leq PPCNG \\ \Delta E_{pi} &\leq PPENG_i \\ \Delta E_{gi} &\leq PGENG_i \\ \frac{\lambda_{pi}}{\lambda_{gi}} &< \eta \end{aligned}$$

where

$$PPENG_i = \int_{U_{i-1}-pgcap}^{U_i-pgcap}\Phi_{i-1}(x)dx - \int_{U_{i-1}}^{U_i}\Phi_{i-1}(x)dx$$

$$PGENG_i = \int_{U_{i-1}}^{U_i}\Phi_{i-1}(x)dx - \int_{U_{i-1}+pgcap}^{U_i+pgcap}\Phi_{i-1}(x)dx$$

PPCNG: the upper reservoir size of pumped-storage units (MWh),
η: efficiency of pumped-storage unit,
ε_p: a set of thermal units degenerated by economic operation of pumped-storage unit.

However, it is difficult to analyze this problem with optimization techniques. In convolution methods, this economic justification is evaluated by repeated searching for if an optimal loading point is there, the marginal cost of the stored energy becomes equal to that of discharged energy. The optimal loading point can be found easily using *PPENG* and *PGENG* defined in (5.28).

6) *Maintenance schedule* – The maintenance schedule of generators has a great effect on supply reliability. It is usually performed in the spring or autumn because peak load occurs in the summer in Korea. There are two methods to consider. The one is that LDC averagely rises as much as the capacity of maintenance, the other is that install capacity (IC) averagely rises as much as the capacity of maintenance. However, it is assumed that maintenance is averagely performed during all times in both methods. There is a difference between both methods and the real maintenance schedule. Therefore, we tried to find a method that can correspond with the real maintenance schedule. In this chapter, it is assumed that the maintenance is ideally performed at a base-load. LDC rises as much as the capacities of maintenance at a base load, as shown in Figure 5.25.LDC means original LDC and MLDC means adjusted

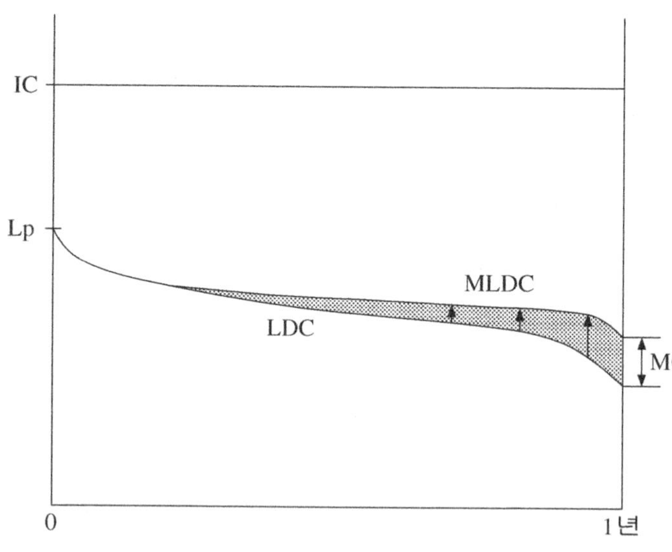

Figure 5.25 LDC rises as much as the capacities of maintenance at a base-load.

LDC considered the maintenance of generators in Figure 5.25. The capacity of maintenance can be expressed as

$$MENG = C_i \times MD_i \times 24 \text{ MWh} \tag{5.29}$$

where
$MENG$: total capacity of maintenance (energy) (MWh),
C_i: the capacity of generator i (MWh),
MD_i: the days of maintenance of generator i (day).

7) *Flow chart* – Figure 5.26 is the overall flow chart of PRASim in order to calculate the power and reliability index using FHT.

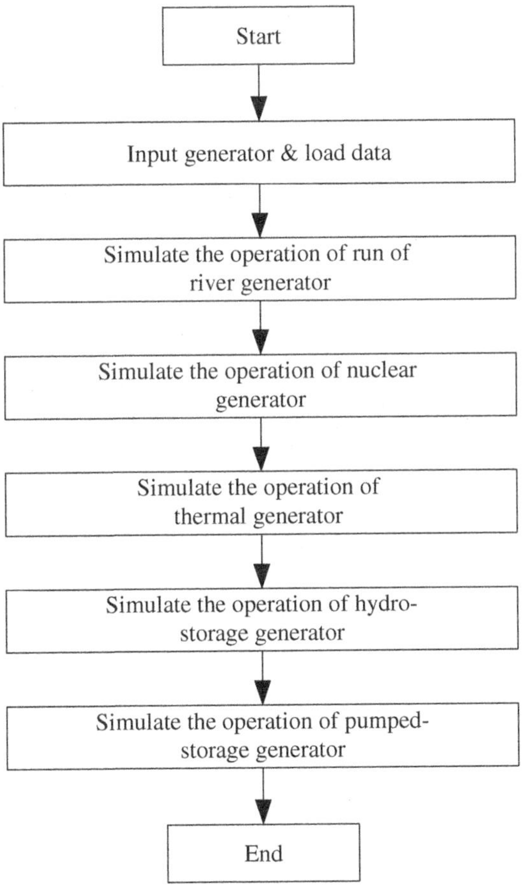

Figure 5.26 Overall flow chart of PRASim.

5.5 Supply Reserve Rate

As a probabilistic reliability index, the LOLE is a difficult concept compared to deterministic reliability indices such as supply reserve rate (SRR) or installed reserve rate (IRR). While the relationship between deterministic reliability indices, SRR or IRR, and probabilistic reliability index LOLE cannot be formulated analytically, a simulation method is available. Using the fast Hartley Transformation (FHT), the PRASim (clouding online simulation system: http://pwrsys.gsnu.ac.kr/comm/freeboard/) is developed based on an exact method, Booth-Baleriaux method, and approximation methods, Cumulant method and Mixture of Normals Approximation method (MONA). Using the probabilistic reliability evaluation and the PRASim, the relationship between LOLE and SRR is developed in simulation.

This section introduces the characteristics of the relationship between probabilistic reliability (LOLE) and the deterministic reliability (SRR) for the year 2008 in Korea power system. The system consists of 153 generators and the total installed capacity is 68,559[MW]. Hydro units are considered as one equivalent generator whose capacity is 1,516[MW] and its FOR and yearly capacity factor are assumed 0.04297 and 20[%], respectively. One equivalent pumped-storage hydro plant is assumed with installed capacity of 3,900[MW]. The input data of nuclear and thermal generators are summarized in Tables 5.10 and 5.11.

The load duration curve of the year 2008 is used, as shown Figure 5.27.

The value of LOLE corresponding to each SRR can be obtained by varying loads. The value of SRR meeting the LOLE criterion is 14.9[%], as shown in Figure 5.28.

Figure 5.29 shows the relationship between SRR and LOLE for varying FOR of nuclear plants from 0[%] to 100[%]. When the FOR of nuclear plants increases by 100[%], the value of SRR meeting the LOLE criterion should increase from 14.9[%] to 18.2[%]. It means that the value of elasticity between the FOR of nuclear plants and the SRR of whole system can be obtained as follows:

Table 5.10 The input data of generators

Type	The number of generators	Capacity [MW]
Nuclear	20	17716
Oil	25	4799
Gas	64	20163
Coal	44	20465
Total	153	63143

Table 5.11 The FOR of generators

Type	Min [MW]	Max [MW]	FOR
Nuclear	0	400	0.10771
	400	800	0.10625
	800	1000	0.09686
	1000	9999999	0.11807
Oil	0	100	0.03156
	100	200	0.04996
	200	300	0.07508
	300	400	0.06257
	400	600	0.0317
	600	800	0.02853
	800	1000	0.02906
	1000	999999	0
Gas	0	100	0.03799
	100	200	0.04023
	200	300	0.02644
	300	400	0.05043
	400	600	0.04463
Coal	100	200	0.03538
	200	300	0.04425
	300	400	0.05962
	400	600	0.05309
	600	800	0.04532
	800	1000	0.04244
	1000	999999	0.06155

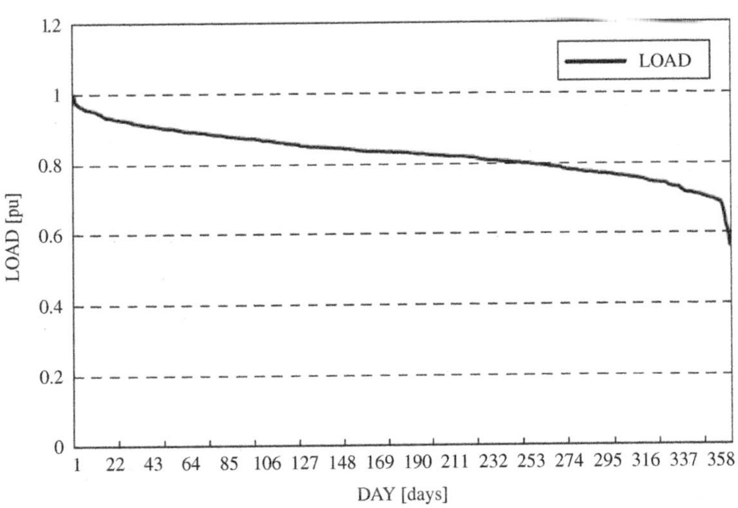

Figure 5.27 Load duration curve of the year 2008.

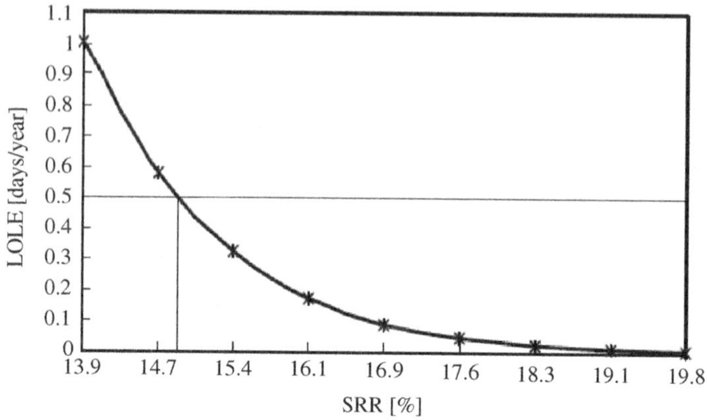

Figure 5.28 Relationship between SRR and LOLE by varying loads in the year 2008.

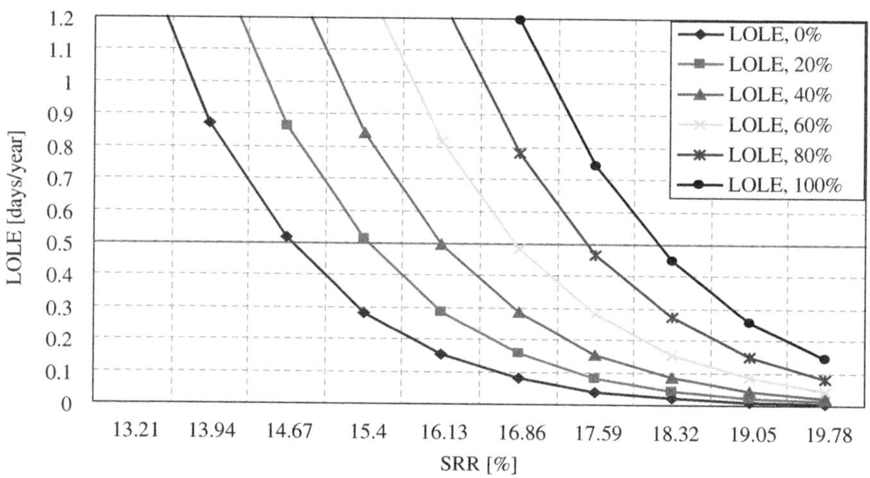

Figure 5.29 Relationship between SRR and LOLE for varying FOR of nuclear plant.

$$\frac{\Delta SRR}{\Delta FOR} = \frac{3.5}{100} \tag{5.30}$$

The relationship for the year 2010 in Korea power system and the peak load of 62,852[MW] has been simulated. The value of SRR meeting the LOLE criterion is 14.5[%], as shown in Figure 5.30.

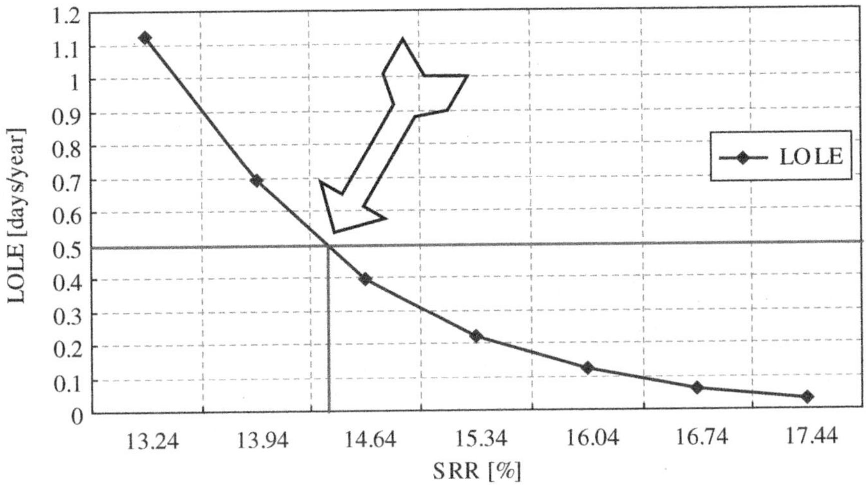

Figure 5.30 Relationship between SRR and LOLE for varying loads in the year 2010.

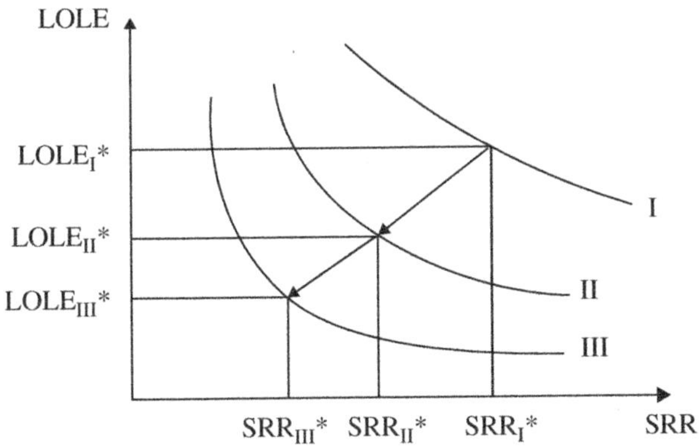

Figure 5.31 Relationship between SRR and LOLE for increasing system size.

From these simulated results, the obtained relationship between SRR and LOLE according to increasing system size in Korea is shown in Figure 5.31. As the system size increases, from I to II to III, LOLE* and SRR* is decreased. Finally, the reliability criteria in Korea is summarized as in Table 5.12.

The SSR can serve as useful guide in preventing the shortage of the installed capacity or oversupply when planning a power generation. Korea has been adopting Cost Based Pool (CBP). It is a transitional pool. So, it fundamentally has issues,

Table 5.12 The adequacy of SRR using PRASim in year 2008 and 2010

Year	LOLE Criterion [Days/year]	The adequacy of SRR [%]
2008	0.5	14.9
2010	0.5	14.5

especially inadequacy is shown when selecting power generation price, because the price of power generation depends on installed capacity regardless of whether a generator is operated or not. The need to study for complementing it is gathering strength. The method described in this chapter to find the adequacy of SRR through the relationship between LOLE and SRR can contribute in determining power generation price.

6

Decision Maker's Satisfaction Using Fuzzy Set Theory

6.1 Introduction

Electricity utilization is environmentally benign and as a form of energy carrier, electricity is clean and safe. It causes no pollution or environmental emissions at the point of end user. But electricity production can cause local and regional environmental impact and may also have long-lasting detrimental global consequences. Some of these impacts like the emissions of sulfur oxides (SO_x), nitrogen oxides (NO_x), and solid particulates, which all have detrimental air quality implications, can be controlled by investing in technology and abatement facilities. These measures can control and reduce such emissions significantly.

There is a growing demand for electricity, high system reliability, and for high-quality power at a minimized production cost. Restructuring of the electricity market has been moving markets from being monopolistic to being competitive. This calls for discontinuing the vertically integrated utility structure and forming a generation company (GENCO), a transmission company (TRANSCO), and a distribution company (DISCO). The goal of a GENCO is to meet the growing load demand at a reasonable price, while enhancing environmental cleanness and providing a reliable quality electric energy source.

The cost of construction of the new power plants, difficulty in placing them unstable security of fuels, the uncertainty of the long-term load forecast, and the environmental constraints make the generation planning problem quite a challenge. Under these circumstances, decision makers are faced with a difficult task. As a result, power system planners are challenged with the task of finding new methods to help the decision makers analyze the energy and power system on the basis of the relevant data for their dynamics and reasonable decision making. Several simple techniques are available for solving the generation mix problem under uncertainty without considerations of emission constraints.

Probabilistic Power System Expansion Planning with Renewable Energy Resources and Energy Storage Systems, First Edition. Jaeseok Choi and Kwang Y. Lee.

Published 2022 by John Wiley & Sons, Inc.

This chapter proposes a fuzzy dynamic programming-based approach for the multi-stage long-term (multi-year) generation mix considering emission constraints under uncertain circumstances. Essentially, the theory consists of the imposition of the framework of a fuzzy decision on the dynamic programming concept. The method can accommodate an arbitrary shape of membership functions and the operation of the pumped-storage hydro plants. Economics, reliability, air pollution, and load uncertainties are evaluated from the membership functions given at all states. Bellman-Zadeh's maximization decision process is used to search for the optimal solution. Also, the 3-dimensional lattice in dynamic programming is used in order to consider load uncertainty [107].

6.2 Fuzzy Dynamic Programming

The fuzzy decision D resulting from q fuzzy goals $G_1, \ldots, G_q$ and p fuzzy constraints $C_1, \ldots, C_p$ is their intersection:

$$D = \left(\bigcap_{i=1}^{q} G_i \right) \cap \left(\bigcap_{j=1}^{p} C_j \right) \tag{6.1}$$

Also, its membership function μ_D resulting from fuzzy goals and constraints is defined by

$$\mu_D(x) = \min \left[\min_{i=1\sim q} \mu_{G_i}, \min_{j=1\sim p} \mu_{C_j} \right] \tag{6.2}$$

The fuzzy mathematical programming problem consists of finding the maximum of the fuzzy decision D:

$$\mu_D(x^*) = \max \mu_D(x) \tag{6.3}$$

where

x^*: optimal decision solution.

For N optimal decisions, (6.3) can be rewritten in a vector form as

$$\mu_D\left(x_1^*, x_2^*, \ldots, x_N^*\right) = \max_{x_1, \ldots, x_N} \mu_D(x_1, x_2, \ldots, x_N) \tag{6.4}$$

To solve this problem using dynamic programming, the principle of optimality is applied to (6.4) as following:

$$\begin{aligned}\mu_D\left(x_1^*, x_2^*, \ldots, x_N^*\right) &= \max_{x_2, \ldots, x_N} \left[\max_{x_1} \left\{ \min \left(\mu_D(x_1), \mu_{F2}(x_2), \ldots, \mu_{F(N-1)}(x_{N-1}), \mu_{FN}(x_N) \right) \right\} \right] \\ &= \max_{x_1, \ldots, x_N} \left[\min \left\{ \mu_D\left(x_1^*\right), \mu_{F2}(x_2), \ldots, \mu_{F(N-1)}(x_{N-1}), \mu_{FN}(x_N) \right\} \right]\end{aligned} \tag{6.5}$$

where

$F_i = G_i \cup C_i$
max: abbreviation for maximize
min: abbreviation for minimize

We can rewrite (6.5) as

$$\mu_D\left(x_1^*, x_2^*, ..., x_N^*\right) = \max_{x_n, ..., x_N}\left[\min\left\{\mu_D\left(x_1^*, x_2^*, ..., x_{n-1}^*\right), \mu_{Fn}(x_n), ..., \mu_{FN}(x_N)\right\}\right] \tag{6.6}$$

Using the principle of optimality by Bellman and Zadeh for an optimal decision of fuzzy set theory, (6.6) yields the familiar general recurrence equation:

$$\mu_D\left(x_1^*, x_2^*, ..., x_n^*\right) = \max_{x_n}\left[\min\left\{\mu_D\left(x_1^*, x_2^*, ..., x_{n-1}^*\right), \mu_{Fn}(x_n)\right\}\right] \tag{6.7}$$

6.3 Best Generation Mix

6.3.1 Problem Statement

This problem can be defined as to determine the generation mix under the following assumptions:

1) The annual loads are given.
2) The number of generators is not that of units but that of types.
3) Nuclear power plants can perform load following.

The system for the proposed method can be modeled as shown in Figure 6.1.

In this chapter, it is assumed that planning for hydro generator construction is performed separately. In the actual test system, the basic resources, which are reserves of hydro power plants, are limited. Therefore, the choice of hydro plant construction is not much and non-flexible. Under this assumption and assuming that the fuel consumption rate, ρ_i (Ton/kWh) and gases density, σ_{in} of type i in year n (ppm/Ton) are constant, the best generation mix (BGM) problem is formulated as follows:

6.3.2 Objective Functions

1) *The economic criterion* (C) – The economic criterion in the best generation mix is the minimization of the sum of the construction cost and the fuel cost as

$$\text{Minimize } Z = \sum_{n=1}^{N}\sum_{i=1}^{NG} K_{cin} d_{in} \alpha_i \Delta x_{in} + \sum_{n=1}^{N}\sum_{i=1}^{NG} K_{fin} f_{in} y_{in} = F(\Delta x_{in}, y_{in}) \tag{6.8}$$

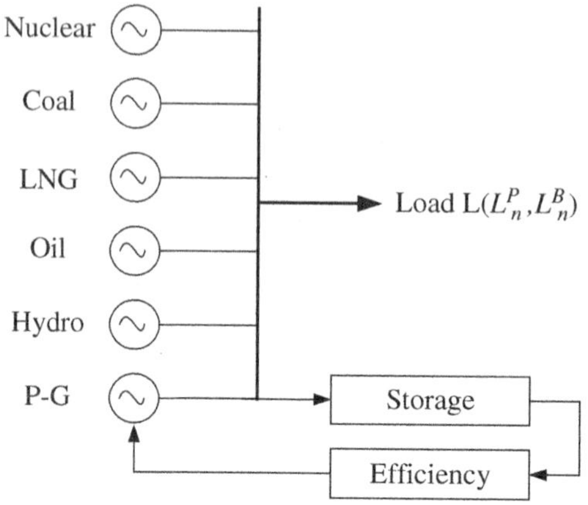

Figure 6.1 Model for the propose method.

where

i: the unit type number (1 for nuclear, 2 for coal, 3 for LNG, 4 for oil, and 5 for pumped-storage generators are specified in this chapter),

N: number of total planning years,

NG: number of unit types,

$$K_{cin} = [(1 + e_{ci})/(1 + r)]^{n\Delta T}$$

$$K_{fin} = [(1 + e_{fi})/(1 + r)]^{n\Delta T}$$

e_{ci}: apparent escalation rate of construction materials of type-i unit,

e_{fi}: apparent escalation rate of fuel of type-i unit,

r: discount rate,

ΔT: step size in years,

d_{in}: construction cost of type-i unit in year n.

f_{in}: marginal fuel cost of type-i unit in year n ($/MWh),

α_i: annual expense rate of type-i unit,

Δx_{in}: construction capacity of type-i unit in year n (MW),

y_{in}: generated energy from type-i unit in year n (MWh).

2) *Reliability criterion* (R) – In this study, the supply reserve rate (SRR) index is used for the reliability criterion. Therefore, the reliability criterion is formulated as

$$\text{Maximize } Z_{2n} = \left(\sum_{i=1}^{NG} x_{in} - L_{pn}\right)/L_{pn} \quad n = 1 \sim N \tag{6.9}$$

where

$$x_{i(n+1)} = x_{in} + \Delta x_{in}$$

3) *Air pollution criterion (A)* – The air pollution index can be formulated as the summation of the CO_2 emissions of fossil-fired units:

$$\text{Minimize } Z_{3n} = \sum_{i=1}^{NG} C_{in}\rho_i y_{in} : n = 1 \sim N \tag{6.10}$$

where
C_{in}: CO_2 density of type-i unit in year n (ppm/Ton),
ρ_i: fuel consumption rate of type-i unit (Ton/MWh).

6.3.3 Constraints

For year n,

1) *Power supply constraint:*

$$\sum_{i=1}^{NG} x_{in} \geq L_{pn}(1 + R_n) \quad n = 0 \sim N \tag{6.11}$$

where
R_n: SRR in year n (p.u.).

2) *Constraint for peak load uncertainty:*

$$L_{pn} \cong L_{pfn} \tag{6.12}$$

where
L_{pfn}: forecasted peak load for year n.

3) *Relation constraints between new plants and existing plants:*

$$x_{in+1} = x_{in} + \Delta x_{in} \quad i = 1 \sim NG, \quad n = 1 \sim N \tag{6.13}$$

4) *Unit generation limit constraints:*

$$x_{\min in} \leq x_{in} \leq x_{\max in} \tag{6.14}$$

5) *SO_x and NO_x emission constraints (fossil-fired units):*

$$\sum_{i=1}^{NG} S_i\rho_i y_{in} \leq SO_{x_n}^{\max} \quad n = 1 \sim N \tag{6.15}$$

$$\sum_{i=1}^{NG} N_i\rho_i y_{in} \leq NO_{x_n}^{\max} \quad n = 1 \sim N \tag{6.16}$$

where

S_i, N_i: SO_x and NO_x emission densities, respectively, of type-i unit [ppm/Ton],
$SO_{x_n}^{\max}, NO_{x_n}^{\max}$: Maximum permissible amount of SO_x and NO_x emissions of type-i unit [Ton].

6.3.4 Membership Functions

1) *The membership function for economic criterion (C)*

$$\mu_c\{x(t-1), u(t)\} = \begin{cases} 1 & : \Delta C(\cdot) \le 0 \\ e^{-W_C \Delta C\{x(t-1), u(t)\}} & : \Delta C(\cdot) > 0 \end{cases} \tag{6.17}$$

where
$\mu_c(\cdot)$: membership function for economic fuzzy set,

$$\Delta C(\cdot) = \{C(x(t)) - C_{asp}(t)\}/C_{asp}(t),$$

$C(x(t))$: cost index at stage t,
$C_{asp}(t)$: aspiration level for cost at stage t,
W_C: weighting factor of the economic membership function,
$x(t+1)$: state parameter at stage $t+1$ (time period) $(=x(t)+\Delta x(t))$ (MW),
$\Delta x(t)$: new construction at stage t (time period) (MW),
$u(t)$: control parameter at stage t $(=\Delta x(t))$ (MW)

2) *Membership function for the reliability criterion*

$$\mu_R\{x(t-1), u(t)\} = \begin{cases} 1 & : \Delta R(\cdot) \ge 0 \\ e^{W_R \Delta R\{x(t-1), u(t)\}} & : \Delta R(\cdot) < 0 \end{cases} \tag{6.18}$$

where
$\mu_R(\cdot)$: membership function of reliability (SRR) fuzzy set,

$$\Delta R(\cdot) = \{R(x(t)) - R_{asp}(t)\}/R_{asp}(t),$$

$R(x(t))$: reliability (supply reserve rate) index at stage t,
$R_{asp}(t)$: aspiration level for reliability (SRR) at stage t,
W_R: weighting factor of reliability membership function.

3) *Membership function for the air pollution criterion (considering CO_2 only)*

$$\mu_A\{x(t-1), u(t)\} = \begin{cases} 1 & : \Delta A(\cdot) \le 0 \\ e^{-W_A \Delta A\{x(t-1), u(t)\}} & : \Delta A(\cdot) > 0 \end{cases} \tag{6.19}$$

where
$\mu_A(\cdot)$: membership function for CO_2 air pollution fuzzy set,

$$\Delta A(\cdot) = \{A(x(t)) - A_{asp}(t)\}/A_{asp}(t),$$

$A(x(t))$: air pollution index at stage t,
$A_{asp}(t)$: aspiration level for air pollution at stage t,
W_A: weighting factor of air pollution membership function.

4) *Membership function for the forecasted peak load*

$$\mu_L(x(t-1), u(t)) = \begin{cases} 1 + \dfrac{1}{W_L(\Delta L(\cdot)/2\sigma_{+p.u}(t))^2} & : \Delta L \geq 0 \\ 1 + \dfrac{1}{W_L(\Delta L(\cdot)/2\sigma_{-p.u}(t))^2} & : \Delta L < 0 \end{cases} \tag{6.20}$$

where
$\mu_L(\cdot)$: membership function for a fuzzy set of the forecasted peak load,

$$\Delta L(\cdot) = \{L_P(x(t)) - L_{pfasp}(t)\}/L_{pfasp}(t),$$

$L_P(x(t))$: the forecasted peak load index at stage t,
$L_{pfasp}(x(t))$: aspiration level for the forecasted peak load at stage t,

$$\sigma_{p.u}(t) = \sigma(t)/L_{pfasp}(t),$$

$\sigma_+(t)$: right-hand side standard deviation of $L_{pfasp}(t)$ (Figure 6.4c),
$\sigma_-(t)$: left-hand side standard deviation of $L_{pfasp}(t)$ (Figure 6.4c),
W_L: weighting factor of the membership function of the forecasted peak load.

6.3.5 The Proposed Fuzzy Dynamic Programming-Based Solution Procedure

Considering (6.1), fuzzy dynamic programming can be formulated as follows:

$$D = C \cap R \cap A \cap L \tag{6.21}$$

where

C, R, A, and L are economics, reliability, air pollution, and forecasted load fuzzy sets, respectively.

Using (6.7), the following recursive algorithm for (6.21) can be derived:

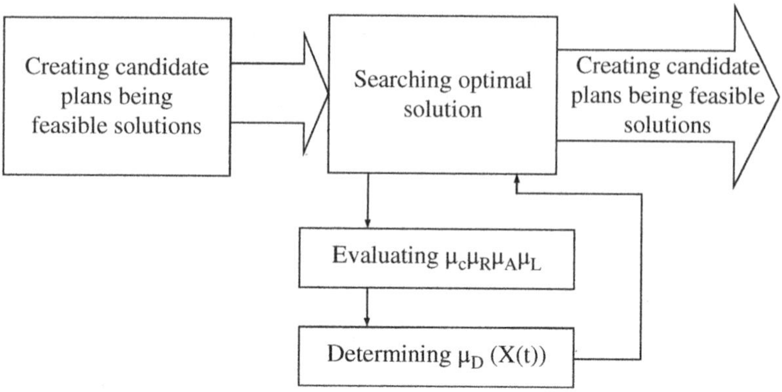

Figure 6.2 Procedure for analyzing the generation mix problem using fuzzy dynamic programming.

$$\mu_D(x(t)) = \max\left[\min\left\{\mu_C(x(t-1), u(t)), \mu_A(x(t-1), u(t)), \mu_R(x(t-1), u(t)), \mu_L(x(t-1), u(t)), \mu_D(x(t-1))\right\}\right] \tag{6.22}$$

$$u_{min}(t) \leq u(t) \leq u_{max}(t)$$

where

$$x(t) = x(t-1) + u(t),$$
$$\mu_D(x(0)) = 1.0,$$

max[min]: maximize[minimum{·}].

The solution procedure for solving the generation mix using the proposed method is shown in Figures 6.2 and 6.3.

6.4 Case Study

In order to demonstrate the effectiveness of the proposed method, the best generation mix (BGM) problem is solved on the KEPCO system. The system consists of six types of generators over 20 years (2006–2026). Year 2006 is assumed to be the initial year. The step size of planning years (stage) is assumed to be five years ($\Delta T = 5$). The maximum, minimum loads, and hydro capacities in standard years are listed in Table 6.1. The characteristics of the economic data are summarized in Table 6.2 and the capacity factor of the gas emissions' input data of generating units are shown in Table 6.3.

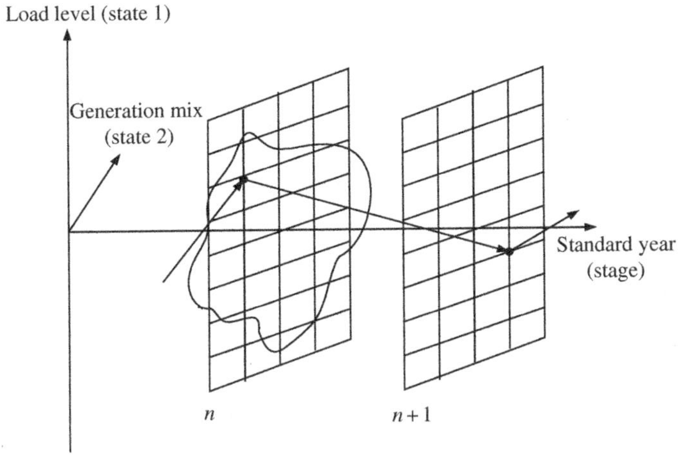

Figure 6.3 Procedure for searching solution of generation mix on 3-D lattice.

Table 6.1 Maximum load, minimum load, and hydro plant at standard years.

Years	Peak load L_P (MW)	Base load L_B (MW)	Hydro (MW)	LEP (10^3 Ton)
2006	48 108	30 340	1800	—
2011	57 340	34 200	2000	4500
2016	69 500	42 500	2200	5500
2021	78 200	47 500	2400	6500
2026	87 000	53 500	2600	7500

Table 6.2 The characteristics of economic input data.

Gen. type	Initial year capacity (MW)	Fixed charge (10^5 won/kW)	A_{ER} of fixed charge (%)	Marginal fuel cost (Won/kWh)	A_{ER} of fuel cost (%)	Annual cost rate (%)
Nucl.	16 715	145.0	2	6.8	1	19
Coal	17 465	100.0	1	13.8	1	17
LNG	14 313	85.0	1	21.5	1	17
Oil	4308	75.0	1	120.0	4	17
P-G	2000	45.0	1	0.0	0	13

where A_{ER} means the apparent escalation rate and the discount rate is assumed as 10%.

Table 6.3 The capacity factors and the gases' emission data of plants.

Gen. type	Capacity factor (%)	Fuel consumption rate (ρ) (Ton/MWh)	Density (σ) (ppm/ton) CO_2, SO_x, NO_x
Nucl.	80	—	
Coal	70	0.4030	700, 450, 500
LNG	65	0.0500	450, 200, 300
Oil	55	0.0234	600, 200, 100
P-G	30	—	

Table 6.4 Yearly total generation capacities (MW).

Year	Generation capacity (MW)
2006	56 601
2011	67 800
2016	88 500
2021	93 000
2026	105 000

In this study, it is assumed that the yearly total generation capacities in the years are shown in Table 6.4. Table 6.5 shows the generation mix on the KEPCO system capacity in the initial year (2006).

The maximum permissible limits on emissions are given in Table 6.6. The aspiration level of the total cost, reliability, and CO_2 in years/stage and weighting factors of membership function are shown in Tables 6.7 and 6.8. The aspiration levels of load of years/stages are the same as the peak loads, *Lp* in Table 6.1.

The membership functions for fuzzy sets of cost and reliability are shown in Figure 6.4. The membership function pattern of CO_2 air pollution is the same as that of cost.

6.4.1 Results and Discussion

The following four test cases were simulated:

1) Case C0: crisp and not considering emission constraints using crisp,
2) Case F0: fuzzy but considering emission constraints,

Table 6.5 Generation mix and capacity in the initial year (2006).

Gen. type	Generation mix (%)	Capacity (MW)
Nucl.	29.5	16 715
Coal	30.9	17 465
LNG	25.3	14 313
Oil	7.6	4308
P-G	3.5	2000
Hyd.	3.2	1000
Total	100	56 601

Table 6.6 Maximum permissible limitation/criterion of the SO_x and NO_x emission constraints (10^3 Ton/yr).

Gases	2011	2016	2021	2026
SO_x	40	40	40	40
NO_x	40	40	40	40

Table 6.7 Aspiration level of total cost, reliability, and CO_2 in years/stage.

Aspiration	2011	2016	2021	2026
Cost (10^{12} won)	65	120	155	182
CO_2 (10^3 Ton/yr)	125	140	155	170
Reliability (%)	10	10	10	10

Table 6.8 Weighting factors of membership function.

	Weighting factors of membership function
Cost	10.0
Reliability	15.0
Load	10.0
CO_2	5.0

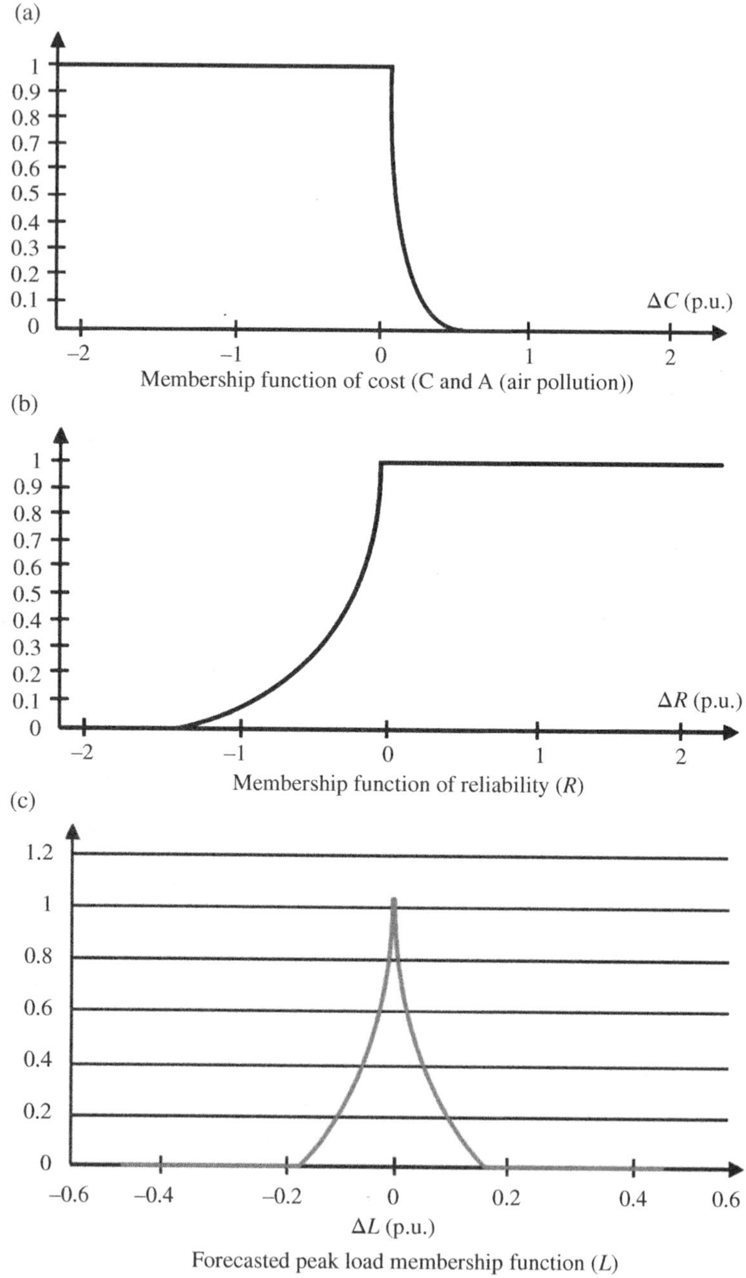

Figure 6.4 Membership functions for fuzzy sets of cost and reliability.

Table 6.9 Best generation mix for the conventional crisp method (not considering CO_2) (%).

	Crisp conventional method (Case C0) (minimize total cost)			
Gen. type	**2011**	**2016**	**2021**	**2026**
Nucl.	28.5	23.5	23.5	22.0
Coal	39.4	48.5	48.5	47.0
LNG	14.5	17.0	18.5	19.0
Oil	10.4	3.0	2.3	2.0
P-G	3.2	4.7	4.5	7.8
Hyd.	4.0	3.3	2.7	2.2

Table 6.10 Best generation mix for the conventional fuzzy method (not considering CO_2) (%).

	Fuzzy conventional method (Case F0) (base case)			
Gen. type	**2011**	**2016**	**2021**	**2026**
Nucl.	23.5	23.5	23.5	46.5
Coal	44.4	48.5	48.5	21.5
LNG	14.5	17.0	18.5	19.0
Oil	10.4	3.0	2.3	2.0
P-G	3.2	4.7	4.5	8.8
Hyd.	4.0	3.3	2.7	2.2

3) Case F1: the fuzzy base case considering emission constraints, and
4) Case F2: the fuzzy case considering CO_2 emission constraint with −10% deceased CO_2 aspiration level.

Test results are shown in Tables 6.9–6.12, respectively. Using the proposed method (case F1) considering CO_2 emission constraints produced the BGM results that are different from those of the conventional crisp and fuzzy methods. The results indicate that the percentage of nuclear power plants increased while that of the coal power plants is decreased. Table 6.10 shows the fact that stricter CO_2 emission constraint increases the nuclear power plant mix and reduces the coal plant mix.

Table 6.11 Best generation mix for the proposed fuzzy method (considering CO_2) (%).

	Fuzzy proposed method (Case F1) (CO_2 reinforcement mix)			
Gen. type	**2011**	**2016**	**2021**	**2026**
Nucl.	46.5	48.5	48.5	48.5
Coal	17.4	23.5	23.5	23.5
LNG	14.5	17.0	18.5	19.0
Oil	10.4	3.0	2.3	2.0
P-G	7.2	4.7	4.5	4.8
Hyd.	4.0	3.3	2.7	2.2

Table 6.12 Best generation mix for the other proposed fuzzy method (considering CO_2 with −10% deceased CO_2 aspiration level) (%).

	Fuzzy proposed method (Case F2) (CO_2 10% more reinforcement mix)			
Gen. type	**2011**	**2016**	**2021**	**2026**
Nucl.	46.5	48.5	48.5	48.5
Coal	17.4	23.5	23.5	23.5
LNG	14.5	17.0	18.5	19.0
Oil	10.4	3.0	2.3	2.0
P-G	7.2	4.7	4.5	4.8
Hyd.	4.0	3.3	2.7	2.2

Table 6.13 shows the satisfaction levels of the simulated results. The table indicates that stricter CO_2 emission constraint lowers the satisfaction level of the decision maker as a result. As an option to improve her/his level of satisfaction, the decision maker may consider purchasing CO_2 emission allowances from the market.

6.5 Conclusion

This chapter proposes a fuzzy dynamic programming-based solution approach for the long-term generation mix with multi-stages (years) considering air pollution constraints on CO_2 emissions. The effectiveness of the proposed approach is

Table 6.13 Satisfaction levels for cases.

	Constraints	Satisfaction level
Case C0	Crisp and minimize cost	—
Case F0	Fuzzy and not considering CO_2	1.0000
Case F1	Fuzzy and considering	0.8984
Case F2	Fuzzy and −10% Aasp	0.5093

demonstrated by using it to determine the best generation mix problem of the KEPCO-system, which contains nuclear, coal, LNG, oil, and pumped-storage hydro plant in multi-years. A result of the CO_2 emission constraint is the recommendation of the nuclear or LNG power plant construction when available. This, in turn, increases the total cost of the BGM. Although this chapter is focused on the effect of CO_2 emissions on the best generation mix, the proposed approach can be generalized to handle other emission constraints such as those on SO_x and NO_x. Also, the proposed approach can accommodate the operation of pumped-storage generation which has a relationship with the operation of nuclear power plant in load following.

7

Best Generation Mix Considering Air Pollution Constraints

7.1 Introduction

There is a global trend toward liberalization and privatization of the electricity supply industry. This is coupled with growing environmental awareness and increasing concern on dwindling fossil energy sources. Electricity is the indispensable form of energy in modern societies. In this chapter, a new approach for the long-term generation mix with multi-criteria considering air pollution constraints, which are not only SO_2 and NO_X but also CO_2 emission limitations, under the uncertain circumstances is proposed using linear programming. The effectiveness of the proposed approach is demonstrated by applying to the best generation mix problem of a power system, which contains nuclear, coal, LNG, oil, and pumped-storage hydro plant in multi-years. This case study in particular is mainly focused on CO_2 emission limitation effect in the best generation mix. The method can accommodate the operation of the pumped-storage generator which has a relationship with operation of nuclear power plant with some requirement for load following.

This chapter illustrates the flexible generation mix problem for year 2030 in Korea, called vision 2030. The proposed approach provides a flexible solution rather than a plan that might be too robust [108].

7.2 Concept of Flexible Planning

Recently, the uncertainties of external conditions for power system scheduling problem are becoming unmanageable due to demand growth, primary energy resource variations, and reliability of energy supply system and energy security. The uncertainties require a more flexible system for preventing power system from the unmanageable collapse. Figure 7.1 means that the flexible planning and

Probabilistic Power System Expansion Planning with Renewable Energy Resources and Energy Storage Systems, First Edition. Jaeseok Choi and Kwang Y. Lee.

Published 2022 by John Wiley & Sons, Inc.

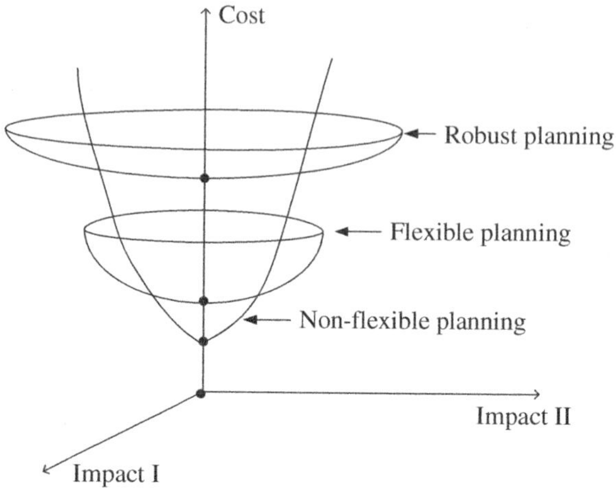

Figure 7.1 Concept of flexibility. *Source*: Based on Takahashi et al. [120] and Arakawa et al. [121].

operation for power system is one which "although not necessarily gives the optimum solution for the basic forecasted conditions, yet can keep the reasonable scheduling solution from being significantly worsened by any assumed changes in the surrounding situations" [120, 121].

The useful methods for flexible generator mix problem and flexible maintenance scheduling problem using fuzzy LP, DP, IP, and search method have been developed in the past. Some cases of application of fuzzy set theory for the flexible planning and operation of power systems are introduced in this chapter. It is expected that more flexible solution can be obtained with the proposed methods because the fuzzy set theory that can reflect the subjective decision of a decision maker is used in these studies. This chapter uses fuzzy set theory in order to obtain a flexible long-term generation mix solution in Korea power system considering CO_2 emission constraint.

7.3 LP Formulation of the Best Generation Mix

7.3.1 Problem Statement

This problem can be defined as to determine the generation mix under the following assumptions:

1) The annual loads are given.
2) The number of generators is not that of units but that of types.

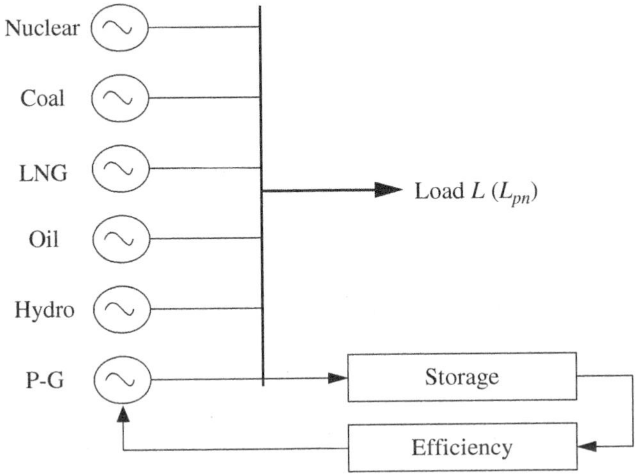

Figure 7.2 A system model for the proposed method.

Nuclear power plants are able to perform load following.

The system for the proposed method can be modeled as shown in Figure 7.2.

In this study, it is assumed that the hydro generator construction is separately planned from that of the other kinds of generation units. In the actual system, as the basic resources on reserve, the hydro power plants have limitations in the country. Therefore, the hydro plant construction is not much available and non-flexible. Under the assumption, the best generation mix problem is formulated in the following.

7.3.2 Objective Functions

1) *The Economic Criterion* – The economic criterion in the best generation mix is the minimization of the sum of the construction cost and the fuel cost as

$$\text{Minimize } Z = \sum_{n=1}^{N}\sum_{i=1}^{NG} K_{cin} d_{in} \alpha_i \Delta x_{in} + \sum_{n=1}^{N}\sum_{i=1}^{NG} K_{fin} f_{in} y_{in}$$
$$= F(\Delta x_{in}, y_{in})$$

where

i: unit type number (1 for nuclear, 2 for coal, 3 for LNG, 4 for oil, and 5 for pumped-storage generators),

N: total number of study years,

NG: number of unit types,

$K_{cin} = ((1 + e_{ci})/(1 + r))^{n\Delta T}$,

$K_{fin} = ((1 + e_{fi})/(1 + r))^{n\Delta T}$,
e_{ci}: apparent escalation rate of construction materials of the ith-unit,
e_{fi}: apparent escalation rate of fuel of ith-unit,
r: discount rate,
$\Delta T_{k,k-1}$: step size years of study years from k to $k - 1$,
d_{in}: construction cost of the ith-unit in the n year,
f_{in}: marginal fuel cost of the ith-unit in the n year (Won/MWh),
α_i: annual expenses rate of the ith-unit,
Δx_{in}: construction capacity of the ith-unit in the n year (MW),
y_{in}: generation capacity of the ith-unit in the n year (MWh).

2) *Constraints*

i) *Installed capacity constraint*

$$\sum_{i=1}^{NG} (x_{in-1} + \Delta x_{in}) + HYD \geq L_{pn}(1 + R_n) \quad n = 1 \sim N$$

where
R_n: supply reserve rate in the nth year. (p.u.),
HYD_n: capacity of hydro generator in the nth year. It is assumed that the HYD_n is given in this study.

ii) *Energy constraint of demand*

$$\sum_{i=1}^{NG} y_{in} \geq (L_{pn} + L_{Bn}) \times 8760/2 + V_n - HYD_n \times 8760 \times CF_H$$

where
L_{Pn}: peak load at the nth year,
L_{Bn}: base load at the nth year,
V_n: the added demand energy is caused by pumped-storage generator,
CF_H: average capacity factor of hydro generator.

iii) *Production energy constraint of generation system*

$$y_{in} \leq (x_{in-1} + \Delta x_{in}) \times 8760 \times CF_i \quad i = 1 \sim NG, \;\; n = 1 \sim N$$

where
CF_i: average capacity factor of the ith-unit.

iv) *Capacity constraint in the initial year*

$$x_{io} = EX_i \quad i = 1 \sim NG$$

where
EX_i: capacity of the ith existing unit.

v) *Constraint of mutual relationship between the existing generator capacity and the new generator capacity (state equation)*

$$x_{in} = x_{in-1} + \Delta x_{in} \quad i = 1 \sim NG, n = 1 \sim N$$

vi) *Energy constraint of LNG thermal plant*

$$y_{3n} \geq LEP_{min_n}/\rho_3 \quad n = 1 \sim N$$

where
LEP_{min_n}: LNG thermal generator production energy for LNG minimum due to consumption in the nth year.

vii) *Constraints of reservoir capacity of pumped-storage generator*

$$y_{5n} = \eta_g \times V_n$$

where
η_g: efficiency of pumped-storage generator.

viii) *No load following power constraints of nuclear power plant*

$$(x_{1n} - x_{5n}) \leq L_{Bn} \quad n = 1 \sim N$$

ix) *Upper-lower constraints of new unit capacity*

$$\Delta x_{in} \leq \Delta X_{max_{in}} \quad i = 1 \sim NG, \quad n = 1 \sim N$$

where $X_{min\ in}$ and $X_{max\ in}$ are the minimum and maximum capacities of the new unit at the nth year (period), respectively.

x) *CO_2 air pollution constraint*

$$\sum_{i=1}^{NG} \xi_i y_i \leq CO_{2MAX_n}$$

where
CO_{2MAX_n}: maximum quantity of CO_2 permitted in the nth year (Ton/yr),
ξ_i: fuel consumption rate of the ith-unit (Ton/MWh).

xi) *SO_X air pollution constraint*

$$\sum_{i=1}^{NG} SO_{X_{in}} \xi_i y_n \leq SO_{XMAX_n}$$

where
SO_{Xin}: SO_X density of the i-unit in the nth year (ppm/Ton),
SO_{XMAX_n}: maximum quantity of SO_X permitted in the nth year (Ton/yr).

xii) *NO_X air pollution constraint*

$$\sum_{i=1}^{NG} NO_{Xin}\xi_i y_n \le NO_{XMAX_n}$$

where
NO_{Xin}: NO_X density of the i-unit in the nth year (ppm/Ton),
NO_{XMAX_n}: maximum quantity of NO_X permitted in the nth year (Ton/yr).

7.4 Fuzzy LP Formulation of Flexible Generation Mix

7.4.1 The Optimal Decision Theory by Fuzzy Set Theory

The fuzzy decision D resulting from q fuzzy goals $G_1, \ldots, G_q$ and p fuzzy constraints $C_1, \ldots, C_p$ is the intersection of them;

$$D = \left(\bigcap_{i=1}^{q} G_i\right) \cap \left(\bigcap_{j=1}^{p} C_j\right) \tag{7.1}$$

and also its membership function μ_D resulting from fuzzy goals and constraints is defined by

$$\mu_D(x) = \min\left[\min_{i=1\sim p} \mu_{G_i}, \min_{j=1\sim q} \mu_{C_j}\right] \tag{7.2}$$

where

min: abbreviation of minimum.

The fuzzy mathematical programming problem consists of finding the maximum of the fuzzy decision D.

$$\mu_D(x^*) = \max \mu_D(x) \tag{7.3}$$

where

x^*: optimal decision solution max is an abbreviation of maximum.

The vector (7.3) can be rewritten as

$$\mu_D(x_1^*, x_2^*, \ldots, x_N^*) = \max_{x_1, \ldots, x_N} \mu_D(x_1, x_2, \ldots, x_N) \tag{7.4}$$

Figure 7.3 shows the concept for optimal decision on fuzzy sets, which is proposed by Bellman and Zadeh.

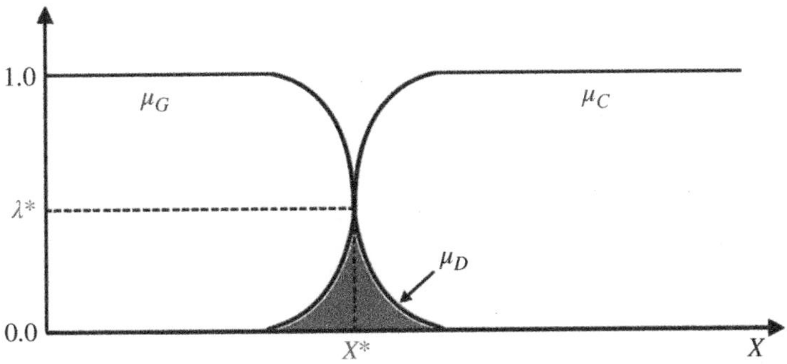

Figure 7.3 Concept for optimal decision on fuzzy sets. *Source*: Modified from Bellman and Zadeh [159].

7.4.2 The Function of Fuzzy Linear Programming

Fuzziness of Cost : $Z \lesssim Z_0$

The satisfaction level in the best generation mix is flexible because the total cost is uncertain.

Fuzziness of Reliability criterion : $R_n \gtrsim R_{0n}$

The reliability criterion in the long-term-based best generation mix may be flexible than entirely strict.

Using parameter λ, therefore, which physically means satisfaction level, the best generation mix problem can be formulated as follows:

Objective: Maximize λ

Constraints: Subject to $F(\Delta x_{in}, y_{in}) + \Delta Z_0\, \lambda = Z_0 + \Delta Z_0$

$$\left(\left(\sum_{i=1}^{NG}(x_{in-1} + \Delta x_{in}) + HYD_n\right) - L_{pn}\right)/L_{pn} - \Delta R_o \lambda = R_o - \Delta R_o$$

$$\sum_{i=1}^{NG} y_{in} \geq (L_{pn} + L_{Bn}) \times 8760/2 + V_n - HYD \times 8760 \times CF_H$$

$$y_{in} \leq (x_{in-1} + \Delta x_{in}) \times 8760 \times CF_i$$

$$x_{i0} = EX_i$$

$$x_{in} = x_{in-1} + \Delta x_{in}$$

$$y_{3n} \geq LEP_{min_n}/\rho_3$$

$$y_{5n} = \eta_g \times Vn$$

$$(x_{1n} - x_{5n}) \leq L_{Bn}$$

$$\Delta x_{in} \leq \Delta X_{max\ in}$$

$$\sum_{i=1}^{NG} \xi_i y_i \leq CO_{2MAX_n}$$

$$\sum_{i=1}^{NG} SO_{Xin} \xi_i y_n \leq SO_{XMAX_n}$$

$$\sum_{i=1}^{NG} NO_{Xin} \xi_i y_n \leq NO_{XMAX_n}$$

7.5 Case Studies

In order to demonstrate the effectiveness of the proposed method, the generation mix is performed on Korea power system, which consists of six types of generators over 25 years (2006–2030). It is assumed that the initial year is 2006. The step size of planning year is assumed as five years ($\Delta T = 5$). The maximum load, minimum load, and hydro capacity in standard years are listed in Table 7.1. The characteristics and economic data are summarized in Tables 7.2 and 7.3, respectively.

The maximum new construction capacity is limited by limitation of human and equipment resources, etc. It is assumed, as shown in Table 7.3, that the maximum and minimum capacities per stage/year of new generator are the same for all stages/years in this study. In addition, in this study, the maximum permissible limitation of air pollution emission was considered as shown in Table 7.4.

Table 7.1 Maximum load, minimum load, and hydro plant in standard years.

Years	Peak load LP (MW)	Base load LB (MW)	Hydro (MW)	LEP (10^3 Ton)
2006	58 990	35 394	1800	—
2011	65 940	39 564	2000	4500
2016	70 050	42 030	2200	5500
2021	74 000	44 400	2400	6500
2026	77 000	46 200	2600	7500
2030	80 000	48 000	2600	7500

Table 7.2 The characteristics and economic data.

Gen. type	Initial capacity (MW)	Fixed charge (10^5 won/kW)	A_{ER} of fixed charge (%)	Marginal fuel cost (Won/kW)	A_{ER} of fuel cost (%)
Nucl.	17 716	144.4	3.5	4	1
Coal	18 465	79.7	3.4	17	1
LNG	17 437	61.4	3.3	67	1
Oil	4686	153.2	3.3	87	4
P-G	3300	63.4	3.5	0	0

Annual cost rate (%)	Capacity factor (%)	Fuel consumption rate (Ton/MWh)	Density (ppm/Ton) CO_2, SO_x, NO_x
19	90	—	
17	90	0.2300	700, 450, 500
17	60	0.1100	450, 200, 300
17	55	0.2000	600, 200, 100
13	30	—	

Where A_{ER} means the apparent escalation rate and the discount rate is assumed as 7.5%.

Table 7.3 Maximum and minimum of capacity per stage year of new generators (MW).

Gen. type	ΔX_{max}	X_{max}
Nuclear	5000	30 000
Coal	5000	30 000
LNG	5000	30 000
Oil	500	2500
P-G	1000	5000

Table 7.4 Maximum permissible limitation of air pollution emission (10^3 Ton/yr).

Air pollution	2011	2016	2021	2026	2030
CO_2	40	40	40	40	40
SO_2	40	40	40	40	40
NO_X	40	40	40	40	40

7.5.1 Results by Non-Fuzzy Model

The simulated results in the cases, base case not considering air pollution criterion and case considering air pollution criterion, are shown in Tables 7.5 and 7.6, respectively. The proposed fuzzy-set theory-based best generation mix, which is simulated under considering air pollution criterion is different from the generation mix using conventional method. The results yield that the mix of nuclear power plants is increasing and that of coal power plants is decreasing. Figures 7.4 and 7.5 show total capacity and percent ratio results for conventional method and proposed method considering air pollution constraints.

Table 7.5 Best generation mix for the conventional methods (%).

	Conventional method (Base case: Case 0)					
Gen. type	**2006**	**2011**	**2016**	**2021**	**2026**	**2030**
Nucl.	27.94	28.71	32.48	35.51	35.05	33.74
Coal	29.12	29.65	27.50	25.73	25.40	24.44
LNG	27.50	27.75	25.74	24.08	23.77	25.58
Oil	7.39	5.92	5.49	5.14	5.07	4.88
P-G	5.20	5.43	6.21	6.91	7.90	8.65
Hyd.	28.4	2.53	2.58	2.63	2.81	2.71

Table 7.6 Best generation mix for the proposed methods (%).

	Proposed method – APC reinforcement mix (Case 1)					
Gen. type	**2006**	**2011**	**2016**	**2021**	**2026**	**2030**
Nucl.	27.94	28.71	32.47	35.62	35.16	33.91
Coal	29.12	29.65	27.53	25.71	25.38	24.48
LNG	27.50	27.75	25.73	24.03	23.71	25.37
Oil	7.39	5.92	5.49	5.13	5.06	4.88
P-G	5.20	5.43	6.21	6.89	7.88	8.65
Hyd.	28.4	2.53	2.58	2.63	2.81	2.71

where APC means the air pollution (CO_2, SO_2, and NO_X) constraints.

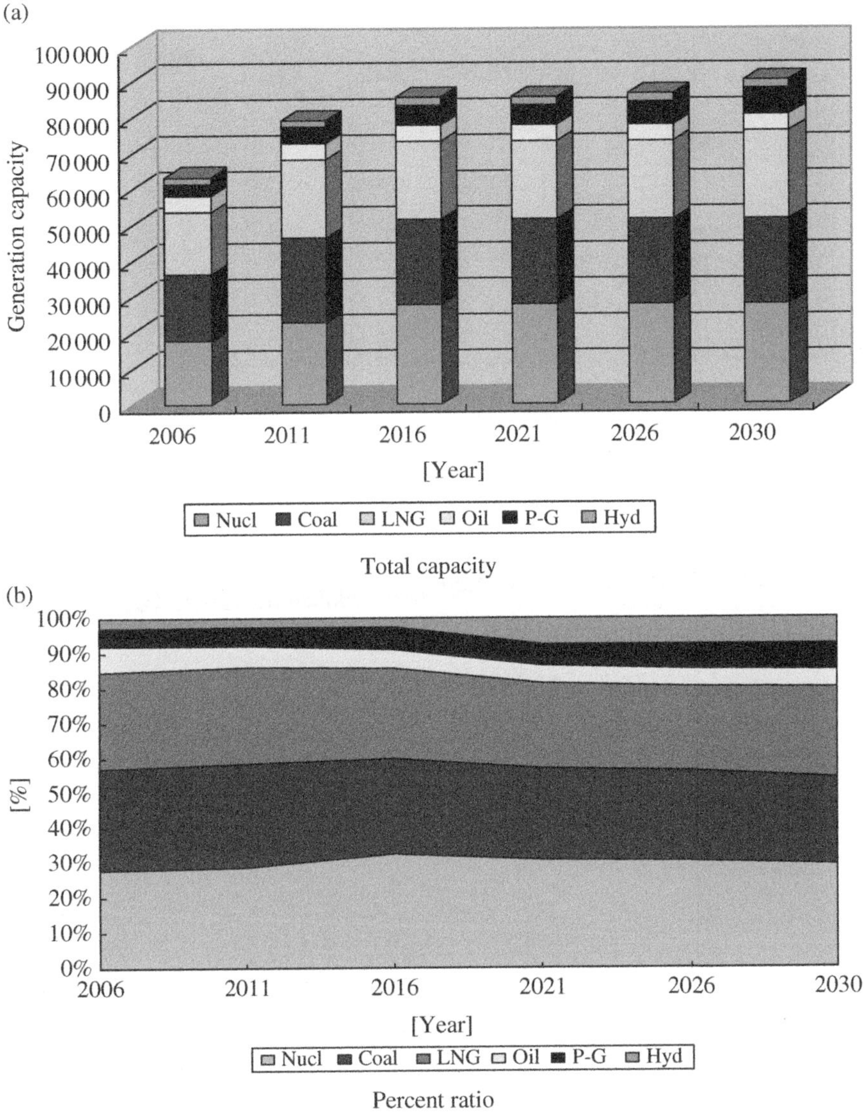

Figure 7.4 Best generation mix by conventional method (Case 0).

Table 7.6 and Figure 7.5 show the results of sensitivity analysis of permissible maximum for CO_2 constraint. As the air pollution constraints are considered, the mix of nuclear power plant is growing up and the total cost is increasing.

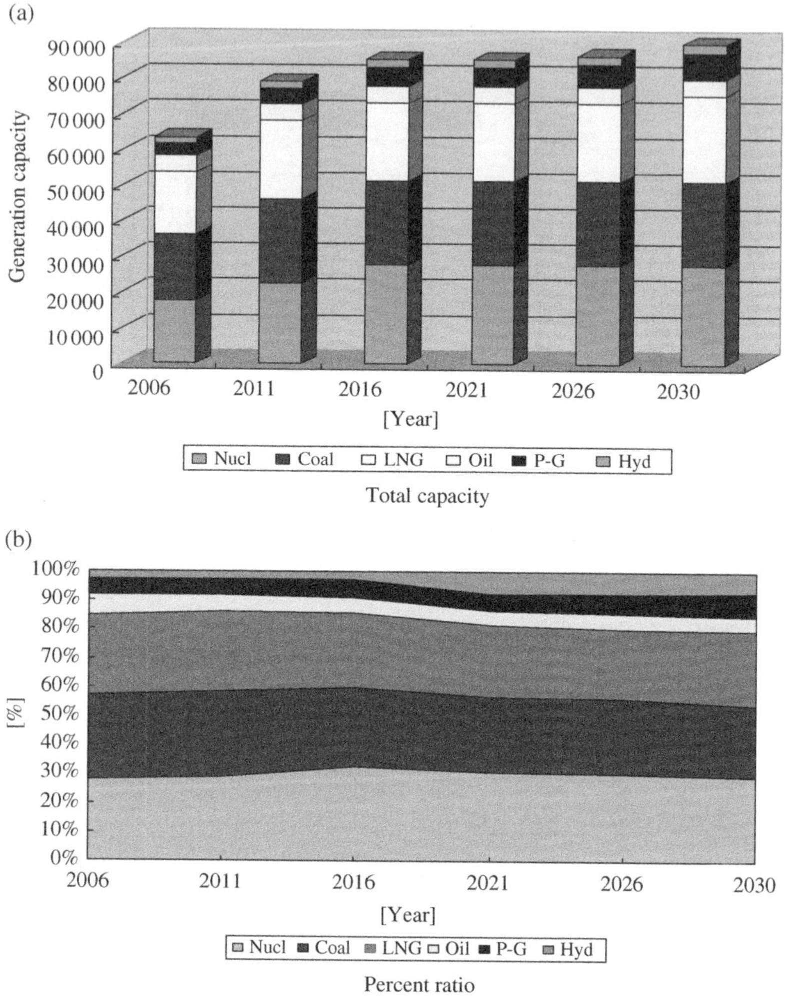

Figure 7.5 The best generation mix by proposed method with CO_2 air pollution constraint (Case 1).

7.5.2 Results by Fuzzy Model

The proposed fuzzy-set theory-based best generation mixes which are simulated under considering air pollution criterion are different from the generation mix using conventional method. The results are shown in Table 7.7, where satisfaction

Table 7.7 Cost variation according to changing of the CO_2 gas air pollution constraints (Billion Won).

	Permissible maximum for CO_2 constraint × 10^6 Ton/yr	Construction cost	Operating cost	Total cost
Case 0	—	4475.10	11 238.99	15 714.09
Case 1	40	4498.82	12 018.50	16 517.32
Case 2	35	4531.21	12 468.79	17 000.00
Case 3	30	4563.62	12 919.07	17 482.69
Case 4	25	4596.02	13 369.35	17 965.37
Case 5	20	4638.19	13 826.55	18 464.74

Table 7.8 Best generation mix for APC reinforcement mix using fuzzy set theory (%).

	Proposed method – APC reinforcement mix					
Gen. type	**2006**	**2011**	**2016**	**2021**	**2026**	**2030**
Nucl.	27.94	28.71	32.47	35.62	35.16	33.91
Coal	29.12	29.65	27.53	25.71	25.38	24.48
LNG	27.50	27.75	25.73	24.03	23.71	25.37
Oil	7.39	5.92	5.49	5.13	5.06	4.88
P-G	5.20	5.43	6.21	6.89	7.88	8.65
Hyd.	28.4	2.53	2.58	2.63	2.81	2.71

level (Z_0) is 16.5×10^3 Billion Won and ΔZ_0 is 10% of Z_0. The results of the best generation mix for APC reinforcement mix using fuzzy-set theory are shown in Table 7.8. Table 7.9 shows the comparison of costs and satisfaction levels of the best generation mix in three cases (Figure 7.6).

Table 7.10 and Figure 7.7 show the relationship between Z_0 and satisfaction level for various aspiration targets using fuzzy-set theory. As the satisfaction level increases by 0.1 point, the total cost increases about 150 (Billion Won), where the abbreviation, APC means air pollution constraint of CO_2, SO_X, and NO_X. The result informs about the incremental rate value of cost in order to make one unit higher satisfaction level.

Table 7.9 Total cost evaluation of the best generation mix in the three cases (Billion Won).

	Construction cost	Operation cost	Total cost	Satisfaction level
Conventional method (Case 1)	4498.82	12 018.50	16 517.32	—
Mix without APC (Case FA1)	4475.10	11 238.99	15 714.09	0.98
Mix with APC (Case FB4)	4498.82	12 018.50	16 517.32	0.49

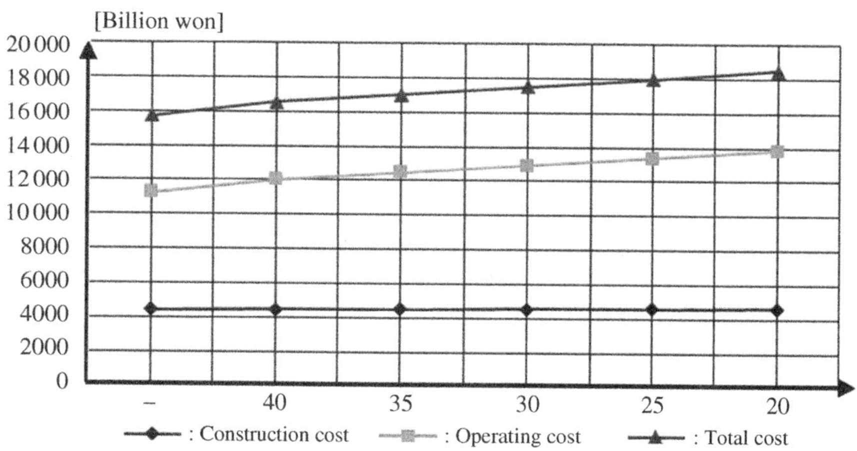

Figure 7.6 Construction, fuel, and total costs' variation according to gas emission constraints.

7.6 Conclusion

In this chapter, an alternative approach for the long-term generation mix with multi-criteria considering air pollution constraints, which are not only SO_2 and NO_X but also CO_2 emission limitations, is proposed using linear programming. The effectiveness of the proposed approach is demonstrated in the best generation mix problem of Korea power system, which contains nuclear, coal, LNG, oil, and pumped-storage hydro plant in multi-years. This chapter presents the solution to the generation mix problem for year 2030 in Korea. The result is the generation mix of nuclear 33.9%, coal 24.5%, LNG 25.4%, and

Table 7.10 Satisfaction-level variation according to changing of cost aspiration level.

	Cases	Aspiration level z_0 (10^3 Billion Won)	Satisfaction level λ
APC not considered cases	Fuzzy A1	16.5	0.98
	Fuzzy A2	16.4	0.92
	Fuzzy A3	16.3	0.86
	Fuzzy A4	16.2	0.80
APC considered cases	Fuzzy B1	17.4	1.00
	Fuzzy B2	17.1	0.84
	Fuzzy B3	16.8	0.67
	Fuzzy B4	16.5	0.49

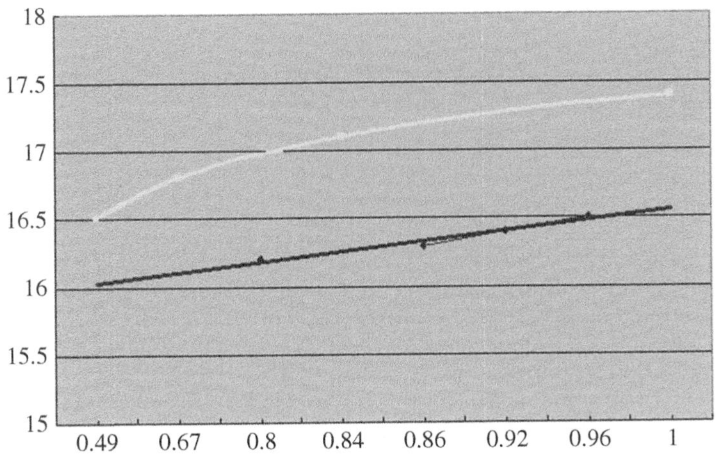

Figure 7.7 Satisfaction-level variation according to changing of cost aspiration level.

pumped-storage generator 8.7%. The method can accommodate the operation of the pumped-storage generator which has an interesting relationship with operation of nuclear power plant with some requirement for load following. The case study in this chapter is mainly focused on the effect of CO_2 emission limitation in the best generation mix. The CO_2 air pollution constraint is more strict, and the nuclear or LNG power plant construction is recommended as shown in the case study, although the total cost is increasing. So, the increasing cost is called the cost for air pollution constraints. The cost for air pollution constraints is 0.82 Won/kWh. Although this is only an example, it provides important data in the carbon dioxide emission trading market. From sensitivity analysis between satisfaction level and the cost, decision makers can obtain information of the incremental rate value of cost in order to make higher satisfaction level by one unit.

8

Generation System Expansion Planning with Renewable Energy

8.1 Introduction

The use of renewable energy sources as a generation resource in electric power systems is growing rapidly due to its advantages that are minimal pollution, non-depletion, and low operating cost. Wind turbine generator (WTG) and solar cell generator (SCG) are the fastest-growing energy sources on a percentage basis. In the past 20 years, for example, the global wind energy capacity has increased five hundredfold – from 3.5 GW in 1994 to almost 50 GW by the end of 2004 and 1560 GW by 2013. As a result, it is becoming more important to analyze the best generation mix (BGM) including WTG and SCG.

This chapter proposes a new approach to solve the long-term generation mix considering renewable energy resources, namely WTGs and SCGs with multi-criteria considering CO_2, SO_x, and NO_x emission constraints, under the uncertainties, is proposed using linear programming. This chapter focuses on the development of tools to analyze the effect of WTGs and SCGs on the best generation mix (BGM) using fuzzy set theory. The method accommodates sensitivity analysis of capacity factors (CFs) of WTG and SCG [36, 38, 39, 129]. The effectiveness of the proposed approach is demonstrated by applying to solve the flexible generation mix problem on the Korea power system for the year 2030. The test system contains nuclear, coal, LNG, oil, pumped-storage hydro plant, WTG, and SCG in multi-year period.

Probabilistic Power System Expansion Planning with Renewable Energy Resources and Energy Storage Systems, First Edition. Jaeseok Choi and Kwang Y. Lee.

Published 2022 by John Wiley & Sons, Inc.

8.2 LP Formulation of the Best Generation Mix

8.2.1 Problem Statement

This BGM problem is to determine the generation mix under the following assumptions:

1) The annual loads are known.
2) The number of generators is not that of units but that of types.
3) Nuclear power plants are able to perform load following.
4) It is assumed that the yearly CFs of WTG and SCG are given.

For the proposed method, the system can be modeled as shown in Figure 8.1. It is considered to be composed of Nuclear = 1, Coal = 2, LNG = 3, Oil = 4, PSH (pumped storage hydro) = 5, WTG = 6, SCG = 7 for convenience in this chapter.

Because wind speed and solar radiation do not maintain a specified stable level and cannot be controlled by the operator, the WTG and SCG have resource supply uncertainties rather than generator outages. Figure 8.2 shows the uncertainties of WTG and SCG plants are different from those of conventional plants. Therefore, the CFs of the WTG and SCG depend entirely on the resource supply with an associated ambiguity. In this chapter, it is assumed that the CFs are available. Therefore, the energy based on the specified CF should be consumed mandatory as is the case with energy-limited generator and hydro generator.

In this study, it is assumed that the hydro generator construction is separately planned from that of the other kinds of generating units. In actual systems, the water resources, which are reserves, of the hydro power plants have limitation in the country. Therefore, the choice of hydro plant construction is not much

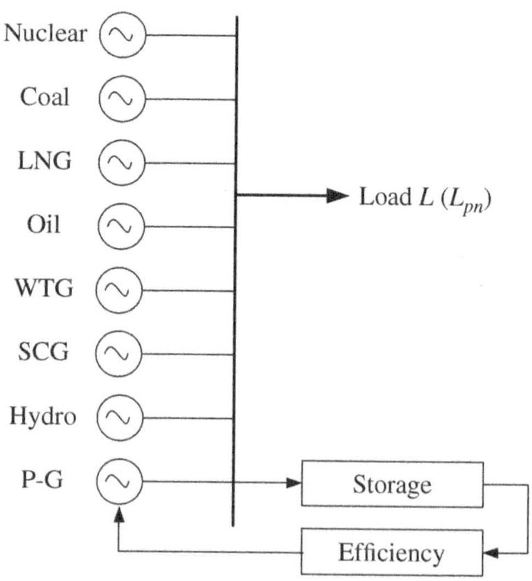

Figure 8.1 A system model for use by the proposed method.

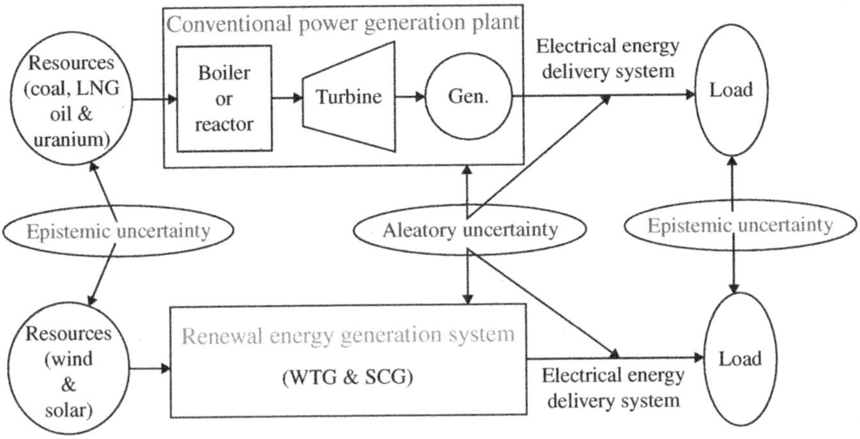

Figure 8.2 Uncertainties of the renewable energy power and conventional power plants.

and non-flexible. Under these assumptions, the BGM problem is formulated as follows:

8.2.2 Objective Function and Constraints

1) *The Economic Criterion* – The economic criterion in the BGM is to minimize the sum of the construction cost and the fuel cost as

$$\begin{aligned}\text{Minimize } Z &= \sum_{n=1}^{N}\sum_{i=1}^{NG} K_{cin} d_{in} \alpha_i \Delta x_{in} + \sum_{n=1}^{N}\sum_{i=1}^{NG} K_{fin} f_{in} y_{in} \\ &= F(\Delta x_{in}, y_{in})\end{aligned} \tag{8.1}$$

where

i: unit type number (1 for nuclear, 2 for coal, 3 for LNG, 4 for oil, and 5 for pumped-storage generators are specified in this chapter),

N: number of total study stage year,

NG: number of unit type,

$K_{cin} = ((1+e_{ci})/(1+r))^{n\ \Delta T_{k,k-1}}$,

$K_{fin} = ((1+e_{fi})/(1+r))^{n\ \Delta T_{k,k-1}}$,

e_{ci}: apparent escalation rate of construction materials of unit i,

e_{fi}: apparent escalation rate of fuel of unit i,

r: discount rate,

$\Delta T_{k,k-1}$: step size years of study years from k to $k-1$,

d_{in}: construction cost of unit i in year n,

f_{in}: marginal fuel cost of unit i in year n (Won/MWh),

α_i: annual expenses rate of unit i,

Δx_{in}: construction capacity of unit i in year n (MW),

y_{in}: generation capacity of unit i in year n (MWh)

2) *Constraints*

i) *Installed capacity constraint*

$$\sum_{i=1}^{NG}(x_{in-1} + \Delta x_{in}) + HYD \geq L_{pn}(1 + R_n) \quad n = 1 \sim N \tag{8.2}$$

where
R_n: supply reserve rate in the year n (p.u.),
HYD_n: capacity of hydro generator in year n. In this study, it is assumed that the HYD_n is given

ii) *Energy constraint of demand*

$$\sum_{i=1}^{NG} y_{in} \geq (L_{pn} + L_{Bn}) \times 8760/2 + V_n - HYD_n \times 8760 \times CF_H \tag{8.3}$$

where
L_{Pn}: peak load in year n,
L_{Bn}: base load in year n,
V_n: the added demand energy caused by pumped-storage generator,
CF_H: average CF of hydro generator

iii) *Production energy constraint of generation system*

$$\begin{aligned} y_{in} &\leq (x_{in-1} + \Delta x_{in}) \times 8760 \times CF_i \quad i = 1 \sim 5, \quad n = 1 \sim N \\ y_{in} &= (x_{in-1} + \Delta x_{in}) \times 8760 \times CF_i \quad i = 6 \text{ and } 7, \quad n = 1 \sim N \end{aligned} \tag{8.4}$$

where
CF_i: average CF of the ith-unit

iv) *Capacity constraint in the initial year*

$$x_{io} = EX_i \quad i = 1 \sim NG \tag{8.5}$$

where
EX_i: capacity of the existing unit i

v) *Constraint of mutual relationship between the existing generator capacity and new generator capacity (state equation)*

$$x_{in} = x_{in-1} + \Delta x_{in} \quad i = 1 \sim NG, \quad n = 1 \sim N \tag{8.6}$$

vi) *Energy constraint of LNG thermal plant*

$$y_{3n} \geq LEP_{min_n}/\rho_3 \quad n = 1 \sim N \tag{8.7}$$

where
LEP_{min_n}: minimum LNG thermal energy in year n.

vii) *Constraints of reservoir capacity of pumped-storage generator*

$$y_{5n} = \eta_g \times V_n \tag{8.8}$$

where
η_g: efficiency of pumped-storage generator

viii) *No load-following power constraints of nuclear power plant*

$$(x_{1n} - x_{5n}) \leq L_{Bn} \quad n = 1 \sim N \tag{8.9}$$

ix) *Upper-lower constraints of new unit capacity*

$$\Delta X_{min_{in}} \leq \Delta x_{in} \leq \Delta X_{max_{in}} \quad i = 1 \sim NG, \ \ n = 1 \sim N \tag{8.10}$$

where
$X_{min_{in}}$ and $X_{max_{in}}$ are the minimum and the maximum capacities of new units at year n, respectively.

x) CO_2 *constraint*

$$\sum_{i=1}^{NG} \xi_i y_i \leq CO_{2MAX_n} \tag{8.11}$$

where
ξ_i: fuel consumption rate of the ith unit (Ton/MWh),
CO_{2MAX_n}: maximum quantity of CO_2 permitted in year n (Ton/yr)

xi) SO_X *constraint*

$$\sum_{i=1}^{NG} SO_{X_{in}} \xi_i y_n \leq SO_{XMAX_n} \tag{8.12}$$

where
$SO_{X_{in}}$: SO_X density of the ith-unit in year n (ppm/Ton),
SO_{XMAX_n}: maximum quantity of SO_X permitted in year n (ppm/yr)

xii) NO_X *constraint*

$$\sum_{i=1}^{NG} NO_{Xin} \xi_i y_n \leq NO_{XMAX_n} \tag{8.13}$$

where
NO_{Xin}: NO_X density of the ith-unit in year n (ppm/Ton),
NO_{XMAX_n}: maximum quantity of NO_X permitted in year n (ppm/yr)

8.3 Fuzzy LP Formulation of Flexible Generation Mix

8.3.1 The Optimal Decision Theory by Fuzzy Set Theory

The fuzzy decision D resulting from q fuzzy goals $G_1,\dots, G_q$ and p fuzzy constraints $C_1,\dots, C_p$ is the intersection of them

$$D = \left(\bigcap_{i=1}^{q} G_i\right) \cap \left(\bigcap_{j=1}^{p} C_j\right) \tag{8.14}$$

and also its membership function μ_D resulting from fuzzy goals and constraints is defined by

$$\mu_D(x) = \min\left[\min_{i=1\sim q} \mu_{G_i}, \min_{j=1\sim p} \mu_{C_j}\right] \tag{8.15}$$

where

min: abbreviation for minimum.

The fuzzy mathematical programming problem consists of finding the maximum of the fuzzy decision D:

$$\mu_D(x^*) = \max \mu_D(x) \tag{8.16}$$

where

x^*: optimal decision solution, max is an abbreviation for maximum.

The (8.16) can be rewritten in a vector form as

$$\mu_D\left(x_1^*, x_2^*, \dots, x_N^*\right) = \max_{x_1,\dots,x_N} \mu_D(x_1, x_2, \dots, x_N) \tag{8.17}$$

Figure 8.3 shows the concept for optimal decision on fuzzy sets, which is proposed by Zadeh [156, 157].

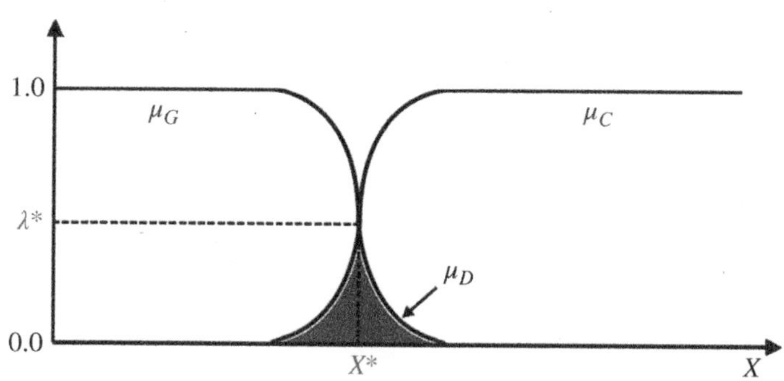

Figure 8.3 Concept for optimal decision on fuzzy sets. *Source:* Based Zadeh [156] and Lee and El-Sharkawi [157].

8.3.2 The Function of Fuzzy Linear Programming

$$\text{Fuzziness of Cost}: Z \lesssim Z_0$$

The satisfaction level in the BGM is flexible because the total cost is unknown.

$$\text{Fuzziness of reliability criterion}: R_n \gtrsim R_{0n}$$

The reliability criterion in the long-term-based BGM may be flexible than fixed. Using parameter λ, therefore, which means satisfaction level physically, the BGM problem can be formulated as follows:

1) *Objective – Maximize* λ
2) *Constraints – Subject to*

$$F(\Delta x_{in}, y_{in}) + \Delta Z_0\,\lambda = Z_0 + \Delta Z_0$$

$$\left(\left(\sum_{i=1}^{NG}(x_{in-1} + \Delta x_{in}) + HYD_n\right) - L_{pn}\right)/L_{pn} - \Delta R_o\lambda = R_o - \Delta R_o$$

$$\sum_{i=1}^{NG} y_{in} \geq \left(L_{pn} + L_{Bn}\right) \times 8760/2 + V_n - HYD \times 8760 \times CF_H$$

$$y_{in} \leq (x_{in-1} + \Delta x_{in}) \times 8760 \times \; CF_i$$

$$x_{i0} = EX_i$$

$$x_{in} = x_{in-1} + \Delta x_{in}$$

$$y_{3n} \geq LEP_{min_n}/\rho_3$$

$$y_{5n} = \eta_g \times Vn$$

$$(x_{1n} - x_{5n}) \leq L_{Bn}$$

$$\Delta x_{in} \leq \Delta X_{max\,in}$$

$$\sum_{i=1}^{NG} \xi_i y_i \leq CO_{2MAX_n}$$

$$\sum_{i=1}^{NG} SO_{X_{in}} \xi_i y_n \leq SO_{XMAX_n}$$

$$\sum_{i=1}^{NG} NO_{X_{in}} \xi_i y_n \leq NO_{XMAX_n}$$

8.4 Case Studies

The BGM was performed on Korea power system to demonstrate the effectiveness of the proposed method. The test system consists of eight types of generators over 22 years (2008–2030) with the initial year being 2008. The maximum load, minimum load, and the hydro capacity in standard years are listed in Table 8.1. Generator characteristics and economic data are summarized in Table 8.2.

The maximum new construction capacity is limited by the limitation of human and equipment resources. It is assumed, as shown in Table 8.3, that the maximum and minimum capacities per stage year of new generators are the same for all stages/years for each type. Table 8.4 shows the maximum permissible limitation of air pollution emission.

8.4.1 Test Results

The proposed fuzzy set theory-based BGM was simulated considering WTG and SCG and air pollution criterion. Aspiration level (Z_0) of total cost (budget) was specified as 27 500 Billion Won. The BGM result is shown in Table 8.5. The satisfaction level for the BGM was found to be 0.832. The result indicates an increase in the mix of nuclear power plants and renewable energy generators and a decrease in that of coal power plants. Figure 8.4 is a graphical representation of the results in percent ratio.

8.4.2 Sensitivity Analysis

8.4.2.1 Capacity Factor of WTG and SCG

The CFs of WTG and SCG in the BGM is important in order to determine their mix. The CFs of WTG and SCG depend on the resources which are wind speed and solar radiation, respectively. The CF has ambiguity. The CFs of WTG and SCG are

Table 8.1 Maximum load, minimum load, and hydro plant at standard years.

Years	Peak load L_P (MW)	Base load L_B (MW)	Hydro (MW)	LNG (10^3 Ton)
2008	58 990	35 394	1800	—
2013	65 940	39 564	2000	4500
2018	70 050	42 030	2200	5500
2023	74 000	44 400	2400	6500
2028	77 000	46 200	2600	7500
2030	80 000	48 000	2600	7500

Table 8.2 Generator characteristics and economic data.

Gen. type	Initial capacity (MW)	Fixed charge (10^3 Won/kW)	[a]A_{ER} of fixed charge (%)	Marginal fuel cost (Won/kWh)	[a]A_{ER} of fuel cost (%)
Nucl.	17 716	2122	3.5	3.730	1
Coal	18 465	1145	3.4	17.794	1
LNG	17 437	741	3.3	65.327	1
Oil	4686	1881	3.3	86.987	4
P-G	3300	915	3.5	0	0
WTG	20	4770	−4.5	0	0
SCG	10	7500	−6.5	0	0

Annual cost rate (%)	Capacity factor (%)	Fuel consumption rate (Ton/MWh)	Density (ppm/Ton) CO_2, SO_X, NO_X
19	90	—	
17	90	0.2300	—, 450, 500
17	60	0.1100	—, 200, 300
17	55	0.2000	—, 200, 100
13	30	—	
10	10	0	
10	15	0	

[a] A_{ER} means the apparent escalation rate (CER) and the discount rate is assumed as 7.5%.

Table 8.3 Maximum and minimum capacity per stage year of new generators (MW).

Generation type	ΔX_{max}	ΔX_{min}
Nuclear	5000	0
Coal	5000	0
LNG	5000	30 000 × 5.25[a]
Oil	500	0
P-G	1000	0
WTG	500~3000	100
SCG	500~3000	100

[a] The limited energy of LNG.

Table 8.4 Maximum permissible limitation of air pollution emission (CO_2:[10^6 Ton/yr], SO_X and NO_X: [10^9 ppm/year]).

Air pollution	2013	2018	2023	2028	2030
CO_{2MAX}	40	40	40	40	40
SO_{XMAX}	40	40	40	40	40
NO_{XMAX}	40	40	40	40	40

Table 8.5 Best generation mix considering renewable energy generators using the proposed method (%).

	Base case: Case F1					
Gen. type	2008	2013	2018	2023	2028	2030
Nucl.	27.93	29.40	31.83	33.64	36.65	36.68
Coal	29.11	30.37	32.02	30.61	28.93	28.82
LNG	27.49	25.19	22.35	20.01	18.91	18.84
Oil	7.39	6.07	5.38	4.82	4.55	4.54
P-G	5.20	5.57	4.94	4.42	4.18	4.16
WTG	0.03	0.67	0.71	0.74	0.80	0.89
SCG	0.02	0.14	0.24	3.30	3.47	3.55
Hyd.	2.84	2.59	2.53	2.47	2.53	2.52
Total	100	100	100	100	100	100

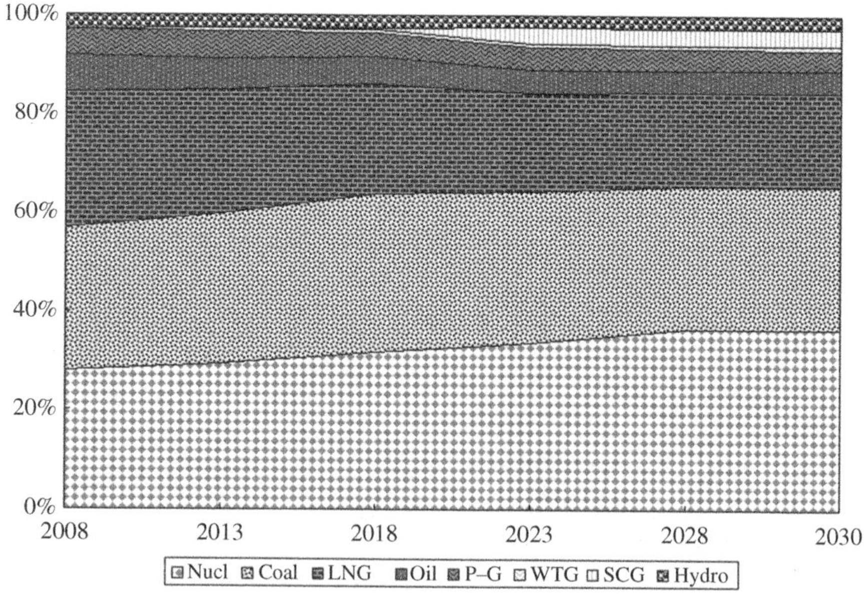

Figure 8.4 Best generation mix by the proposed method (Case F1).

usually in the ranges of 8–16% and 14–18%, respectively. Tables 8.6 and 8.7 show the BGM results at $CF_{WTG} = 12\%$ and $CF_{WTG} = 14\%$, respectively, while $CFSCG = 15\%$. Tables 8.8 and 8.9 show the cost evaluation results for the various CFs of WTG and SCG.

From the results, as the CF of the renewable energy generators increases, the total cost decreases because the operating cost of conventional generator decreases relatively even if the construction cost increases. The total costs per unit CF decreased from the highest CFs of WTG and SCG yield about 20 billion (Won/%) and 27 billion (Won/%), respectively. The result describes the economic

Table 8.6 Best generation mix for CF_{WTG} = 12% using fuzzy set theory (%).

	Case CF3 (CF_{WTG} = 12%)					
Gen. type	**2008**	**2013**	**2018**	**2023**	**2028**	**2030**
Nucl.	27.93	29.39	31.82	33.61	36.71	36.76
Coal	29.11	30.36	32.02	30.60	28.99	28.88
LNG	27.49	25.21	22.37	20.02	18.96	18.89
Oil	7.39	6.06	5.38	4.81	4.56	4.54
P-G	5.20	5.56	4.94	4.42	4.19	4.17
WTG	0.03	0.67	0.71	0.78	0.84	0.93
SCG	0.02	0.14	0.24	3.30	3.22	3.31
Hyd.	2.84	2.59	2.53	2.47	2.53	2.52

Table 8.7 Best generation mix for CF_{WTG} = 14% using fuzzy set theory (%).

	Case CF4 (CF_{WTG} = 14%, CF_{SCG} = 15%)					
Gen. type	**2008**	**2013**	**2018**	**2023**	**2028**	**2030**
Nucl.	27.93	29.39	31.82	33.61	36.71	36.76
Coal	29.11	30.34	31.78	30.56	28.95	28.84
LNG	27.49	25.26	22.15	20.04	18.99	18.92
Oil	7.39	6.06	5.31	4.81	4.55	4.54
P-G	5.20	5.56	4.88	4.41	4.18	4.16
WTG	0.03	0.67	1.72	3.85	3.74	3.82
SCG	0.02	0.14	0.24	0.32	0.40	0.49
Hyd.	2.84	2.59	2.53	2.47	2.53	2.52

Table 8.8 Cost variations corresponding to changes in the CF of WTG (Billion Won).

	CF_{WTG} (%)	Construction cost	Operating cost	Total cost	Satisfaction level
Case CF2	8	8355.0	19 618.5	27 973.5	0.828
Case CF1	10	8417.9	19 543.8	27 961.8	0.832
Case CF3	12	8420.9	19 519.5	27 940.4	0.840
Case CF4	14	8458.4	19 437.5	27 895.9	0.856
Case CF5	16	8565.0	19 259.0	27 824.0	0.882

Table 8.9 Cost variations corresponding to changes in the CF of SCG (Billion Won).

	CF_{SCG} (%)	Construction cost	Operating cost	Total cost	Satisfaction level
Case CF6	14	8444.9	19 541.4	27 986.3	0.823
Case CF1	15	8417.9	19 543.8	27 961.8	0.832
Case CF7	16	8396.5	19 541.7	27 938.2	0.840
Case CF8	17	8352.5	19 563.9	27 916.5	0.849
Case CF9	18	8693.8	19 186.1	27 879.9	0.862

credit of renewable energy resource in the BGM problem. Finally, the higher CF makes the satisfaction level to increase because the total cost is decreased. Apparent escalation rate (e_c) of construction materials of WTG and SCG is reflected in the total cost in (8.1).

It is expected that the apparent escalation rate (CER, e_c) of construction material cost of the WTG and SCG will be decreased as a result of technology development that has been taking place. Therefore, the negative CER may be considered. Specially, the CER of SCG will be decreased more rapidly rather than the CER of the WTG. For example, Tables 8.10 and 8.11 show the best generation mix result at $e_{cSCG} = -7.0\%$ and $e_{cSCG} = -5.5\%$. Tables 8.12 and 8.13 show the cost results for the various CERs of WTG and SCG. As the CER of renewable generators increases, the total cost decreases because the construction cost of renewable energy generator decreases regardless of whether the operating cost of conventional generators increases or stays fixed. It is noted that the WTG and SCG can win conventional generators in mix from $e_{cWTG} = -4.0\%$ and $e_{cSCG} = -6.0\%$, respectively as shown in Tables 8.10–8.13. The total costs per unit ER decreased by the lower CER of WTG and SCG yield about 10 billion (Won/%) and 50 billion (Won/%).

Table 8.10 Best generation mix for e_{cSCG} = −7.0% using the proposed approach (%).

	Case F12 (e_{cSCG} = −7.0%)					
Gen. type	**2008**	**2013**	**2018**	**2023**	**2028**	**2030**
Nucl.	27.93	29.38	31.83	33.61	36.71	36.77
Coal	29.11	30.35	31.97	30.61	29.00	28.89
LNG	27.49	25.76	22.87	20.46	19.38	19.31
Oil	7.39	6.06	5.38	4.81	4.56	4.54
P-G	5.20	5.56	4.94	4.42	4.19	4.17
WTG	0.03	0.16	0.25	0.33	0.41	0.50
SCG	0.02	0.14	0.24	3.30	3.22	3.31
Hyd.	2.84	2.59	2.53	2.47	2.53	2.52

Table 8.11 Best generation mix for e_{cSCG} = −5.5% using the proposed approach (%).

	Case F10 (e_{cSCG} = −5.5%)					
Gen. type	**2008**	**2013**	**2018**	**2023**	**2028**	**2030**
Nucl.	27.93	29.43	31.85	34.85	37.62	38.33
Coal	29.11	30.39	32.03	31.24	30.22	29.79
LNG	27.49	25.14	22.31	20.68	19.37	19.09
Oil	7.39	6.07	5.38	4.99	4.67	4.61
P-G	5.20	5.57	4.94	4.58	4.29	4.23
WTG	0.03	0.67	0.71	0.77	0.82	0.90
SCG	0.02	0.14	0.24	0.33	0.41	0.50
Hyd.	2.84	2.59	2.53	2.47	2.53	2.52

Table 8.12 Total cost evaluation corresponding to changes in the CER of WTG (Billion Won).

	e_{cWTG} (%)	Construction cost	Operating cost	Total cost	Satisfaction level
Case F10	−3.5	8360.6	19 610.7	27 971.3	0.829
Case F11	−4.0	8423.6	19 543.8	27 967.5	0.830
Case F1	−4.5	8417.9	19 543.8	27 961.8	0.832
Case F12	−5.0	8412.5	19 543.8	27 956.3	0.834
Case F13	−5.5	8407.3	19 543.8	27 951.1	0.836

Table 8.13 Total cost evaluation corresponding to changes in the CER of SCG (Billion Won).

	e_{cSCG} (%)	Construction cost	Operating cost	Total cost	Satisfaction level
Case F10	−5.5	8385.6	19 613.4	27 999.0	0.818
Case F11	−6.0	8454.4	19 534.1	27 988.5	0.822
Case F1	−6.5	8417.9	19 543.8	27 961.8	0.832
Case F12	−7.0	8391.8	19 543.8	27 935.6	0.842
Case F13	−7.5	8372.4	19 534.8	27 907.2	0.852

The results describe the economic credits of renewable energy resources in the viewpoint of construction material cost. The lower escalation rate makes the satisfaction level increase because total cost is decreased.

8.5 Conclusion

This chapter has proposed a fuzzy LP-based approach for the long-term best generation mix with multi-criteria considering renewable energy generators such as WTGs and SCGs and air pollution constraints. The proposed method can accommodate sensitivity analysis of CF and CER of WTG and SCG with uncertainties.

The effectiveness of the proposed approach is demonstrated by applying it to the best multi-year generation mix problem of the Korea power system for the year 2030. The test system contains nuclear, coal, LNG, oil, pumped-storage hydro, WTG, and SCG plants. The economic credits of renewable energy resource in the best generation mix problem were proposed in the viewpoint of CF and CER of WTG and SCG. It is noted that WTG and SCG are competitive generator types in the mix and WTG and SCG can win the conventional generators in mix from $e_{cWTG} = -4.0\%$ and $e_{cSCG} = -6.0\%$, respectively for the Korea power system. The proposed model and method are useful tools for various economics evaluation in the BGM considering renewable energy generators and emission constraints.

9

Reliability Evaluation for Power System Planning with Wind Generators and Multi-Energy Storage Systems

This chapter presents a new methodology for a probabilistic power system reliability evaluation using a Monte Carlo simulation (MCS) with multi-energy storage system (ESS) installed at wind farms for planning ESS-integrated power system. A large-scale wind turbine generator (WTG) creates significant power fluctuations and should affect the stability, frequency control, and then the reliability of the power system. High penetration of wind farms can result in unacceptable variations in the frequency and voltage in the power system. The impact of significant power fluctuation of the WTG can, however, be reduced by installing an ESS. The proposed model can facilitate the reliability analysis and evaluation of the contribution of each ESS installed at multiple wind farms integrated into a power system. The presented simulation method can also be used to assess an appropriate capacity of the ESS in the power system resulting from a sensitivity analysis. A case study demonstrates the proposed model and methodology on a power system with a similar size to the one in Jeju Island, Korea.

9.1 Introduction

A reasonable level of continuity in an electric power supply is indispensable for a better quality of life and economic advancement. ESSs are being added to power systems with the major objective of mitigating the adverse impacts of variability and uncertainty associated with renewable energy generation (REG). The ESSs are needed in the power system to reduce the high volatility resulting from the increasing penetration of WTG. Rapid changes in WTG lead to instability in customer satisfaction. As a solution, an ESS can be installed in wind farms in connection with a conventional power system. For reference, the current application of ESS in Korea is mainly for frequency regulation. In the future, however, ESS is

Probabilistic Power System Expansion Planning with Renewable Energy Resources and Energy Storage Systems, First Edition. Jaeseok Choi and Kwang Y. Lee.

Published 2022 by John Wiley & Sons, Inc.

expected to be used in integrating large installations of renewable energy in addition to the frequency regulation. This will allow the flexibility to manage and control the power supply pattern of different renewables depending on weather, location, and so on. It is therefore expected that the deployment of ESS will continue to increase with the increase in REG. For example, in Korea in 2018, ESS was distributed more than 20 times compared to the previous year. In addition, the use of ESS for renewable energy is estimated to have increased about 16 times. Here, the rapid increase is attributed to the financial incentives, such as the weighting of renewable energy certificates (RECs).

It is therefore important to analyze the influence of ESS on the power-supply reliability. Reliability evaluation of single-ESS coordinated with single WTG using MCS was studied by Hu, Karki, and Billinton in 2009 [31]. A similar study has been done by Gao and Wang [32]. Additionally, reliability evaluation of ESS with renewable distributed generation in microgrids was studied by Borges and Cantarino [33]. However, these studies were on reliability evaluation for single-ESS coordinated with single WTG. Nowadays, however, multi-ESS coordinated with multi-WTGs are in operation in the actual system.

This is achieved through the development of a new methodology embedded in a simulation program to analyze the power system reliability. An MCS method was used to incorporate the dependency and correlation between the various system variables and characteristics in the operation of an ESS and WTG. The proposed algorithm is applicable as both a next-generation energy technology and conversion technology for feasibility analysis.

The integration of multiple wind farms in a power system causes large fluctuations in power generation that are highly uncertain. The resulting net load can therefore exhibit a random and erratic pattern leading to challenges in preparing, committing, and dispatching appropriate capacities and reserves to provide a timely balance in supply and demand. For example, the California independent system operator (CAISO) recently analyzed that the daytime load declines sharply due to an increase in the amount of renewable energy generation, which was called a duck curve (Figure 9.1) [40, 41]. If such a phenomenon becomes severe, it warns the overall system to be vulnerable to potential blackout situations. Japan has recently curtailed an increase in REG due to the concern that the rapidly increasing REG may create difficulties in maintaining the stability of the power supply. The utilization of ESS is widely perceived as a potential solution to mitigate the challenges of large-scale integration of wind farms in power systems.

Therefore, the objective of this chapter is to develop a new methodology for probabilistic reliability evaluation for the power system when multi-ESSs are installed at multi-wind farms. First of all, in order to develop a new methodology, a new model is proposed. Furthermore, the optimal operation equation of multi-ESS coordinated with multi-WTG is newly formulated in the perspective of

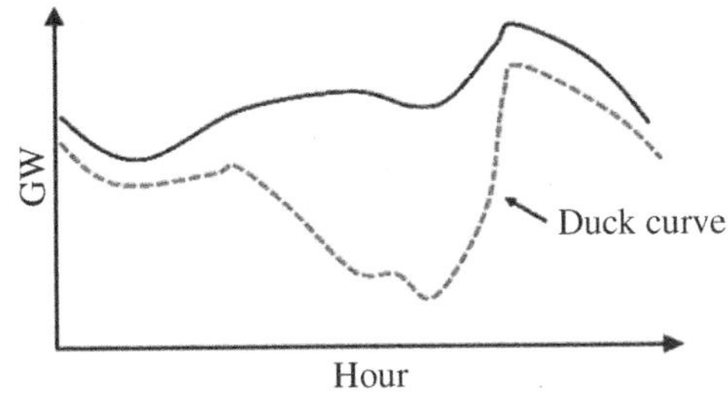

Figure 9.1 An example of duck curve. *Source:* Modified from California ISO – Fast Facts, CAISO. Available at http://www.caiso.com/Documents/Flexibleresourceshelprenewables_FastFacts.pdf.

reliability maximization. The method is verified by application to an actual power system on the Jeju Island in Korea. For reference, it is assumed that wind farm is equivalent to WTG in this chapter for convenience.

9.2 Probabilistic Reliability Evaluation by Monte Carlo Simulation

9.2.1 Probabilistic Operation Model of Generator 1

The MCS method uses a stochastic model and a probabilistic method to quantify the level of reliability while incorporating the inherent uncertainties in the system. A deterministic model uses an analytical method, but the stochastic nature of the system variables and the dependencies renders it almost impossible to find the solution. Therefore, the solution is obtained using a probabilistic method through a simulation by repeatedly and numerically generating a series of random numbers [34, 42].

For the reliability evaluation by the MCS method, random numbers are generated as shown in (9.1) and (9.2) using the Forced Outage Rate (FOR) of the generator. The artificial operation history of the generators having a long history should be reconstructed by using Mean Time to Failure (MTTF) and Mean Time to Repair (MTTR), which are two variables constituting FOR. For reference, the Time to Repair (TTR) must be reconfigured sequentially after the Time to Failure (TTF).

$$TTF_{i,k} = -MTTF_i \ln U_k = -\frac{1}{\lambda_i} \ln U_k \tag{9.1}$$

$$TTR_{i,k} = -MTTR_i \ln U'_k = -\frac{1}{\mu_i} \ln U'_k \tag{9.2}$$

where

$TTF_{i,k}$: TTF of the ith generator at the kth state (hours),
$TTR_{i,k}$: TTR of the ith generator at the kth state (hours),

λ_i: failure rate of the ith generator,
μ_i: repair rate of the ith generator,
U_k, U'_k: two uniformly distributed random numbers between [0,1] at the kth state.

Hence, the total generator output TG_{ck} (MW) of the kth state generator for the MCS is calculated as in (9.3).

$$TG_{ck} = \sum_{i \in NG} G_{i,k} \times ISK_{i,k} \tag{9.3}$$

$$ISK_{i,k} = \begin{cases} 1 & k \in \Omega_{TTF_i} \\ 0 & k \in \Omega_{TTR_i} \end{cases} \tag{9.4}$$

where

NG: generator set,
TG_{ck}: total generation capacity at the kth state (MW),
$G_{i,k}$: the ith generator capacity at the kth state (MW),
$ISK_{i,k}$: the ith generator probability at the kth state,
Ω_{TTFi}: TTF set of the ith generator,
Ω_{TTRi}: TTR set of the ith generator.

9.2.2 Probabilistic Operation Model of Generator 2

Figure 9.2 shows the sum of the virtual generator operating plans in the individual generators and the load. The shaded area in Figure 9.2 denotes the energy not supplied due to capacity shortage leading to load curtailment. The expected energy not supplied (EENS) for a period under study can be computed by

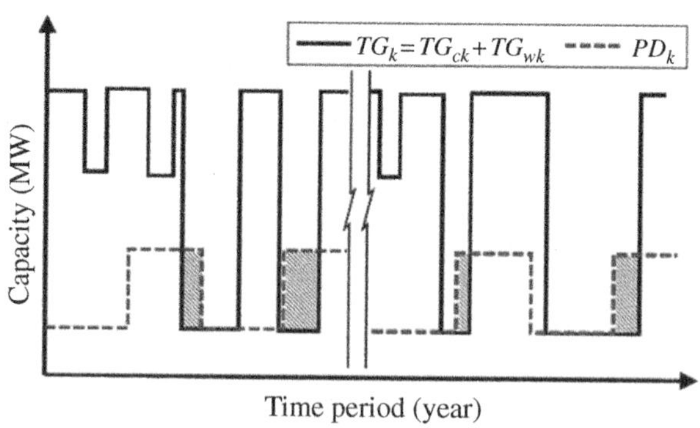

Figure 9.2 An example of load for each state and the conventional generator's probabilistic output (where TG_K: Total Capacity of WTG and CG at the kth state [MW]).

$$EENS = \sum_{k=1}^{NSS} ENS_k/NY \tag{9.5}$$

where

NY: years of MCS and ENS,
NSS: total states (=NY × NS), NS: states of each year in MCS,
ENS: energy not supplied calculated by

$$ENS_k = \begin{cases} SRP_k & SRP < 0 \\ 0 & SRP \geq 0 \end{cases} \tag{9.6}$$

where

SRP: supply reserve power (MW) obtained from

$$SRP_k = TG_{ck} + TG_{wk} - PD_k \tag{9.7}$$

where

TG_{wk}: total capacity of the WTG at the *k*th state (MW).

9.3 Probabilistic Output Prediction Model of WTG

Figure 9.3 shows a typical power curve of a WTG. The theoretical output of the WTG is explained in (9.8). As shown in the theoretical output equation of the WTG, the output greatly depends on the wind speed [37, 44–45].

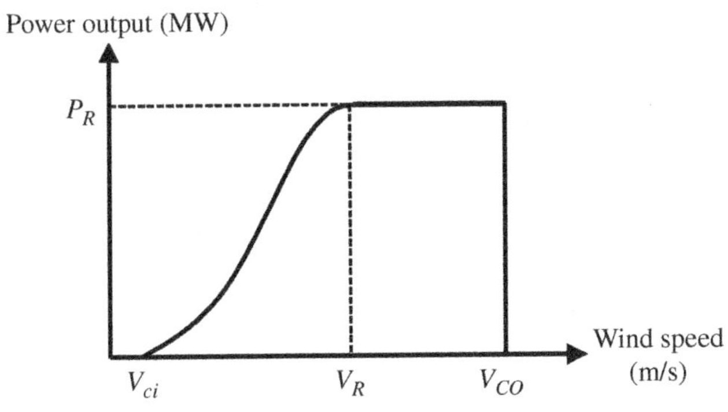

Figure 9.3 Typical output curve of WTG. *Source:* Based on Refs. [37, 43–45].

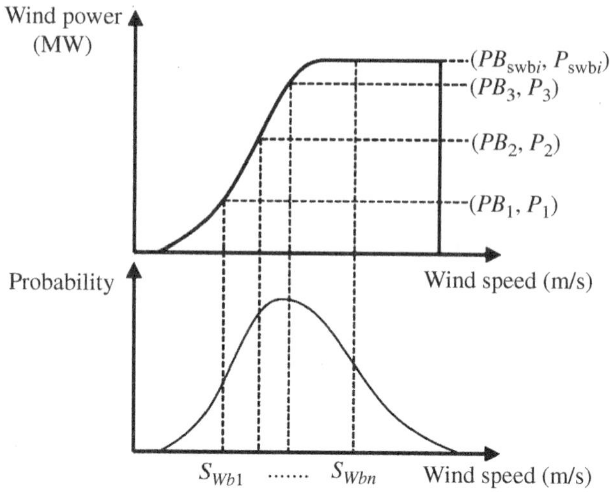

Figure 9.4 Probabilistic output prediction model of an equivalent WTG of a wind farm.

Generally, wind speed changes greatly according to space and time. The distribution of the wind speed is presented as a Weibull probability distribution similar to a normal probability distribution. An output probability distribution function is obtained for a multi-state WTG, which combines a wind power model that considers the uncertainty of the wind power supply with a characteristic model of the WTG output. The results can be used to calculate the expected output using the previous value [46]. Figure 9.4 illustrates the process of acquiring a multi-state wind power model. The wind speed probability density function is combined with the WTG's power curve as shown in Figure 9.4. In this figure, P_{SWbi} and PB_{SWbi} are the power output and its corresponding probability when the wind speed is SW_{bi}, which is the wind speed of the ith wind speed band. Finally, (9.9) is the expected output power EP_{kt} at time t of wind farm k and ρ_{ki} is the set of wind speeds on wind farm k in a time period t [47–50].

$$\begin{aligned} P_{SWbi} &= 0, && 0 \leq SW_{bi} \leq V_{ci} \\ &= P_R\left(A + B \times SW_{bi} + C \times SW_{bi}{}^2\right), && V_{ci} \leq SW_{bi} \leq V_R \\ &= P_R, && V_R \leq SW_{bi} \leq V_{co} \\ &= 0 && V_{co} < SW_{bi} \end{aligned} \tag{9.8}$$

where

V_{ci}: cut-in speed (m/s),
V_R: rated speed (m/s),
V_{co}: cut-out speed (m/s),

P_R: rated power (kW),

P_{SWbi}: output of WTG corresponding to average wind speed SW_{bi} of the ith wind speed band (kW).

$$EP_{WTG_{kt}} = \sum_{i \in \rho_{ki}} (P_{SWbi} \times PB_{SWBbi}) \tag{9.9}$$

9.4 Multi-Energy Storage System Operational Model

This chapter focuses on HLI (Hierarchical Level I). It is assumed that there are multiple wind farms and ESS is installed in each wind farm to perform charging and discharging through an appropriate control method. Figure 9.5 shows the conventional model and Figure 9.6 shows the new proposed model for the power system including multi-WTG and multi-ESS [31, 38, 39]. Figure 9.7 shows the charging and discharging of ESS in the proposed model shown in Figure 9.6.

Here, an operation scenario was developed where $X\%$ is assumed to be the maximum permissible wind power in the percentage of the system load at any point in time. This means the permissible rates will not change even though the load varies over time. The simulation procedure for determining the energy storage state is as follows [34, 36]:

Step 1: The wind speed model is estimated for the selected wind speed regime used by the wind farm. From the output characteristic curve of the WTG, the time series $TG_{wi,k}$ (MW) of the wind power output is estimated from the output characteristic curve of WTG. In addition, the time series $TG_{ci,k}$ (MW) of the conventional generators is determined.

Step 2: The ratio of the wind power that can be supplied and used is estimated according to the load and the ratio of the wind power stored in the ESS. In the simulation process, the surplus wind generation SG_{wk} (MW) and the obligation output ratio can be either (+) or (−) [38, 39].

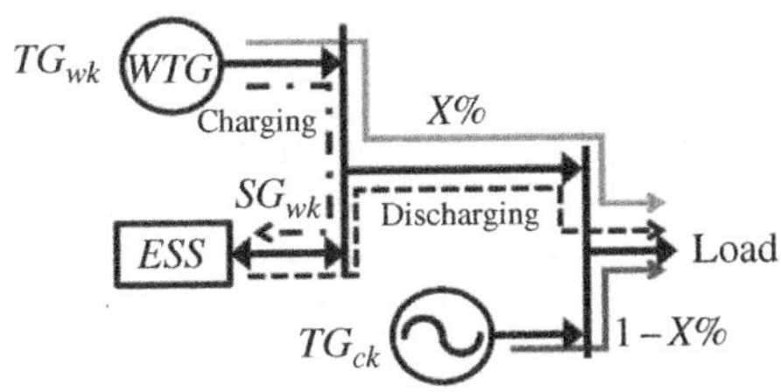

Figure 9.5 Conventional model (single-WTG linked with a single-ESS). *Source:* Based on Refs. [31, 38, 39].

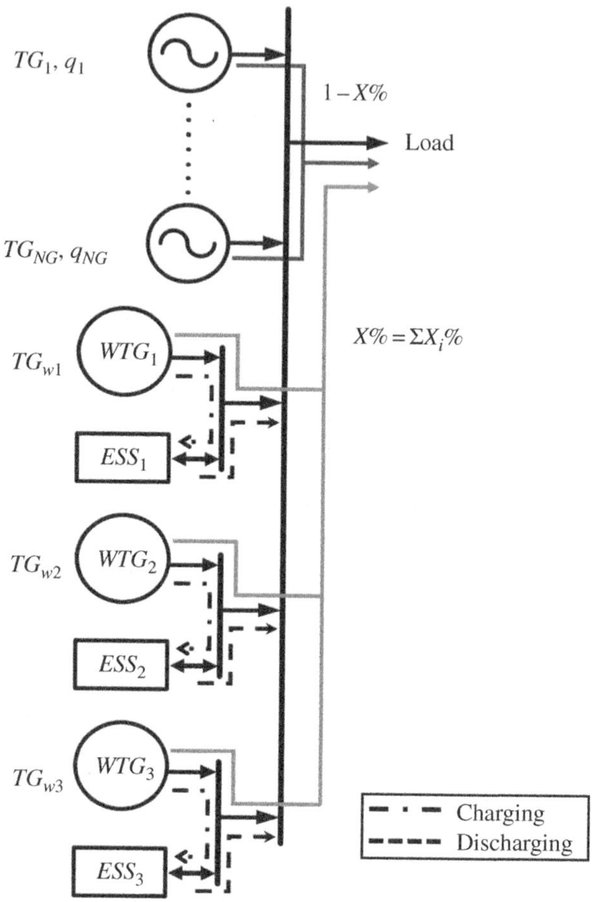

Figure 9.6 Proposed model of multiple WTG by multi-ESS (example: Jeju Island-sized sample system: three wind farms linked with multi-ESSs). *Source:* Based on Refs. [31, 38, 39].

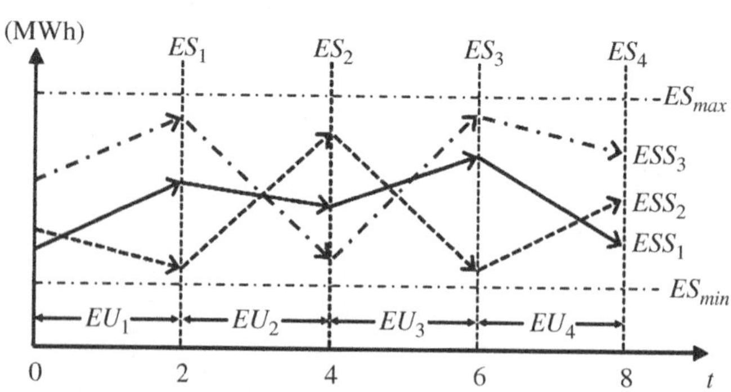

Figure 9.7 ESS state of charge (SOC) transition feature.

$$SG_{wk} = \sum TG_{wi,k} - X_i\% \times L_k \tag{9.10}$$

$$SG_{ck} = \sum TG_{ci,k} - (1 - X_i\%) \times L_k \tag{9.11}$$

where

SG_{wk}: surplus generation of the WTG at the kth state (MW),
SG_{ck}: surplus generation of the CG at the kth state (MW),
$TG_{wi,k}$: total capacity of the ith WTG at the kth state (MW),
$TG_{ci,k}$: total capacity of the ith CG at the kth state (MW),
$X_i\%$: percentage of a wind power dispatch restriction to the load at the ith WTG (p.u.),
L_k: load at the kth state (MW).

It is assumed that if SG_{wk} is positive (+), the WTG cannot exceed $X_i\% \times L_k$ to supply to the load. If SG_{ck} is negative (−), the ESS should be discharged so that even if there is excess output, it shall not charge the ESS.

Finally, $ES_{i,k}$ (MWh) represents the energy stored in the ESS, $E_{Ui,k}$ the energy variation of the ESS, and $TGD_{i,k}$ (MWh) the discharge energy of the ESS time series, which can be obtained as following:

$$ES_{i,k+1} = \begin{cases} ES_{i,k} + SG_{wi,k} \times t_k & SG_{wi,k} \geq 0 \quad \text{and} \quad SG_{ci,k} \geq 0 \\ ES_{i,k} + SG_{ci,k} \times t_k & SG_{wi,k} \geq 0 \quad \text{and} \quad SG_{ci,k} < 0 \\ ES_{i,k} & SG_{wi,k} < 0 \quad \text{and} \quad (SG_{wi,k} + SG_{ci,k}) \geq 0 \\ ES_{i,k} + (SG_{wi,k} + SG_{ci,k}) \times t_k & SG_{wi,k} < 0 \quad \text{and} \quad (SG_{wi,k} + SG_{ci,k}) < 0 \end{cases} \tag{9.12}$$

$$EU_{i,k} = \begin{cases} SG_{wi,k} \times t_k & SG_{wi,k} \geq 0 \quad \text{and} \quad SG_{ci,k} \geq 0 \\ SG_{ci,k} \times t_k & SG_{wi,k} \geq 0 \quad \text{and} \quad SG_{ci,k} < 0 \\ 0 & SG_{wi,k} < 0 \quad \text{and} \quad (SG_{wi,k} + SG_{ci,k}) \geq 0 \\ (SG_{wi,k} + SG_{ci,k}) \times t_k & SG_{wi,k} < 0 \quad \text{and} \quad (SG_{wi,k} + SG_{ci,k}) < 0 \end{cases} \tag{9.13}$$

$$TG_{Di,k} = \begin{cases} 0 & SG_{wi,k} \geq 0 \quad \text{and} \quad SG_{ci,k} \geq 0 \\ -SG_{ci,k} \times t_k & SG_{wi,k} \geq 0 \quad \text{and} \quad SG_{ci,k} < 0 \\ 0 & SG_{wi,k} < 0 \quad \text{and} \quad (SG_{wi,k} + SG_{ci,k}) \geq 0 \\ -(SG_{wi,k} + SG_{ci,k}) \times t_k & SG_{wi,k} < 0 \quad \text{and} \quad (SG_{wi,k} + SG_{ci,k}) < 0 \end{cases} \tag{9.14}$$

9.4.1 Constraints of ESS control ($EU_{i,k}$)

The maximum constraint on charge/discharge control is

$$-EU_{max,i} \leq EU_{i,k} \leq EU_{max,i} \tag{9.15}$$

$$EU_{max,i} = [(ES_{max,i} - ES_{min,i})/TM_{ESS,i}] \times \Delta t \tag{9.16}$$

where

$TM_{ESS,i}$: operation time from $ES_{max,i}$ to $ES_{min,i}$ of the ith ESS (hours).

The maximum permissible energy constraint on charge/discharge in line with the maximum capacity constraint of the ESS is

$$ES_{min,i} \le ES_{i,k} + EU_{i,k} \le ES_{max,i} \tag{9.17}$$

Step 3: The energy $ES_{i,k}$ stored in the ESS is limited by a maximum value ($ES_{max,i}$) and minimum value, as shown in (9.18). The time series of the load are L_k and t_k, where t_k is the ith time interval, and T is the number of annual intervals, which refers to the time interval in the annual simulation. In general, if $t_k = 1$, then $T = 8736$. This value was used in the model.

$$ES_{i,k} = ES_{i,k-1} + EU_{i,k} \tag{9.18}$$

$$ES_{min,i} \le ES_{i,k} \le ES_{max,i} \tag{9.19}$$

where

$ES_{i,k}$: energy stored in ESS of the ith ESS at the kth state (MWh),
$EU_{i,k}$: energy variation of the ith ESS at the kth state (MWh),
$ES_{max,i}$: maximum energy capacity (SOC) of the ith ESS (MWh),
$ES_{min,i}$: minimum capacity (SOC) of the ith ESS (MWh).

9.5 Multi-ESS Operation Rule

The basic purpose of ESS installation is to prevent the power supply from being disrupted by considering reliable operation only. Therefore, to prevent the power supply from being disrupted, the choice is to determine discharging mode of ESS first and then determine charging mode of ESS.

9.5.1 Discharging Mode

Select the largest ESS-discharging capacity first for discharging in the following cases:

- If the total output of the existing generator is less than 1–X%.
- If the sum of the output of wind power and that of the conventional generator is less than the load.

9.5.2 Charging Mode

In the charging mode, when the total of wind power generation exceeds $X\%$, each ESS installed in the WTG is charged with the excess of the WTG output. At this time, the discharge mode of ESS is excluded. Here, the following conditions are considered as a method of appropriately distributing the charged amount.

- Except for the discharge mode of ESS, the rechargeable state of the remaining ESS is searched. And first, charge from the maximum chargeable ESS.

The charging and discharging conditions of the proposed model (Figure 9.6) are summarized in the following Table 9.1 based on the charging/discharging operation mode described above.

However, the difference between the proposed multi-ESS model (Figure 9.6) and the conventional single-ESS model (Figure 9.5) is shown in the black mark (●) in Table 9.1. Existing models contribute in different ways to a single-ESS model (Figure 9.5). The major disadvantage in such models is that an ESS in a discharging mode cannot be charged even if the wind is blowing well and the output of the wind power source is high. However, in the case of the multi-ESS model (Figure 9.6) proposed in this chapter, it has a clear advantage in availing a number of ESS units to charge and store surplus renewable energy that becomes available while other ESS units are in a discharging mode. The proposed model allows the flexibility to optimize the dispatch and operational modes of multiple ESS distributed throughout the grid network.

9.6 Reliability Evaluation with Energy Storage System

This study adopted an MCS method as a stochastic numerical approach for predicting values to reflect the uncertainty of the WTG and a CG. In addition, a sample status was created for the WTG that considers uncertainty by applying a random number. The reliability index was calculated by summing and averaging the

Table 9.1 Charge and discharge conditions.

SG_{wk}	SG_{ck}	SG_k	Charging	Discharging
+	+	+	○	×
+	−	+ (−)	●	○
−	+	+	×	×
−	−	−	×	○

results after obtaining the loss of load expectation (*LOLE* [hours/year]) and expected energy not served (*EENS* [MWh/year]) for each state. For the energy index of reliability (*EIR* [p.u.]) defined from the *EENS*, the higher the value, the higher is the reliability. This is formulated in the following equations:

$$LOLE = \frac{1}{NY} \sum_{k \in \Omega_D} t_k \tag{9.20}$$

$$EENS = \frac{1}{NY} \sum_{k \in \Omega_D} (TG_{Dk} + EU_k) \tag{9.21}$$

$$EIR = 1 - \frac{EENS}{TDE} \tag{9.22}$$

where

NY: years of MCS,
Ω_D: a set of discharge modes. NSS: Total number of states,
EU_k: discharging control energy as a negative value (MWh),
TG_{Dk}: discharge energy indispensable for eliminating any lack of supply for the load (marked as an absolute value) (MWh).

The followings are proposed to express the supply reliability for the ESS. Also, (9.25) represents the contribution of the supply shortage cost and is proposed for an ESS that is installed on a wind farm.

$$\Delta LOLE_{ESS,X\%} = LOLE_{WTG,X\%} - LOLE_{WTG+ESS,X\%} \tag{9.23}$$

$$\Delta EENS_{ESS,X\%} = EENS_{WTG,X\%} - EENS_{WTG+ESS,X\%} \tag{9.24}$$

$$EVU_{ESS,X\%} = EENS_{ESS,X\%} \times IEAR \tag{9.25}$$

where

$LOLE_{WTG,X\%}$: LOLE of WTG,
$EENS_{WTG,X\%}$: EENS of WTG,
$LOLE_{WTG+ESS,X\%}$: LOLE of WTG combined with ESS,
$EENS_{WTG+ESS,X\%}$: EENS of WTG combined with ESS,
$IEAR$: interrupted energy assessment rate (won/kWh).

9.7 Case Studies

9.7.1 Power System of Jeju Island

This study evaluates the reliability in terms of the supply adequacy by applying the developed model to a sample power system integrated with three wind farms

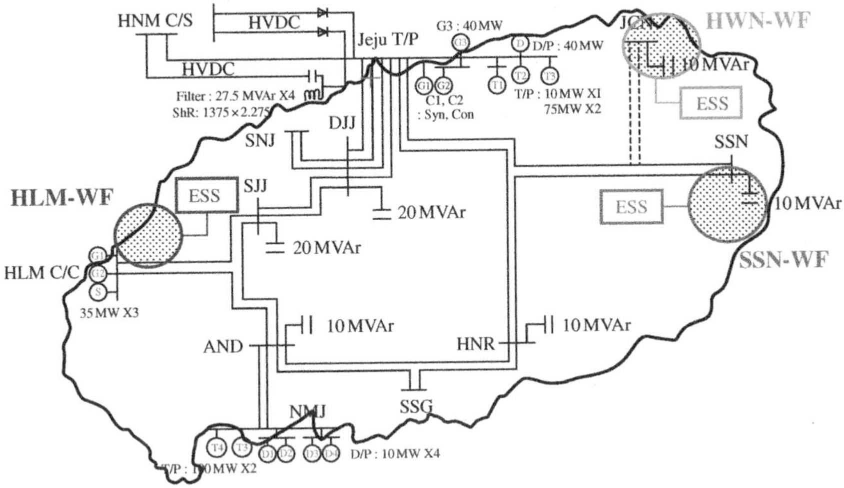

Figure 9.8 Power system of Jeju Island.

Table 9.2 Generator characteristic data in Jeju Island system.

	Name	Type	Capacity (MW)	Num	α (Gcal/ MW²h)	β (Gcal/ MWh)	γ (Gcal/ hour)	Fuel cost (f) ($/Gcal)	FOR
1	HVDC	DC	150	2	0.004	1.512	45.207	43.300	0.028
2	NMJ3	T/P	100	2	0.004	1.512	45.207	43.300	0.012
3	JJU1	T/P	10	1	0.062	2.100	5.971	43.599	0.015
4	JJU2	T/P	75	2	0.003	1.832	30.231	43.599	0.012
5	HLM1	G/T	35	2	0.004	2.401	20.320	77.909	0.013
6	HLM1	S/T	35	1	0.004	2.401	20.320	77.909	0.013
7	JJU3	D/P	40	1	0.025	0.364	28.484	43.599	0.018
8	NMJ1	D/P	10	4	0.006	1.999	1.360	43.300	0.018

connected with ESS, the data from the three main wind farms on Jeju Island: Haengwon (HWN), Seongsan (SSN), and Hanlim (HLM) are used in this study. The Jeju-sized sample power system and three wind farms are shown in Figure 9.8.

The generator data of the CG and WTG are given in Table 9.2.

Tables 9.3 shows the WTG characteristics and wind speed data of HWN, SSN, and HLM wind farms, and the load variation curve used in this case study is shown in Figure 9.9. Table 9.4 shows the ESS specifications for the system. In this study,

Table 9.3 WTG characteristic data and wind speed data of Jeju Island system.

	HWN	SSN	HLM
WTG capacity (MW)	100	60	40
Cut-in speed (V_{ci}) (m/s)	5	5	5
Rated speed (V_R) (m/s)	10	10	10
Cut-out speed (V_{co}) (m/s)	25	25	25
Mean wind speed (m/s)	5.67	5.67	5.67
α (Shape parameter)	3.82	3.82	3.82
β (Scale parameter)	3.45	3.45	3.45

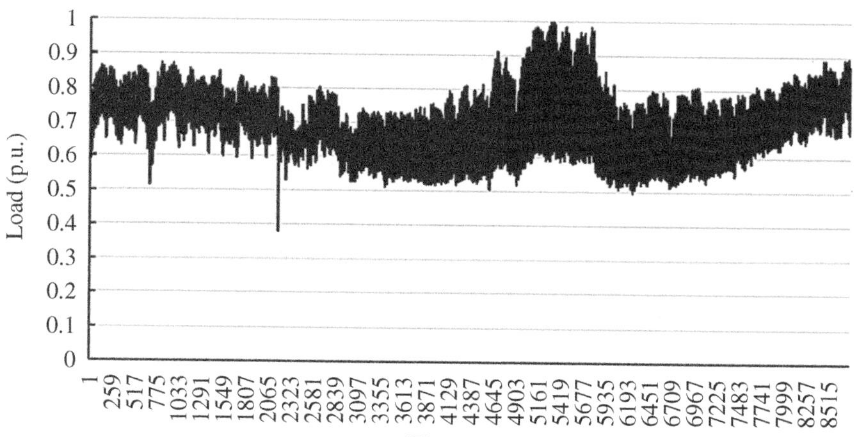

Figure 9.9 Jeju Island load variation curve of the power system.

the peak load is set to 851 MW. For reference, this shows the recent peak load of Jeju Island.

The following three cases are considered (Figure 9.10):

1) *Comparison of reliability indices when a WTG was combined with an ESS (X% = 0.1)* – The reliability evaluation results are shown in Table 9.5. It can be seen that the reliability of the wind-integrated system improves significantly with the installation of the multi-ESS in system B as shown in the table. It can be seen that the improvement in LOLE is over about 94 times and in EENS, is about 150 times better than that of the base system.

Table 9.4 Multi-ESS specifications.

	Max. capacity (ES_M) (MWh)	Min. capacity (ES_m) (MWh)	Time length for charge/discharge (hours)	X% (p.u.)	Initial SOC of ESS (MWh)
HWN ESS_1	50	10	1	0.1	15
SSN ESS_2	30	10	1	0.1	15
HLM ESS_3	20	10	1	0.1	15

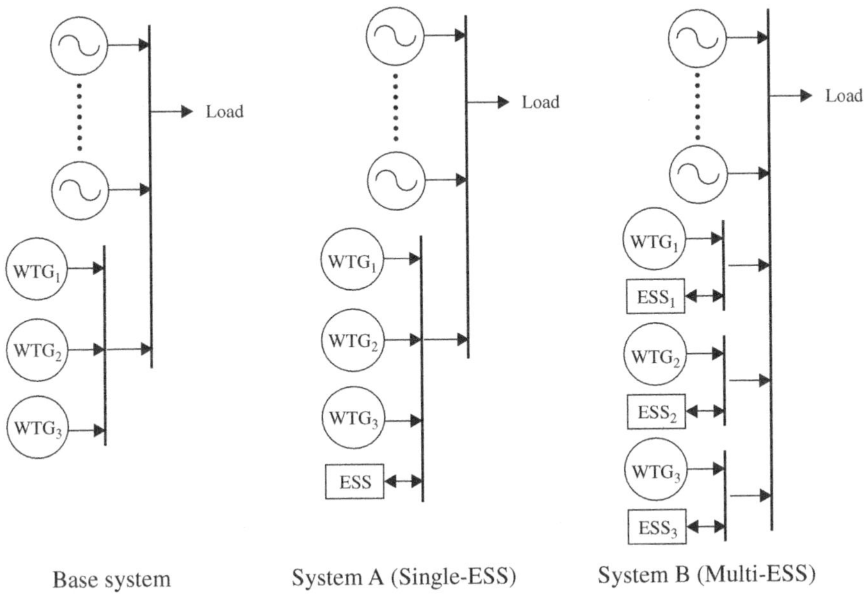

Figure 9.10 Models for a case study on the Jeju Island system. Base system, system A (single-ESS), and system B (multi-ESS).

Table 9.5 Reliability evaluation result of the case.

	Base system (with WTG)	System A (single-ESS)	System B (multi-ESS)
Total capacity (MW)	1045	1145	1145
LOLE (hours/day)	47.33	32.694	0.504
EENS (MWh/day)	2292.0	1864.8	15.2
EIR (p.u.)	0.999 56	0.999 642	0.999 997 1

Table 9.6 Calculation result of reliability in the case of changing the maximum capacity ESS (ES_M).

Max. capacity (ES_M) (MWh)	*LOLE* (hours/year)	*EENS* (MWh/year)	*EIR* (p.u.)
System B.1 (50/30/20)	0.504	15.2	0.999 997 1
System B.2 (100/60/40)	0.037	1.4	0.999 999 7
System B.3 (150/90/60)	0.014	0.4	0.999 999 92
System B.4 (200/120/80)	0.014	0.4	0.999 999 92
System B.5 (250/150/100)	0.014	0.4	0.999 999 92

Also, as shown in Table 9.5, multi-ESS and single-ESS were compared in order to clarify the effect in terms of reliability. In this case, it can be seen that the multi-ESS model (the proposed model like Figure 9.6) has an effect of improving the reliability by about 64 times as much as the single-ESS model (the conventional model like Figure 9.5).

2) *The dependency of the reliability index on the maximum ESS capacity* – For the increased ESS capacity, the reliability was improved by about 14 times when the capacity was increased by 2 times from System B.1 to System B.2. When the maximum capacity was increased to System B.3 and beyond, it was saturated. As shown in Table 9.6, there is adequate reliability capacity in System B.2 or System B.3 with average variances of $ESS_1 = 125$ MWh, $ESS_2 = 75$ MWh, and $ESS_3 = 50$ MWh. This suggests that when the equivalent WTG in each wind farm changes, the appropriate ESS will also change.

9.7.2 Reliability Evaluation of Single-ESS

1) *Reliability indices according to changes in X%* – First, the reliability evaluation is performed for the single-ESS (Figure 9.5). The simulation was conducted by comparing the case between the base system and system A. Here, the capacity of the ESS is set to 100 MW, which is the same as the total capacity of the multi-ESS. Table 9.7 shows the reliability indices according to changes in *X*%, which is the ratio of the maximum allowable output to the load in the WTG. The single-ESS was the least reliable when *X*% was 0. The reason is that there are three WTGs but one ESS, so the ESS does not operate properly. That is, as explained before, the biggest disadvantage is that the ESS in a discharging mode cannot be charged even if the wind is blowing well and the output of the wind power is high.

As shown in Figures 9.11 and 9.12, when the ESS has been installed in the wind farm, the reliability index can deteriorate rapidly as *X*% increases.

Table 9.7 Reliability evaluation result of cases.

X% (p.u.)	Case	*LOLE* (hours/year)	*EENS* (MWh/year)	*EIR* (p.u.)
0	Base system	83.51	4220.4	0.999 189 9
	System A	50.51	3483.7	0.999 331 2
0.1	Base system	47.33	2292.0	0.999 56
	System A	32.694	1864.8	0.999 642
0.2	Base system	46.384	2255.1	0.999 567 1
	System A	41.256	2110.6	0.999 595
0.3	Base system	46.384	2255.1	0.999 567 1
	System A	46.144	2206.6	0.999 576 4
0.4	Base system	46.384	2255.1	0.999 567 1
	System A	46.384	2255.1	0.999 567 1

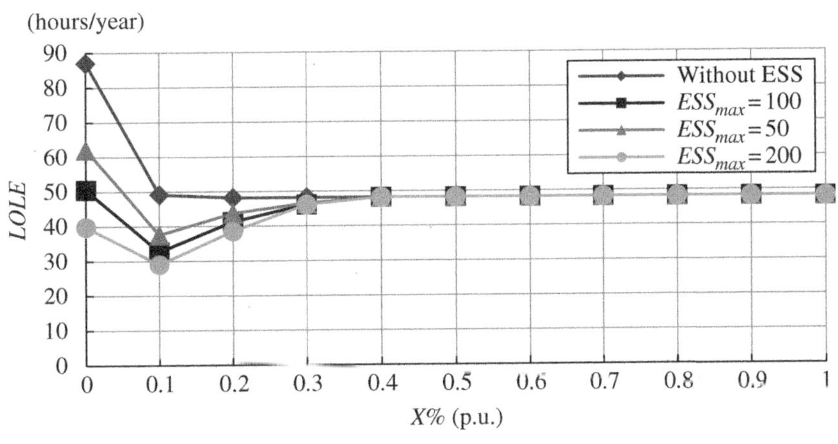

Figure 9.11 Variation of LOLE according to *X*%.

Therefore, $X\% = 0.1$ p.u. is appropriate. If $X\% = 0.4$ p.u., the reliability is the same as the base case where the ESS has not been installed. Thus, the installation of the ESS is meaningless.

2) *Reliability evaluation on changes in X% and ESS capacity* – Figures 9.13 and 9.14 show the changes in reliability indices as defined in (9.23) and (9.24), which show the contribution of ESS on the reliability.

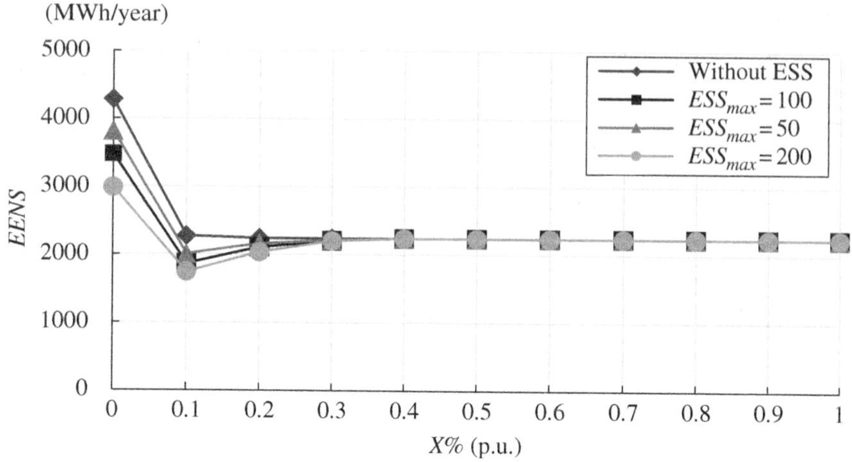

Figure 9.12 Variation of EENS according to *X*%.

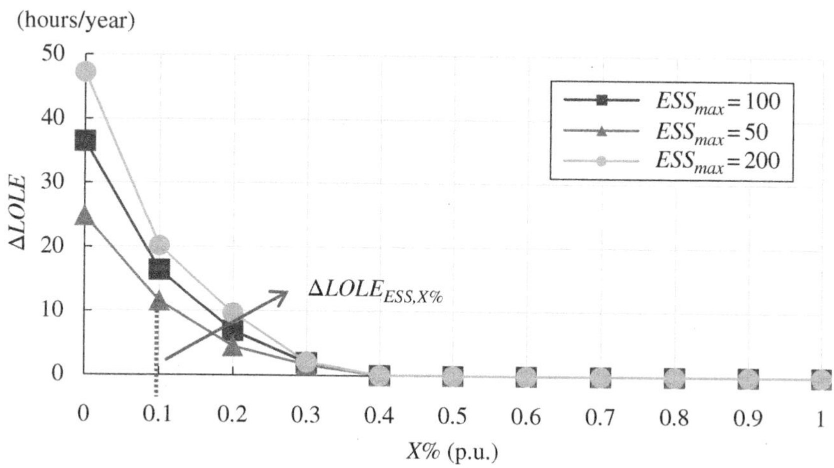

Figure 9.13 Variation of $\triangle LOLE_{ESS,X\%}$ according to *X*% and the ESS capacity.

In Figures 9.13 and 9.14, $\triangle LOLE_{ESS,X}\%$ and $\triangle EENS_{ESS,X\%}$ indicate contributions of ESS on LOLE and EENS, respectively, when $X\% = 0.1$ p.u. As can be seen in Figure 9.13, before and after the installation of ESS, the contribution to the power system reliability is as much as 14.6 hours/year.

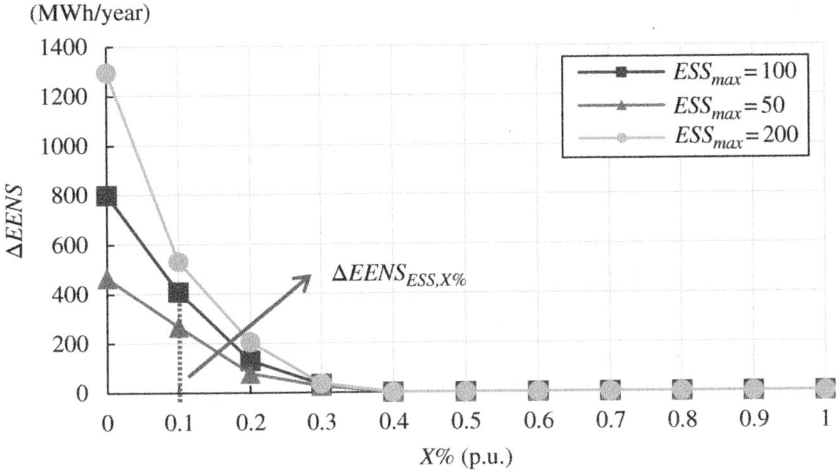

Figure 9.14 Change of $\triangle EENS_{ESS,X\%}$ according to $X\%$ and the ESS capacity.

9.7.3 Reliability Evaluation of Multi-ESS

1) *Reliability indices according to changes in X%* – Second, the reliability evaluation is performed for the multi-ESS (Figure 9.6). The simulation was conducted by comparing the case between the base system and system B. Table 9.8 shows the reliability indices according to changes in $X\%$, which is the ratio of the maximum allowable output to the load in the WTG. The reliability is improved

Table 9.8 Reliability evaluation result of cases.

X% (p.u.)	Case	*LOLE* (hours/year)	*EENS* (MWh/year)	*EIR* (p.u.)
0	Base system	83.51	4220.4	0.999 19
	System B	0.568	10.7	0.999 998
0.1	Base system	47.33	2292.0	0.999 56
	System B	0.504	15.2	0.999 997
0.2	Base system	46.384	2255.1	0.999 567
	System B	5.141	192.7	0.999 963
0.3	Base system	46.384	2255.1	0.999 567
	System B	26.569	1182.5	0.999 773
0.4	Base system	46.384	2255.1	0.999 567
	System B	46.175	2243.3	0.999 569

as $X\%$ becomes lower. The reason is that the ESS is used actively when $X\%$ becomes lower. This is the biggest difference from the single-ESS. The permissible value of $X\%$ depends on the regulating margin and ramping capability of the spinning reserves.

As shown in Figures 9.15 and 9.16, when the ESS has been installed in the wind farm, the reliability index can deteriorate rapidly as $X\%$ increases. Therefore, $X\% = 0.1$ or 0.2 p.u. is appropriate. If $X\% = 0.4$ p.u., the reliability is the same as the base case where the ESS has not been installed. Thus, the installation of the ESS is meaningless. These properties are thought to be useful in deciding the appropriate $X\%$ for a wind farm complex.

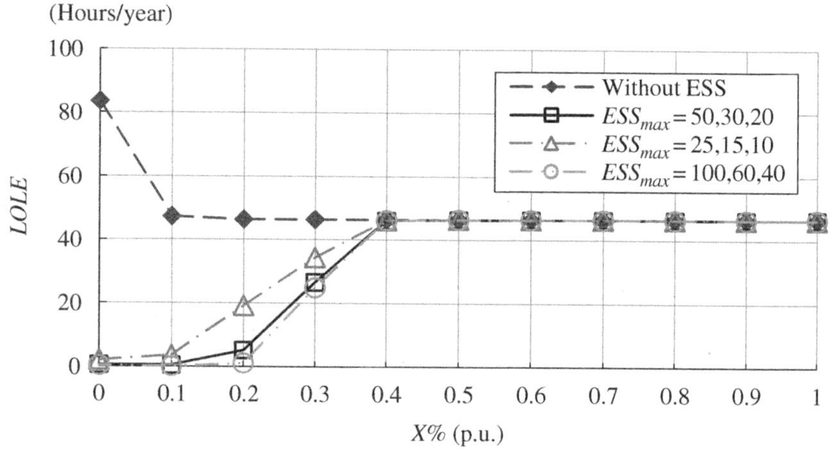

Figure 9.15 Variation of LOLE according to $X\%$.

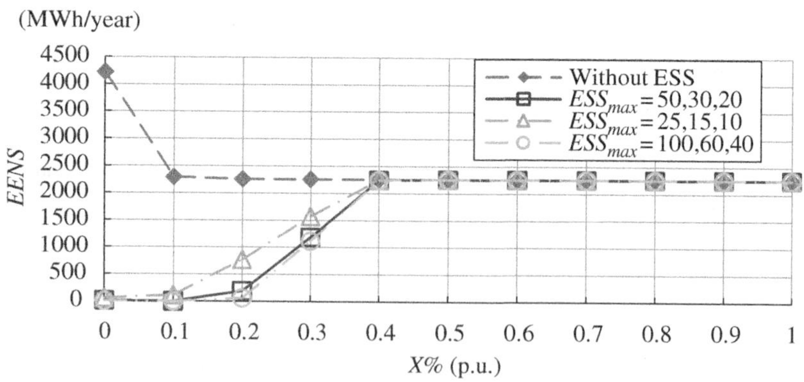

Figure 9.16 Variation of EENS according to $X\%$.

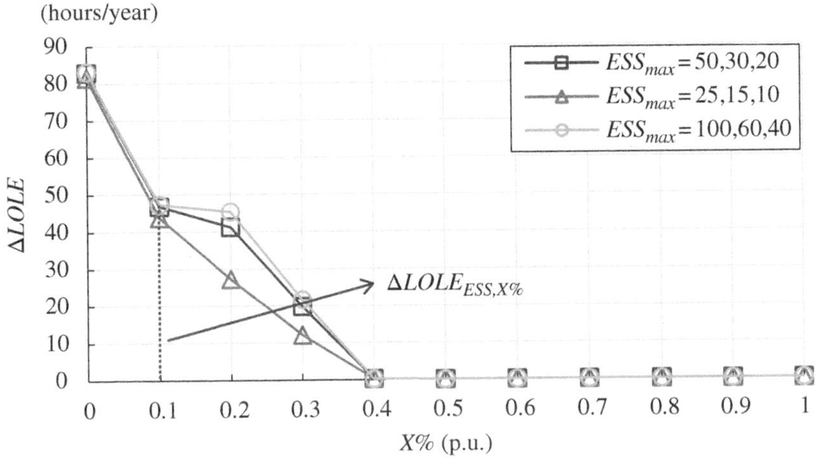

Figure 9.17 Variation of $\triangle LOLE_{ESS,X\%}$ according to $X\%$ and the ESS capacity.

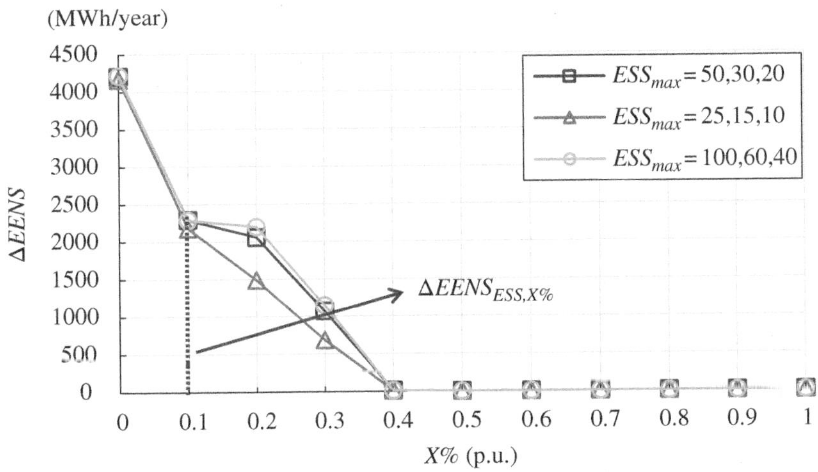

Figure 9.18 Change of $\triangle EENS_{ESS,X\%}$ with $X\%$ and the ESS capacity.

2) *Reliability evaluation on changes of X% and ESS capacity* – Figures 9.17 and 9.18 show the changes in reliability indices as defined in (9.23) and (9.24), which show the contribution of EES to reliability. In Figures 9.17 and 9.18, $\triangle LOLE_{ESS,X\%}$ and $\triangle EENS_{ESS,X\%}$ indicate contributions of ESS on LOLE and EENS, respectively, when $X\% = 0.1$ p.u. As can be seen in Figure 9.17, before and after the installation of ESS, the contribution to the power system reliability is as much as 46.8 hours/year.

Table 9.9 Reliability evaluation result between system A (single-ESS) and system B (multi-ESS).

X% (p.u.)	Case	*LOLE* (hours/year)	*EENS* (MWh/year)	*EIR* (p.u.)
0	System A	50.51	3483.7	0.999 331
	System B	0.568	10.7	0.999 998
0.1	System A	32.694	1864.8	0.999 642
	System B	0.504	15.2	0.999 997
0.2	System A	41.256	2110.6	0.999 595
	System B	5.141	192.7	0.999 963
0.3	System A	46.144	2206.6	0.999 576
	System B	26.569	1182.5	0.999 773
0.4	System A	46.384	2255.1	0.999 567
	System B	46.175	2243.3	0.999 569

9.7.4 Comparison of System A and System B

To delve into the effects of multi-ESS on the reliability where an ESS is installed for each of the multiple WTG sources, the multi-ESS is compared with the single-ESS where one ESS is installed for multiple WTGs as in Figure 9.5.

Table 9.9 compares the calculated results between system A (single-ESS) and system B (multi-ESS). Here, the capacity of the WTG is shown in Table 9.3, the

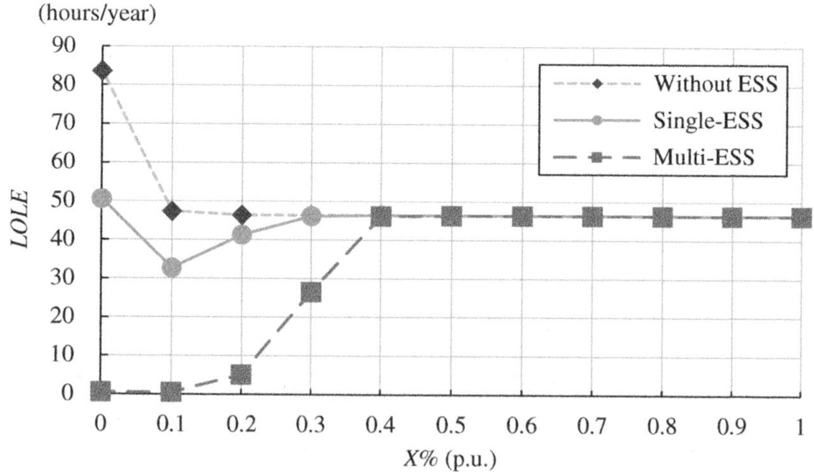

Figure 9.19 Variation of *LOLE* between single-ESS and multi-ESS according to *X%*.

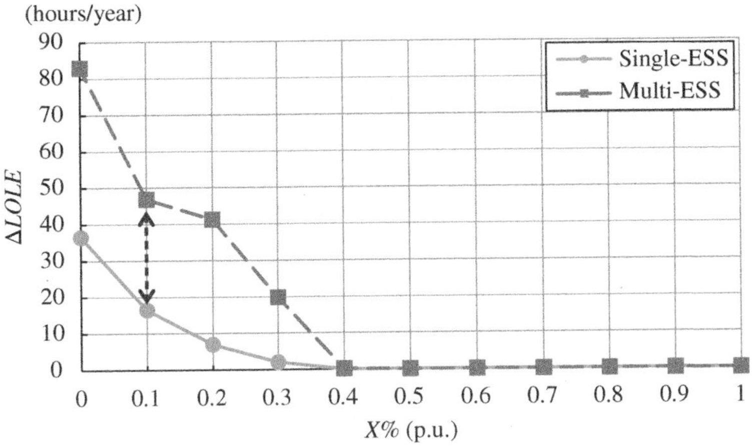

Figure 9.20 Variation of $\triangle LOLE_{ESS,X\%}$ between single-ESS and multi-ESS according to $X\%$ and the ESS capacity.

capacity of the single-ESS is 100 MW and the multi-ESS has a total of 100 MW (HWN $ESS_1 = 50$ MW, SSN $ESS_2 = 30$ MW, HLM $ESS_3 = 20$ MW).

As Figures 9.19 and 9.20 shows, multi-ESS contributes more to the system reliability than the single-ESS. For example, when $X\%$ is 0.1, the comparison of the two systems in terms of LOLE index shows that the multi-ESS ($LOLE = 46.8$ hours/year) is contributing about three times more than the single-ESS ($LOLE = 16.4$ hours/year).

9.8 Conclusion

This study presented the basic model and algorithm for the reliability evaluation of a power system involving ESS and WTG. The main points of this study are summarized as follows:

First of all, the effectiveness of the model was verified through a reliability evaluation conducted on a sample power system of Jeju Island. A large-scale WTG creates huge power fluctuations and may be dangerous for the stability of the power supply. However, this danger can be reduced by installing an ESS.

Second, the models could also help determine the proper capacity of an ESS and use it efficiently through a sensitivity analysis. The ESSs are very useful for compensating the output of renewable energy sources, which have uncertainty in terms of cost and efficiency.

Third, the scale of the optimum permitted output of renewable energy sources, such as wind power and solar power, can then be selected based on reliability. This provides valuable input to operation planning to increase the efficiency and reliability of renewable energy integration in a power system.

Finally, the proposed models can be used to facilitate the reliability analysis of ESSs and make planning decisions to improve the economic efficiency of a power system.

The proposed algorithm described in the Appendix provides a very essential concept for ESS expansion planning problem because reliability evaluation of ESS is the basic and essential element for optimal power system (including smart grid and microgrid) expansion planning, although the capital cost is still not considered.

9.A Appendices

9.A.1 Single-ESS Model

In order to demonstrate the developed algorithm for ESS operation in the perspective of reliability, sample input data and calculation process are presented in Tables 9.A.1–9.A.3, where $X\%$ (p.u.) and ESS charging time (TM_{ESS}) are assumed as 20% (0.2 p.u.) and two hours, respectively.

Therefore, maximum charging and discharging energies (maximum control energy) of ESS at all states using (9.16) can be calculated.

1) *Operation Method of Single-ESS Model*

Step 1: Time interval between states is assumed as two hours. Table 9.A.1, therefore, means one-day operation. As the sample system is a very simple problem, load (L_k), WTGs' generation (TG_{wk}) and CGs' generation (TG_{ck}) at $\#k$ state are assumed arbitrary for convenience. In actual system operation, this input data should be generated by Monte Carlo method using random number considering FOR of generators and uncertainties of wind speed should be given. Summation power (TG_k) of WTGs and CGs is calculated.

Step 2: In here, It is assumed that permissible WTG generation ($X\%$) is limited as 0.2 pu of load at each state. As it is, WTG is generated within 20% of load and CGs should be generated over 80% of the load. Therefore, surplus generation of wind generators (SG_{wk}) (9.10) and surplus generation of CGs (SG_{ck}) (9.11) at $\#k$ state can be calculated. For example, the first state is calculated as follows. And, the summation (SG_k) is also obtained.

$$SG_{w1} = TG_{w1} - X\% \times L_1 = 30 - (100 \times 0.2) = 10\ \text{MW} \tag{9.A.1}$$

$$SG_{c1} = TG_{c1} - (1 - X\%) \times L_1 = 95 - (100 \times 0.8) = 15\ \text{MW} \tag{9.A.2}$$

$$SG_1 = TG_{w1} + TG_{c1} = 25\ \text{MW} \tag{9.A.3}$$

Table 9.A.1 Calculation of single-ESS operation in the sample system.

k	L_k (MW)	TG_{wk} (MW)	TG_{ck} (MW)	TG_k (MW)	SG_{wk}(MW)	SG_{ck}(MW)	SG_k (MW)	C_k/D_k (MWh)	TG_{Dk}(MWh)	EU_k(MWh)	ES_{k+1}(MWh)	EENS(MWh)	LOLE (hours)
1	100	30	95	125	30–20 = +10	95 − 80 = +15	25	(20/0)	0	10 × 2 = 20	10 + 20 = 30		
2	100	25	75	100	25 − 20 = +5	75 − 80 = −5	0	(0/−10)	5 × 2 = 10	−5 × 2 = −10	30 − 10 = 20		
3	100	15	90	105	15 − 20 = −5	90 − 80 = +10	5	(0/0)	0	0	20		
4	100	15	75	90	15 − 20 = −5	75 − 80 = −5	−10	(0/−20)	10 × 2 = 20	−10 × 2 = −20	20 − 15 = 5	20 − 15 = 5	1 × 2 = 2
5	110	30	105	135	30 − 22 = +8	105 − 88 = +17	25	(16/0)	0	8 × 2 = 16	5 + 16 = 21		
6	110	30	85	115	30 − 22 = +8	85 − 88 = −3	5	(0/−6)	3 × 2 = 6	−3 × 2 = −6	21 − 6 = 15		
7	110	15	105	120	15 − 22 = −7	105 − 88 = +17	10	(0/0)	0	0	15		
8	110	15	85	100	15 − 22 = −7	85 − 88 = −3	−10	(0/−20)	10 × 2 = 20	−10 × 2 = −20	15 − 10 = 5	20 − 10 = 10	1 × 2 = 2
9	105	25	95	120	25 − 21 = +4	95 − 84 = +11	15	(8/0)	0	4 × 2 = 8	5 + 8 = 13		
10	105	25	70	95	25 − 21 = +4	70 − 84 = −14	−10	(0/−28)	14 × 2 = 28	−14 × 2 = −28	13 − 8 = 5	28 − 8 = 20	1 × 2 = 2
11	90	30	80	110	30 − 18 = +12	80 − 72 = +8	20	(24/0)	0	12 × 2 = 24	5 + 24 = 29		
12	100	25	80	105	25 − 20 = +5	80 − 80 = 0	5	(10/0)	0	5 × 2 = 10	29 + 10 = 39		
	1140	255	960	1215								35 (MWh/day)	6 (hours/day)

where

L_k, TG_{wk}, TG_{ck}, TG_k: input value assumed for sample system.

Table 9.A.2 WTG specification.

	#1	#2	#3	Total
Capacity (MW)	10	10	10	30

Table 9.A.3 Single-ESS specification.

	Max. capacity (ES_{max}) (MWh)	Min. capacity (ES_{min}) (MWh)	Time length for charge/discharge ($\triangle t$) (hours)	X% (p.u.)	Initial SOC of ESS (ES_0) (MWh)
ESS	50	5	2	0.2	10

Step 3: ESS chargeable/dischargeable energy (SOC) (C_k [Charging]/D_k [Discharging]) can be decided. Because wind surplus generation at first state ($k = 1$) is "+," the surplus energy is chargeable in ESS. But, although the surplus energy of CG exists, it is not chargeable. In conclusion, the chargeable energy is calculated as following with consideration of two hours' time (TM_{ESS}).

$$C_1\ (= 10 \times 2) = 20\,\text{MWh}/D_1(= 0\,\text{MWh}) = 20/0 \tag{9.A.4}$$

As this case has no outage time and energy, and also, discharging is not necessary, therefore, discharge energy (TG_{Dk}) is zero "0."

Step 4: Considering the maximum control energy limitation 45 MWh, ESS is charged. As this state is satisfied with ($EU_1 < EU_{max}$) condition, ESS can be charged. Therefore, the next state SOC is calculated.

$$ES_1 = ES_0 + EU_1 = 10 + 20 = 30\,\text{MWh} \tag{9.A.5}$$

2) *Reliability Evaluation for Single-ESS Model*

From Table 9.A.1, reliability indices are calculated as (9.A.6)–(9.A.8). The SOCs of ESS at states are demonstrated in Figure 9.A.1.

$$LOLE = 3 \times 2 = 6\ \text{hours/day} \tag{9.A.6}$$

$$EENS = 35\ \text{MWh/day} \tag{9.A.7}$$

$$\text{Frequency of outage} = 3\ \text{occ/day} \tag{9.A.8}$$

Times of Charging/Discharging are described in Table 9.A.4.

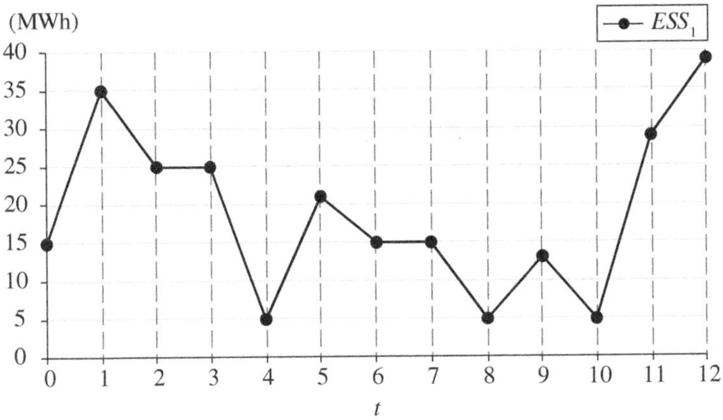

Figure 9.A.1 Single-ESS charge/discharge state (SOC) transition.

Table 9.A.4 ESS charged and discharged energy and frequency.

	Total energy (MWh)	Times
Charging	78	5
Discharging	84	5
Total	162	10

9.A.2 Multi-ESS Model

In order to introduce the new operation methodology of multi-ESS model for convenience, very similar data specification of single-ESS is applied to the multi-ESS model. The summation of multi-ESS is the same as that of single-ESS as like in Table 9.A.5. And also, charging time (TMESS) is the same as that of single- ESS (2 hours).

Table 9.A.5 Multi-ESS specification.

	Max. capacity (ES_{max})(MWh)	Min. capacity (ES_{min})(MWh)	Time length for charge/discharge ($\triangle t$)(hours)	$X\%$(pu)	Initial SOC of ESS (ES_0)(MWh)
ESS_1	20	2	2	0.2	4
ESS_2	15	2	2	0.2	3
ESS_3	15	1	2	0.2	3
Total	50	5	2	0.2	10

Using the maximum control energy (chargeable or dischargeable energies) (9.16) of three ESSs, the maximum control energies (SOC) of three ESSs at #1 state can be calculated as (9.A.9)–(9.A.11),

$$EU_{max,1} = [(20-2)/2] \times 2 = (18/2) \times 2 = 18\,\text{MWh} \tag{9.A.9}$$

$$EU_{max,2} = [(15-2)/2] \times 2 = (13/2) \times 2 = 13\,\text{MWh} \tag{9.A.10}$$

$$EU_{max,3} = [(15-1)/2] \times 2 = (14/2) \times 2 = 14\,\text{MWh} \tag{9.A.11}$$

9.A.3 Operation of Multi-ESS Models

1) **Operation of *Multi-ESS model (Algorithm)*** – Operation method and process of multi-ESS model is very similar to that of single-ESS model. However, the difference between single ESS-ESS and multi-ESS is that the decision of charge/discharge depends on each ESS in the multi-ESS model. **For the multi-ESS model problem, as it is, we have questions like "How do we charge (or discharge) surplus energies of WTG into (from) the available ESSs?"** We can propose several optimal dispatch methods for this problem. Two operational methods are proposed here. Two new proposed methods based on processing and results are presented in Tables 9.A.6 and 9.A.7:

Method 1: Energy Rate Dispatch Method (ERDM)
Method 2: Maximum First Priority Method (MFPM)

Additionally, a more detailed algorithm for the multi-ESS model is introduced.

Step 1: Load (L_k), WTG generation (TG_{wk}), and CG generation (TG_{ck}) are used alike with input data of the single-ESS model sample system.

Step 2: In the multi-ESS operation, it is assumed that permissible WTG generation ($X\%$) is limited to 0.2 p.u. of load at each state. As it is, WTG is generated within 20% of load and CGs should be generated over 80% of the load. Therefore, surplus generation of wind generators (SG_{wk}) (9.10) and that of CGs (SG_{ck}) (9.11) at #k state can be calculated. For example, the first state is calculated as follows. And, the summation (SG_k) is also obtained. So far, the calculation is the same as the single-ESS calculation.

Step 3: Available quantities of (C_k [Charging]/D_k [Discharging]) of ESS is calculated.

Step 4: Considering charging/discharging of ESS, WTG surplus energies should be charged. And also, considering the maximum control quantity (7), dispatchable energies 18, 13, and 14 MWh for three ESSs are calculated. Therefore, this is smaller than the limit of the available energy. This is satisfied with a chargeable condition constraint ($EU_{1,1} < EU_{max,1}$, $EU_{2,1} < EU_{max,2}$, $EU_{3,1} < EU_{max,3}$).

Using the above dispatchable algorithm, three ESSs can be charged/discharged.

Table 9.A.6 Calculation of multi-ESS operation in sample system (method 1).

		TG_{wk} (MW)										EU_{AXk} (MWh)			$EU_{i,k}$ (MWh)			ES_k (SOC) (MWh)				
k	L_k (MW)	#1	#2	#3	TG_{ck} (MW)	TG_k (MW)	SG_{wk} (MW)	SG_{ck} (MW)	SG_k (MW)	C_k/D_k (MWh)	TG_{Dk} (MWh)	$c_{max1,k}$ $z_{1,k}$	$c_{max2,k}$ $z_{2,k}$	$c_{max3,k}$ $z_{3,k}$	#1	#2	#3	#1	#2	#3	EENS (MWh)	LOLE (hours)
1	100	10	10	10	95	125	$30-20=+10$	$95-80=+15$	+25	(20/0)	0	(20/16)	(20/12)	(20/12)	8	6	6	4	3	3		
		30										−2	−1	−2				12	9	9		
2	100	10	10	5	75	100	$25-20=+5$	$75-80=-5$	0	(10/−10)	10	(20/8)	(20/6)	(10/6)	−10	5	5	2	14	14		
		25										−10	−7	−8								
3	100	5	5	5	90	105	$15-20=-5$	$90-80=+10$	+5	(0/0)	0	(10/18)	(10/1)	(10/1)	0	0	0	2	14	14		
		15										0	−12	−13								
4	100	5	5	5	75	90	$15-20=-5$	$75-80=-5$	−10	(0/−20)	20	(10/18)	(10/1)	(10/1)	0	−7	−13	2	7	1		
		15										0	−12	−13								
5	110	10	10	10	105	135	$30-22=+8$	$105-88=+17$	+25	(16/0)	0	(20/18)	(20/8)	(20/14)	7	3	6	9	10	7		
		30										0	−5	0								
6	110	10	10	10	85	115	$30-22=+8$	$85-88=-3$	+5	(16/−6)	6	(20/11)	(20/5)	(20/8)	10	−6	6	19	4	13		
		30										−7	−8	−6								
7	110	5	5	5	105	120	$15-22=-7$	$105-88=+17$	+10	(0/0)	0	(10/1)	(10/11)	(10/2)	0	0	0	19	4	13		
		15										−17	−2	−12								
8	110	5	5	5	85	100	$15-22=-7$	$85-88=-3$	−10	(0/−20)	20	(10/1)	(10/11)	(10/2)	−10	−2	−8	9	2	5		
		15										−17	−2	−12								
9	105	10	10	5	95	120	$25-21=+4$	$95-84=+11$	+15	(8/0)	0	(20/11)	(20/13)	(10/10)	3	3	2	12	5	7		
		25										−7	0	−4								

(*Continued*)

Table 9.A.6 (Continued)

		TG_{wk} (MW)										EU_{AXk} (MWh)			$EU_{i,k}$ (MWh)			ES_k (SOC) (MWh)				
k	L_k (MW)	#1	#2	#3	TG_{ck} (MW)	TG_k (MW)	SG_{wk} (MW)	SG_{ck} (MW)	SG_k (MW)	C_k/D_k (MWh)	TG_{Dk} (MWh)	$c_{max1,k}$ $z_{1,k}$	$c_{max2,k}$ $z_{2,k}$	$c_{max3,k}$ $z_{3,k}$	#1	#2	#3	#1	#2	#3	EENS (MWh)	LOLE (hours)
10	105	10	5	10	70	95	25 − 21 = +4	70 − 84 = −14	−10	(8/−28)	28	(20/8)	(10/10)	(20/8)	−10	−3	−6	2	2	1	9	1 × 2 = 2
		25										−10	−3	−6								
11	90	10	10	10	80	110	30 − 18 = +12	80 − 72 = +8	+20	(24/0)	0	(20/18)	(20/13)	(20/14)	10	6	8	12	8	9		
		30										0	0	0								
12	100	5	10	10	80	105	25 − 20 = +5	80 − 80 = 0	+5	(10/0)	0	(10/8)	(20/7)	(20/6)	4	3	3	16	11	12		
		25										−10	−6	−8								
	1240	280			1040	1320															9 (MWh/day)	2 (hours/day)

where

L_k, TG_{wk}, TG_{ck}, TG_k: input value assumed for the sample system.

Table 9.A.7 Calculation of multi-ESS operation in sample system (Method 2).

K	L_k (MW)	TG_{wk} (MW) #1	#2	#3	TG_{ck} (MW)	TG_k (MW)	SG_{wk} (MW)	SG_{ck} (MW)	SG_k (MW)	C_k/D_k (MWh)	TG_{Dk} (MWh)	EU_{MAXk} (MWh) $c_{max1,k}$ $z_{1,k}$	$c_{max2,k}$ $z_{2,k}$	$c_{max3,k}$ $z_{3,k}$	$EU_{i,k}$ (MWh) #1	#2	-#3	ES_k (SOC) (MWh) #1	#2	#3	EENS (MWh)	LOLE (hours)
1	100	10	10	10	95	125	30 − 20 = +10	95 − 80 = +15	+25	(20/0)	0	(20/16)	(20/12)	(20/12)	16	0	4	4	3	3		
		30										−2	−1	−2				20	3	7		
2	100	10	10	5	75	100	25 − 20 = +5	75 − 80 = −5	0	(10/−10)	10	(20/0)	(20/12)	(10/8)	−10	10	0	10	13	7		
		25										−19	−1	−6								
3	100	5	5	5	90	105	15 − 20 = −5	90 − 80 = +10	+5	(0/0)	0	(10/10)	(10/2)	(10/8)	0	0	0	10	13	7		
		15										−8	−11	−6								
4	100	5	5	5	75	90	15 − 20 = −5	75 − 80 = −5	−10	(0/−20)	20	(10/10)	(10/2)	(10/8)	−8	−11	−1	2	2	6		
		15										−8	−11	−6								
5	110	10	10	10	105	135	30 − 22 = +8	105 − 88 = +17	+25	(16/0)	0	(20/18)	(20/13)	(20/9)	16	0	0	18	2	6		
		30										0	0	−5								
6	110	10	10	10	85	115	30 − 22 = +8	85 − 88 = −3	+5	(16/−6)	6	(20/2)	(20/13)	(20/9)	−6	13	3	12	15	9		
		30										−16	0	−5								
7	110	5	5	5	105	120	15 − 22 = −7	105 − 88 = +17	+10	(0/0)	0	(10/8)	(10/0)	(10/6)	0	0	0	12	15	9		
		15										−10	−13	−8								
8	110	5	5	5	85	100	15 − 22 = −7	85 − 88 = −3	−10	(0/−20)	20	(10/8)	(10/0)	(10/6)	−7	−13	0	5	2	9		
		15										−10	−13	−8								
9	105	10	10	5	95	120	25 − 21 = +4	95 − 84 = +11	+15	(8/0)	0	(20/15)	(20/13)	(10/6)	8	0	0	13	2	9		
		25										−3	0	−8								

(Continued)

Table 9.A.7 (Continued)

K	L_k (MW)	TG_{wk} (MW) #1	#2	#3	TG_{ck} (MW)	TG_k (MW)	SG_{wk} (MW)	SG_{ck} (MW)	SG_k (MW)	C_k/ D_k (MWh)	TG_{Dk} (MWh)	EU_{MAXk} (MWh) $c_{max1,k}$ $z_{1,k}$	$c_{max2,k}$ $z_{2,k}$	$c_{max3,k}$ $z_{3,k}$	$EU_{i,k}$ (MWh) #1	#2	-#3	ES_k (SOC) (MWh) #1	#2	#3	EENS (MWh)	LOLE (hours)
10	105	10	5	10	70	95	25 − 21 = +4	70 − 84 = −14	−10	(8/−28)	28	(20/7)	(10/13)	(20/6)	−11	8	−8	2	10	1	9	1 × 2 = 2
		25										−11	0	−8								
11	90	10	10	10	80	110	30 − 18 = +12	80 − 72 = +8	+20	(24/0)	0	(20/18)	(20/8)	(20/14)	18	0	6	20	10	7		
		30										0	−8	0								
12	100	5	10	10	80	105	25 − 20 = +5	80 − 80 = 0	+5	(10/0)	0	(10/0)	(20/5)	(20/8)	0	2	8	20	12	15		
		25										−18	−8	−6								
	1240	280			1040	1320															9 (MWh/ day)	2 (hours/ day)

where
L_k, TG_{wk}, TG_{ck}, TG_k: input value assumed for sample system.

Method 1: Energy Rate Dispatch Method (ERDM)

$$EU_{1,1} = \frac{C_{max \times 1,1}}{C_{max \times 1,1} + C_{max \times 2,1} + C_{max \times 3,1}} \times C_1 = \frac{16}{16+12+12} \times 20 = 8\,\text{MWh} \quad (9.A.12)$$

$$\text{EU}_{2,1} = \frac{C_{\text{max} \times 2,1}}{C_{\text{max} \times 1,1} + C_{\text{max} \times 2,1} + C_{\text{max} \times 3,1}} \times C_1 = \frac{12}{16+12+12} \times 20 = 6\,[\text{MWh}] \quad (9.A.13)$$

$$EU_{3,1} = \frac{C_{max \times 3,1}}{C_{max \times 1,1} + C_{max \times 2,1} + C_{max \times 3,1}} \times C_1 = \frac{12}{16+12+12} \times 20 = 6\,\text{MWh} \quad (9.A.14)$$

Therefore, SOCs of ESS are calculated as follows:

$$ES_1 = ES_{1,0} + EU_{1,1} = 4 + 8 = 12\,\text{MWh} \quad (9.A.15)$$

$$ES_2 = ES_{2,0} + EU_{2,1} = 3 + 6 = 9\,\text{MWh} \quad (9.A.16)$$

$$ES_3 = ES_{3,0} + EU_{3,1} = 3 + 6 = 9\,\text{MWh} \quad (9.A.17)$$

The process is presented in the first row in Table 9.A.6.

Method 2: Maximum First Priority Method (MFPM)

Table 9.A.7 presents the operation processed results of the multi-ESS sample system using Method 2, Maximum First Priority Method (MFPM). The surplus WTG energy 20 MWh is distributed according to the priority of the maximum chargeable energy rate. At #1 state, each maximum chargeable energy for three ESSs is as follows:

$$C_{max \times 1,1} = 16\,\text{MWh}, C_{max \times 2,1} = 12\,\text{MWh}, C_{max \times 3,1} = 12\,\text{MWh} \quad (9.A.18)$$

Therefore, #1 ESS is selected first. The available minimum chargeable energy 16 MWh of #1 ESS is charged into the first selected #1 ESS. For convenience, we make a decision of selection rule that the remaining energy 4 MWh (=20–16 MWh) is charged into #3 ESS. For this problem, actually, the maximum chargeable energy of #2 ESS is the same as that of #3 ESS. But, it is assumed that #3 ESS is selected for convenience in the next ESS. Several optimal selected methods can be proposed at this time. This is marked in the first row in Table 9.A.7.

$$EU_{1,1} = 16\,\text{MWh}, EU_{2,1} = 0\,\text{MWh}, EU_{3,1} = 4\,\text{MWh} \quad (9.A.19)$$

Therefore, states (SOC) of the multi-ESS are calculated as

$$ES_1 = ES_{1,0} + EU_{1,1} = 4 + 16 = 20\,\text{MWh} \tag{9.A.20}$$

$$ES_2 = ES_{2,0} + EU_{2,1} = 3 + 0 = 3\,\text{MWh} \tag{9.A.21}$$

$$ES_3 = ES_{3,0} + EU_{3,1} = 3 + 4 = 7\,\text{MWh} \tag{9.A.22}$$

2) Reliability Evaluation of *Multi-ESS Model*
   From Tables 9.A.6 (Method 1) and 9.A.7 (Method 2), reliability indices are calculated as (9.A.23)–(9.A.25). The reliability of both methods is exactly the same for this sample system.

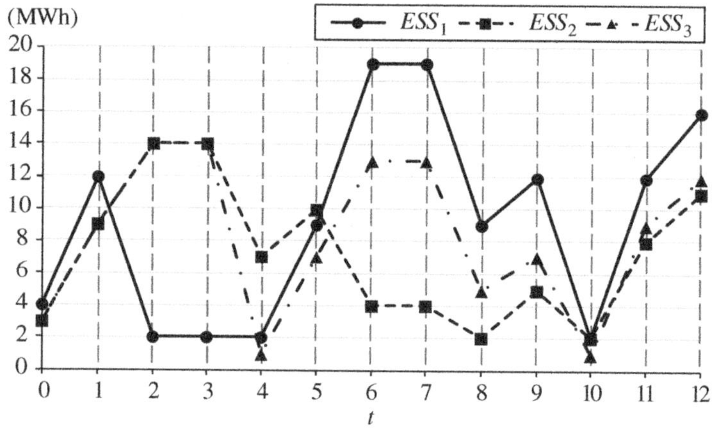

Figure 9.A.2 ESS charge/discharge state (SOC) transition (Method 1).

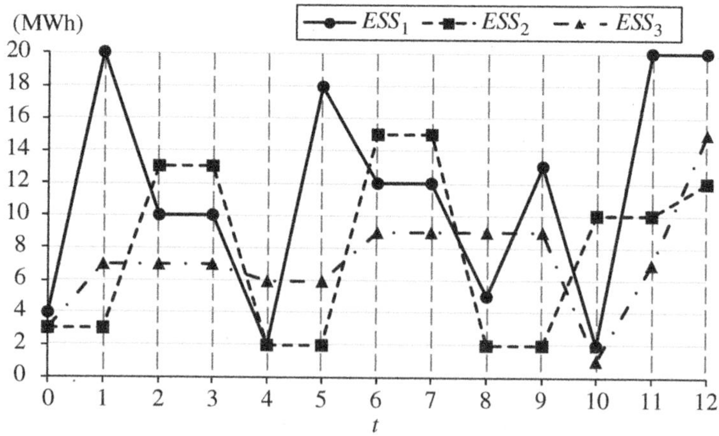

Figure 9.A.3 ESS charge/discharge state (SOC) transition (Method 2).

Table 9.A.8 Energy and frequency of charging and discharging ESS.

	Total energy (MWh)	Frequency
Charging	112	8
Discharging	84	5
Total	196	13

The SOCs of ESS at states for two methods are demonstrated in Figures 9.A.2 and 9.A.3. They have a difference in the operation and final SOC as can be seen in detail in Tables 9.A.6 and 9.A.7, and Figures 9.A.2 and 9.A.3. In terms of SOC at the last state, the last SOC of Method 2 (total final SOC = 20 + 12 + 15 = 47 MWh) is larger than that of Method 1 (total final SOC = 16 + 11 + 12 = 39 MWh). We are able to conclude that Method 2 is a better ESS operational method than Method 1 in terms of the power system reliability for this sample system (Table 9.A.8).

$$LOLE = 1 \times 2 = 2\,\text{hours/day} \tag{9.A.23}$$

$$EENS = 9\,\text{MWh/day} \tag{9.A.24}$$

$$\text{Frequency of outage} = 1\,\text{occ/day} \tag{9.A.25}$$

9.A.4 A Comparative Analysis of Single-ESS and Multi-ESS Models

The appendix proposes an algorithm of WTG-integrated ESS operation in the viewpoint of power system reliability. This sample study can be manual computation. And also, the sample system describes the comparison between the single-ESS and multi-ESS model. Table 9.A.9 presents the comparison results. This comparison result says that the multi-ESS system is surely better than the single ESS system. The installation cost (capital cost) problem is not considered here. The frequency (times) of charging/discharging of ESS is described in Table 9.A.10.

Table 9.A.9 Reliability results of single-ESS and multi-ESS models.

	Single-ESS	Multi-ESS
LOLE (hours/day)	6	2
EENS (MWh/day)	35	9
Frequency of loss (occ/day)	3	1

Table 9.A.10 Total energy and frequency of ESS on both models.

	Single-ESS	Multi-ESS (average)
Total energy for charge/discharge (MWh)	162	196 (65.3)
Frequency	10	13 (4.3)

where Average: total of amount (Total energy of charge/discharge)/number of ESS.

Table 9.A.11 Amount of final energy charged to the ESS (SOC).

		Multi-ESS	
	Single-ESS	Method 1	Method 2
Amount of final energy charged to the ESS (MWh)	39	39	47

Finally, the last state's total SOCs of ESS for three cases, which are the single-ESS, the multi-ESS method 1 and the multi-ESS method 2 are compared in Table 9.A.11. The multi-ESS system is the more efficient system for using surplus wind energy from the results in terms of the power system reliability. And also, Method 2 is a better operational algorithm than Method 1 for this sample system in terms of the last SOC maximum operation.

Remarks: Please be careful that the installation cost (capital cost [construction cost]) is ignored in the sample problem. For ESS expansion problem, the capital cost should be considered. The capital cost-considered algorithm development will be studied more in future in order to obtain the optimal ESS expansion planning.

10

Genetic Algorithm for Generation Expansion Planning and Reactive Power Planning

10.1 Introduction

The generation expansion planning (GEP) and reactive power planning problems are nonlinear dynamic optimization problems that can only be fully solved by complete enumeration, a process that is computationally impossible for realistic planning problems. Therefore, modern heuristic approaches, such as genetic algorithm (GA), are well suited for these complex planning problems. GA is a search algorithm based on the hypothesis of natural selections and natural genetics [51]. Recently, a global optimization technique using GA has been successfully applied to various areas of power system such as economic dispatch [52, 53], unit commitment [54, 55], reactive power planning [56–58], power plant control [59, 60], and GEP [14]. The GA-based approaches have several advantages. Naturally, they can not only treat the discrete variables but also overcome the dimensionality problem. In addition, they have the capability to search for the global optimum or quasi-optimums within a reasonable computation time. However, there exist some structural problems in the conventional GA, such as premature convergence and duplications among strings in a population as generation progresses [51].

This chapter introduces two approaches in enhancing the capability of GA for power system planning. First, an improved genetic algorithm (IGA), which can overcome the shortfalls of the conventional GA to some extent, is introduced for GEP [14]. Second, a synthetic optimization procedure is presented by combining two optimization methods together, random search and optimization algorithm, for reactive power generation planning problem [57].

Probabilistic Power System Expansion Planning with Renewable Energy Resources and Energy Storage Systems, First Edition. Jaeseok Choi and Kwang Y. Lee.

Published 2022 by John Wiley & Sons, Inc.

10.2 Generation Expansion Planning

The GEP is one of the most important decision-making activities in electric utilities. Least-cost GEP is to determine the minimum-cost capacity addition plan (i.e., the type and number of candidate plants) that meets forecasted demand within a prespecified reliability criterion over a planning horizon.

A least-cost GEP problem is a highly constrained nonlinear discrete dynamic optimization problem that can only be fully solved by complete enumeration in its nature [7, 11, 61]. Therefore, every possible combination of candidate option over a planning horizon must be examined to get the optimal plan, which leads to the computational explosion in a real-world GEP problem.

To solve this complicated problem, a number of methods have been successfully applied during the past decades. Masse and Gilbrat [8] applied a linear programming approach that necessitates the linear approximation of an objective function and constraints. Bloom [9] applied a mathematical programming technique using a decomposition method and solved it in a continuous space. Park, Lee, and Youn [10] applied Pontryagin's maximum principle whose solution also lies in a continuous space. Although the above-mentioned mathematical programming methods have their own advantages, they possess one or both of the following drawbacks in solving a GEP problem. That is, they treat decision variables in a continuous space. And there is no guarantee to get the global optimum since the problem is not mathematically convex. Dynamic programming (DP)-based framework is one of the most widely used algorithms in GEP [7, 11, 12, 61, 62]. However, the so-called "curse of dimensionality" has interrupted the direct application of the conventional full DP in practical GEP problems. For this reason, WASP [11] and EGEAS [7] use a heuristic tunneling technique in the DP optimization routine where users prespecify states and successively modify tunnels to arrive at a local optimum. David and Zhao developed a heuristic-based DP [12] and applied the fuzzy set theory [62] to reduce the number of states. Recently, Fukuyama and Chiang [13] and Park, Park and Won [63] applied GA to solve sample GEP problems, and showed promising results. However, an efficient method for a practical GEP problem that can overcome a local optimal trap and the dimensionality problem simultaneously has not been developed yet.

In this chapter, an IGA, which can overcome the aforementioned problems of conventional GA to some extent, is introduced [14]. The IGA incorporates the following two main features. First, an artificial creation scheme for an initial population is devised, which also takes the random creation scheme of the conventional GA into account. Second, a stochastic crossover strategy is developed. In this scheme, one of the three different crossover methods is randomly selected from a biased roulette wheel where the weight of each crossover method is determined through pre-performed experiments. The stochastic crossover scheme is similar to the stochastic selection of reproduction candidates from a mating pool. The results of the IGA are compared with those of the conventional simple GA, the full DP, and the tunnel-constrained DP employed in WASP.

10.3 The Least-Cost GEP Problem

Mathematically, solving a least-cost GEP problem is equivalent to finding a set of optimal decision vectors over a planning horizon that minimizes an objective function under several constraints. The GEP problem to be considered is formulated as follows [10]:

$$\min_{U_1,\ \dots,\ U_T} \sum_{t=1}^{T} \{ f_t^1(U_t) + f_t^2(X_t) - f_t^3(U_t) \} \tag{10.1}$$

$$\text{s.t.} \quad X_t = X_{t-1} + U_t (t = 1, \dots, T) \tag{10.2}$$

$$\text{LOLP}(X_t) < \varepsilon \qquad (t = 1, \dots, T) \tag{10.3}$$

$$\underline{R} \le R(X_t) \le \overline{R} \qquad (t = 1, \dots, T) \tag{10.4}$$

$$\underline{M_t^j} \le \sum_{i \in \Omega_j} x_t^i \le \overline{M_t^j} \qquad (t = 1, \dots, T \text{ and } j = 1, \dots, J) \tag{10.5}$$

$$0 \le U_t \le \overline{U_t} \qquad (t = 1, \dots, T) \tag{10.6}$$

where

T: number of periods (years) in a planning horizon,
J: number of fuel types,
Ω_j: index set for j-th fuel-type plant,
X_t: cumulative capacity (MW) vector of plant types in year t,
x_t^i: cumulative capacity (MW) of i-th plant type in year t,
U_t: capacity addition (MW) vector by plant types in year t,
$\overline{U_t}$: maximum construction capacity (MW) vector by plant types in year t,
u_t^i: capacity addition (MW) of i-th plant in year t,
LOLP(X_t): loss of load probability (LOLP) with X_t in year t,
$R(X_t)$: reserve margin with X_t in year t,
ε: reliability criterion expressed in LOLP,
$\overline{R}, \underline{R}$: upper and lower bounds of reserve margin,
$\overline{M_t^j}, \underline{M_t^j}$: upper and lower bounds of j-th fuel type in year t,
$f_t^1(U_t)$: discounted construction costs ($) associated with capacity addition U_t in year t,
$f_t^2(X_t)$: discounted fuel and O&M costs ($) associated with capacity X_t in year t,
$f_t^3(U_t)$: discounted salvage value ($) associated with capacity addition U_t in year t.

The objective function is the sum of tripartite discounted costs over a planning horizon. It is composed of discounted investment costs, expected fuel and O&M costs, and salvage value. To consider investments with longer lifetimes than a planning horizon, the linear depreciation option is utilized [11]. In this paper, five types of constraints are considered. (10.2) implies state equation for dynamic planning problem [56], and (10.3) and (10.4) are related with the LOLP reliability criteria and the reserve margin bands, respectively. The capacity mixes by fuel types are considered in (10.5). Plant types give another physical constraint in (10.6), which reflects the yearly construction capabilities.

Although the state vector, X_t, and the decision vector, U_t, have dimensions of MW, we can easily convert those into vectors which have information on the number of units in each plant type. This mapping strategy is very useful for GA implementation of a GEP problem such as encoding and treatment of inequality (10.6), and illustrated in the following equations:

$$X_t = \left(x_t^1, \ldots, x_t^N\right)^T \quad \rightarrow \quad X'_t = \left({x'}_t^1, \ldots, {x'}_t^N\right)^T \tag{10.7}$$

$$U_t = \left(u_t^1, \ldots, u_t^N\right)^T \quad \rightarrow \quad U'_t = \left({u'}_t^1, \ldots, {u'}_t^N\right)^T \tag{10.8}$$

where

N: number of plant types including both existing and candidate plants,
X'_t: cumulative number of units by plant types in year t,
U'_t: additional number of units by plant types in year t,
${x'}_t^i$: i-th plant type's cumulative number of units in year t,
${u'}_t^i$: i-th plant type's additional number of units in year t.

10.4 Simple Genetic Algorithm

Basically, GA is a search mechanism based on the hypothesis of natural selection [51, 157]. The GA is an artificial optimization scheme that emulates the hypothetical adaptive nature of natural genetics. It provides solutions by generating a set of chromosomes referred to as a generation. Each string (chromosome) has its own fitness measure that reflects how well a creature can survive under surrounding environments. The new generation of strings is provided through three major genetic operations – reproduction, crossover, and mutation, which provide a powerful global search mechanism. Reproduction is a process in which individual strings are copied into a mating pool according to their fitness values. Crossover, the most important genetic operator, is a structured recombination operation.

In the classical one-point crossover, a random position in a string is chosen and all characters to the right of this position are swapped. Mutation, the secondary operator in GA, is an occasional random alteration of the value of a string position. Variations of the simple GA for power system applications can be found in the Refs. [13, 14, 52–60, 63]. The improvements on the conventional GA will be described in the subsequent sections.

The simple GA consists of a population of bit strings transformed by three genetic operations: selection, crossover, and mutation. Each string represents a possible solution, with each substring representing a value for a variable of interest. The algorithm starts from an initial population generated randomly. A new generation is generated by using the genetic operations considering the fitness of a solution that corresponds to the objective function for the problem. The fitness of solutions is improved through iterations of generations. When the algorithm converges, a group of solutions with better fitness is obtained, and the optimal solution is obtained.

10.4.1 String Representation

String representation is an important factor in solving the planning problem using the SGA. In order to accommodate different representations of object parameters, i.e., the investment variables, the following representation method is used.

A string consists of sub-strings; the number of sub-strings is equal to the number of total candidate buses, for example, for adding capacitors or inductors in VAR planning. Each sub-string is in the form of binary corresponding to the amount of investment. This binary representation is certainly not unique but it is simple to implement. For example, the 3-bit binary for a unit can represent $2^3 = 8$ different amounts of installation for generator, capacitive, or inductive VAR. On the other hand, the 4-bit binary represents $2^4 = 16$ choices.

10.4.2 Genetic Operations

1) *Initial population generation* – Initial population of binary strings is created randomly. Each of the three strings represents one feasible solution satisfying constraints.
2) *Fitness evaluation* – The solution strings and each candidate solution is tested in its environment. The fitness of each candidate solution is evaluated through some appropriated measure such as the inverse of the cost function Z. The algorithm is driven toward maximizing this fitness.
3) *Selection and reproduction* – Selection and reproduction create a new population from the old population. A set of old strings is selected to reproduce a set of new strings according to the probability determined by the simulated spin of a

weighted roulette wheel. The roulette wheel is biased with the fitness of each of the solution candidates. The wheel is spun N times where N is the number of strings in the population.

4) *Crossover* – Crossover is performed on two strings at a time that are selected from the population at random. It involves choosing a random position in the two strings and swapping the bits that occur after this position. Crossover can occur at a single position (single crossover). Crossover can be performed in different methods. Two different means are used in this chapter: tail-tail crossover and head-tail crossover.

   The tail-tail crossover tends to change less significant bits. On the other hand, the head-tail crossover gives more chances of changes by changing more significant bits. The crossover methods can be changed during iterations: the head-tail crossover can be used in early generations and then switched to tail-tail crossover in later generations for fine-tuning.
5) *Mutation* – Mutation is performed sparingly, typically every 100–1000 bit transfers from crossover, and it involves selecting a string at random as well as a bit postion at random and changing it from a 1 to a 0 or vice versa. It is used to escape from a local minimum. After mutation, the new generation is completed, and the procedure begins again with the fitness evaluation of the population.

10.5 Improved GA for the Least-Cost GEP

10.5.1 String Structure

Since it is convenient to use integer values for GA implementation of a GEP problem, the reordered structure of (10.8) by plant types covering a planning horizon is used for encoding of a string as shown in (10.9). Here, each element of a string (i.e., $\hat{U}'^{n} = \left(u'^{n}_{1}, u'^{n}_{2}, \dots, u'^{n}_{T}\right)^{T}$ for $n = 1, \dots, N$) corresponds to a substring and its structure is depicted in Figure 10.1.

$$\hat{U}' = \left(u'^{1}_{1}, u'^{1}_{2}, \dots, u'^{1}_{T}, \dots, u'^{n}_{1}, u'^{n}_{2}, \dots, u'^{n}_{T}, \dots, u'^{N}_{1}, u'^{N}_{2}, \dots, u'^{N}_{T}\right)^{T} = \left(\hat{U}'^{1}, \dots, \hat{U}'^{n}, \dots, \hat{U}'^{N}\right)^{T} \tag{10.9}$$

10.5.2 Fitness Function

The objective function or cost of a candidate plan is calculated through the probabilistic production costing and the direct investment costs' calculation [7, 11]. The fitness value of a string can be evaluated using the following [51, 59]:

$$f = \frac{\alpha}{1 + J} \tag{10.10}$$

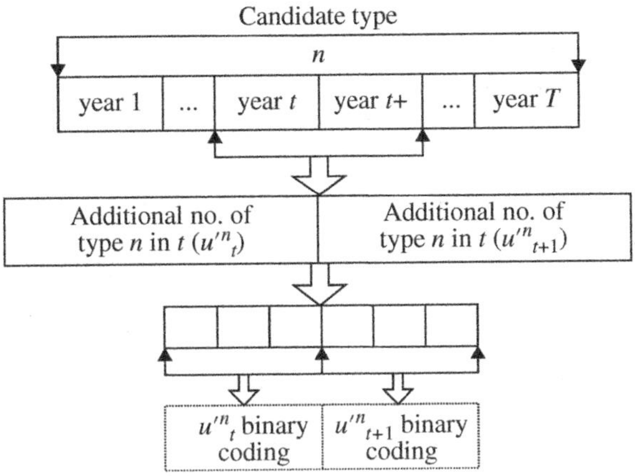

Figure 10.1 A substring structure.

where

α: constant,
J: objective function of (10.1).

However, this simple mapping occasionally brings about a premature convergence and duplications among strings in a population since strings with higher fitness values dominate the occupation of a roulette wheel.

To ameliorate these problems, the following modified fitness function, which normalizes the fitness value of strings into real numbers within [0,1], is used in this paper [51].

$$f'(i) = \frac{f(i) - f_{min}}{f_{max} - f_{min}} \tag{10.11}$$

where

$f(i)$: fitness value of string i using (10.10),
f_{max}, f_{min}: maximum and minimum fitness values in a generation,
$f'(i)$: modified fitness value of string i.

10.5.3 Creation of an Artificial Initial Population

It is important to create an initial population of strings spread out throughout the whole solution space, especially in a large-scale problem. One alternative method could be to increase the population size, which yields a high computational

Table 10.1 Example of a look-up table with three plant types for three planning years.

	Type 1 (upper limit: 3 units/year)			Type 2 (upper limit: 3 units/year)			Type 3 (upper limit: 5 units/year)		
m	Year 1	Year 2	Year 3	Year 1	Year 2	Year 3	Year 1	Year 2	Year 3
1	00	00	00	00	00	00	000	000	000
2	01	01	01	01	01	01	001	001	001
3	10	10	10	10	10	10	010	010	010
4	11	11	11	11	11	11	011	011	011
5	00	00	00	00	00	00	100	100	100
6	01	01	01	01	01	01	101	101	101
7	10	10	10	10	10	10	000	000	000
8	11	11	11	11	11	11	001	001	001
9	00	00	00	00	00	00	010	010	010
10	01	01	01	01	01	01	011	011	011
11	10	10	10	10	10	10	100	100	100
12	11	11	11	11	11	11	101	101	101
Generated string 1: 011100101011000101010									

burden. This chapter suggests a new artificial initial population (AIP) scheme, which also takes the random creation scheme of the conventional GA into account. The procedures are illustrated in the following and in Table 10.1:

Step 1. Generate all possible binary seeds of each plant type considering (10.6). For example, if i-th plant type has an upper limit of 3 units per year, then generate four possible binary seeds (i.e., 00,01,10,11).

Step 2. Find the least common multiple (LCM) m from the numbers of the binary seeds of all types, and fill m binary seeds in a look-up table for all plant types and planning years. For example, if three plant types have upper limits of 3, 3, and 5 units per year, respectively, then the numbers of binary seeds are 4, 4, and 6, and m becomes 12.

Step 3. Select an integer within $[1, m]$ at random for each element $u_t'^n$ of a string in (10.9). Fill the string with the corresponding binary digits and delete it from the look-up table. Repeat until m different strings are generated.

Step 4. Check the constraints of (10.3), (10.4), and (10.5). If a string satisfies these constraints for all years, then it becomes a member of an initial population. Otherwise, the only parts of the string that violate the constraints in year t are generated at random until they satisfy the constraints. Go to step 3 n times for $n{\cdot}m$ less than P, where P is the number of strings in a population and n is an arbitrary positive integer.

Step 5. The remaining *P-n·m* strings are created using uniform random variables with binary number {0,1}. Go to step 4 to check the constraints and regenerate them if necessary. This process is repeated until all strings, which satisfy the constraints, are generated.

This AIP is based on both artificial and random selection schemes, which allows all possible string structures to be included in an initial population.

10.5.4 Stochastic Crossover, Elitism, and Mutation

Most of the GA works are based on Goldberg's simple genetic algorithm (SGA) framework [51]. This paper proposes two different schemes for genetic operation: a stochastic crossover technique and the application of elitism. The stochastic crossover scheme covers three different crossover methods; 1-point crossover, 2-point crossover, and 1-point substring crossover as illustrated in Figure 10.2. Each crossover method has its own merits. The 1-point substring crossover can provide diverse bit structures to search solution space; however, it easily destroys the string structure that may have partial information on the optimal structure (Figure 10.3).

Although the 1- and 2-point crossovers cannot explore solution space as widely as the above crossover, the probability of destroying an already-found partial optimal structure is very low. The stochastic crossover strategy is similar to the process of stochastic selection of reproduction candidates from a mating pool. That is, one of the three different crossover methods is selected from a biased roulette wheel, where each crossover method has a roulette wheel slot-sized according to its performance. The weight for each crossover method has been determined.

The second feature lies in the application of elitism [55]. The roulette wheel selection scheme gives a reproduction opportunity to a set of recessive members

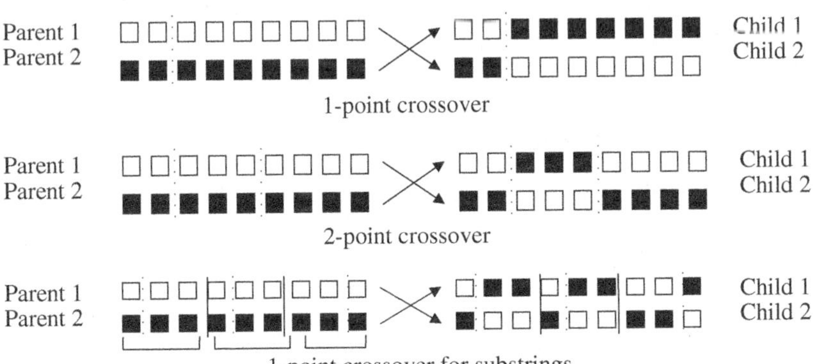

Figure 10.2 Three different crossover methods used.

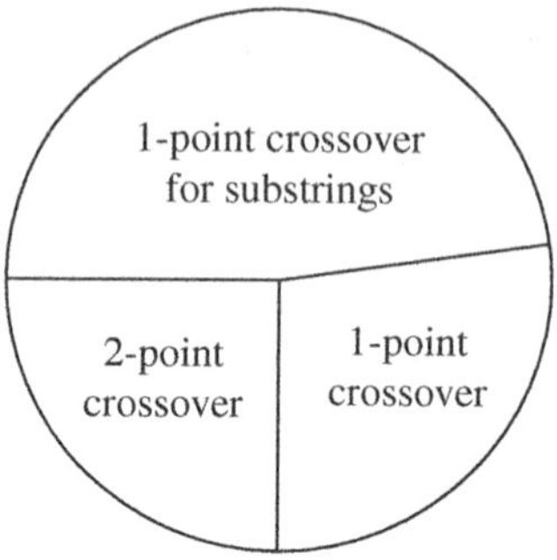

Figure 10.3 Roulette wheel for stochastic selection of crossover method.

and might not give the set of dominant strings (i.e., an elite group) a chance to reproduce. Furthermore, the application of genetic operations changes the string structures of the fittest solutions. Thus, the best solutions in the current generation might not appear in the next generation. To circumvent these problems, an elite group is directly copied into the next generation. We have applied the conventional mutation scheme [51] where it performs bit-by-bit for the strings undergone the stochastic crossover operator. However, the mutation procedure is not applied to the set of dominant strings to preserve the elitism. After genetic operations, we check all strings whether they satisfy the constraints of (10.3)–(10.5) or not. If any string violates the constraints of (10.3)–(10.5), only the parts of the string that violate the constraints in year t are generated at random until they satisfy the constraints as described in the AIP scheme.

10.6 Case Studies

The IGA, SGA, tunnel-constrained dynamic programming (TCDP) employed in WASP, and full dynamic programming (DP) were implemented using the FORTRAN77 language on an IBM PC/Pentium (166 MHz) computer.

10.6.1 Test Systems' Description

The IGA, SGA, TCDP, and DP methods have been applied in two test systems: Case 1 for a power system with 15 existing power plants, 5 types of candidate options and a 14-year study period, and Case 2 for a real-scale system with a 24-year study period. The planning horizons of 14 and 24 years are divided into 7 and 12 stages (two-year intervals), respectively. The forecasted peak demand over the study period is given in Table 10.2.

Tables 10.3 and 10.4 show the technical and economic data of the existing plants and candidate plant types for future additions, respectively.

Table 10.2 Forecasted peak demand.

Stage (year)	0 (1996)	1 (1998)	2 (2000)	3 (2002)	4 (2004)	5 (2006)	6 (2008)
Peak (MW)	5000	7000	9000	10 000	12 000	13 000	14 000
Stage (year)		7 (2010)	8 (2012)	9 (2014)	10 (2016)	11 (2018)	12 (2020)
Peak (MW)		15 000	17 000	18 000	20 000	22 000	24 000

Table 10.3 Technical and economic data of the existing plants.

Name (fuel type)	No. of units	Unit capacity (MW)	FOR (%)	Operating cost ($/kWh)	Fixed O&M cost ($/kW-Mon)
Oil #1 (Heavy Oil)	1	200	7.0	0.024	2.25
Oil #2 (Heavy Oil)	1	200	6.8	0.027	2.25
Oil #3 (Heavy Oil)	1	150	6.0	0.030	2.13
LNG G/T #1 (LNG)	3	50	3.0	0.043	4.52
LNG C/C #1 (LNG)	1	400	10.0	0.038	1.63
LNG C/C #2 (LNG)	1	400	10.0	0.040	1.63
LNG C/C #3 (LNG)	1	450	11.0	0.035	2.00
Coal #1 (Anthracite)	2	250	15.0	0.023	6.65
Coal #2 (Bituminous)	1	500	9.0	0.019	2.81
Coal #3 (Bituminous)	1	500	8.5	0.015	2.81
Nuclear #1 (PWR)	1	1000	9.0	0.005	4.94
Nuclear #2 (PWR)	1	1000	8.8	0.005	4.63

Table 10.4 Technical and economic data of candidate plants.

Candidate type	Construction upper limit	Capacity (MW)	FOR (%)	Operating cost ($/kWh)	Fixed O&M cost	Capital cost ($/kW)	Life time (years)
Oil	5	200	7.0	0.021	2.20	812.5	25
LNG C/C	4	450	0	0.035	0.90	500.0	20
Coal (Bitum.)	3	500	9.5	0.014	2.75	1062.5	25
Nuc (PWR)	3	1000	9.0	0.004	4.60	1625.0	25
Nuc (PHWR)	3	700	7.0	0.003	5.50	1750.0	25

10.6.2 Parameters for GEP and IGA

There are several parameters to be predetermined, which are related to the GEP problem and GA-based programs. In this chapter, we use 8.5% as a discount rate, 0.01 as LOLP criteria, and 15 and 60% as the lower and upper bounds for reserve margin, respectively. The considered lower and upper bounds of capacity mix are 0

Table 10.5 Parameters for IGA implementation.

Parameters	Values
• Population size	300
• Maximum generation	300
• Probabilities of crossover and mutation	0.6, 0.01
• Number of elite strings	3 (1%)
• Weights of 1-point, 2-point, and 1-point Crossover for Substrings in a Biased Roulette Wheel	0.15; 0.15; 0.70

Table 10.6 Results obtained by each crossover method.

	Objective function in million dollars (Errors against optimal solution, %)		
Crossover method	*PC* = 0.6	*PC* = 0.7	*PC* = 0.8
One-point crossover	5035.53 (0.59%)	5013.50 (0.15%)	5057.30 (1.02%)
Two-point crossover	5034.89 (0.57%)	5032.98 (0.54%)	5034.89 (0.57%)
One-point substring crossover	5012.53 (0.13%)	5012.46 (0.13%)	5010.63 (0.09%)
DP	5006.19		

and 30% for oil-fired power plants, 0 and 40% for LNG-fired, 20 and 60% for coal-fired, and 30 and 60% for nuclear, respectively.

Parameters for IGA are selected through experiments. Especially, the dominant parameters such as crossover probabilities and weights for crossover techniques are determined empirically from a test system with a six-year planning horizon with other data being the same as Cases 1 and 2 (Table 10.5).

To decide the weight of each crossover method in a biased roulette wheel for stochastic crossover, nine experiments are performed by changing the probability of crossover from 0.6 to 0.8, and the results are compared with the optimal solution obtained by the full DP as shown in Table 10.6.

Among the three crossover methods, the 1-point substring crossover showed the best performance in every case. Thus, we set the 1-point substring crossover with the biggest weight, and others with an equal smaller weight.

To determine the weight of each crossover method in a biased roulette wheel, 18 simulations were performed with different weights and crossover probabilities as shown in Table 10.7.

Table 10.7 Results obtained by stochastic crossover method.

Weights	Objective function in million dollars (Errors against optimal solution, %)		
	PC = 0.6	PC = 0.7	PC = 0.8
0.05:0.05:0.90	5007.40[a] (0.02%)	5010.63 (0.09%)	5007.40 (0.02%)
0.10:0.10:0.80	**5006.19 (0.00%)**	5010.63 (0.09%)	5012.37 (0.12%)
0.15:0.15:0.70	5007.40 (0.02%)	**5006.19 (0.00%)**	**5006.19 (0.00%)**
0.20:0.20:0.60	**5006.19 (0.00%)**	**5006.19 (0.00%)**	5011.79 (0.11%)
0.25:0.25:0.50	**5006.19 (0.00%)**	5007.40 (0.02%)	5018.37 (0.24%)
0.30:0.30:0.40	**5006.19 (0.00%)**	5012.46 (0.13%)	5007.40 (0.02%)

[a] The solution with objective function as 5007.40 million dollars is the second-best solution found by dynamic programming.

Among 18 simulations, we have found the optimal solution 7 times and the second-best solution 4 times. Furthermore, the optimal or the second-best solution is found by applying the stochastic crossover technique when the probability of crossover is 0.6. Also, when the weight of 1-point substring crossover is 0.7 and weights for others are 0.15, it always found optimal or the second-best optimal solution. Therefore, we have set the weights in the stochastic crossover technique as 0.15:0.15:0.70 among the three crossover methods. This choice has resulted in the robustness of the stochastic crossover method.

10.6.3 Numerical Results

The developed IGA was applied to two test systems and compared with the results of DP, TCDP, and SGA. Throughout the tests, the solution of the conventional DP is regarded as the global optimum and that of TCDP as a local optimum. Both the global and a local solution can be obtained in Case 1; however, only a local solution can be obtained by using TCDP in Case 2 since the "curse of dimensionality" prevents the use of the conventional DP.

Figure 10.4 illustrates the convergence characteristics of various GA-based methods in Case 1. It also shows the improvement of IGA over SGA. The IGA employing the stochastic crossover scheme (IGA2) has shown better performance

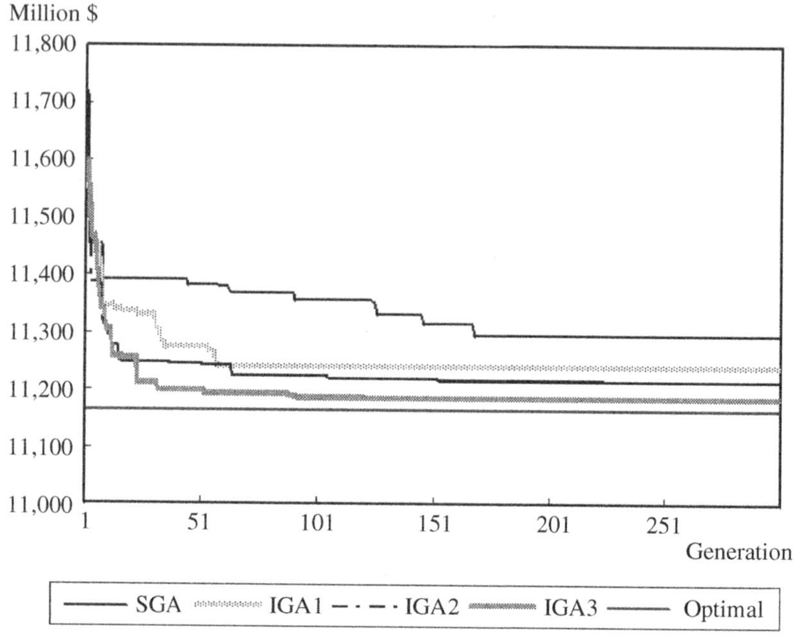

Figure 10.4 Convergence characteristics of IGA method in Case 1 system.

Table 10.8 Summary of the best results obtained by each solution method.

Solution method		Cumulative discounted cost (10^6 $)	
		Case 1 (14-year study period)	Case 2 (24-year study period)
DP		11 164.2	Unknown
TCDP		11 207.7	16 746.7
SGA		11 310.5	16 765.9
IGA	IGA1	11 238.3	16 759.2
	IGA2	11 214.1	16 739.2
	IGA3	11 184.2	16 644.7

than the IGA using the AIP scheme (IGA1). By considering both schemes simultaneously (IGA3), the performance is significantly enhanced.

Table 10.8 summarizes the costs of the best solution obtained by each solution method. In Case 1, the solution obtained by IGA3 is within 0.18% of the global solution costs while the solutions by SGA and TCDP are within 1.3 and 0.4%,

Table 10.9 Cumulative number of newly introduced plants in Case 1 and Case 2 by IGA3.

Type (year)	Oil (200 MW)	LNG C/C (450 MW)	Coal (500 MW)	PWR (1000 MW)	PHWR (700 MW)
1998	3 (5)[a]	2 (1)	2 (3)	0 (1)	2 (0)
2000	5 (6)	3 (1)	5 (6)	0 (1)	4 (1)
2002	5 (7)	3 (1)	5 (6)	0 (2)	4 (1)
2004	8 (10)	7 (3)	6 (7)	0 (2)	4 (1)
2006	10 (12)	10 (3)	6 (7)	0 (2)	6 (2)
2008	10 (13)	10 (3)	6 (9)	0 (2)	6 (2)
2010	10 (13)	10 (3)	6 (9)	0 (2)	6 (4)
2012	14	11	8	1	7
2014	17	14	8	1	7
2016	19	15	10	1	9
2018	19	17	10	3	9
2020	20	18	12	3	9

[a] The figures within parentheses denote the results of IGA3 in Case 1.

respectively. In Case 1 and Case 2, IGA3 has achieved a 0.21 and 0.61% improvement of costs over TCDP, respectively. Although SGA and IGAs have failed in finding the global solution, all IGAs have provided a better solution than SGA. Furthermore, solutions of IGA3 are better than that of TCDP in both cases, which implies that it can overcome a local optimal trap in a practical long-term GEP. Table 10.9 summarizes generation expansion plans of Case 1 and Case 2 obtained by IGA3.

The execution time of GA-based methods is much longer than that of TCDP. That is, IGA3 requires approximately 3.7 and 6 times of execution time in Case 1 and Case 2, respectively. However, it is much shorter than the conventional DP. Figure 10.5 shows the observed execution time of IGA3 and DP as the stages are expanded. The execution time of IGA3 is almost linearly proportional to the number of stages while that of DP exponentially increases. In the system with 11 stages, it takes over nine days for DP, and requires about 1.2 million of array memories to obtain the optimal solution while it takes only 11 hours by IGA3 to get the near optimum.

The proposed method definitely provides quasi-optimums in a long-term GEP within a reasonable computation time. Also, the results of the proposed IGA method are better than those of TCDP employed in the WASP, which is viewed as a very powerful and computationally feasible model for a practical long-term GEP problem. Since a long-range GEP problem deals with a large amount of investment, a slight improvement by the proposed IGA method can result in substantial cost savings for electric utilities.

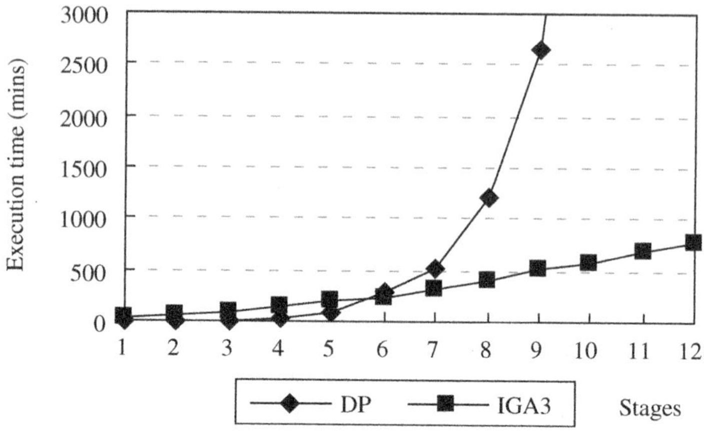

Figure 10.5 Observed execution time for the number of stages.

10.6.4 Summary

The IGA has been successfully applied to long-term GEP problems. It provided better solutions than conventional SGA. Moreover, by incorporating all the improvements (IGA3), it was found to be robust in providing quasi-optimums within a reasonable computation time and yield better solutions compared to the TCDP employed in WASP. Contrary to the DP, the computation time of the proposed IGA is linearly proportional to the number of stages.

The developed IGA method can simultaneously overcome the "curse of dimensionality" and a local optimum trap inherent in GEP problems. Therefore, the proposed IGA approach can be used as a practical planning tool for a real-system scale long-term GEP.

10.7 Reactive Power Planning

The reactive power, or VAR, planning problem is a nonlinear optimization problem. Its main objective is to find the most economic investment plan for new reactive sources at selected load buses which will guarantee proper voltage profile and the satisfaction of operational constraints. Usually, the planning problem is divided into operational and investment planning subproblems. In the operational planning problem, the available shunt reactive sources and transformer tap-settings are optimally dispatched at a minimal operation cost. In the investment planning problem, new reactive sources are optimally allocated over a planning horizon at a minimal total cost (operational and investment).

During the past decade, there has been a growing concern in power systems about reactive power operation and planning. The recent approaches to the VAR planning problem are becoming very sophisticated in minimizing installation cost and for the efficient use of VAR sources to improve system performance. Various mathematical optimization formulations and algorithm have been developed in most cases, by using nonlinear [64], linear [65], or mixed integer programming [66], and decomposition method [67–70]. More recently, simulated annealing [71] and GA [9, 72] have also been used. With the help of powerful computers, it is now possible to do a large amount of computation in order to achieve a global optimal solution instead of a local optimal solution.

Simulated annealing method is a random search method. Hsiao et al. [70] provided an approach for the simulated annealing method using the modified fast decoupled load flow. However, only the new configuration (VAR installation) is checked with the load flow and the existing resources such as generators and regulating transformers are not fully exploited. SGA method is a powerful optimization technique analogous to the natural genetic process in biology. Theoretically, this technique converges to the global optimum solution with probability one, provided that certain conditions are satisfied. The SGA method is known as a robust optimization method. It is useful especially when other optimization methods fail in finding the optimal solution. However, it often requires too many repeated computations in obtaining final results.

In order to obtain a good result for reactive power generation planning problem, a synthetic optimization procedure is presented by combining two optimization methods together with random search and optimization algorithm.

This chapter presents an improved method of operational and investment planning by using an SGA combined with the successive linear programming method. The Benders cuts are constructed during the SGA procedure to enhance the robustness and reliability of the algorithm. The method takes advantage of both the robustness of the SGA and the accuracy of the conventional optimization method.

The proposed VAR planning approach is in the form of a two-level hierarchy. In the first level, the SGA is used to select the location and the amount of reactive power sources to be installed in the system. This selection is passed on to the operation optimization subproblem in the second level in order to solve the operational planning problem. It is a common practice to use a successive linear programming (LP) formulation to improve the computation speed and to enhance the computation accuracy; the LP method is fast and robust. The operational planning problem is decoupled into coupled real (P) and reactive (Q) power optimization modules; and the successive linearized formulation of the P-Q optimization modules speeds up computation, and allows the LP to be used in finding the solution of the nonlinear problem [73]. The dual variables in the LP are transferred from the

P-Q optimization modules to the SGA module in the first level to set up the Benders cut for investment planning. This hierarchical optimization approach allows the SGA to obtain correct VAR installations, and at the same time, satisfy all the operational constraints and the requirement of minimum operation cost.

10.8 Decomposition of Reactive Power Planning Problem

The reactive power planning problem is to determine the optimal investment of VAR sources over a planning horizon. The cost function to be minimized is the sum of the operation cost and the investment cost. The investment cost is the cost to install new shunt reactive power compensation devices for the system. The fuel cost for generation is the only operation cost to be considered in this chapter.

10.8.1 Investment-Operation Problem

The reactive power planning problem involves both operation and investment costs, and it can be written in the following form:

$$\min_{Y,U} f(Y.U) = L_o(Y) + L_u(U) \tag{10.12a}$$

subject to

$$G_1(Y,U) \leq 0 \tag{10.12b}$$

$$G_2(U) \leq 0 \tag{10.12c}$$

where,

$Y = [P^T, V^T, N^T]$: vector of operational variables
P: vector of real power generations,
V: vector of bus voltage magnitudes,
N: vector of tap-settings,
U: vector of investment variables
$L_o(Y)$: operation cost
$L_u(U)$: investment cost
$G_1(Y, U)$: constraint involving both Y and U
$G_2(U)$: constraint involving U only.

Here, (10.12a) consists of investment and operation cost, and (10.12b) are coupled constraints for operation and investment variables. It includes load flow balance and other important operational constraints, on the other hand, (10.12c) includes constraints relative to only investment variables.

10.8.2 Benders Decomposition Formulation

The operation cost is nonlinear and the investment cost can be assumed to be linear with respect to the amount of newly added reactive power compensation. According to this assumption, the minimization problem (10.12) can be expressed in a nonlinear programming formulation:

$$\min_{Y,U} C^T U + f(Y) \tag{10.13a}$$

subject to

$$H(Y) + BU \leq b_1 \tag{10.13b}$$

$$DU \leq b_2 \tag{10.13c}$$

where

Y: vector of operation variables,
U: vector of investment variables,
C: vector of cost coefficients,
B,D: matrices of constraints,
f,H: cost and constraint functions, respectively.
b_1, b_2: vector of constraints.

Because of the structure of the constraints, it is quite natural to consider a two-level hierarchical approach to solve the problem. That is, use the SGA to select the device and amount, and use an optimization method to obtain optimal results under the given installation.

In this chapter, the generalized Benders decomposition (GBD) method [74] is used in the SGA module in setting up a Benders cut in order to improve the convergence characteristics. The procedure is as follows:

i) Assuming a feasible investment U, the feasible decision Y is obtained by solving the Y (operation) subproblem:

$$\min_{Y} f(Y) \tag{10.14a}$$

subject to

$$H(Y) \leq b_1 - BU \tag{10.14b}$$

ii) Having found optimal Y from the first stage, the decision for the feasible investment U is obtained by solving the U (investment) subproblem:

$$\min_{Y,U} Z = C^T U + \sigma \tag{10.15a}$$

subject to

$$DU \leq b_2 \tag{10.15b}$$

$$W(U) \leq \sigma \tag{10.15c}$$

where σ is low limit variable, and $W(U)$ is called the Benders cut and is a function which supplies information concerning the capacity decision U in terms of the operation feasibility. Then the problem would determine a solution (U, Y) that would minimize the global function (10.13a).

The Benders decomposition method builds the function $W(U)$ based on the solution of the Y subproblem. In nonlinear optimization, $W(U)$ can be determined if we observe that the simplex multiplier vector associated with the first stage (Y subproblem) is the basic feasible solution for the dual problem. Therefore,

$$W(U) = v\left(U^k + \lambda^k\left(BU^k - BU\right)\right) \tag{10.16}$$

where $v(.)$ is the optimal operation cost with the installation of U^k.

The dual solution λ^k is the simplex multiplier associated with the constraint in the operation subproblem, where k is the iteration number. Since the revised simplex method is used for solving the operation subproblem, λ^k is obtained as a by-product and new constraints, each corresponding to a different investment installation, are established.

From (10.15a) to (10.15c), it can be seen that λ_i^k presents the change of the cost caused by the unit change in investment for unit i. If $\lambda_i^k > 0$ then $U_i > U_i^k$ is helpful in generating a new member of the population, which may decrease the total cost Z. If only one constraint is considered, a decreasing direction similar to the steepest descent method can be found. The Benders cut is viewed as a coordinator between investment and operation subproblems, and the GBD method iterates between the two. At each iteration, a new constraint is added into $W(U)$ to form a new constraint set.

10.9 Solution Algorithm for VAR Planning

The planning methodology developed in the chapter is simulated for the reactive power planning problem. The problem is decomposed into investment and operation subproblems and solved iteratively until convergence [68].

The operation subproblem is again decomposed into economic real (P) and reactive (Q) power dispatch problems to minimize the fuel cost function [73, 74]. In the P-module optimal values of real power generation, and in the Q-module, the optimal values of bus voltage magnitudes and transformer tap-settings are obtained. In addition, the optimal values of reactive power dispatched by the generators and compensators are also obtained.

In each population, total operating and investment costs are calculated for each investment. The fitness is simply the inverse of this total cost. The ratio of the average fitness and the maximum fitness of the population is computed and generation is repeated until:

$$\frac{\text{Average fitness}}{\text{Maximum fitness}} \geq AP$$

where AP is a given number that represents the degree of satisfaction. If the convergence has been reached at given accuracy, then optimal values for investment are found. Other criteria, such as the difference between the maximum and minimum fitnesses and the rate of increase in maximum fitness, can also be used as the stopping criteria. Another possibility is to stop the algorithm at some fitness number of generations and designate the result as the best fit from the population.

The iterative process is as follows:

Step 1. Initial population generation – compute the fitness of each member according to operation subproblem results.
Step 2. Generate new population – typical SGA methods, reproduction crossover and mutation, are used. The Benders cut is used on a subset of strings to obtain one new and better member of the population.
Step 3. Compute the fitness of the new generation.
If convergence condition is satisfied, stop computation. Otherwise, return to Step 2 and begin a new generation.

The most important step is Step 2. A new population is generated according to the fitness of the old population through the simulated spin of a weighted wheel in the SGA [51]. Some modifications are made to the SGA for our planning problem, resulting in a modified SGA (MSGA):

1) In the GBD, the iteration procedure is an alternate computation between investment and operation till the convergence is reached. The Benders cuts are selected and constructed from old population. It is used to obtain a new member of population. The number of cuts can be adjusted as a part of the procedure. Some better-fitted strings and some worse fitted strings are selected to construct the cuts. The Benders cut helps in narrowing down the space of possible solutions, and thus speeds up the convergence.
2) An abandoning rate is considered in giving up some poor alternatives by assorting the fitness of the alternatives.
3) Different crossovers are also considered, that is, the tail-tail crossover and head-tail crossover, and the crossover position is selected randomly. The head-tail can also be used in producing new strings from two identical parents.

In the original SGA, only the fitness value resulting from the operation subproblem is used to generate a new generation. However, the new population generated only by its fitness is random and blind. By using the Benders cut, which makes use of both the dual variable information and the cost function, a new and better string can be found. If this new string is a good one (it may be the best one), i.e. it has a higher fitness value, it will survive to the next generation. Otherwise, it will likely die afterward. In this method, the robust characteristics of the SGA can still be maintained; at the same time, it increases the chance to find the optimal result earlier. The Benders cut can be set up without difficulty because all variables are made available when the operation optimization subproblem is solved.

10.10 Simulation Results

The systems tested and described here are: The 6- and 30-bus networks. The emphasis is on the effectiveness of the technique and validity of results. The following parameters are used for SGA program:

Mutation rate: 0.01
Crossover rate: 1.0
Abandoning rate: 0.9
Parameter resolution: 3 bits per substring

10.10.1 The 6-bus System

The 6-bus system given in [73] is considered, which has two generators at buses 1 and 2, and two load buses, 4 and 6, are used for shunt reactive compensation. The initial load flow results show that with no reactive compensation, there are under-voltages at load buses 3 through 6. Thus, the reactive power supply from generators is not adequate to maintain the required voltage profile.

Table 10.10 shows the maximum, minimum, and average fitness obtained by the simple genetic method (SGA) and modified simple genetic method (MSGA). Both methods give the same final results. However, the MSGA method needs less iteration than the SGA method.

Table 10.10 The fitness values for the 6-bus system.

Method	Gen.		Min.	Avg.	Max.
SGA	11	Initial	0.2526	0.2543	0.2559
		Final	0.2556	0.2558	0.2561
MASG	9	Initial	0.2526	0.2543	0.2559
		Final	0.2547	0.2556	0.2561

Table 10.11 Summary of results for the 6-bus system.

	Limits		Initial	Test 1	Test 2	Test 3
Var.	Low	Upper	State	Result	Result	Result
V_1	1.0	1.1	1.00	1.1	1.065	1.032
V_2	1.0	1.15	1.00	1.15	1.150	1.150
V_3	0.9	1.0	0.78	0.932	0.948	0.983
V_4	0.9	1.0	0.88	0.966	0.995	0.995
V_5	0.9	1.0	0.82	0.968	0.995	0.996
V_6	0.9	1.0	0.87	0.946	0.979	0.995
P_1	10.0	100.0	94.8	53.68	52.67	52.40
P_2	50.0	100.0	50.0	100.0	100.0	100.0
Q_1	−20.	100.0	53.84	67.535	33.06	14.097
Q_2	−20.	100.0	25.19	18.047	12.14	8.930
C_4	0.0	30.0	0.0	0.0	20.0	25.0
C_6	0.0	30.0	0.0	0.0	20.0	30.0
Total cost ($)			6.19.53	397.78	391.35	390.78
Losses (MW)			9.83	18.68	17.70	17.39

Test 1: Operation optimization without investment.
Test 2: Operation optimization with investment by Benders decomposition method only.
Test 3: Operation optimization by using MSGA.

After the reactive power planning is completed, the total reactive power compensation is summarized in Table 10.11. It is observed that the voltage profile is within the operating range of 0.90–1.15 p.u. Both voltage limits are satisfied. The total cost is decreased from $619.53 to $390.78, a decrease of 36.9%. The final operation cost for the optimization without capacitor investment is $397.78, higher than the optimal result. The column for Test 2 is corresponding to the result obtained by using the Benders decomposition method. The total cost is also higher than that of the modified SGA method.

10.10.2 IEEE 30-bus System

For the IEEE 30-bus system [75], there are six generator buses. Seven buses are selected to add capacitors. Each candidate bus has 3 bits for parameter resolution and it can represent eight different values for installation. The length of the chromosome string is 21 bits and the population size is 25.

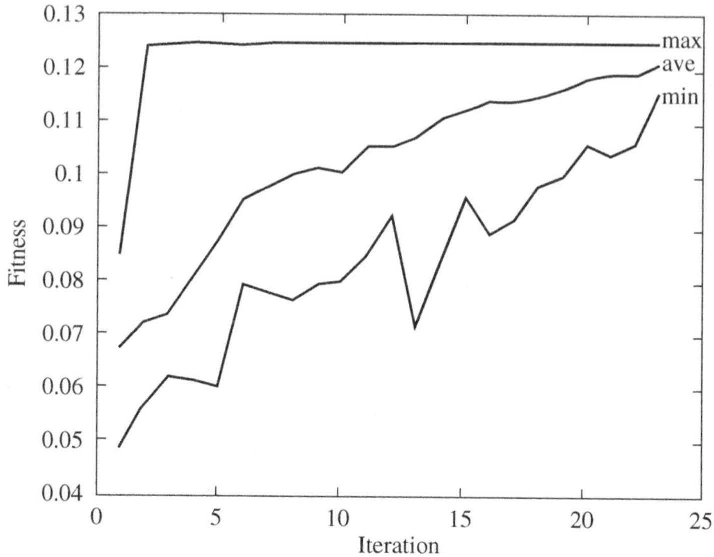

Figure 10.6 MSGA iteration result.

Initial optimization is run for operational variables. The result shows that the system can maintain all operation constraints without any new capacitor installed, but at a higher cost. In order to test the effectiveness of the program, high unit installation cost is used. It was anticipated that the SGA method should find an optimal result after certain generations and in which case, the additional installation should be zero.

Figure 10.6 shows the iteration result for the test case using the MSGA with Benders cut added. There were a total of 264 crossovers and 104 discards for some string with bed fitness values. Figure 10.7 shows the iteration result for the test case using only the SGA, where there were 325 crossovers and 141 discards.

It can be seen that when the Benders cuts are added for the MSGA, only two generations are needed to find the optimal result. After that, the optimal results are still maintained during later iterations. As indicated in Figure 10.7, the SGA method needs 18 generations to find the final result. Due to random search, the optimal result can only be reached after a considerable number of iterations. The convergence procedure is slower than the MSGA method.

10.10.3 Summary

The voltage profile throughout the planning period was improved from the undervoltages seen in the initial load flow, to the required operation range. It was also seen that new shunt capacitors are installed at or near the load buses that exhibit

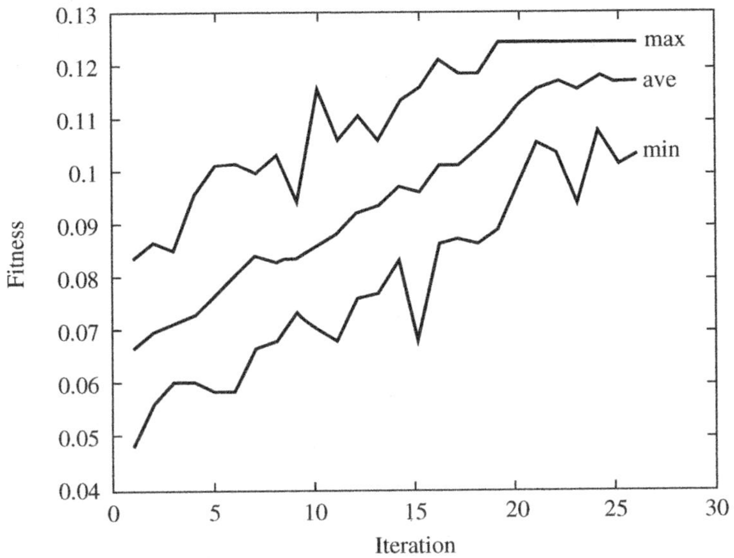

Figure 10.7 The SGA iteration result.

under-voltage violation. Test shows that the MSGA method is robust in algorithm and gives good results which include the global minimum as a solution.

The SGA needs a higher cpu time compared with an analytical optimization method. However, the SGA is flexible, robust, and easy for modification. There are no need of assumptions for linearity, convexity, and so on. As it is shown that the method can be easily combined with other methods. The heuristic experience can be added on without difficulty. With the help of high speed computers, using an efficient optimization method for operation subproblem, and making use of the parallel nature of the SGA, the MSGA promises as a useful tool for planning problems.

10.11 Conclusion

This chapter introduced an IGA for a long-term least-cost GEP problem. The IGA includes several improvements such as the incorporation of an AIP scheme, a stochastic crossover technique, elitism, and scaled fitness function. The IGA has been successfully applied to long-term GEP problems. It provided better solutions than the conventional SGA. Moreover, by incorporating all the improvements, it was found to be robust in providing quasi-optimums within a reasonable computation time and yield better solutions compared to the TCDP employed in WASP.

Contrary to the DP, computation time of the proposed IGA is linearly proportional to the number of stages. The developed IGA method can simultaneously overcome the "curse of dimensionality" and a local optimum trap inherent in GEP problems. Therefore, the IGA approach can be used as a practical planning tool for a real-system scale long-term GEP.

A synthetic method of reactive power planning is presented. Different from the conventional SGA, which mainly uses the objective function for its fitness evaluation, the approach presented in this chapter, MSGA, makes use of not only the objective function but also the dual variable information. The SGA is a random search algorithm and useful in finding the global optimal solution. The new formulation of the Benders method for investment-operation decomposition improves the robustness of the random algorithm. Test shows that the MSGA method is robust in algorithm and gives good results which include the global minimum as a solution.

The SGA needs a higher cpu time compared with an analytical optimization method. However, the SGA is flexible, robust, and easy for modification. There is no need of assumptions for linearity, convexity, and so on. As it is shown that the method can be easily combined with other methods. The heuristic experience can be added on without difficulty. With the help of high-speed computers, using an efficient optimization method for operation subproblem, and making use of the parallel nature of the SGA, the MSGA promises as a useful tool for planning problems.

References

1 Wang, X. and McDonald, J.R. (1944). *Modern Power System Planning. McGraw-Hill.*

2 Sullivan, R.L. (1977). *Power System Planning. McGraw-Hill.*

3 Schlabbach, J., Rofalski, K.-H., and Schlabbach, J. (2008). *Power System Engineering: Planning, Design, and Operation of Power Systems and Equipment. Wiley* [Structures, substations, equipment design].

4 Elkarmi, F., Shikhah, N.A., and Elkarmi, F. (2012). *Power System Planning Technologies and Applications. IGI Global.* [Technical planning at micro level].

5 Sepasian, S.H. and Sadegh, M. (2011). *Electric Power System Planning.* Springer.

6 Ankita and Sharma (2015). *Power System Planning. Genius Publications.* [finite element methods for structures].

7 M. C. Caramanis, F.C. Schweppe, and R.D. Tabors (August 1982). Electric Generation Expansion Analysis System. EPRI Report EL-2561, Vol. 1 and 2.

8 Masse, P. and Gilbrat, R. (1957). Application of linear programming to investments in the electric power industry. *Management Science* 3 (2): 149–166.

9 Bloom, J.A. (1982). Long-range generation planning using decomposition and probabilistic simulation. *IEEE Transactions on Power Apparatus and Systems* PAS-101 (4): 797–801.

10 Park, Y.M., Lee, K.Y., and Youn, L.T.O. (1985). New analytical approach for long-term generation expansion planning based maximum principle and Gaussian distribution function. *IEEE Transactions on Power Apparatus and Systems* 104: 390–397.

11 Jenkins, R.T. and Joy, D.S. (1974). Wien automatic system planning package (WASP) - an electric utility optimal generation expansion planning computer code. *Oak Ridge National Laboratory*, Oak Ridge, Tennessee.

12 David, A.K. and Zhao, R. (1989). Integrating expert systems with dynamic programming in generation expansion planning. *IEEE Transactions on Power Systems* 4 (3): 1095–1101.

13 Fukuyama, Y. and Chiang, H. (1996). A parallel genetic algorithm for generation expansion planning. *IEEE Transactions on Power Systems* 11 (2): 955–961.

Probabilistic Power System Expansion Planning with Renewable Energy Resources and Energy Storage Systems, First Edition. Jaeseok Choi and Kwang Y. Lee.

Published 2022 by John Wiley & Sons, Inc.

14 Park, J.-B., Park, Y.-M., Won, J.-R., and Lee, K.Y. (2000). An improved genetic algorithm for generation expansion planning. *IEEE Transactions on Power Systems* 15 (3): 916–922.

15 Billinton, R. (1970). *Power System Reliability Evaluation*. New York: Gordon and Breach.

16 Billinton, R., Ringlee, R.J., and Wood, A.T. (1973). *Power system Reliability Calculations*. MIT.

17 Finger, S. (February 1979). Electric power system production costing and reliability analysis including hydroelectric, storage and time-dependent power plant. *Technical Report 79-006*, MIT Energy Lab.

18 Chung, T.S., Li, Y.Z., and Wang, Z.Y. (2004). Optimal generation expansion planning via improved genetic algorithm approach. *Electrical Power and Energy Systems* 26: 655–659.

19 Sage, A.P. (1968). *Optimum Systems Control*. Englewood Cliffs, NJ: Prentice-Hall.

20 J. B. Rosen(1960). Nonlinear programming. Part I. Linear constraints. *The Japan Society for Industrial and Applied Mathematics* 3: 181–217.

21 Booth, R.R. (1972). Optimal Generation Planning Considering Uncertainty. *IEEE Transactions on Power Systems* PAS-91: 70–77.

22 Baleriaux, H. (1967). Simulation de 1'exploitation d'unpare de machines thermiques de production d'electricite couple a des stations de pompage. *Revue E(edition S.R.B.E.), S.A. EBES, Brussels, Belgium* 5 (7): 1–24.

23 Youn, L.T.O. (1984). Optimal generation expansion planning. Ph.D. thesis, University of Houston, Houston, TX, pp. 22–30.

24 Brian Manhire (May 1980). Probabilistic simulation of multiple energy storage devices for production cost calculations. EPRI Report, EA-1411, Vol. 1–2, TSA 78-804, pp. 4–2, 4–28, and 4–33.

25 Noonan, F., and Giglio, R.J. (1974). A mathematical programming model for long-range planning of electric power genration. *Presented at the ORSA/TIMS Puerto Rico Meeting*, pp. 24–29.

26 Meslier, F. (1975). Simulation of the optimal management of a thermal and hydraulic generation system (GRETA model). *IEEE Summer Meeting*: 55–60.

27 Synthetic electric utility systems for evaluating advanced technologies. EPRJ Report No. EM-285, Section 4, February 1977.

28 Choi, J., Tran, T.T., Kwon, J. et al. (2008). Nodal probabilistic production cost simulation considering transmission system unavailability. *IET Generation, Transmission and Distribution* 2 (1): 34–42.

29 Waterworth, D.A., Arnold, C.P., and Watson, N.R. (1998). Reliability assessment technique for HVdc systems. *IPENZ Transactions* 25 (1).

30 Swaitti, O. (2007). *Assessing the Impacts of Increasing Penetration of HVDC Lines on Power System Reliability. KTH Royal Institute of Technology.*

31 Hu, P., Karki, R., and Billinton, R. (2009). Reliability evaluation of generating systems containing wind power and energy storage. *Journal of IET Generation, Transmission & Distribution* 3 (8): 783–791.

32 Gao, Z.Y., and Wang, P. (2010). Reliability evaluation of power systems with WTGs and energy storage. *IPEC 2010 Conference Proceedings*, Singapore, pp. 654–659.

33 Borges, C.L.T. and Cantarino, E. (2011). Microgrids reliability evaluation with renewable distributed generation and storage system. *The 18th IFAC World Congress*, Milano, Italy, pp. 11695–11700.

34 Billinton, R. and Li, W.Y. (1994). *Reliability Assessment of Electric Power Systems Using Monte Carlo Methods. Plenum Press.*

35 Choi, J., Mount, T., and Thomas, R. (2007). Transmission expansion planning using contingency criteria. *IEEE Transactions on Plasma Science* 22 (4): 2249–2261.

36 Karki, R., Hu, P., and Billinton, R. (2006). A simplified wind power generation model for reliability evaluation. *IEEE Transactions on Energy Conversion* 21 (2): 533–540.

37 Weisser, D. (2003). A wind energy analysis of grenada: an estimation using the 'Weibull' density function. *Renewable Energy* 28 (11): 1803–1812.

38 Oh, U.J., Lee, Y.C., Lim, J.T. et al. (2015). Reliability evaluation with wind turbine generators and an energy storage system for the Jeju island power system. *Journal of KIEE* 64 (1): 1–7.

39 Oh, U.J., Lee, Y.C., Choi, J.S. et al. (2016). Development of reliability contribution function of power system including wind turbine generators combined with battery energy storage system. *Journal of KIEE* 65 (3): 371–381.

40 California ISO – Fast Facts. (2016). http://www.caiso.com/Documents/FlexibleResourcesHelpRenewables_FastFacts.pdf#search=California%20ISO%20%E2%80%93%20Fast%20Facts (accessed 11 November 2011).

41 Denholm, P., O'Connell, M., Brinkman, G., and Jorgenson, J. (2015). Overgeneration from solar energy in California: a field guide to the duck chart. *NREL*: 1–46.

42 Talat, L.A. (2012). Bulk electric system reliability evaluation with wind turbine and energy storage. *Journal of Engineering Sciences, Assiut University* 40 (3): 885–898.

43 Choi, J.S., Park, J.J., Cho, K.H., Oh, T.G., and Shahidehpour, M. (2010). Probabilistic reliability evaluation of composite power systems including wind turbine generators. *PMAPS 2010*, Grand Copthorne Waterfront Hotel, Singapore, pp. 802–807.

44 Seguro, J.V. and Lambert, T.W. (2000). Modern estimation of the parameters of the weibull wind speed distribution for wind energy analysis. *Journal of Wind Engineering and Industrial Aerodynamics* 85 (1): 75–84.

45 Cho, K.H., Park, J.J., Oh, T.G. et al. (2011). Probabilistic reliability evaluation of composite power systems including wind turbine generators. *Journal of International Council on Electrical Engineering* 1 (3): 274–280.

46 Cho, K.H., Park, J.J., and Choi, J.S. (2012). Probabilistic reliability based grid expansion planning of power system including wind turbine Generators. *Journal of Electrical Engineering and Technology* 7 (5): 698–704.

47 Park, Y.G., Kim, C.W., and Park, J.B. (2016). MILP-based dynamic efficiency scheduling model of battery energy storage systems. *Journal of Electrical Engineering and Technology* 11 (5): 1063–1069.

48 Choi, J.S., Lim, J.T., and Lee, K.Y. (2013). DSM considered probabilistic reliability evaluation and an information system for power systems including wind turbine generators. *IEEE Transactions on Smart Grid* 4 (1): 425–432.

49 Do, D.P.N., Lee, Y.C., and Choi, J.S. (2016). Hourly average wind speed simulation and forecast based on ARMA model in Jeju island, Korea. *Journal of Electrical Engineering and Technology* 11 (6): 1548–1555.

50 Hwang, M., Chun, Y.H., Park, J.W., and Kang, Y.C. (2015). Dynamic droop-based inertial control of a wind power plant. *Journal of Electrical Engineering and Technology* 10 (3): 1363–1369.

51 Goldberg, D.E. (1989). *Genetic Algorithms in Search, Optimisation and Machine Learning*. Massachusetts: Addison-Wesley Publishing Company Inc.

52 Walters, D.C. and Sheble, G.B. (1993). Genetic algorithm solution of economic dispatch with valve point loading. *IEEE Transactions on Power Systems* 8 (3): 1325–1332.

53 Chen, P.H. and Chang, H.C. (1995). Large-scale economic dispatch by genetic algorithm. *IEEE Transactions on Power Systems* 10 (4): 1919–1926.

54 Dasgupta, D. and McGregor, D.R. (1994). Thermal unit commitment using genetic algorithms. *IEE Proceedings of Generation, Transmission and Distribution* 141 (5): 459–465.

55 Sheble, G.B., Maifeld, T.T., Brittig, K., and Fahd, G. (1996). Unit commitment by genetic algorithm with penalty methods and a comparison of Lagrangian search and genetic algorithm-economic dispatch algorithm. *International Journal of Electrical Power & Energy Systems* 18 (6): 339–346.

56 Iba, K. (1994). Reactive power optimization by genetic algorithm. *IEEE Transactions on Power Systems* 9 (2): 685–692.

57 Lee, K.Y., Bai, X., and Park, Y.M. (1995). Optimization method for reactive power planning using a genetic algorithm. *IEEE Transactions on Power Systems* 10 (4): 1843–1850.

58 Lee, K.Y. and Yang, F.F. (1998). Optimal reactive power planning using evolutionary algorithms: a comparative study for evolutionary programming, evolutionary strategy, genetic algorithm, and linear programming. *IEEE Transactions on Power Systems* 13 (1): 101–108.

59 Dimeo, R. and Lee, K.Y. (1995). Boiler-Turbine control system design using a genetic algorithm. *IEEE Transactions on Energy Conversion* 10 (4): 752–759.

60 Zhao, Y., Edwards, R.M., and Lee, K.Y. (1997). Hybrid feedforward and feedback controller design for nuclear steam generators over wide range operation using genetic algorithm. *IEEE Transactions on Energy Conversion* 12 (1): 100–106.

61 Nakamura, S. (1984). A review of electric production simulation and capacity expansion planning programs. *Energy Research* 8: 231–240.

62 David, K. and Zhao, R. (1991). An expert system with fuzzy sets for optimal planning. *IEEE Transactions on Power Systems* 6 (1): 59–65.

63 Park, Y.M., Park, J.B., and Won, J.R. (1996). A genetic algorithms approach for generation expansion planning optimization. *Proceedings of the IFAC Symposium on Power Systems and Power Plant Control,* Pergamon, UK, pp. 257–262.

64 Billington, R. and Sachdev, S.S. (1973). Optimum network VAR planning by nonlinear programming. *IEEE Transactions on Power Apparatus and Systems* PAS-92: 1217–1225.

65 Hegdt, G.T. and Grady, W.M. (1983). Optimal var siting using linear load flow formulation. *IEEE Transactions on Power Apparatus and Systems* PAS-102 (5): 1214–1222.

66 Aoki, K., Fan, M., and Nishikori, A. (1988). Optimal var planning by approximation method for recursive mixed integer linear planning. *IEEE Transactions on Power Systems* PWRS-3 (4): 1741–1747.

67 Lee, K.Y., Ortiz, J.L., Park, Y.M., and Pond, L.G. (1986). An optimization technique for reactive power planning of sub-transmission network under normal operation. *IEEE Transactions on Power Systems* PWRS-1: 153–159.

68 Mangoli, M.K., Lee, K.Y., and Park, Y.M. (1993). Optimal long-term reactive power planning using decomposition technique. *Electric Power Systems Research*: 41–52.

69 Nadira, R., Lebow, W., and Usoro, P. (1987). A decomposition approach to preventive planning of reactive volt-ampere (VAR) source expansion'. *IEEE Transactions on Power Systems* PWRS-2 (1).

70 Hsiao, Y.-T., Liu, C.-C., Chiang, H.-D., and Chen, Y.-L. (1993). A new approach for optimal VAR sources planning in large scale electric power systems. *IEEE Transactions on Power Systems* 8 (3): 988–996.

71 Reeves, R.C. (1993). *Modern Heuristic Techniques for Combinatorial Problems", An Imprint of John Wiley & Son, Inc.* New York: Halsted Press.

72 Miranda, V., Ranito, J.V., and Proenca, L.M. (1994). Genetic algorithms in optimal multistage distribution network planning. *IEEE PES Winter Meeting*, WM 229-5-PWRS.

73 Mangoli, M.K., Lee, K.Y., and Park, Y.M. (1993). Optimal real and reactive power control using linear programming. *Electric Power Systems Research*: 1–10.

74 Geoffrion, M. (1972). Generalized Benders decomposition. *Journal of Optimization Theory and Applications* 10 (4): 237–261.

75 Lee, K.Y., Park, Y.M., and Ortiz, J.L. (1985). United approach to optimal real and reactive power dispatch. *IEEE Transactions on Power Apparatus and Systems* PAS-104: 1147–1153.

76 Jenkins, R.T., and Vorce, T.C. (1977). Use of Fourier series in the power system probabilistic simulation. *Proceedings of the 2nd WASP Conference*, pp. 35–44.

77 Anderson, D. (1972). Model for determining least-cost investments in electricity supply. *Bell Journal of Economics and Management Science* 3: 267–299.

78 Adams, N., Belgari, F., Laughton, M.A., and Mitra, G. (1972). Mathematical programming systems in generation, transmission and distribution planning. *Proceedings of PSCC*, No. 1, Vol. 13, Grenoble, France.

79 Bessiere, F. (1966). Le Modele' lnvestissements 85' d' electrocite de Franc. *Proceedings of PSCC*, Stockholm, Sweden.

80 Phillips, D., Jenkin, F.P., Pritchard, J.A.T., and Rybicki, K. (1969). A mathematical model for determining generating plant mix. *Proceedings of PSCC*, Rome, Italy.

81 Evans, R.A. (1973). A quadratic programming algorithm for optical generation planning. *lEE PES Winter Meet*, Paper C 73 112-0, New York.

82 Peschon, J., and Henault, P.H. (1970). Long-term power system expansion planning by dynamic programming and production cost simulation. *Proceedings of IEEE Symposium on Adaptive Processes*, Decision and Control, Paper 12.

83 Oatman, E.N. and Hamant, L.J. (1973). A dynamic approach to generation expansion planning. *IEEE Transactions on Power Apparatus and Systems* PAS-92: 1888–1897.

84 de France, E. (1976). La Nouvelle Version du MNI. *Working Paper M.* 333.

85 de France, E. (1976). Une Representation Simplifies de Ia Gestion Annulle de Moyens de Production. *Etudes Economique Generales Working Paper M.* 334.

86 Rau, N.S., Toy, P., and Schenk, K.F. (1980). Expected energy production costs by the method of moments. *IEEE Transactions on Power Apparatus and Systems* PAS-99 (5): 1908–1915.

87 Stremel, J.P. (1981). Sensitivity study of the cumulant method of calculating generation system reliability. *IEEE Transactions on Power Apparatus and Systems* PAS-100 (2): 771–778.

88 Duran, H. (1984). A recursive approach to the cumulant method of calculating reliability and production cost. *Presented at the IEEE/PES 1984 Summer Meeting*, Paper No. 84SM 547-6, Seattle, WA, 15–20 July 1984.

89 Lee, K. Y., and Youn, L.T.O. (1983). Optimal long term generation investment planning. *Proc. l983 IEEE Region 5 Conference*, pp. l08–ll3.

90 Papoulis, A. (1979). *Probability, Random Variables, and Stochastic Processes*, 189–199. McGraw-Hill Kogakusha, LTD.

91 Happ, H.H. (1977). Optimal power dispatch: a comprehensive survey. *IEEE Transactions on Power Apparatus and Systems* PAS-96: 841–854.

92 Alsac, O., Bright, J., Prais, M., and Stott, B. (1990). Further development in LP-based optimal power flow. *IEEE Transactions on Power Systems* 5 (3): 697–711.

93 Hobson, E. (1980). Network constrained reactive power control using linear programming. *IEEE Transactions on Power Apparatus and Systems* PAS-99 (4): 1040–1047.

94 Fernandes, R., Lange, F., Burchettr, R. et al. (1983). Large scale reactive power planning'. *IEEE Transactions on Power Apparatus and Systems* PAS-102 (5): 1083–1088.

95 Happ, H.H. and Wirgau, K.A. (1978). Static and dynamic var compensation in system planning. *IEEE Transactions on Power Apparatus and Systems* PAS-97 (5): 1564–1578.

96 Hughes, A., Jee, G., Hsiang, P. et al. (1981). Optimal power planning. *IEEE Transactions on Power Apparatus and Systems* PAS-100: 2189–2196.

97 Shoults, R.R. and Chen, M.S. (1976). Reactive power control by least square minimization. *IEEE Transactions on Power Apparatus and Systems* PAS-95: 397–405.

98 Lebow, W.M., Mehra, R.K., Nadira, R., Rouhani, R., and Usoro, P.B. (1984). Optimization of reactive volt-ampere (VAR) sources in system planning. EPRI Report, EL-3729, Project 2109-1, Vol. 1.

99 Fatih Birol (2019). *World Energy Outlook 2019, Executive Summary*. International Energy Agency, IEA Publications.

100 Covarrubias, A.J. (1979). Expansion planning for electric power systems", International Atomic Energy Agency – Viennna. *IAEA BULLETIN* 21 (2–3): 55–64.

101 Chuang, A.S., Wu, F., and Varaiya, P. (2001). A game-theoretic model for generation expansion planning: problem formulation and numerical comparisons. *IEEE Transactions on Power Systems* 16 (4): 885–891.

102 IAEA (1984). *Expansion Planning for Electrical Generating Systems: A Guidebook.* Vienna: *International Atomic Energy Agency.*

103 Zhu, J. and Chow, M.-y. (1997). A review of emerging techniques on optimal generation expansion planning. *IEEE Transactions on Power Systems* 12 (4).

104 Das, P.K., Chanda, R.S., and Bhattacharjee, P.K. Combined generation and transmission system expansion planning using implicit enumeration and linear programming technique. *IE(I) Journal -EL* 86: 110–116.

105 Joao, C., Henggeler Antunes, C., Gomesmartins, A., and Almeida, A.T. (1995). A multiple objective linear programming model for power generation expansion planning. *International Journal of Energy Research* 19: 419–432.

106 Park, J., Jeong, S., Choi, J., Yun, Y., Cha, J., and Yoon, Y. (2008). A study on the relationship between probabilistic and deterministic reliabilities in Korea. *PMAPS2008*, Rincón, Puerto Rico.

107 Choi, J., Lee, S., Kang, H.I., Song, K., and Namkung, J.Y. (1994). The construction of the multi-criteria generation mix by the Fuzzy dynamic programming. *Proceedings of International Conference on Neural Information*, Seoul, October 1994, pp. 890–895.

108 Jeong, S., Choi, J., Kim, J., Lee, Y., El-Keib, A.A., and Shahidehpour, M. (2008). Flexible best generation mix for Korea Power System considering CO_2 constraint – vision 2030. *IEEE PES GM 2008*, Pittsburg, PA.

109 Park, J., Wu, L., Choi, J., Cha, J., El-Keib, A.A., and Watada, J. (2009). Fuzzy theory-based best generation mix considering renewable energy generators. *Fuzzy-IEEE ICC Jeju*, Jeju Island, Korea.

110 Fukuyama, Y. and Ueki, Y. (1995). Application of parallel genetic algorithms to generation expansion planning using parallel processors. *Electrical Engineering in Japan* 115 (6): 71–81.

111 Walters, D.C., and Sheble, G.B. (1995). Genetic algorithm to optimal reactive power dispatch including voltage-dependent load models. *IFAC International Symposium on Large Scale Systems*, London, July 1995.

112 Levin, N. and Zahavi, J. (1984). Optimal mix algorithm with existing units. *IEEE Transaction 1984* PAS-103 (5): 954–962.

113 Levin, N. and Zahavi, J. (1985). Optimal mix algorithm with limited-energy plants. *IEEE Transactions 1985* PAS-104 (5): 1131–1139.

114 Dhar, S.B. (1979). Power system long-range decision analysis under fuzzy environment. *IEEE Transactions on Power Apparatus and Systems* PAS-98 (2).

115 Wong, K.P. (1993). Artificial intelligence and neural network application in power systems. *IEE Second International Conference on advances in Power System Control*, Operation and Management, Hong Kong, December 1993.

116 Stremel, J.P., Jenkins, R.T., Babb, R.A., and Bayless, W.D. (1980). Production costing using the cumulant method of representing the equivalent load curve. *IEEE* PAS-99 (5): 1947–1956.

117 Bracewell, R.N. (1984). The fast hartley transform. *Proceedings of the IEEE* 72 (8): 1010–1017.

118 Bracewell, R.N. (1986). *The Hartley Transform. Oxford University Press.*

119 Yasuda, K., Nishiya, K., and Hasegawa, J. (1988). Optimal generation expansion planning with electric energy storage systems. *IECON*, pp. 550–555.

120 Takahashi, K. (1988). Power systems flexibility, principles and means, available methods at the planning stage. *CIGRE SC-37*, Brussels.

121 Arakawa, F. (1989). Energy security assessment quantative analysis for flexible system planning. *CIGRE SC-37*, Florence.

122 Tanabe, R., Yasuda, K., and Yokoyama, R. (1992). Determination of robust generation mix considering uncertainties. *JIEE, Japan* 112-B (4): 331–338.

123 Jenkins, N., Allan, R., Crossley, P. et al. (2000). Embedded generation. *IET Power and Energy Series* 31: 31–38.

124 Divan, D., Luckjiff, G.A., Freeborg, J.W., Brumsickle, W., Bielinski, D., and Grant, T. (2002). I-Grid™: a new paradigm in distribution grid power quality and reliability monitoring. *Proceedings of EPRI PQA North America Conference*, Portland, OR.

125 Dy-Liacco, T.E. (1999). Enabling technologies for operation and real-time control in a liberalized environment. *EPRI Second European Conference*, 1999, pp. 2–4.

126 Billinton, R., and Gan, L. (2000). Wind power modeling and application in generating adequacy assessment. *Proceedings of the 14th Power Systems Computation Conference*, Sevilla, Spain.

127 Billinton, R. and Gao, Y. (2008). Multi-state wind energy conversion system models for adequacy assessment of generating systems incorporating wind energy. *IEEE Transactions on Energy Conversion* 23 (2): 163–170.

128 Park, J., Wu, L., Choi, J. et al. (2009). Probabilistic production cost credit evaluation of wind turbine generators. *IJICIC (International Journal of Innovative Computing, Information and Control)* 5 (11(A)): 3637–3646.

129 Billinton, R., and Huang, D. (2008). Aleatory and epistemic uncertainty considerations in power system reliability evaluation. *PMAPS 2008*, Puerto Rico.

130 Choi, J., Kim, H., Cha, J., and Billinton, R. (2001). Nodal probabilistic congestion and reliability evaluation of a transmission system under deregulated electricity market. *IEEE PES GM2001*, Vancouver, Canada.

131 Choi, J., Billinton, R., and Futuhi-Firuzabed, M. (2005). Development of a nodal effective load model considering of transmission system element unavailabilities. *IEE Proceedings of Generation Transmission Distribution* 152 (1): 79–89.

132 Choi, J., Mount, T., Thomas, R., and Billinton, R. (2006). Probabilistic reliability criterion for planning transmission system expansions. *IET Generation, Transmission and Distribution* 153 (6): 719–727.

133 Liang, W., Park, J., Choi, J., and El-Keib, A.A. (2009). Probabilistic reliability evaluation of power systems including wind turbine generators considering wind speed correlation. *KIEE, Journal of Electrical Engineering & Technology* 4 (4): 514–519.

134 Wu, L., Park, J., Choi, J., El-Keib, A.A., Shahidehpour, M., and Billinton, R. (2009). Probabilistic reliability evaluation of power systems including wind turbine generators using a simplified multi-state model: a case study. *IEEE PES GM*, Calgary.

135 Park, J., Cho, K., Oh, T., Choi, J., and El-Keib, A.A. (2010). Evaluation of economic impacts of wind turbine and solar cell generators on power system considering CO_2 emission. *PMAPS 2010*, Grand Copthorne Waterfront Hotel, Singapore.

136 Choi, J. (1990). An efficient algorithm for probabilistic operating simulation. Ph.D. thesis, Korea University: Seoul, Korea.

137 Tran, T., El-Keib, A., Thomas, R. et al. (2005). A method for transmission system expansion planning considering probabilistic reliability criteria. *IEEE Transactions on Plasma Science* 20 (3): 1406–1415.

138 Alguacil, N., Motto, A.L., and Conejo, A.J. (2003). Transmission expansion planning: a mixed-integer LP approach. *IEEE Transactions on Power Systems* 18 (3): 1070–1077.

139 Romero, R., Rochay, C., Mantovani, M., and Mantovani, J.R.S. (2007). Evaluation of hybrid models for multistage transmission system planning. *Revista Controle & Automação* 18 (1): 106–114, Janeiro, Fevereiro e Março.

140 Kaltenbach, J.-C., Peschon, J., and Gehrig, E.H. (1970). A mathematical optimization technique for the expansion of electric power transmission systems. *IEEE Transactions on Power Apparatus and Systems* PAS-89 (1): 113–119.

141 Dusonchet, Y.P. and El-Abiad, A. (1973). Transmission planning using discrete dynamic optimizing. *IEEE Transactions on Power Apparatus and Systems* PAS-92 (4): 1358–1371.

142 Shayeghi, H., Mahdavi, M., and Haddadian, H. (2008). DCGA based-transmission network expansion planning considering network adequacy. *International Journal of Information Technology* 4 (4): 299–304.

143 Choi, J., Cha, J., Watada, J., Mount, T.D., and Thomas, R. (2007). Transmission system expansion planning considering outage cost. *ICICIC 2007*, Kumamoto, Kusyu, Japan.

144 Huiling (Michelle) Bao (2007). Application of the jint deterministic-probabilistic (D-P) creiteria to bulk electric system planning. A thesis for mater of science in the Department of Electrical Engineering University of Saskatchewan Saskatoon.

145 Jaeseok, C., El-Keib, A.A., and Tran, T. (2005). A fuzzy branch and bound-based transmission system expansion planning for the highest satisfaction level of the decision maker. *IEEE Transactions on Plasma Science* 20 (1): 479–484.

146 Tran, T., Kwon, J., Choi, J., El-Keib, A.A., and Watada, J. (2007). Application of fuzzy set theory to transmission system expansion planning. *ISME2007*, Waseda University, Kitakyushu, Japan.

147 Choi, J., Thomas, R., Wang, Z., Rahim, A., El-Keib, A., and Billinton, R. (2005). A study on probabilistic optimal reliability criterion determination in composite power system expansion planning. *IEEE PES GM2005*, San Francisco, CA.

148 Park, J., Choi, J., Jung, Y., Yun, Y., and Cha, J. (2009). Determination of a deterministic reliability criterion for composite power system expansion planning. *IEEE PES GM2009*, Calgary, AB.

149 Wang, H., and Redfern, M.A. (2010). The advantages and disadvantages of using HVDC to interconnect AC networks. *UPEC2010*.

150 Tjernberg, L.B. (2008). Introduction to HVDC technology for reliable electrical power systems. *IEEE Xplore*.

151 Li, W. (2006). Expected energy not served (EENS) study for vancouver island transmission reinforcement project. *Report-BCTC*.

152 Wang, T. and Guan, L. (2014). Improvements on the HVDC reliability assessment index system considering preventive maintenance. *Eksploatacja I Niezawodnosc – Maintenance and Reliability* 16 (4): 629–636.

153 HVDC: technology for energy efficiency and grid reliability. *ABB*. https://studylib.net/doc/18268252/hvdc--technology-for-energy-efficiency-and-grid-reliability.

154 Brand, L., de Silva, R., Bebbington, E., and Chilukuri, K. (2014). Grid west project HVDC technology review. *PSC*.

155 Quaddus, M.A. and Goh, T.N. (1985). Electric power generation expansion: planning with multiple objectives. *Applied Energy* 19 (4): 301–319.

156 Zadeh, L.A. (1979). Fuzzy sets and information granularity, memorandum No. UCB/ERL M79/45, Electronics Research Laboratory, University of California, Berkeley, CA, 19 July 1979.

157 Lee, K.Y. and El-Sharkawi, M.A. (eds.) (2008). *Modern Heuristic Optimization Techniques: Theory amd Applications to Power Systems*. Wiley-IEEE Press.

158 Bellman, R. and Programming, D. (1957). *A Rand Corporation Research Study*. Princeton, NJ: Princeton University Press. ISBN: 0-691-07951-X.

159 Bellman, R.E. and Zadeh, L.A. (1970). Decision-making in a fuzzy environment. *Management Science* 17 (4): B-141–B-164.

Part II

Transmission System Expansion Planning

11

Transmission Expansion Planning Problem

This chapter addresses the long-term transmission expansion planning. The transmission expansion planning algorithm is based on the linear power flow and failure probabilities for the operation cost model and the maximum principle for the long-term investment model. The optimization problem consists of the long-term problem to determine the annual investment and the yearly problem to determine the optimal operation. The long-term problem is formulated as a Hamiltonian minimization problem and the yearly problem is formulated as a linear programming problem [1].

11.1 Introduction

The power system expansion planning problem is concerned with obtaining expansion plans which dictate what new generation and transmission facilities to add, when to add, and where. The objective is to select the most economical and reliable expansion plans in order to meet future power demand at minimum cost and maximum reliability over a long period of time, usually 5–30 years, subject to a multitude of technical, economical, environmental, legal, and political constraints. The practical planning problem is an extremely difficult and complex design problem. Costs and constraints cannot easily be quantified and are difficult to predict. Since planning decisions involve considerable investment and operating costs and commit the utility company to at least 10 years into the future, it is important to have flexible decision strategies for adding the right equipment, at the right time, and in the right location so as to achieve the above objective in the presence of the uncertainties.

At present, most power system expansion planning methods are based on a sequence of simulation studies such as load flow, stability, reliability, and economic studies. Clearly, this approach requires a large number of simulation runs

Probabilistic Power System Expansion Planning with Renewable Energy Resources and Energy Storage Systems, First Edition. Jaeseok Choi and Kwang Y. Lee.

Published 2022 by John Wiley & Sons, Inc.

and a combination of experience and judgment by the system planner to determine a satisfactory (not necessarily optimum) plan. Since the cost of power system equipment is high and the operational and reliability requirements are severe, most of the current effort is concentrated on more accurate mathematical models and efficient computational analysis methods. Because of the size and complexity of the problem, there is no one method which can solve the whole expansion problem. Therefore, the problem is normally decomposed into two major long-range expansion problems, namely optimal generation expansion planning and optimal transmission expansion planning.

The first problem of generation expansion planning was the thrust of EPRI RP1529-Electric Generation Expansion Analysis System (EGEAS) [2]. Independently, this problem has also been studied by the authors resulting in an efficient optimal long-term generation expansion methodology [3–5]. The method has reduced the computational requirement considerably by developing an accurate analytical production cost model based on the probabilistic assumption on the load and plant outages.

Following the similar philosophy of the probabilistic model and with the goal of reducing computational requirement, this book also covers the second problem, namely the optimal transmission expansion planning problem. In the first problem of generation expansion, only the overall volume and the type (nuclear, coal, oil, gas turbine, etc.) of generation are determined. The second problem is to distribute each type of generation within the planning area and determine the optimal location, size, and required transmission network with minimum cost. Toward this goal, the authors have developed a methodology for optimal reactive power planning and operation [6, 7].

11.2 Long-Term Transmission Expansion Planning

The long-range transmission network planning problem seeks a series of economical and reliable network expansion plans which will accommodate the future generation and load growth over the planning period. The basic objective is to find the type, location, and timing of the network additions (or changes) at minimum cost subject to reliability, environmental, legal, and political constraints. Since many factors are difficult to quantify and predict, this problem is extremely difficult to formulate mathematically.

Currently, the transmission network expansion optimization problem objectives have concentrated on seeking a series of economical and reliable transmission expansion plans for a given future generation and load profiles, existing transmission network, and future right-of-way constraints. The goal is to obtain new transmission networks which minimize the cost of network changes under the right-of-way constraints and the reliability constraints.

The simplest power system model which recognizes the electrical characteristics is defined by the d.c. load flow model [8, 9].

The design parameters for this model are the network elements in terms of susceptance, B, per line or transformer. It should be noted, however, that this gives a nonlinear model since the power flows throughout the network are nonlinear functions of B.

In the planning problem, the system variables are the generation capacities, P_G, transmission network element (line or transformer) impedances, Z, (or susceptance, B, and the forecast load demand profile P_l). The design parameters are B for transmission planning while the other system variables are assumed to be specified or their expected values are known. There are a number of constraints imposed on the design parameters:

1) The design parameters are discrete variables since the transmission elements B can only be added in discrete steps.
2) The design parameters are a function of time over the planning period. Since, in general, the planning decisions are only made a discrete-time steps, the design parameters are also discrete-time variables, i.e. $B(t)$, where t denotes the time index.
3) The design parameters are constrained as to the maximum number of lines per right-of-way.
4) The design parameters B are stochastic variables because of the uncertainty in the load, operating conditions, and the state of the system, such as outages, fuel availability, etc.

The cost $C(B)$, which is minimized, consists of capital investment cost $C^{'}$, operating costs $C^{''}$, and sometimes shortage costs $C^{'''}$.

There are various investment cost models, $C^{'}(B)$, available but the one that recognizes the cost per mile plus terminal cost is the most desirable. One planning method [10] uses a cost-capacity model per right-of-way, which takes into account the right-of-way width and the type of construction used. It indicates that the investment cost of line additions does depend on other decisions of online additions if these lines use the same right-of-way. Although nonlinear cost-susceptance models have been used, there seem to be little reported on them. With regard to operating costs, $C^{''}(B)$, they are usually neglected in the transmission expansion problem. In order to include them, one needs a model as to line loading, cost of maintenance, etc.

There are several planning algorithms available for the solution of the above problem, each based on a different interpretation of the system model and the choice of the design objective. None of these algorithms contains all the above model requirements. The linear programing approach uses a linear cost model and the transportation-type model [11–13]. The nonlinear model uses either nonlinear or linear cost model, but the system model is the d.c. load flow model [14, 15].

Most of the work done has concentrated on obtaining efficient algorithms which can expand large networks for the static case, i.e. for a particular planning year and for a given generation load profile (usually for peak demand). There are several time-dependent planning algorithms such as dynamic programming, but they have been applied to small problems and usually require prefiltering to reduce the number of possible network configurations [16–18].

As mentioned earlier, the expansion planning problem faces the uncertainties of future load, transmission line outages, and generation failures. A statistical planning method was developed which guarantees a specified level of probability of supplying the uncertain bus loads [18]. The approach, however, does not consider the uncertainties of transmission lines probabilistically. Its use was limited to static planning and the time-dependent planning was not included.

Another important transmission planning formulation is the so-called Mexico model developed by the EDG (Electricite de France) in France. It also uses the d.c. load flow method and minimizes the sum of the fuel cost and the failure cost. The model, however, has a disadvantage that it does not handle investment from a long-term perspective. The linear programming formulation of the model requires a great deal of memory and computational time. Moreover, the investment problem is treated as a maximal flow problem that does not give accurate line flows.

11.3 Yearly Transmission Expansion Planning

The optimal long-term transmission expansion planning problem is divided into two subtasks: the short-term, or yearly, transmission planning problem, and the long-term transmission planning problem. The yearly transmission planning problem is to determine an optimal transmission operation plan for a single year. The load forecast for each region, the overall volume of each type of generation facilities to be installed, and a list of candidate plant sites are given. Then the yearly planning problem is to find the optimal distribution of generating units and the optimal operation of the transmission system.

11.3.1 Power Flow Model

Although a.c. power flow gives accurate operation cost, d.c. power flow is viewed as sufficient for planning purpose because of its simplicity and much less computation time. Consider a general representation of a.c. power flow given by the following equations:

$$P_i = V_i \sum_{k=1}^{NB} V_k [G_{ik} \cos(\theta_i - \theta_k) + B_{ik} \sin(\theta_i - \theta_k)] \tag{11.1}$$

$$Q_i = V_i \sum_{k=1}^{NB} V_k [G_{ik} \sin(\theta_i - \theta_k) - B_{ik} \cos(\theta_i - \theta_k)] \tag{11.2}$$

where

P_i: real power of bus i,
Q_i: reactive power of bus i,
V_i: voltage magnitude of bus i,
θ_i: voltage phase tangle of bus i,
G_{ik}: real part of element (i, k) of bus admittance matrix,
B_{ik}: imaginary part of element (i, k) of bus admittance matrix,
NB: total number of bus.

For d.c. power flow model, we normally make the following assumptions:

i) $V_i \simeq 1$ for all i,
ii) $\sin(\theta_i - \theta_k) \simeq \theta_i - \theta_k, \quad \cos(\theta_i - \theta_k) \simeq 1$,
iii) $G_{ik} = 0$.

Then the a.c. power flow (11.1) is simplified to yield the following d.c. power flow equation:

$$P_i = \sum_{k=1}^{NB} B_{ik}(\theta_i - \theta_k), \quad i = 1, \ldots, NB \tag{11.3}$$

where B_{ik} is the line susceptance between bus i and k. The d.c. power flow equation can be written in a matrix form:

$$P_{bus} = [A][S]\ \theta \tag{11.4}$$

where

P_{bus}: $NB \times 1$ vector of bus real power,
θ: $NB \times 1$ vector of bus voltage angles,
$[A]$: $NB \times NL$ bus incidence matrix,
$[S]$: $NL \times NB$ line incidence matrix,
NL: total number of lines.

The line incidence matrix is defined by

$$[S] = \left[S_1^T\ S_2^T\ \cdots\ S_{NL}^T\right]^T \tag{11.5}$$

where

$$S_l = b_l[0 \;\cdots\; \overset{i}{0}\; 1\; 0 \;\cdots 0\; \overset{k}{-1}\; 0 \;\cdots] \tag{11.6}$$

which is a $1 \times NB$ row vector, and b_l is the lth line susceptance or primitive admittance b_l between bus i and k.

Using the line incidence matrix, the line flow can be easily obtained as

$$F_l = [S]\ \theta \tag{11.7}$$

where P_l is a $NL \times 1$ line flow vector.

Finally, the bus admittance matrix can also be expressed as

$$[Y_{bus}] = [A][S] \tag{11.8}$$

which is an $NL \times NB$ matrix.

11.3.2 Optimal Operation Cost Model

The optimal operation is to minimize the operation cost under the given constraints. The operation cost can be made of fuel cost and failure cost. The failure cost represents supply shortage cost on the customer side due to load shedding resulting from the lack of transmission capability. As in the case of generation planning [5], the failure cost can be represented as a supply-shortage marginal cost, which increases as a function of the unserved power. For the purpose of simplicity, we assume that it is a linear function of the unserved power.

The optimal operation is then formulated as an optimization problem that minimizes the sum of operation cost:

$$\min_{P_G, P_L} \quad P_0 = \beta_0^T P_G^T + \alpha^T \left(P_L - P_L^d\right) \tag{11.9}$$

subject to

$$\underline{P}_G \le P_G \le \overline{P}_G \tag{11.10}$$

$$P_L^d \le P_L \le 0 \tag{11.11}$$

$$-\overline{P}_l \le P_l \le \overline{P}_l \tag{11.12}$$

where

P_G: bus power generation vector, $NB \times 1$,
P_L: bus load vector, $NB \times 1$,
β_0: fuel cost vector, $NB \times 1$,
α: supply-shortage cost vector, $NB \times 1$,

P_l: line flow vector $NL \times 1$,
P_L^d: bus load demand vector, $NB \times 1$,
$\overline{P}, P$: upper tower limits of P.

The above model is the basis for the Mexico model. This mode, however, has two significant drawbacks: (i) it is necessary to invert the bus admittance matrix $[Y_{bus}]$ and (ii) it is difficult to obtain marginal cost, which are very important information in transmission planning.

A new approach eliminates these drawbacks. First, we choose the voltage phase angle θ as variables in optimization rather than the bus power. Secondly, when the load and generation coexist at the same bus, they are combined so that the net power is represented either as a generation or load.

Using these assumptions and (11.4)–(11.7), the optimization problem (11.9)–(11.12) can be transformed into the following equivalent problem:

$$\min_{\theta} \; B = \beta^T [A][S]\theta - \alpha^T P_L^d \tag{11.13}$$

subject to

$$\underline{P}_B \leq [A][S]^{\theta} \leq \overline{P}_B \tag{11.14}$$

$$-\overline{P}_l \leq [S]^{\theta} \leq \overline{P}_l \tag{11.15}$$

where

$$\beta = \beta_0 + \partial \tag{11.16}$$

$$\overline{P}_B = \begin{cases} \overline{P}_G & \text{for generator bus,} \\ 0 & \text{for load bus,} \end{cases} \tag{11.17}$$

$$\underline{P}_B = \begin{cases} \underline{P}_G & \text{for generator bus,} \\ P_L^d & \text{for load bus,} \end{cases} \tag{11.18}$$

The above formulation has many advantages. Since the variable for optimization is the voltage angle θ, the optimization problem becomes a linear programming problem. The initial estimate for the variable is trivial, i.e. $\theta = 0$ and thus the number of iterations is reduced. There is no need to invert the bus admittance matrix $[Y_{bus}]$ which otherwise is a major computational burden. The sparsity of matrices $[A]$ and $[S]$ are preserved and, therefore, a sparsity technique can be used to reduce memory size and computation time. Finally, due to its linear formulation, the marginal costs can be obtained analytically, which is a significant advantage over conventional transmission planning techniques.

Since the operation model (11.13)–(11.15) is in a linear formulation, a linear programming technique can be conveniently used to obtain a stable solution. However, for this problem, we use the Gradient Projection Method which was developed and used efficiently for the generation planning problem [3, 4]. Since there are a large number of constraints, this Gradient Projection Method is particularly useful when combined with the sparsity technique. This will be more significant when we consider the long-term transmission planning problem, where the memory size and computation time are critical.

11.3.3 Probability of Line Failures

The long-term transmission planning problem should not only consider the normal operation but also include contingencies due to changes in the system, e.g. generator failures, line failures, and load uncertainties. The generator failures were already treated in terms of expected power output in the generation planning problem. Similarly, the load uncertainties were also treated in terms of the expected or mean load curves. Since, in our case, there is no need for convolution or deconvolution, the equivalent load duration curves can also be used for simplicity. In transmission planning, therefore, it only remains to quantify line failures in order to obtain the expected value of operation cost.

In general, most of the line failures are single line outages and it is very rare to have multiple line outages. Therefore, it is sufficient to consider only single- and double-line outages without loss of generality.

Let's define the following definitions:

q_l: failure rate of line l,
Prob(0, 0): probability of no line outage,
Prob(l, 0): probability of single line outage of line l,
Prob(l, k): probability of double line outage of lines l and k.

Then, these probabilities can be computed as

$$\text{Prob}(0,0) = \prod_{l=1}^{NL}(1-q_l) \triangleq q_0' \tag{11.19}$$

$$\text{Prob}(l,0) = \frac{q_l}{1-q_l}\prod_{l=1}^{NL}(1-q_l) = \frac{q_l}{1-q_l}q_0' \tag{11.20}$$

$$\text{Prob}(l,k) = \frac{q_l q_k}{(1-q_l)(1-q_k)}\prod_{l=1}^{NL}(1-q_l) = \frac{q_l q_k}{(1-q_l)(1-q_k)}q_0' \tag{11.21}$$

However, the probability of space should satisfy the following condition:

$$1 \triangleq \text{Prob}(0,0) + \sum_{l=1}^{NL} \text{Prob}(l,0) + \sum_{l=1}^{NL} \sum_{k=l+1}^{NL} \text{Prob}(l,k) = Q_T \tag{11.22}$$

Therefore, the probabilities (11.19)–(11.21) are normalized as

$$\text{Prob}(0,0) = Q_0 \tag{11.23}$$

$$\text{Prob}(l,0) = \frac{q_l}{1-q_l} Q_0 \tag{11.24}$$

$$\text{Prob}(l,k) = \frac{q_l q_k}{(1-q_l)(1-q_k)} Q_0 \tag{11.25}$$

where

$$Q_0 = Q_0'/Q_{T'}$$

11.3.4 Expected Operation Cost

Once the failure probabilities (11.23)–(11.25) are available, the expected value of operation cost can be computed. The failure of a network is reflected in the line incidence matrix $[S]$ in the operation model (11.13)–(11.15), and the line incidence matrix $[S]$ is a function of the primitive admittance vector b.

The line primitive admittance vector b changes when a failure occurs; thus, it becomes a random vector $b(\omega)$ where ω denotes a random event. Therefore, the line incidence matrix can be represented by

$$[S] = [S(b)] = [S(\omega)] \tag{11.26}$$

where b is the $NL \times l$ random primitive admittance vector.

The optimal operation cost can also be represented as a random variable

$$f(\omega) \triangleq \min_{\theta} \{B^T[A][S]\theta - a^T P_L^d\} \tag{11.27}$$

such that

$$\underline{P}_B \leq [A][S]^\theta \leq \overline{P}_B \tag{11.28}$$

$$-\overline{P}_l \leq [S]^\theta \leq \overline{P}_l \tag{11.29}$$

where, $f(\omega)$ is the optimal operation cost for an event ω.

When $\text{Prob}(\omega_k)$ denotes the failure probability for an event ω by summing the costs for all possible events multiplied by corresponding probabilities, the *expected operation cost* can be computed as

$$E\{ f(\omega)\} \triangleq \sum_{k=1}^{W} \text{Prob}(\omega_k) f(\omega_k) \tag{11.30}$$

where W is the total number of events.

11.3.5 Annual Expected Operation Cost

The operation pattern can be grouped into daily, weekly, monthly, and yearly operations, each of which requires different load curves. However, these different load curves can be handled conveniently when an effective load duration curve is used. For practical purpose, we choose monthly or seasonal load duration curves since they also contain daily and weekly load curves.

By dividing a monthly load duration curve into a number of time-bands and computing the expected operation cost for each time-band, the total yearly expected operation cost can be computed as

$$F_i \triangleq \sum_{s=1}^{S} n_s \sum_{t=1}^{T} \tau_t \sum_{k=1}^{W} \text{Prob}(\omega_k) f_\Delta[b(\omega_k)] \tag{11.31}$$

where

$\Delta = i, s, t$: index of year, season, and time-band, respectively,

$f_\Delta[b(\omega_k)]$: the optimal operation cost for event ω_k for year i, season s, and time-band t,

τ_t: length of time-band

T: number of time-bands in a load duration curve,

n_s: number of monthly load cycles in season S,

S: number of seasons in a year,

ω: number of events.

The optimal operation cost f_Δ can be computed in the optimization problem (11.27)–(11.29) for each event by using the gradient projection method, then (11.31) gives the gradient projection method and (11.31) gives the yearly expected operation cost, which will be asked to the investment cost for long-term planning problem.

11.4 Long-Term Transmission Planning Problem

The long-term transmission planning problem is to determine an optimal transmission expansion plan over several years of the planning horizon. The data required for the short-term planning problem is also required for the long-term planning problem. Then the long-term planning problem is to determine the optimal distribution of new generating units and the addition of new transmission lines during a planning horizon with minimum cost.

11.4.1 Long-Term Transmission Planning Model

The long-term investment problem in the transmission facility is to minimize investment cost while maximizing its utilization. Thus, the optimal investment model is to minimize the sum of total investment cost and operation cost over the planning horizon. The model will be more realistic if the salvage value is also included in the total cost in order to consider the value of the facility beyond the end of the planning horizon.

Since the life of a transmission line is over 30 years, an added line is to remain in the system longer than a typical planning horizon. Therefore, the investment model should be treated as a *dynamic system*, as in the case of the generation planning problem. This also makes it possible to formulate the problem as an optimal control problem and use Pontryagin's minimum principle to obtain solution easily [3–5].

For simplicity, we first assume that an investment in the transmission line is made at the beginning of each year and it will be operational at the period of the same year, and the equivalent investment cost and operation cost are converted to the present values.

Then the long-term transmission planning problem is to minimize the present value of total annual costs:

$$G_i(y_i, u_i) = C_i(u_i) + F_i(x_i, u_i) \tag{11.32}$$

$$C_i(u_i) = c_i u_i \tag{11.33}$$

where

C_i: present value of capital investment cost for year i,
F_i: present value of operation cost for year i,
c_i: present value of unit capacity cost for year i,
x_i: existing primitive admittance vector Δb for year i,
u_i: added primitive admittance vector Δb for year i.

Additionally, a salvage cost $S(X_I)$ can be added to represent the cost at the terminal year I.

The long-term transmission planning problem is now stated formally as

$$\min_{u_i} \; G = \sum_{i=1}^{I} G_i(x_i, u_i) + S(X_I) \tag{11.34}$$

such that

$$x_{i+1} = x_i + u_i, \quad x_I = \text{specified} \tag{11.35}$$

subject to

$$u_i^{min} \leq u_i \leq u_i^{max} \tag{11.36}$$

$$c_i^T u_i \leq c_i^{max} \tag{11.37}$$

where

$i = 1, 2, \ldots, I$: index of years in the planning horizon.

The problem (11.34)–(11.37) is a typical stochastic optimization problem. In view of Eqs. (11.27)–(11.29) and (11.34)–(11.37), the problem is formulated as a linear problem. This is a very desirable feature for the complex long-term planning problem since any linear programming technique can be used for a guaranteed solution, and, moreover, computation line can be reduced significantly following the following solution technique.

11.4.2 Solution Technique for the Planning Problem

The (11.35) is a dynamic state equation describing the state of installed transmission facilities each year, that is, the sum of existing facilities and newly added facilities is the installed facilities for the next year. Considering the newly added facilities as decision or control variables, the long-term planning problem is now formulated as an "optimal control problem."

As was proposed in the generation planning problem, Pontryagin's minimum principle [3–5] can be used. Then the optimal control is solved by solving the following Hamiltonian minimization problem:

$$\min_{u_i} \; H_i(x_i, u_i, \lambda_{i+1}) \triangleq G_i(x_i, u_i) + \lambda_i^T (x_i + u_i) \tag{11.38}$$

such that

$$x_{i+1} = x_i + u_i, \quad x_I = \text{specified} \tag{11.39}$$

$$\lambda_i^T = \lambda_{i+1}^T + \frac{\partial H_i}{\partial x_i}, \quad \lambda_I^T = \frac{\partial S}{\partial x_I} \tag{11.40}$$

where

λ_i: costate vector for year i, $NL \times 1$,
x_i: state vector for year i, $NL \times 1$,
u_i: control vector for year i, $NL \times 1$.

As in the case of the generation planning problem, the long-term transmission planning problem for a number of years is reduced to the Hamiltonian minimization problem for a single year. This resulted in a significant reduction in the number of decision variables u_i from $NL \times 1$ to NL in a minimization routine, where I is normally 10–30 years.

Note that the Hamiltonian minimization problem (11.38)–(11.40) is in a linear formulation. Therefore, a solution is guaranteed and computation time is much shorter compared to a nonlinear formulation. It is up to the planner to use his own preferred minimization package. In our study, we use the Gradient Projection Method which is developed and proven to be very effective for this type of problem [3, 4].

The costate (11.40) requires computing the marginal value of operation cost, $\partial F_i/\partial x_i$, with respect to the existing facility. Similarly, the Hamiltonian minimization problem (11.38) requires computing the marginal value of investment cost, $\partial C_i/\partial u_i$, with respect to the newly added facility. These marginal values are not only required to find the optimal solution, but also provide a planner with useful technical and marginal information. These can be found in explicit forms from the operation and investment models.

Finally, it will be useful to note one practical problem. The line additions, u_i, are normally in the discrete multiples of some base units rather than continuous variables, and often, the unit size is very large. Thus, the integer programming method can be applied to this problem. However, the integer programming method is very difficult to apply to a large-scale system and is not used in practice.

In order to overcome this problem, the following approximate method is applied. In each iteration, when the optimization results in an investment of a line capacity, more than half of the base unit is invested. On the other hand, when the investment is called for less than half of the base unit, no investment is made.

Although this procedure may not be giving the exact solution that can be obtained by the integer programming method, its error is expected to be small because the procedure is iterated several times.

11.5 Case Study

A hypothetical five-bus transmission system shown in Figure 11.1 is considered for testing the proposed algorithm. The system consists of 2 generating plants, 3 load centers and 11 existing transmission lines (in solid lines). The right-of-way constraint is assumed to allow 8 more transmission lines for future expansion (shown in dotted lines).

The initial line data are shown in Table 11.1 where the existing lines between buses 3 and 4 are combined as an equivalent line 7, and, similarly, between buses 4 and 5 as line 5. The bus power data is given in Table 11.2 where each load bus is

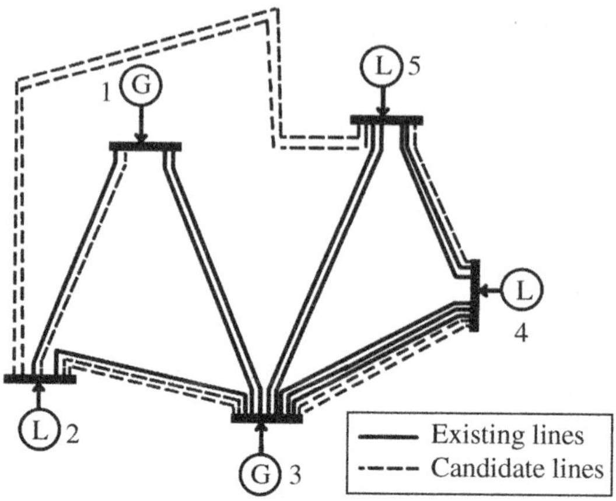

Figure 11.1 Sample system.

Table 11.1 Initial line data.

Line	From bus	To bus	Admittance	Capacitive coeff.	Failure rate
1	1	2	59.30	0.44	0.015
2	1	2	0.00	0.44	0.015
3	1	3	30.96	0.30	0.015
4	1	3	15.38	0.30	0.015
5	4	5	43.86	0.80	0.015
6	4	5	0.00	0.80	0.015
7	4	3	46.19	0.60	0.015
8	4	3	0.00	0.60	0.015
9	4	3	0.00	0.50	0.015
10	2	5	0.00	0.55	0.015
11	2	3	51.60	0.55	0.015
12	2	3	0.00	0.55	0.015
13	2	3	0.00	0.55	0.015
14	5	3	87.72	0.65	0.015
15	5	3	55.54	0.65	0.015
16	2	5	0.00	0.50	0.015

Table 11.2 Bus power data.

[100 MVA Base]

Year	Bus / Time band	1	2	3	4	5
1	1	100.0	−51.5	100.0	−4.0	−51.5
	2	100.0	−45.5	100.0	−3.0	−45.5
	3	100.0	−40.5	100.0	−2.0	−40.5
2	1	100.0	−60.0	100.0	−10.0	−60.0
	2	100.0	−50.0	100.0	−9	−50.0
	3	100.0	−40.0	100.0	−7.0	−40.0
3	1	100.0	−70.0	100.0	−30.0	−70.0
	2	100.0	−60.0	100.0	−20.0	−60.0
	3	100.0	−50.0	100.0	−15.0	−50.0
4	1	100.0	−80.0	150.0	−50.0	−90.0
	2	100.0	−70.0	150.0	−40.0	−80.0
	3	100.0	−60.0	150.0	−40.0	−80.0
5	1	150.0	−100.0	150.0	−80.0	−120.0
	2	150.0	−90.0	150.0	−70.0	−100.0
	3	150.0	−80.0	150.0	−60.0	−110.0
6	1	150.0	−120.0	200.0	−130.0	−150.0
	2	150.0	−100.0	200.0	−110.0	−130.0
	3	150.0	−90.0	200.0	−100.0	−110.0

represented by high, medium, and low demand load levels. The length of time-band for each load level in the yearly load duration curve is shown in Table 11.3. The load is expected to increase for each year as shown in Table 11.2 and generating capacities are also increased in later years. The financial data are given in Tables 11.4 and 11.5 and the data on candidate transmission lines for future expansion are given in Table 11.6.

Table 11.3 Yearly load duration.

(Unit : Hour)

Time band	1	2	3
Duration hour	2000	4000	2760

Table 11.4 Computed fuel and supply-shortage costs.

(Unit : $/kWh)

Coefficients \ Bus	1	2	3	4	5
Fuel cost	0.09	0.0	0.08	0.0	0.0
Supply-shortage cost	0.0	1.827	0.0	1.827	1.827

Table 11.5 Investment cost limits.

(Unit : K$)

Year	1	2	3	4	5
Investment cost	1000.0	1000.0	1500.0	1500.0	1000.0

The results of the optimal investment algorithm for a five-year planning problem are shown in Table 11.7. During the first year, the candidate line 3 is added in double between buses 3 and 4. This is due to its low investment cost and low marginal cost. Similarly, the candidate line 5 is added in double between buses 2 and 3 during the second year. Although both lines 3 and 5 have the same investment costs, line 3 is added first since its marginal cost is lower.

During the third year, the candidate line 4 is added in double between buses 2 and 5 due to its low investment cost, while the candidate line 2 is added during the third and fourth years due to its higher investment cost. Similarly, the candidate line 6 is added between buses 2 and 3 during the fourth and fifth years and the candidate line 8 is added between buses 1 and 2 during the last year.

Table 11.6 Candidate line data.

Candidate line	Line number	Life (year)	Investment cost (K$/℧)	Unit cap. (℧)	Investment limit (℧)
1	6	30	50	15	30
2	8	30	40	15	30
3	9	30	30	15	30
4	10	30	30	10	20
5	12	30	30	15	30
6	13	30	40	15	30
7	16	30	360	15	30
8	2	30	40	15	30

Table 11.7 Optimal investment results.

Candidate line \ Year	1	2	3	4	5
1					
2			15.0	15.0	
3	30.0				
4			20.0		
5		30.0		15.0	15.0
6					
7					
8					15.0

The cost comparison is given in Table 11.8 for the cases of with and without investment. It should be noted that the difference in annual operation cost becomes significant in later years due to the load growth.

The algorithm has demonstrated its effectiveness by allowing investment to take place either to increase the effective line capacities of the existing facilities or to

Table 11.8 Operating and investment costs.

(Unit: K$)

Year	1	2	3	4	5
Operation cost w/o investment	68.47	104.4	307.1	579.7	1618.0
Operation cost with investment	68.47	104.2	267.5	298.0	1353.0
Investment cost	900	900	1400	1200	900

install new lines between buses where there was no line initially (e.g. between buses 2 and 5). The algorithm has converged in 3 iterations for the five-year planning problem and the CPU time was 20 minutes 55 seconds in the Hp-1000 microcomputer. Therefore, it is expected that the computation will be much faster in a large computer even for a larger transmission system.

11.6 Conclusion

An operation cost model is developed using the voltage angle as the optimization variable and an analytical expression for the investment marginal cost is developed. This makes it possible to apply the maximum principle for long-term transmission planning. The sample study has shown the effectiveness of this approach. Some of the conclusions are summarized below:

1) An operation cost model is developed using the voltage angle as the optimization variable. This makes the power flow equation simple and the initial guess for the solution trivial, resulting in reduced computer ion time.
2) The model eliminates the need for inverting the bus admittance matrix; thus, the memory requirement and computation time are reduced.
3) The use of supply-shortage cost eliminates the power balance equation, which will later be satisfied through the optimization process.
4) The investment marginal cost is developed in the form of the sensitivity matrix as a function of the dual values in linear programming formulation. This makes it possible to apply the maximum principle and solve the Hamiltonian optimization problem.

5) The use of the maximum principle has transformed the multi-year planning problem into yearly Hamiltonian optimization problems and resulted in a reduction in memory requirement.
6) The use of line failure probabilities in the operation cost model has eliminated the need for Monte Carlo method and thus of the repetitive computations.

One optimization method, the gradient projection method, was used for both the operation cost optimization and the investment cost optimization.

12

Models and Methodologies

12.1 Introduction

The major premise for transmission expansion planning (TEP) is that demand forecasting and generation expansion schedule for the study period are given. The objective of TEP is to establish an optimal counterplan of electric energy delivery in the economic and electrotechnical sense to meet the demand forecast and the generation expansion planning (GEP). The plan involves the generation site, line capacity, location, and time of transmission line addition. Some aspects to be considered in the TEP are economy in terms of trade-off between investment cost and operation cost, stability of the power system, and reliability and quality of the service. In solving the TEP problem, constraints are fund-raising, right-of-way, electromechanical limits in terms of voltage, conductor size, line overload, short current, stability, and load shedding, etc. However, there are difficulties in handling the TEP problem such as the large-scale problem of the system with various decision variables, the burden of decision making, a great deal of uncertainty since deterministic thinking is not adequate, and unresolved technical problems such as quantifying reliability, stability, etc.

12.2 Transmission System Expansion Planning Problem

A composite power system that includes generation and transmission facilities is shown in Figure 12.1, where TS refers to the transmission system, NG is the number of generators, ${}_k\Phi_0$ is the inverted load duration curve at load point k, and NL is the number of load points. Table 12.1 shows two main kinds of optimal model for transmission system expansion planning problem [19].

Probabilistic Power System Expansion Planning with Renewable Energy Resources and Energy Storage Systems, First Edition. Jaeseok Choi and Kwang Y. Lee.

Published 2022 by John Wiley & Sons, Inc.

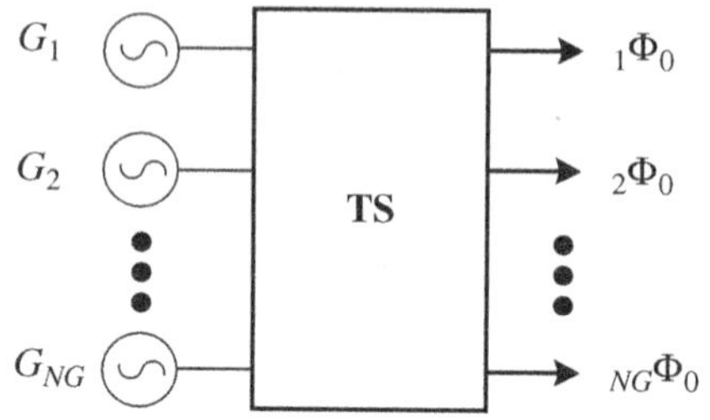

Figure 12.1 A composite power system including the transmission system.

Table 12.1 Two kinds of optimal model for transmission system expansion planning problem.

	Model I	Model II
Objective function	Minimize z = Investment cost + Operating cost	Minimize z = Investment cost + Operating cost + Outage cost
Main constraint	Nodal and system reliability criterion ($LOLE_k^*$ and $LOLE_{sys}^*$)	—
Method for assessing outage cost	—	$\sum_{k\in\Omega_{bus}} (IEAR_k \times EENS_k)$ Where, k = Bus number Ω_{bus} = Bus set
Remarks	Optimal reliability criterion $LOLE_k^*$ and $LOLE_{sys}^*$ are necessary	Program assessing outage cost is necessary

12.3 Cost Evaluation for TEP Considering Electricity Market

The objective of TEP is to establish an optimal counter plan of electric energy delivery in economic and electro-technical sense to meet the demand forecast and the generation expansion planning. The plan involves generation site, line capacity, location and time of transmission line addition. Some aspects to be considered in the TEP are economy in terms of trade-off between investment cost and operation cost, stability of the power system, and reliability and quality of the service.

The prime mover behind the TEP is the profitability of the investment based on economic evaluation. For a given demand forecasting over a long-term planning horizon, the TEP is to provide means of delivering energy established by generation expansion planning (GEP). The delivery of energy is dictated by the electricity market, which involves short-term development scenarios, market pricing, fuel

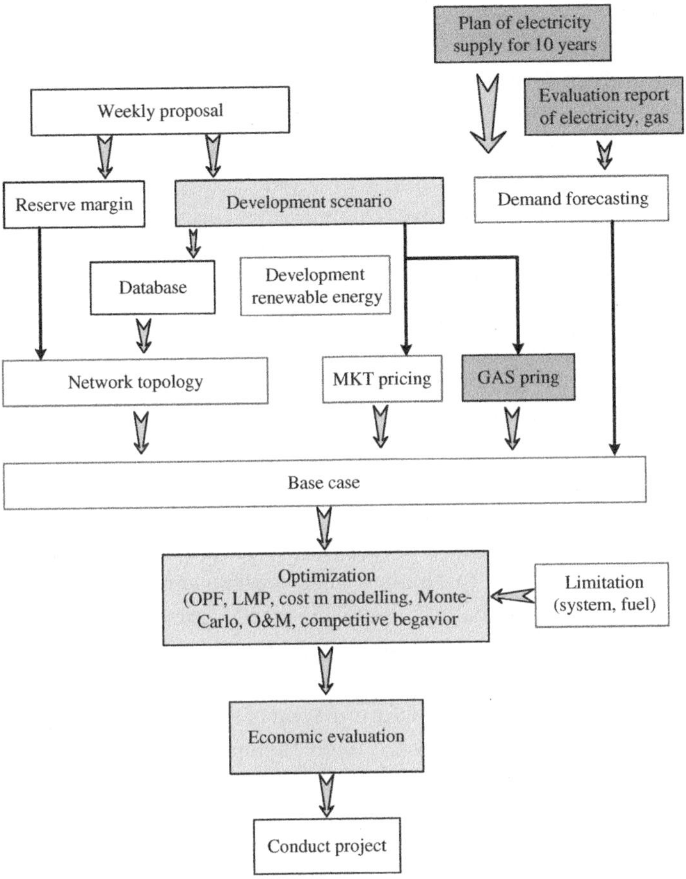

Figure 12.2 Example: economic evaluation process for grid expansion planning considering electricity market.

pricing, optimization, and economic evaluation. An example of the economic evaluation process is shown in Figure 12.2.

12.4 Model Development History for TEP Problem

The long-term transmission expansion problem is a stochastic decision problem that consists of determining the time, the location, and the type of transmission lines to be built.

After Garver's paper that was published in 1970, much research has been done on the field of transmission network expansion planning (TNEP) problem. Some of them were related to the problem-solution method. Some others proposed different approaches for the solution of this problem considering various parameters such as uncertainty in demand, reliability criteria, and economic factors. Also, some of them investigated this problem and GEP together. Recently, different methods such as greedy randomized adaptive search procedure (GRASP), Benders decomposition, a Benders hierarchical decomposition approach (HIPER), branch and bound algorithm, sensitivity analysis, genetic algorithm, simulated annealing and Tabu search have been proposed for the solution of static transmission network expansion planning (STNEP) problem.

This problem arises either in centralized or competitive frameworks. In practice, further simplification is needed and it consists of assuming a given voltage level for both static and dynamic approaches. Notwithstanding, finding a solution tool that guarantees optimality, for either instance of the TEP problem, remains an open issue because of the combinatorial nature of the problem, which renders its solution difficult for medium-sized problems. For this reason, some researchers have applied various heuristic techniques to obtain a good feasible solution. Among those techniques are simulated annealing, neural networks, game theory, and genetic algorithms. The application of mixed-integer programming techniques using Benders decomposition is combined with some heuristics. However, these results hinge on a modification of the natural formulation of the original problem. A more rigorous decomposition scheme that is suitable for Benders decomposition was reported [20].

12.5 General DC Power Flow-Based Formulation of TEP Problem

$$\text{Min } v = \sum_{(i,j)} c_{ij} n_{ij} \tag{12.1}$$

s.t.

$$Sf + g = d \tag{12.2}$$

$$f_{ij} - \gamma_{ij}\left(n_{ij}^{0} + n_{ij}\right)\left(\theta_i - \theta_j\right) = 0 \tag{12.3}$$

$$\left|f_{ij}\right| \le \left(n_{ij}^{0} + n_{ij}\right)\overline{f_{ij}} \tag{12.4}$$

$$0 \le g \le \overline{g}$$
$$0 \le n_{ij} \le \overline{n_{ij}}$$
$$(i,j) \in \Omega$$

where

S: Branch-node incidence matrix.
F: Active power matrix in each corridor.
G: Generation vector.
D: Demand vector.
N: Number of network buses.
θ: Phase angle of each bus.
g_{ij}: Total susceptance of circuits in corridor $i - j$.
n_{ij}^0: Number of initial circuits in corridor $i - j$.
$\overline{n}_{ij}$: Maximum number of constructible circuits in corridor $i - j$.
$\overline{f}_{ij}$: Maximum of transmissible active power through corridor $i - j$ which will have two different rates according to the voltage level of candidate line.
n_{ij}: Number of new circuits in corridor $i - j$.
Ω: Set for all corridors.
c_{ij}: Cost of a circuit added to the right-of-way $i - j$.
n_{ij}: The number of circuits added in the right-of-way $i - j$.
n_{ij}^0: the number of lines in the base case.

Constraint (12.2) represents the conservation of power in each node for Kirchhoff's Current Law (KCL) constraint in the equivalent DC network. Constraint (12.3) represents the Kirchhoff's Voltage Law (KVL). Moreover, these constraints are nonlinear [21].

12.5.1 Linear Programming

We now require that the reliability constraints of (12.5) be satisfied when the branch capacities are changed from γ_k to $\gamma_k + \Delta\gamma_k$ to accommodate the node injections $g_i\ (t + \Delta t)$. First, the outage tests are conducted for the initial configuration (with capacities γ_k) and the new node injections $g_i\ (t + \Delta t)$. We obtain the quantities Ψ_k^* using (12.6) and compute the Jacobian matrix B.

$$\Psi_k^* \le \overline{\overline{\Psi}}_k \quad k = 1, \ldots, r \tag{12.5}$$

$$\Psi_k^* = \max_{m \in [1, \ldots, M]} \left|\Psi_k^{(m)}\right| \quad k = 1, \ldots, r \tag{12.6}$$

where

Ψ_k^* is the maximum voltage phase-angle difference that can ever appear in branch k during the outage tests.

Then we solve the linear programming (LP) problem:

$$\text{minimize } \Delta C = \sum_{m=1}^{r} C_m \Delta\gamma_m$$

Subject to

$$-\overline{\overline{\Psi}}_k \leq \Psi_k^* + \sum_{m=1}^{r} \frac{\partial \varphi_k}{\partial r_m} \Delta\gamma_m \leq \overline{\overline{\Psi}}_k, \quad k = 1, \ldots, r$$

$$\Delta\gamma_m \geq q_m, \quad m = 1, \ldots, r$$

where the unknowns are the $\Delta\gamma_m$. The lower bounds q_m to the $\Delta\gamma_m$ have either zero value (if no decrease of capacity is allowed) or negative values, which can be freely chosen. Once a solution $\Delta\gamma$ has been found by the LP, the power flow and overload computations are repeated for the new capacities $\gamma_m + \Delta\gamma$. Since the higher-order terms have been omitted in (12.7), some of the largest overloads are sometimes not completely eliminated, as they should be if (12.7) were rigorous.

$$\Delta\Psi_i = \sum_{j=1}^{r} \frac{\partial \varphi_i}{\partial \gamma_j}(\gamma)\Delta\gamma_j, \quad i = 1, \ldots, r \tag{12.7}$$

If those remaining violations are very small, they may be ignored. Otherwise, corrections to the $\Delta\gamma_k$ can be determined by a new application of the LP method and so on until reliability is satisfactory. Convergence can be shown experimentally [23]. A simplified flowchart of computational procedure using the LP is given in Figure 12.3.

12.5.2 Dynamic Programming

Dynamic programming (DP) is widely recognized as a very powerful technique for solving variational problems and it is readily applicable to problems discrete in nature. The TEP problem is of this kind since the objective of the optimization is the complete expansion schedule. However, the complete problem cannot be readily solved by DP for two reasons: (i) the tremendous number of decision alternatives; and (ii) the state dependency of the reliability constraints (i.e., adding new capacity to a branch changes the reliability constraints for the remainder of the

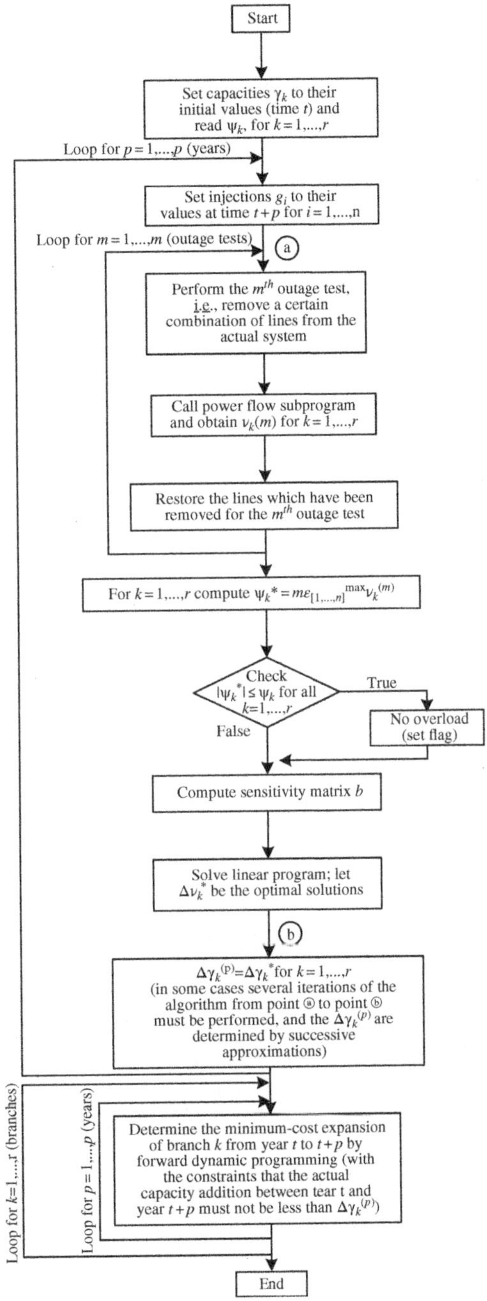

Figure 12.3 Simplified flow chart of computational procedure using linear programming.

system; the lower bounds to branch capacities depend on one another, and their relationship is not simple). There is some hope that these difficulties will eventually be overcome by using recent developments to the theory of DP, but considerable work still remains to be done until that approach can be practical here.

It is, however, possible to use DP in conjunction with the idea of the preceding section for determining a close-to-optimal sequence of investment decisions during a period of P years in a much more accurate manner than with the look-ahead factor method. Given the branch capacities $\Delta\gamma_k$ in year t, we use the LP model to compute the minimum-cost capacity changes required for the direct transition from the node injections in year t to the node injections in year $t+p$, for $p = 1, 2, \ldots, P$. Let $\Delta\gamma_k^{(P)}$ be the solutions of the respective linear programs. The expansion of each branch k during the P-year period is now optimized separately via forward DP with the constraint that the capacity addition in branch k between year t and year $t+p$ must be larger than $\Delta\gamma_k^{(P)}$, $p = 1, 2, \ldots, P$. The objective is the minimization of the present worth of investment expenses during this P-year period. The choice of P is empirically limited to a small number of years, generally less than five or six, for otherwise, the required $\Delta\gamma_k^{(P)}$ become too large for the linearized sensitivity (12.7) to remain valid. A less accurate, but more manageable approach, is used to estimate capacity requirements beyond time $t+P$ and until the end T of the planning period, where T is usually in the order of 20–30 years. Thus, the planning extends over a 20- or 30-year period, as generally recommended. However, the first five or six years are treated with more accuracy than the following ones [22, 24]. A simplified flowchart of computational procedure using the DP is given in Figure 12.4.

12.5.3 Integer Programming (IP)

1) *The Objective Function:* The conventional definition of cost in TEP is to minimize the total construction costs plus the operating costs. This chapter proposes a new definition of the total cost (C^T) that includes the construction cost (CC) associated with investing in new transmission lines, the operation/production cost (OC), and the standby cost (SC) of generators as expressed in (12.8). In the case of a competitive electricity market environment, a grid owner, or independent system operator (ISO), or grid reliability committee will want to consider the problem of

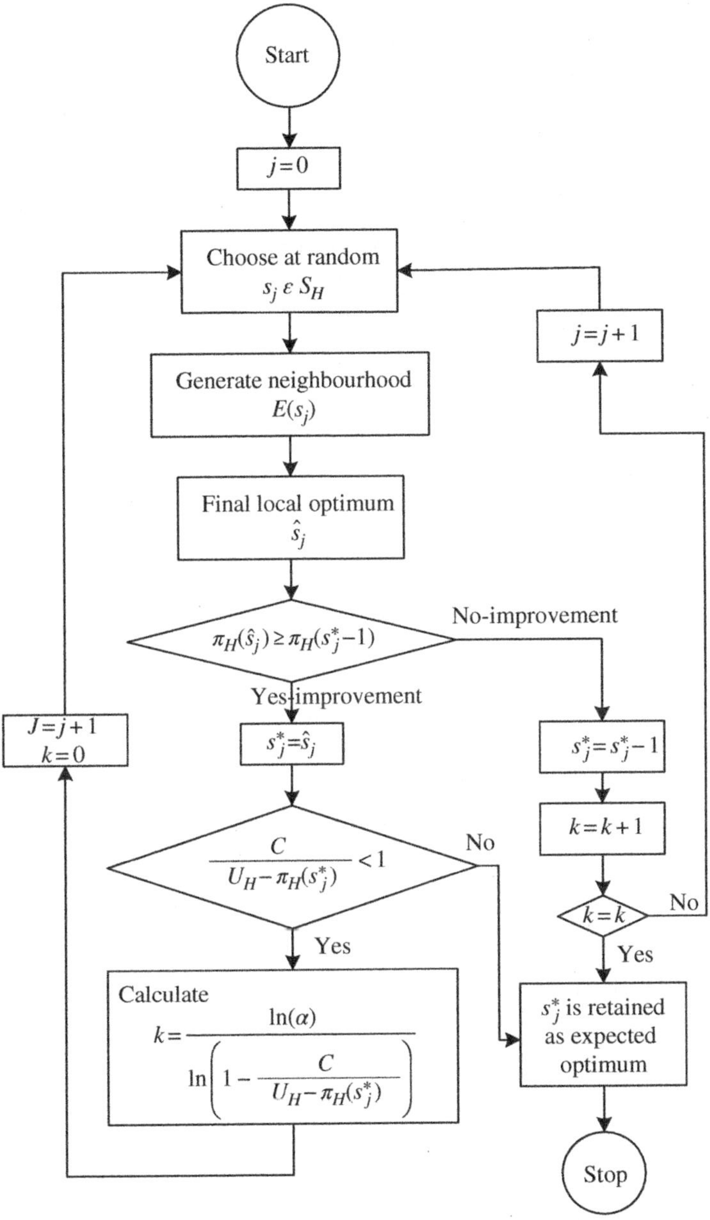

Figure 12.4 Flowchart of TEP using dynamic programming.

searching for a better plan from the point of view of a generation company (GENCO) as well as the grid owners. This chapter considers production cost instead of the transmission line loss cost for the *OC*. A network model and a network flow method that neglects line loss are used in this chapter. Consequently, the objective is to minimize the total cost given by:

$$C^T = CC + OC + SC \quad (\$) \tag{12.8}$$

2) *The Construction Cost*: The total cost of construction is the summation of the construction costs of new elements. In general, stricter reliability criteria are expected to increase the construction cost because additional transmission lines will be required in order to increase reliability. Therefore, the construction costs of new lines and transformers, *CC*, can be formulated as:

$$CC = \left[\sum_{i=1}^{m(x,y)} C_{(x,y)}^{(i)} U_{(x,y)}^{(i)}\right] \quad (\$) \tag{12.9}$$

where

ρ: is the set of all transformers and transmission lines that are represented as branches (segments connecting nodes) in a network model.

$m(x,y)$: is the number of circuits (ckt) of new candidate branches (segments connecting nodes in a network) connecting buses x and y, which are represented as nodes (points to which branches are connected in a network).

$C_{(x,y)}^{(i)}$: is the sum of construction costs of the 1st ckt through the i-th ckt of the new transformers and lines connecting buses x and y. Specifically, $C_{(x,y)}^{(i)} = \sum_{j=1}^{i} \Delta C_{(x,y)}^{(j)}$.

$\Delta C_{(x,y)}^{(j)}$: is the construction cost of the j-th ckt connecting buses x and y.

$U_{(x,y)}^{(i)}$: is a decision (integer) variable associated with the ckt connecting buses x and y (1 if the 1st ckt to i-th ckt lines are to be constructed, and 0 otherwise).

$$\sum_{i=1}^{m(x,y)} U_{(x,y)}^{(i)} = 1$$

$$P_{(x,y)}^{(i)} = \sum_{j=1}^{i} \Delta P_{(x,y)}^{(j)}$$

$P_{(x,y)}^{(i)}$: is the sum of the capacities of the 1st ckt through the i-th ckt connecting buses x and y.

$\Delta P_{(x,y)}^{(j)}$: is the capacity of the j-th ckt connecting buses x and y.

12.5.4 Genetic Algorithm by Mixed Integer Programming (MIP)

The TEP problem is a mixed-integer nonlinear optimization problem. Due to the consideration of transmission line conditions after expansion from the loading viewpoint, the adequacy of the transmission network is included in STNEP objective function. Therefore, the proposed objective function could be defined as follows [25]:

$$OF = K + \sum_{i,j \in \Omega} C_{ij} n_{ij} - T_A C_A \ \$ \tag{12.10}$$

where

C_{ij}: Construction cost of each line in branch $i-j$
n_{ij}: Number of new circuits in corridor $i-j$
$\boldsymbol{\Omega}$: Set for all corridors
T_A: Required time for missing the expanded network adequacy (in years).
C_A: Annual worth of transmission network adequacy (\$/year). Determination of this is based on the importance of network adequacy for network owners. Naturally, its high quantities lead to having a network with high adequacy and, of course, expensive configuration for expansion.
K: A constant parameter that is large enough to prevent obtaining negative values of an objective function. Several restrictions given to be mathematical solutions are in line with the planning requirements. These constraints are as follows:

$$Sf + g - d = 0 \tag{12.11}$$

$$f_{ij} - \gamma_{ij}\left(n_{ij}^0 + n_{ij}\right)\left(\theta_i - \theta_j\right) = 0 \tag{12.12}$$

$$\left|f_{ij}\right| \le \left(n_{ij}^0 + n_{ij}\right)\overline{f_{ij}} \tag{12.13}$$

$$0 \le n_{ij} \le \overline{n_{ij}} \tag{12.14}$$

$$0 \le g \le \overline{g} \tag{12.15}$$

$$N-1 \ \text{Safe Criterion} \tag{12.16}$$

where

S: Branch-node incidence matrix.
f: Active power matrix in each corridor.
g: Generation vector.
d: Demand vector.
N: Number of network buses.
θ: Phase angle of each bus.
g_{ij}: Total susceptance of circuits in corridor $i-j$.

n_{ij}^0: Number of initial circuits in corridor $i-j$.
$\overline{n}_{ij}$: Maximum number of constructible circuits in corridor $i-j$.
$\overline{f}_{ij}$: Maximum of transmissible active power through corridor $i-j$ which will have two different rates according to the voltage level of candidate line.
n_{ij}: Number of new circuits in corridor $i-j$.
Ω: Set for all corridors.

12.6 Branch and Bound Algorithm

12.6.1 Branch and Bound Algorithm and Flow Chart

The flow chart of the branch and bound algorithm is summarized in Figure 12.5. Details are explained as follows [26]:

1) Check the need for transmission expansion for the system and its possibility of using the candidate lines.
2) Set $j=1$ (initial system), $j_{opt}=0, jmax=0, \lambda_{opt}=0$ and $ENNOD_j=0$. $ENNOD_j$ is 1 means that the #j system is the end node.
3) If $ENNOD_j=1$, the #j system is an end node at which the branch operation of a branch and bound is finished (bound) in the solution graph used to obtain the optimal solution and there is no need to consider any of the other graphs following this system. Go to step 15.
4) Calculate the minimum cut-set using the maximum flow method for #j system (#j solution in the solution graph.)
5) Select the #i branch/line of the candidate branches/lines set (S_j) involved in the minimum cut-set and add to the #j system. In what follows, the new system is named the #ji system.
6) If the #ji system is already considered in the solution graph, go to step 14.
7) Calculate the total cost $C_{ji}^T = C_j^T + C\left(P_{(x,y)}^{(i)}\right)$ for the #ji system and evaluate the cost membership grade μ_{Cji}.
8) Calculate the minimum cut-set capacity $P_{Cji}(X, \overline{X})$ using the maximum flow/minimum cut-set theorem and evaluate the delivery marginal rate (reliability) member grade μ_{Rji}.
9) Calcuate λ_{ji} minimum $\{\mu_{Cji}, \mu_{Rji}\}$.
10) If $\mu_{Cji} < \lambda_{opt}$ (the current #ji system) with a cost of C_{ji}^T is optimal and the lambdas of the following systems will be lower than the current optimal solution (λ_{opt}), go to step 14.
11) Set $jmax = jmax + 1$.
12) If $\lambda_{ji} > \lambda_{opt}$, set $\lambda_{opt} = \lambda_{ji}$, $j_{opt} = jmax$, go to step 14.

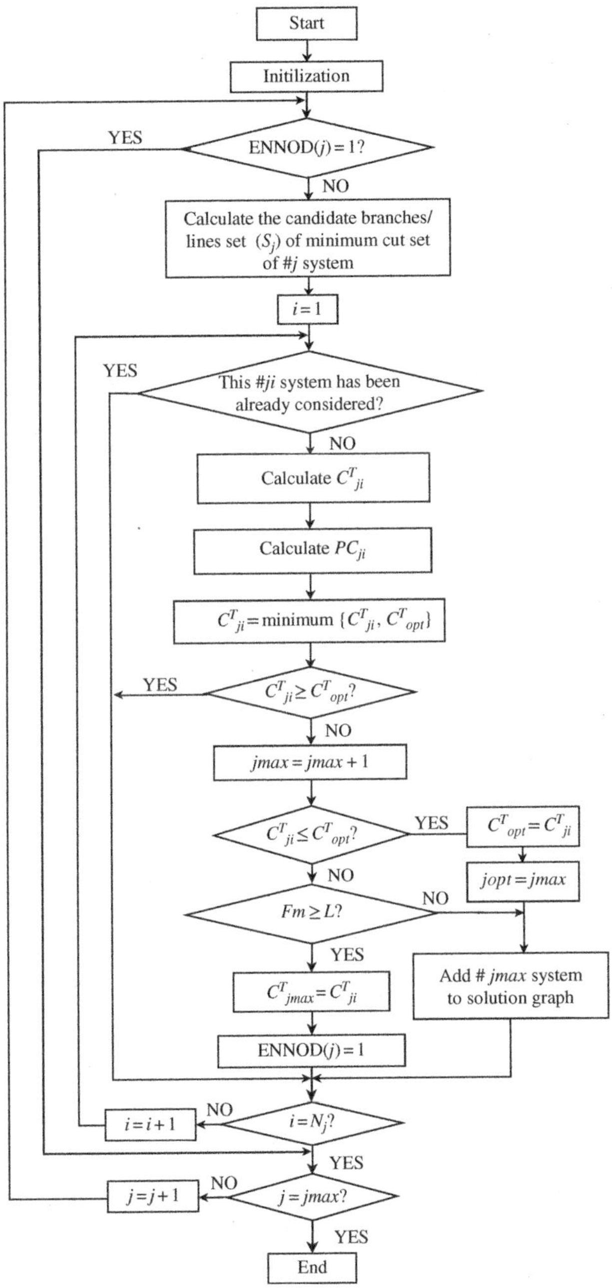

Figure 12.5 Flow chart of TEP using branch and bound.

13) If $\mu_{Cji} < \mu_{Rji}$, set $\lambda_{jmax} = \lambda_{ji}$, $ENNOD_{jmax} = 1$, go to step 15.
14) Add the solution *#jmax(#ji)* to the solution graph.
15) If all the candidate branches/lines in the cut-set S_j have been considered, go to step 16. Otherwise, set $i = i + 1$ and go to step 5.
16) If $j = jmax$, continue. Otherwise, set $j = j + 1$ and go to step 4.
17) For $j = jmax$, the solution graph has been constructed fully and the optimal solution with λ_{opt} being the highest satisfaction level of the decision maker is obtained at step 12.

12.6.2 Sample System Study by Branch and Bound

In this section, a comparison between deterministic and probabilistic reliability criterion approaches for composite power system expansion planning is made by using the branch and bound method.

A two-bus sample is given in Figure 12.6 with input data provided in Table 12.2. The system required reliability criterion is assumed to be LOLE$*$ = 100 h/yr. The load duration curves on a standard day are given in Figure 12.7.

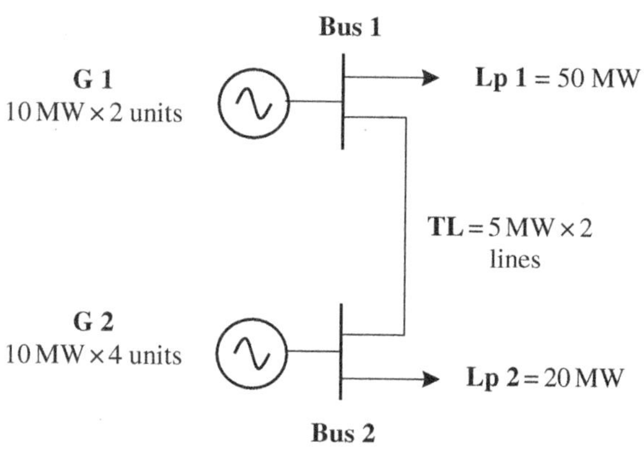

Figure 12.6 Sample study - 2 buses system.

Table 12.2 Candidate generators and transmission lines.

Units	Capacity (MW)	Candidate units	Cost (M$/unit)
G 1	10	3	10
G 2	10	3	8
TL	5	4 (circuits)	2

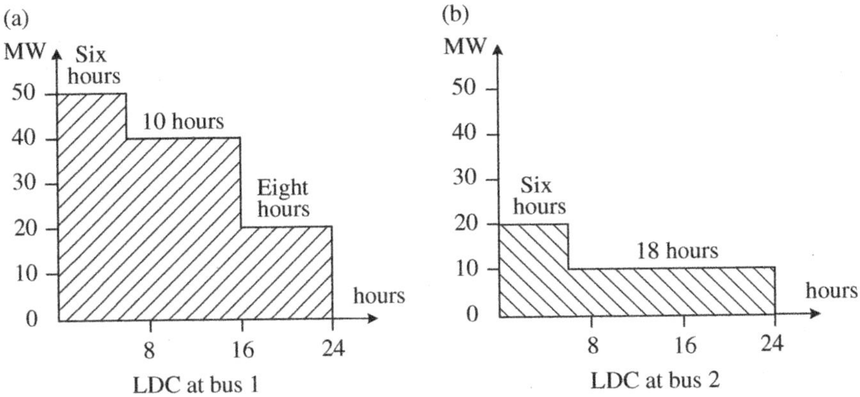

Figure 12.7 Load curves on a standard day.

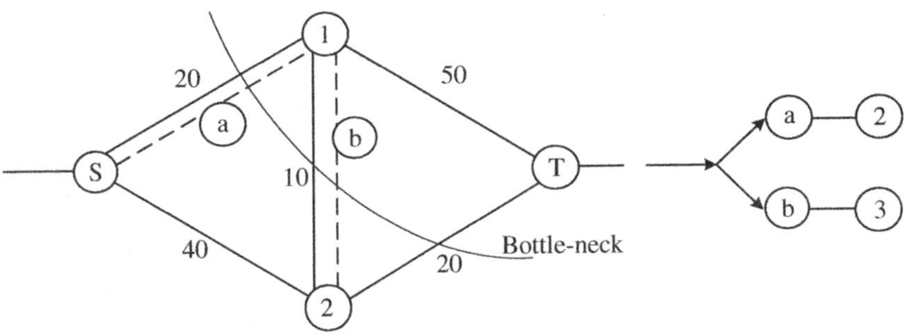

Figure 12.8 Initial system.

Figure 12.8 shows the initial branch diagram for the sample 2-bus system in Figure 12.6. It shows two bottle-neck lines, one between the source to bus-1 and another between bus-1 and bus-2. The corresponding LOLE for this system is 4160 h/yr, which obviously does not meet the reliability requirement. The minimum cut set of these branches, say (a) and (b), are candidate lines. The figure shows two paths: adding new lines to branch (a) or branch (b). It also shows the next step to proceed: To add a new line to branch (a), go to step 2; while to add a new line to branch (b), go to step 3.

Figures 12.9 and 12.10 show the branch diagrams when new lines are added to branches (a) and (b), respectively, where the new investment cost and the LOLE are evaluated. It is noted that the reliability is improved in both steps; however, it falls short of the requirement. Thus, the steps will be repeated.

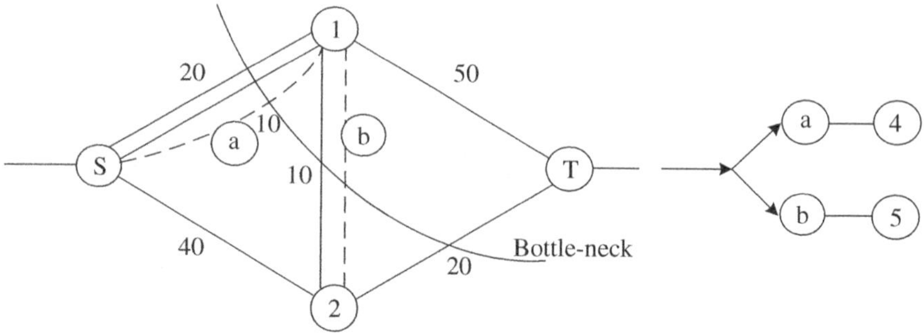

Figure 12.9 Step 2, added line.

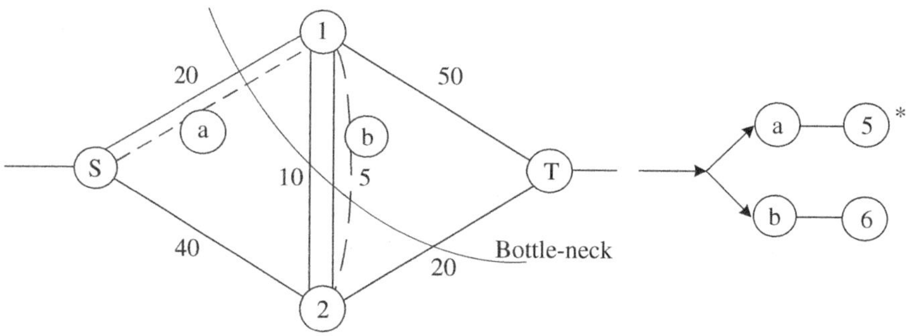

Figure 12.10 Step 3, added line.

Figures 12.11–12.13 show the branch diagrams when new lines are added in steps 4, 5, and 6, respectively, where the new investment cost and the LOLE are evaluated. It is interesting to note that step 4 has the least LOLE but the highest investment cost, while step 6 has the least cost but he highest LOLE. Therefore, the steps need to be repeated.

Figures 12.14 and 12.15 show the branch diagrams when new lines are added in steps 7 and 8, respectively, where the new investment cost and the LOLE are evaluated. Comparing the two cases, step 7 is selected as the optimal solution by the deterministic approach in the branch and bound method, with the optimal solution:

G_1^1: 10 MW, T_{1-2}^1, T_{1-2}^2: 5 MW × 2 circuits

Steps 1 through 8 are summarized in Figure 12.16 as a solution graph for obtaining optimal solution by the branch and bound method using deterministic approach.

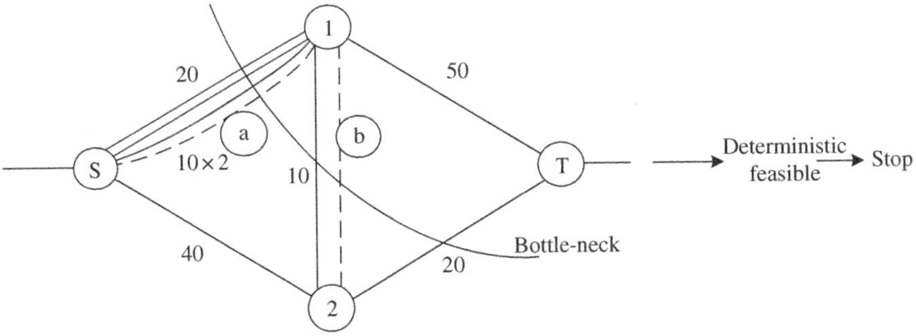

Figure 12.11 Step 4, added 2 lines.

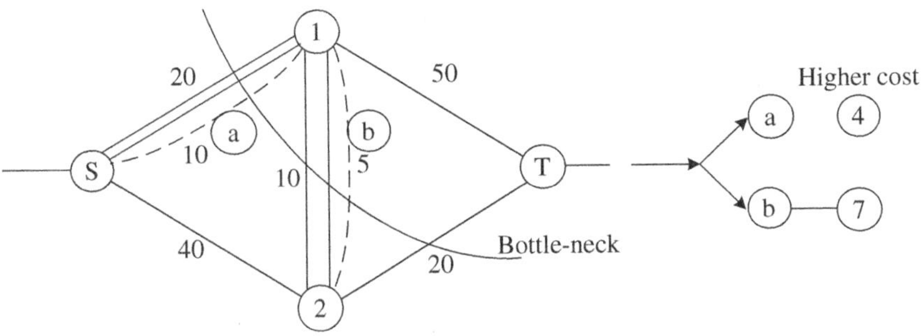

Figure 12.12 Step 5, added 2 lines.

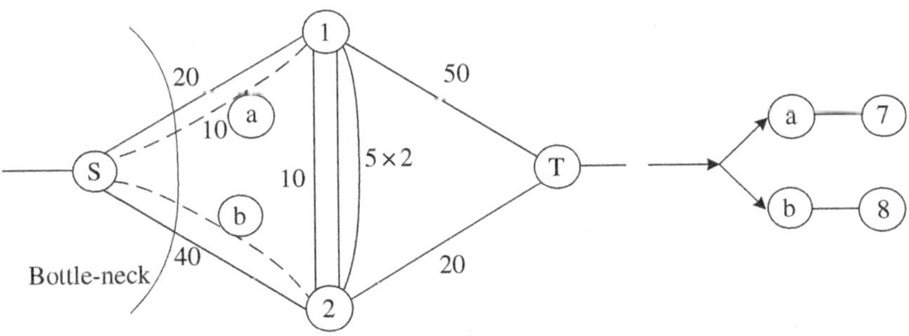

Figure 12.13 Step 6, added 2 lines.

It should be noted that the deterministic approach stopped at steps 4, 7, and 8, and step 7 was selected as the optimal solution. The question remains on steps 4 and 8, which could lead to alternative solutions. This leads to the probabilistic

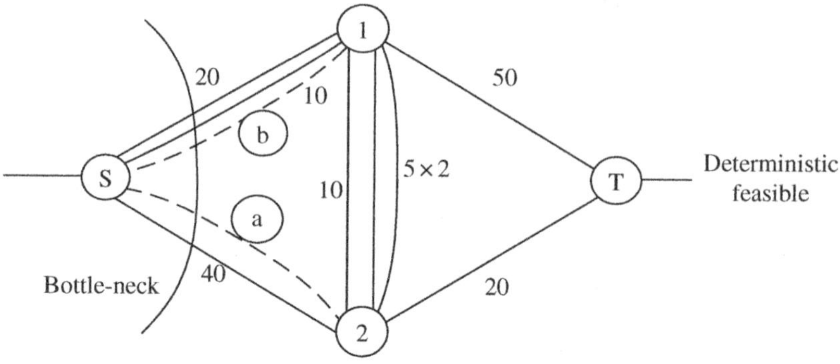

Figure 12.14 Step 7, added 3 lines.

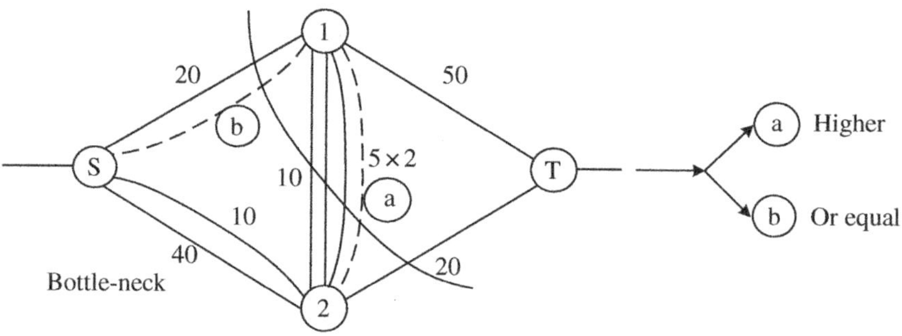

Figure 12.15 Step 8, added 3 lines.

approach by continuing the search at steps 4 and 8. Figures 12.17 and 12.18 show the branch diagrams when additional lines are added in steps 9 and 10, respectively, where the new investment cost and the LOLE are evaluated. While the reliability is improved in step 9, the cost is increased. On the other hand, step 10 provides a significant improvement in reliability at the expense of small additional investment cost. Repeating the step, Figure 12.19 shows the branch diagram when the maximum of four lines are invested in step 11, where the LOLE further improved at the expense of a small investment cost.

Steps 1 through 11 are summarized in Figure 12.20 as a solution graph for obtaining optimal solution by the Branch and Bound Method using the probabilistic approach.

The progression of search for optimal expansion planning is summarized in Table 12.3.

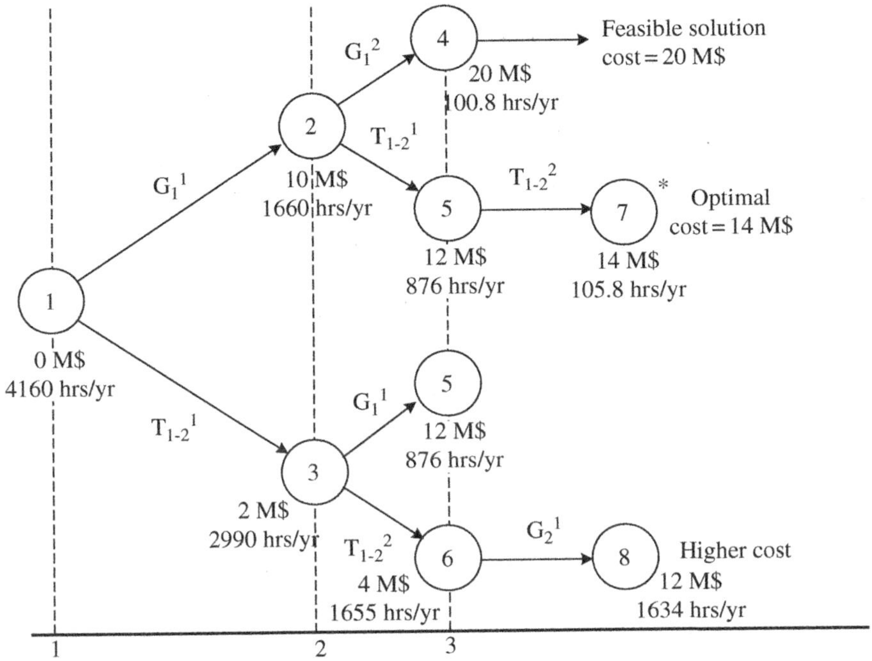

Figure 12.16 Solution graph for obtaining an optimal solution by branch and bound method, deterministic approach.

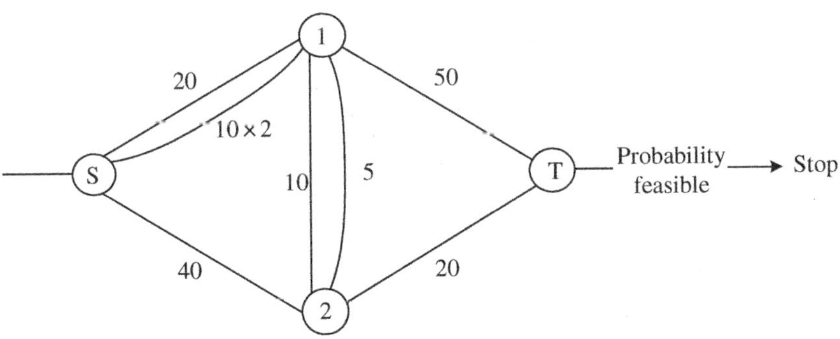

Figure 12.17 Step 9, added 3 lines.

1) *System* – Required Reliability Criteria LOLE∗ = 100 hours/year assumed
2) *Input Data*
3) *Network Branch and Bound*

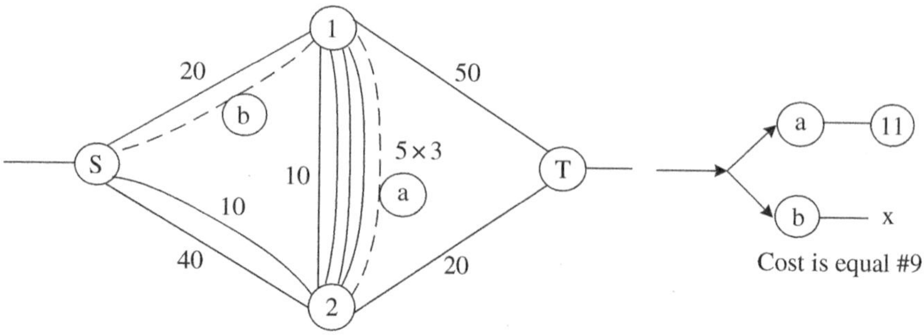

Figure 12.18 Step 10, added 4 lines.

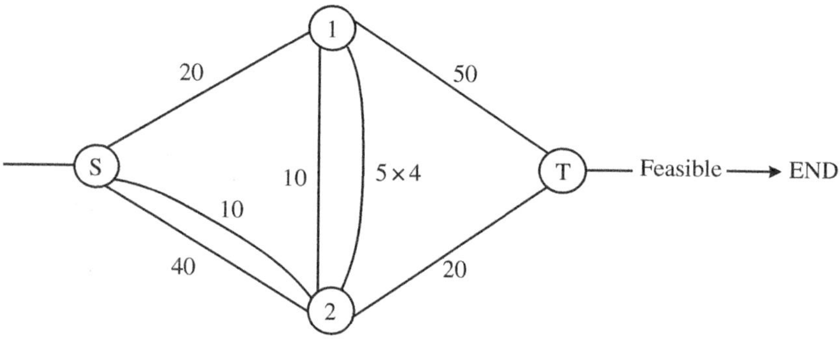

Figure 12.19 Step 11, added 4 lines.

i) *Initial system*
   Cost = 0 M$
   LOLE = 4160 hours/year

ii) *Step 2*
   Cost = 10 M$
   LOLE = 1660 hours/year

iii) *Step 3*
   Cost = 2 M$
   LOLE = 2990 hours/year

iv) *Step 4*
   Cost = 20 M$
   LOLE = 100.8 hours/year

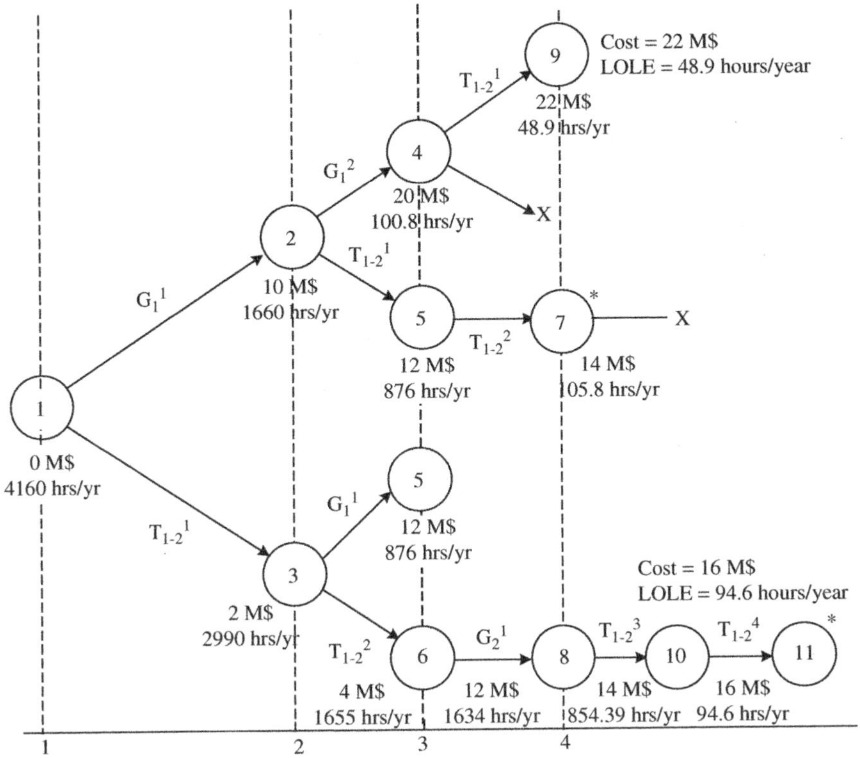

Figure 12.20 Solution graph for obtaining an optimal solution by branch and bound method, probabilistic approach.

v) *Step 5*
   Cost = 12 M$
   LOLE = 876 hours/year

vi) *Step 6*
   Cost = 4 M$
   LOLE = 1655 hours/year

vii) *Step 7*
   Cost = 14 M$
   LOLE = 105.8 hours/year
   Note: Deterministic approached optimal solution and cost are obtained.

viii) *Step 8*
   Cost = 12 M$, LOLE = 1634 hours/year

Table 12.3 Progression of search for optimal expansion plans.

System #	Con. Gens and lines	Cost (M$)	LOLE (hrs/yr)	EENS (MWh/yr)	Connected system #	Remarks
①	Initial system	0	4160	100 555	②③	NF
②	$G_1^{\ 1}$	10	1660	34 002	④⑤	NF
③	$T_{1-2}^{\ \ 1}$	2	2990	66 736	⑤⑥	NF
④	$G_1^{\ 1}, G_1^{\ 2}$	20	100,8	2042	⑨	DF, NPF
⑤	$G_1^{\ 1}, T_{1-2}^{\ \ 1}$	12	876	17 894	⑦	NF
⑥	$T_{1-2}^{\ \ 1}, T_{1-2}^{\ \ 2}$	4	1655	33 849	⑦⑧	NF
⑦*	$G_1^{\ 1}, T_{1-2}^{\ \ 1}, T_{1-2}^{\ \ 2}$	14	105.8	2140	—	DFO, NPF
⑧	$G_2^{\ 1}, T_{1-2}^{\ \ 1}, T_{1-2}^{\ \ 2}$	12	1634	33 298	⑩	NPF
⑨	$G_1^{\ 1}, G_1^{\ 2}, T_{1-2}^{\ \ 1}$	22	48.9	989	END	PF
⑩	$G_2^{\ 1}, T_{1-2}^{\ \ 1}, T_{1-2}^{\ \ 2}, T_{1-2}^{\ \ 3}$	14	854.39	17 331	⑪	NPF
⑪*	$G_2^{\ 1}, T_{1-2}^{\ \ 1}, T_{1-2}^{\ \ 2}, T_{1-2}^{\ \ 3}, T_{1-2}^{\ \ 4}$	16	94.6	1909	END	PFO

* Optimal solution by Branch and Bound Method in the probabilistic approach.

Therefore, #7 new system is an optimal solution by a deterministic approach.

$G_1^{\ 1}$: 10 MW, $T_{1-2}^{\ \ 1}, T_{1-2}^{\ \ 2}$: 5 MW × 2 circuits

ix) *Step 9 ← Step 4*
Cost = 22 M$
LOLE = 48.9 hours/year
Remark: (4)–(b) has a higher cost than #9. Therefore, it is ignored.

x) *Step 10 ← Step 8*
Cost = 14 M$
LOLE = 854.39 hours/year

xi) *Step 11*

Cost = 16 M$, LOLE = 94.6 hours/year

13

Probabilistic Production Cost Simulation for TEP

13.1 Introduction

Deregulation with open access to the transmission system has become a hot issue in electrical energy industries. It has moved the industry from conventional monopolistic electricity markets to competitive markets. In a competitive market, the price of the nodal delivered energy and the nodal quality of electrical energy including voltage quality and reliability of service are the main factors for business success. Appropriate concepts of nodal probability production cost simulation and reliability evaluation for independent transmission system operators under a competitive environment need to be developed. Successful electricity market operation under a deregulated electricity environment depends on the transmission system as well as nodal reliability management. The development of generalized models for evaluating nodal reliability indices and simulating nodal probabilistic energy production and cost considering generator and transmission line uncertainties is an important objective. Nodal reliability indices together with the related information can be used for the management and control of congestion and reliability in both ISO and TRANSCO domains in a deregulated electricity market.

The conventional effective load duration curve (ELDC) for only generation systems, without considering transmission elements, has played an important role in probabilistic production cost simulation and reliability evaluation for generation system expansion planning and supplies some very useful information. This curve has been used widely since Baleriaux and Jamoulle [45] developed the recursive equation in 1967, which incorporate the forced outage rates of generators with the LDC in order to obtain probabilistic production costs and reliability indices. This method was extended by Booth et al. [46]. In 1980, Stremel et al. [47] developed the cumulant method in which the values of a function at every point on the curve can be obtained by using the expansion formula of the Gram–Charlier series while all orders of cumulants of the ELDC are known. Although this approach has

Probabilistic Power System Expansion Planning with Renewable Energy Resources and Energy Storage Systems, First Edition. Jaeseok Choi and Kwang Y. Lee.

Published 2022 by John Wiley & Sons, Inc.

the problem that the reliability index could be negative if a system includes low forced-outage rate generators, the cumulant method has been used widely in application because of its computation efficiency and flexibility. In order to overcome certain faults, the Mixture of Normals Approximation (MONA) method was developed by Gross, Garapic, and McNutt in 1988 [48]. All these methods, however, consider only the forced outage rates of the generating units without considering the uncertainty associated with forced outages of transmission elements. Considerable research has been conducted on the reliability evaluation of composite power systems since systematic methodology and algorithm was developed and first introduced in 1969. Algorithms based on the analytical enumeration method encounter difficulties in computation time in reliability evaluation of large-sized systems due to the many uncertainty states that must be considered. Algorithms based on Monte Carlo simulation have been developed for large-scale system evaluation. The Monte Carlo simulation method is considered to be more flexible for large-scale system applications. More recently, another methodology using probabilistic AC load flow for reliability evaluation of composite power system has been introduced.

This chapter illustrates an extended effective load model for nodal indices of probabilistic production cost simulation and reliability evaluation. The conventional effective load model arises from convolution integral processing of the outage capacity probabilistic distribution functions of the generators and a total load of the power system. The model is used successfully for generation system expansion planning, although it does not consider the capacities and uncertainties of transmission lines. The proposed effective load model is obtained from convolution integral processing of the outage capacity probability distribution function (PDF) of a synthesized fictitious equivalent generator (SFEG). This is designated as the SFEG in this chapter and considers the forced outage rates of generators as well as transmission lines and the original load given at the load point. An improved ELDC based on the extended effective load model is also proposed in this chapter. It is tentatively named the CoMposite power systems Effective Load Duration Curve (CMELDC). It is expected that the CMELDC will be applied to research areas such as reliability evaluation, probabilistic production cost simulation, and analytical outage cost assessment, etc., at the load points in a composite power system. An analytical enumeration method is used in this chapter as the primary focus of this study is to develop the alternative effective load model and clearly review the utility of the proposed model prior to applying it to large power systems.

The algorithm for making the CMELDC, considering the forced outage rates of the generators as well as transmission lines, is illustrated in detail by a numerical example using a simple model. The algorithm is then applied to a test system to demonstrate the effectiveness of the proposed method [27, 28].

13.2 Modeling of Extended Effective Load for Composite Power System

1) *Hierarchical Levels (HL)* – The basic techniques for adequacy assessment can be categorized in terms of their application to segments of a complete power system. These segments are shown in Figure 13.1 and can be defined as the functional zones of generation, transmission, and distribution. The goal of this research is to develop an extended effective load model for CMELDC at a load point in a composite power system (HLII).
2) *Effective Load*
   i) *Effective load at HLI*
   The actual system at HLI in Figure 13.2a can be represented by an equivalent system, as shown in Figure 13.2b. The capacity and forced outage rate of the ith generator are C_i (MW) and q_i, respectively in Figure 13.2a. It is equivalent to considering the capacity and forced outage rate of the generator as C_i

Figure 13.1 Hierarchical-level diagram of a power system.

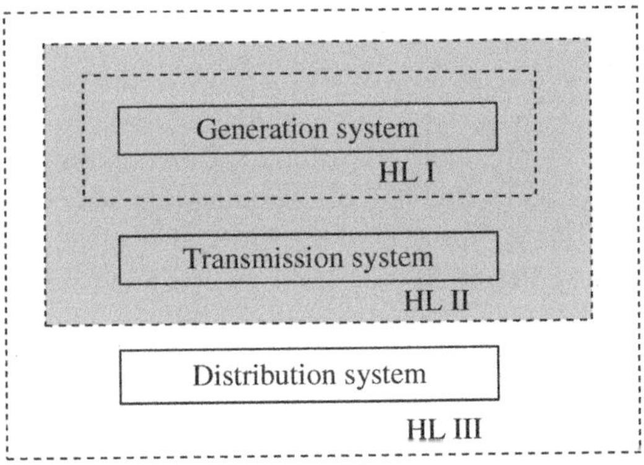

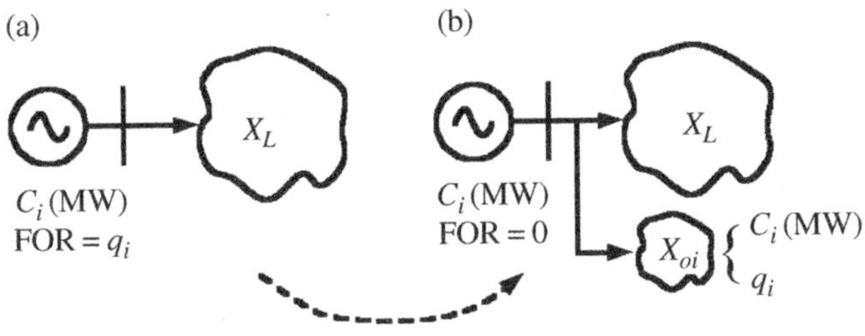

Figure 13.2 Actual and equivalent systems at HLI (for GEP). (a) Actual system. (b) Equivalent system.

(MW) and 0 with the fictitious load added to the original load. The capacity and forced outage rate of the fictitious load for the ith generator are C_i (MW) and qi, respectively as shown in Figure 13.2b. The effective load is defined as the sum of the original load and the probabilistic additional load caused by the forced outage of the generators. The effective load is formulated as shown in (13.1).

$$x_e = x_L + \sum_{i=1}^{NG} x_{oi} \tag{13.1}$$

where

x_e: random variable of the effective load,

x_L: random variable of the original load,

x_{oi}: random variable of the probabilistic load caused by the forced outage of generator ith,

NG: the total number of generators.

The ELDC at HLI can be calculated using (13.2).

$$\Phi_i(x_e) = \Phi_{i-1}(x_e) \otimes f_{oi}(x_{oi}) = \int \Phi_{i-1}(x_e - x_{oi}) f_{oi}(x_{oi}) dx_{oi} \tag{13.2}$$

where

$\otimes$: operator indicating the convolution integral where $\Phi_o(x_e - x_{oi}) = \Phi_o(x_L)$,

$f_{oi}(x_{oi})$: the PDF of the outage capacity of generator ith,

$\Phi_i(x_e)$: the PDF of the ELDC convolved with the PDFs of the outage capacities of all the generators 1st to NGth,

$\Phi_{i-1}(x_e)$: the PDF of the ELDC convolved with the PDFs of the outage capacities of all the generators 1st to $NG - 1$,

$\Phi_{i-1}(x_e - x_{oi})$: the PDF of the integral convolution styled ELDC convolved with the PDFs of the outage capacities of all the generators 1st to NG-1

The PDF of the ELDC convolved with the PDFs of the outage capacities of all the generators 1st to NGth can be obtained using (13.3), where $\Phi_0(x_e)$ is the total synthesized probability and $f_{osNG}(X_{os})$ is a distribution function of the outage capacity of the generation system.

$$\Phi_{NG}(x_e) = \Phi_o(x_e) \otimes f_{osNG}(x_{os}) \tag{13.3}$$

ii) *Effective load at HLII*

An effective load model at each load point of a composite power system considering the forced outage rates of the generating units and the unavailabilities of the transmission system facilities is proposed. The concept of an ELDC at HLII can be defined in a similar way as that at HLI. It is, however, more complex and difficult to obtain than that at HLI. Figure 13.3 presents the main concept of the equivalent system and the effective loads at HLII. CG, CT, q, and q_l in Figure 13.3 are the capacities and forced outage rates of generators and transmission lines, respectively. NT is the number of transmission lines and k and j are the number indices of load point (bus) and system states, respectively. Figure 13.3a is the original composite power system. Using optimal load flow with a specified objective function and operating generators from 1st to tth, it is possible to calculate the maximum arrival power (${}_kAP_{sij}$) at the load point and the state probabilities (${}_kq_{sij}$) for system state j as shown in Figure 13.3b. This can be designated as an SFEG with multi-operating states with forced outage rate ${}_kq_{sij}$ and operating powers ${}_kAP_{sij}$ at the load point.

The capacity of the SFEG comes from the largest maximum arrival power (${}_kAP_{sij}$). The SFEG is similar to Figure 13.3a, the actual system at HLI without the transmission system. The equivalent generator indicates that the generator system contains units from 1st to ith. The ${}_kf_{osi}$ in Figure 13.3b is the outage capacity PDF of the SFEG created by generator units from 1st to ith. The fictitious equivalent generator is designated as the SFEG in the following discussion. The new effective load in HLII can be defined from the sum of the original load and the probabilistic load caused by the forced outage of generators and transmission lines and designated as the SFEG, at each load point as shown in Figure 13.3c. The new effective load is random and can be formulated as shown in (13.4)

$$ {}_kx_e = {}_kx_L + \sum_{j=1}^{NS} kx_{osij} \tag{13.4} $$

where

${}_kx_e$: random variable of the effective load on the composite power system at load point #k,

${}_kx_L$: random variable of the original load at load point #k,

${}_kx_{osij}$: random variable of the probabilistic load caused by the SFEG at load point #k,

J: number of system states,

NS: total number of system states

After loading the generators 1st to ith, the PDF, ${}_k\Phi_i$ of the CMELDC at load point #k can be calculated as shown in (13.5).

$$ {}_k\Phi_i(x_e) = {}_k\Phi_o(x_e) \otimes {}_kf_{osi}(x_{osi}) = \int_k \Phi_o(x_e - x_{osi})\,{}_kf_{osi}(x_{osi})\,dx_{osi} \tag{13.5} $$

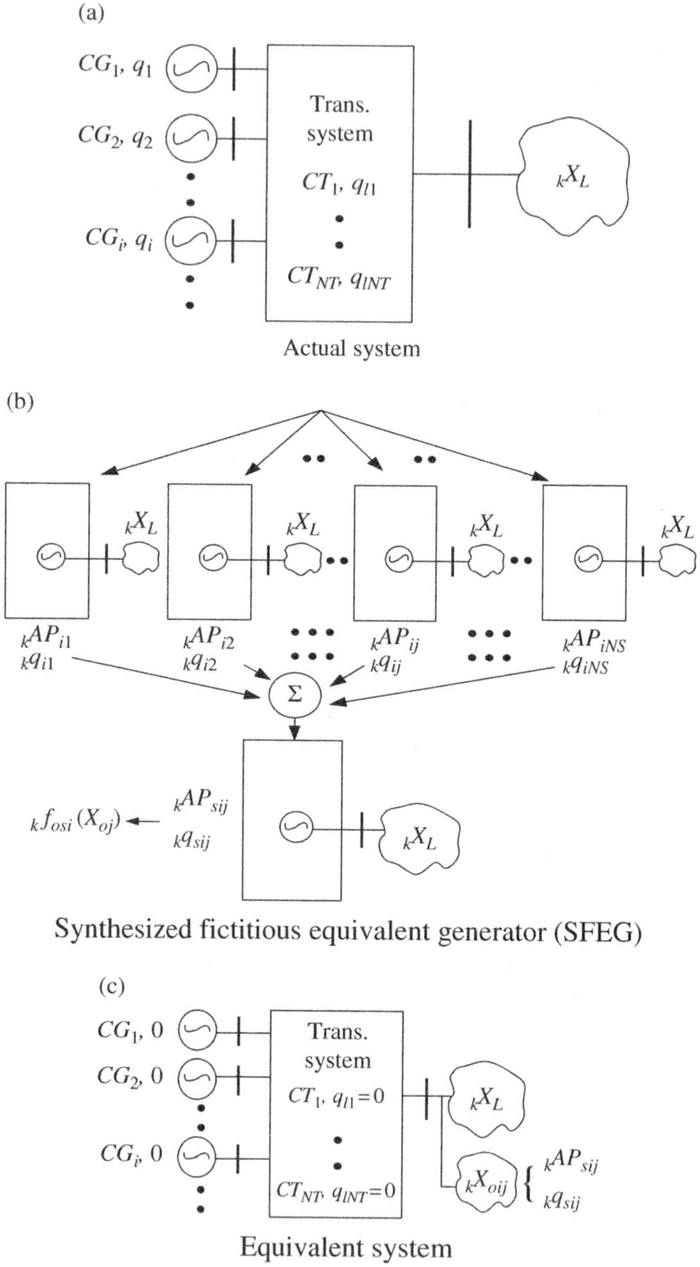

Figure 13.3 Actual and equivalent systems and the effective load model at HLII proposed in this chapter for TEP.

where

$_{k}\Phi_{o}$: inverted LDC at load point #k

$_{k}f_{osi}$: outage capacity PDF of the SFEG operated by generators 1st to ith at load point #k,

13.3 Probability Distribution Function of the Synthesized Fictitious Equivalent Generator

1) *State Probability Calculation* – The studies described in this chapter were conducted using an analytical enumeration approach. Research based on the new effective load model using Monte Carlo simulation and DC load flow has been recently conducted by authors. It is not necessary to consider all the contingency states in a large system as the probabilities of relatively large numbers of generators and transmission lines being failed at the same time are virtually zero. (13.6) can be used in these cases for the state probabilities ($_{k}q_{sij}$) for system state j. The state probabilities become the values of the outage capacity PDF, $_{k}f_{osi}$ of the SFEG created by generator units 1st to ith.

$$q_{sij} = P(e_j)Q(\bar{e}_j) \quad \exists\forall \ n(\bar{e}_j) \le \text{Ncont} \tag{13.6}$$

where

e_j and $\overline{e_j}$: sets of elements on operation and outage, respectively, of system state j,

$n(\bar{e}_j)$: number of elements on outage of set, $\overline{e_j}$,

$P(e_j)$: available probability of set, e_j,

$Q(\bar{e}_j)$: unavailable probability of set, $\overline{e_j}$,

Ncont: number of generator and transmission line contingencies (Ncont = 7 used in this study).

2) *Maximum Arrival Power Evaluation using the Network Flow Method* – Since there are several possible solutions when calculating the load point outage powers for each state, the objective function for minimum outage power must be set up and an optimal solution obtained by the network flow method [38, 39]. The objective function was established to minimize the outage power at a load point as shown in (13.7). AC or DC load flows can be used in order to obtain more accurate maximum arrival power [31–33, 35]. Nonlinear optimal power flow can require considerable computation time, particularly in a large power system. A linearized flow model was used in this case, in order to

illustrate the concept and development of the extended ELDC considering transmission line uncertainties. In this study, transmission line losses are ignored and only effective power is considered for computational convenience using the network flow method [37–39]. In the future, this model will be extended using AC or DC flow to improve accuracy.

i) *Objective function*

$$\text{Minimize } \left\{ \max \left(L_{pk} - P_k \right) / L_{pk} \right\} \quad k \in B_L \tag{13.7}$$

where

L_{pk}: peak load power at load point #k

B_L: set of buses with loads

P_k: control variable meaning effective power flow of load point branches connecting with the terminal (sink) node (therefore, they represent the arrival power ($_kAP$) at load point #k).

ii) *Constraints*

constraint of incident circuit

$$\sum_{l=1}^{NB} a_{il} P_l \leq CG_i \quad i \in B_B \tag{13.8}$$

where

a_{il}: node–branch incidence matrix,

B_B: set of all nodes including all buses, source, and terminal (sink) node,

NB: total number of branches (generator, transmission lines, and load points),

CG_i: generation at bus/node #i (MW) and G_i if #i node is generation plant bus, but zero if it is not,

P_l: control variable meaning effective power flow of all branch #l constraints of transmission line capacity

$$-CT_{l\,max} \leq P_l \leq CT_{l\,max} \quad l \in B_T \tag{13.9}$$

where

CT_{lmax}: capacity of transmission line l (MW),

P_l: control variable meaning effective power flow of transmission line branch #l,

B_T: set of transmission lines

(13.7)–(13.9) can be summarized as shown in (13.10).

Minimize λ

Subject to

$$\begin{cases} \sum_{l=1}^{NB} a_{il} P_l \leq CG_i & i \in B_B \\ -CT_{l\,max} \leq P_l \leq CT_{l\,max} & l \in B_T \\ \left(L_{pk} - P_k\right)/L_{pk} \leq \lambda & k \in B_L \end{cases} \tag{13.10}$$

Using Linear Programming, the maximum arrival power (${}_kAP_{sij}$) at the load points can be obtained by (13.10) at contingency state j. The outage capacity PDF (${}_kf_{osi}$) of the SFEG in (13.5) can be obtained from the state probabilities (${}_kq_{sij}$) of (13.6) and the maximum arrival power (${}_kAP_{sij}$) of (13.10).

13.4 Reliability Evaluation and Probabilistic Production Cost Simulation at Load Points

After loading the generators from 1 to i according to the merit order or bidding order of the electricity market, the reliability indices and the CMELDC (${}_k\Phi_{si}$) at load point #k can be calculated using the convolution integral, (13.5). The reliability indices, Loss of Load Expectation (${}_kLOLE_i$) and Expected Energy Not Supplied (${}_kEENS_i$) can be calculated using (13.11) and (13.12) with ${}_k\Phi_i(x)$ as shown in Figure 13.4. In this figure, L_{pk} and ${}_kAP_i$, respectively on the horizontal axis, express the peak load and the capacity of the SFEG at load point k. It should be noted that ${}_kAP_i$ has noncoherence characteristics and that the CMELDC cannot be recursively obtained, unlike that at HLI, because of capacity limitations of the transmission system. A proof for inducing nodal/bus reliability indices, LOLP and EENS using CMELDC, is given in the Appendix.

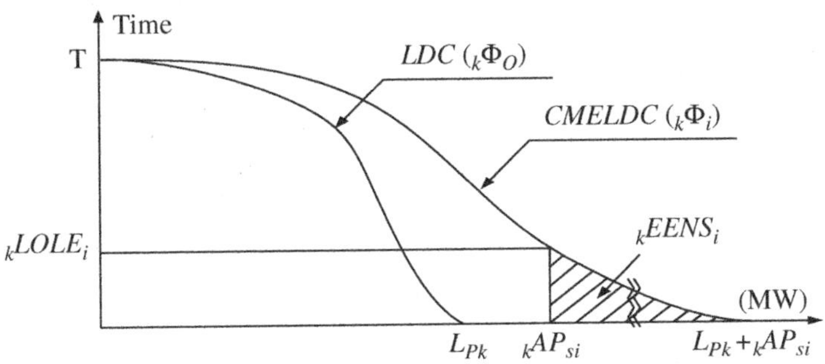

Figure 13.4 Reliability indices and CMELDC at load point k.

$$_{k}LOLE_{i} = {}_{k}\Phi_{si}(x)|_{x={}_{k}AP_{i}} \tag{13.11}$$

$$_{k}EENS_{i} = \int_{_{k}AP_{i}}^{_{k}AP_{i}+Lp_{k}} {}_{k}\Phi_{i}(x)dx \tag{13.12}$$

The nodal probabilistic production energy $_{k}E_{i}$ and probabilistic production cost $_{k}PC_{i}$ of generator i at load point k can be calculated using (13.13) and (13.14).

$$_{k}EENS_{i} = \int_{_{k}AP_{i}}^{_{k}AP_{i}+Lp_{k}} {}_{k}\Phi_{i}(x)dx \tag{13.13}$$

$$_{k}PC_{i} = F_{i}({}_{k}E_{i},\ {}_{k}LOLE_{i-1})\ \ (\$) \tag{13.14}$$

where

$_{k}EENS_{i-1}$: expected energy not supplied before loading generator i,
$_{k}EENS_{i}$: expected energy not supplied after loading generator i,
F_{i}: operating cost function of generator i (\$/h)

If F_i is a linear function of $A_i + B_iP_i$ for a power P_i, the $_{k}PC_{i}$ can be calculated using (13.15), where A_i and B_i are, respectively, the load fuel cost rate and the incremental fuel cost rate of generating unit i.

$$_{k}PC_{i} = A_{i}'(1-FOR_{i})'_{k}LOLE_{i-1} + B_{i}{}'_{k}E_{i} \tag{13.15}$$

The flow chart for nodal probabilistic production cost simulation and constructing the new ELDC based on a new nodal effective load model is shown in Figure 13.5.

13.5 Case Studies

13.5.1 Numerical Calculation of a Simple Example

A simple numerical example is used in order to demonstrate in detail the algorithm for probabilistic production cost simulation and reliability evaluation based on the proposed model. This simple system has two generators, two transmission lines, and a single load as shown in Figure 13.6.

The equipment capacities and forced outage rates and the generator operating cost functions are shown in Table 13.1. Generator 1 is therefore loaded before generator 2. The system load duration curve is shown in Figure 13.7.

1) *Probabilistic Production Cost Simulation of Generator #1*

   Step 1: The associated probabilities in operating only generator #1 are shown in Table 13.2.

   The capacity, CFEG of the SFEG, is calculated as follows:

$$\text{CFEG} = \text{Maximum}\left(AP_j\right) - \text{Minimum}\left(AP_j\right) = 30 - 0 = 30\ \text{MW} \tag{13.16}$$

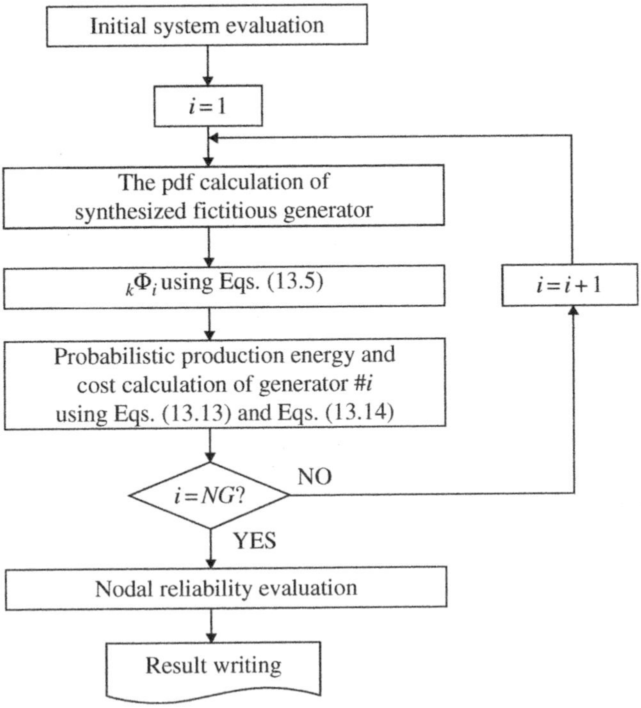

Figure 13.5 The flow chart for nodal reliability evaluation and probabilistic production cost simulation based on CMELDC.

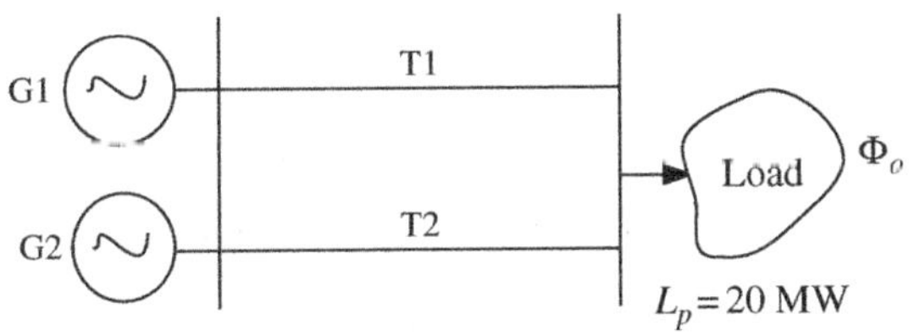

Figure 13.6 Simple system model.

Step 2: The outage capacity PDF of the SFEG is obtained as shown in Table 13.3.

Step 3: In order to calculate the CMELDC, the ${}_k\Phi_1(x_e)$ at the load point after loading generator #1 using (13.5), the outage capacity PDF of the SFEG and the original load duration curve (LDC) are shown in Table 13.4 and Figure 13.8.

Step 4: Integral processing of the PDF of the SFEG and the original LDC function to obtain the CMELDC, ${}_k\Phi_1(x_e)$ is illustrated in Table 13.5, where the LOLE in shaded area is integrated to compute EENS.

Table 13.1 Equipment capacities and forced outage rates and generator operating cost functions.

Elements	Capacity (MW)	FOR	Operating cost functions ($/h)
G1	30	0.1	4 + 2P
G2	10	0.1	4 + 3P
T1	15	0.003	
T2	15	0.003	

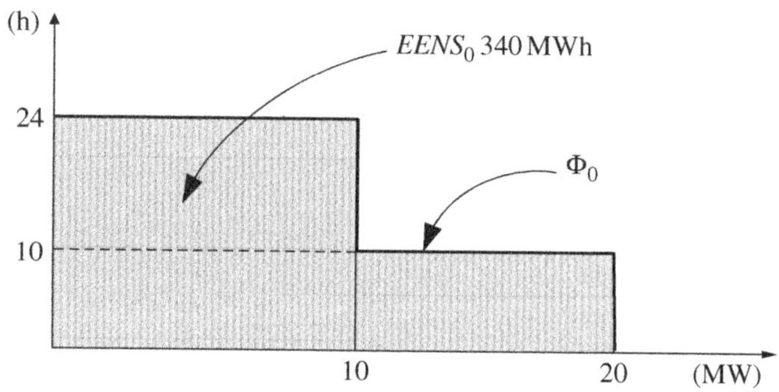

Figure 13.7 System load duration curve.

Table 13.2 State probabilities, available, and outage capacities.

State #	G_1	T_1, T_2	Probability	AP_j
1	0	0	$0.9 \times 0.997^2 = 0.894\,608\,1$	30
2	0	1	$0.9 \times 0.997 \times 0.003 \times 2 = 0.005\,383\,8$	15
3	0	2	$0.9 \times 0.003^2 = 0.000\,008\,1$	0
4	1	0	$0.1 \times 0.997^2 = 0.099\,400\,9$	0
5	1	1	$0.1 \times 0.997 \times 0.003 \times 2 = 0.000\,598\,2$	0
6	1	2	$0.003^2 \times 0.1 = 0.000\,000\,9$	0

AP_j: maximum arrival power (available capacity).

Table 13.3 Probability distribution table of the synthesized fictitious equivalent generator ($_{k}f_{os1j}$).

AP_j	OP_j	pdf ($_{k}f_{os1}$) of SFEG
30	0	0.894 608 1
25	5	0.0
20	10	0.0
15	15	0.005 383 8
10	20	0.0
5	25	0.0
0	30	0.100 008 1
Total		1.000 000 0

Outage capacity, $OP_j = CFEG - AP_j$.

Table 13.4 Distribution of outage capacity PDF of the SFEG and the original LDC.

State #	OP_j	$_{k}f_{os1}$	Load level	$_{k}\Phi_{o}$
7	30	0.100 008 1	−30	24
6	25	0.0	−25	24
5	20	0.0	−20	24
4	15	0.005 383 8	−15	24
3	10	0.0	−10	24
2	5	0.0	−5	24
1	0	0.894 608 1	0	24
			0	24
			5	24
			10	24
			15	10
			20	10

Considering only generator 1, the reliability indices and the probabilistic production energy and cost can be calculated as follows:

Results with only generator 1 operating:

i) *Reliability Evaluation*

$$LOLE_1 = 2.454\,032\,4 \text{ hours/days}$$

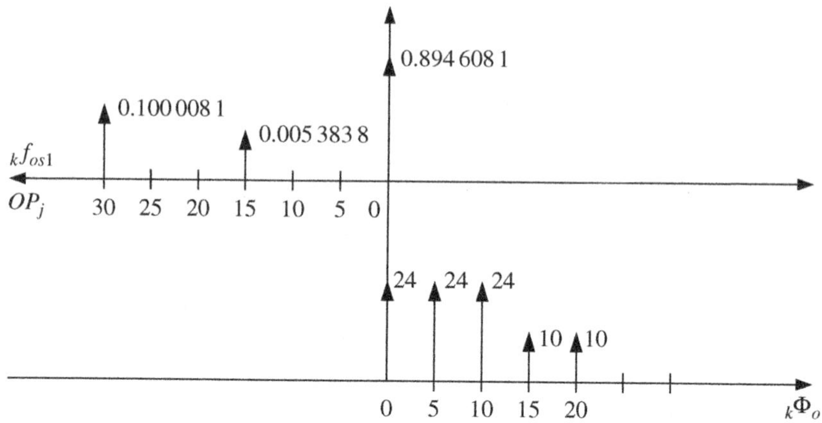

Figure 13.8 Outage capacity PDF of the SFEG and LDC.

Table 13.5 Convolution integral processing [level: (MW)].

State #	1	4	7	Total
Power level Load level	0	15	30	CMELDC ($_k\Phi_1$)
0	21.470 594 4	0.129 211 2	2.400 194 4	24.0
5	21.470 594 4	0.129 211 2	2.400 194 4	24.0
10	21.470 594 4	0.129 211 2	2.400 194 4	24.0
15	8.946 081	0.129 211 2	2.400 194 4	11.400 113 4
20	8.946 081	0.129 211 2	2.400 194 4	11.400 113 4
25		0.129 211 2	2.400 194 4	2.529 305 6
30		0.053 838	2.400 194 4	2.454 032 4
35		0.053 838	2.400 194 4	2.454 032 4(LOLE)
40			2.400 194 4	2.400 194 4
45			1.000 081	1.000 081
50			1.000 081	1.000 081

$$EENS_1 = 2.454\,032\,4 \times 5 + 2.400\,194\,4 \times 5 + 1.000\,081\,0 \times 10 = 34.271\,944 \text{ MWh}$$

ii) *Probabilistic Production Energy and Cost*

Probabilistic Production Energy

$$\begin{aligned} E_1 &= EENS_0 - EENS_1 \\ &= 340 - 34.271\ 944 = 305.728\ \text{ MWh} \end{aligned}$$

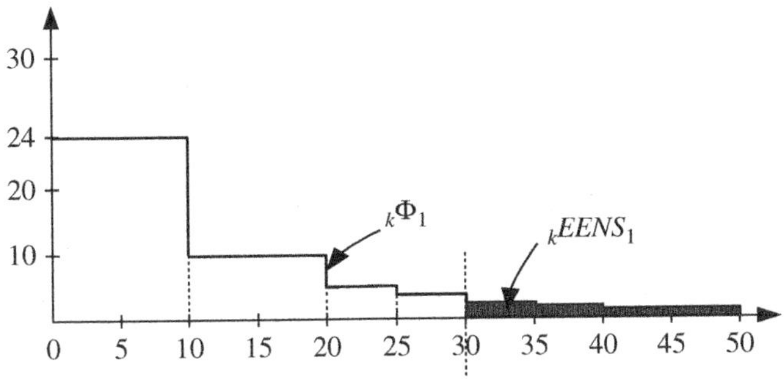

Figure 13.9 The CMELDC for loading generator #1.

Probabilistic Production Cost

$$PC_1 = 4 \times 0.9 \times 24 + 2 \times 305.728\,056$$
$$= 697.856\,112\ \$$$

The CMELDC ($_k\Phi_1$) obtained when loading generator #1 is shown in Figure 13.9.

2) *Probabilistic Production Cost Simulation of Generator #2*

Step 1: The probabilities of all the contingency states when operating generators 1 and 2 are shown in Table 13.6.

The capacity, CFEG of the SFEG as in the former case, can be calculated as follows. The capacity is the same as in the case of loading generator 1 only because of the capacity limitations of the transmission lines.

$$\text{CFEG} = \text{Maximum}\left(AP_j\right) - \text{Minimum}\left(AP_j\right) = 30 - 0 = 30\ \text{MW} \quad (13.17)$$

Step 2: The outage capacity PDF of the SFEG considering the two generators can be obtained as shown in Table 13.7.

Step 3: In order to calculate the CMELDC, the $_k\Phi_2(x_e)$ at the load point after loading generators 1 and 2 using (13.5), the distributions of the outage capacity PDF of the SFEG and the original LDC are shown in Table 13.8 and Figure 13.10.

Step 4: Integral processing of the PDF of the SFEG and the original LDC function in order to obtain the CMELDC, $_k\Phi_2(x_e)$ is illustrated in Table 13.9, where the LOLE in shaded area is integrated to compute EENS.

Considering both generators 1 and 2, the reliability indices and the probabilistic production energy and cost can be calculated as follows:

Results with both generators 1and 2 operating:

i) *Final Reliability Evaluation of the System*

$$LOLE_2 = 1.194\,043\,7\ \text{hours/days}$$

Table 13.6 State probabilities, available, and outage capacities.

State #	T	$G1$	$G2$	Probability (${}_{kJ}f_{os2j}$)	AP_j
1	0	0	0	$0.997\,2 \times 0.92 = 0.805\,147\,29$	30
2	0	1	0	$0.997\,2 \times 0.1 \times 0.9 = 0.089\,460\,81$	10
3	0	0	1	$0.997\,2 \times 0.9 \times 0.1 = 0.089\,460\,81$	30
4	0	1	1	$0.997\,2 \times 0.12 = 0.009\,940\,09$	0
5	1	0	0	$0.003 \times 0.997 \times 2 \times 0.92 = 0.004\,845\,42$	15
6	1	1	0	$0.003 \times 0.997 \times 2 \times 0.1 \times 0.9 = 0.000\,538\,38$	10
7	1	0	1	$0.003 \times 0.997 \times 2 \times 0.9 \times 0.1 = 0.000\,538\,38$	15
8	1	1	1	$0.003 \times 0.997 \times 2 \times 0.12 = 0.000\,059\,82$	0
9	2	0	0	$0.003\,2 \times 0.92 = 0.000\,007\,29$	0
10	2	1	0	$0.003\,2 \times 0.9 \times 0.1 = 0.000\,000\,81$	0
11	2	0	1	$0.003\,2 \times 0.9 \times 0.1 = 0.000\,000\,81$	0
12	2	1	1	$0.003\,2 \times 0.12 = 0.000\,000\,09$	0

Table 13.7 Probabilistic distribution table of the synthesized fictitious equivalent generator (${}_{kJ}fos_{2j}$).

AP_j	OP_j	pdf of SFEG (${}_{kJ}f_{os2}$)	$\Delta {}_{kJ}f_{os} = {}_{kJ}f_{os1} - {}_{kJ}f_{os2}$
30	0	0.894 608 1	0.0
25	5	0.0	0.0
20	10	0.0	0.0
15	15	0.005 383 8	0.0
10	20	0.089 999 19	−0.089 999 19
5	25	0.0	0.0
0	30	0.010 008 91	+0.089 999 19
Total		1.000 000 0	0.0

$$
\begin{aligned}
EENS_2 &= 1.194\,043\,7 \times 5 + 1.140\,205\,7 \times 5 + 0.100\,089\,1 \times 10 \\
&= 12.672\,138\,0\ \text{MWh}
\end{aligned}
$$

ii) *Probabilistic Production Energy and Cost of Generator #2*

Probabilistic Production Energy

Table 13.8 Distribution of outage capacity PDF of the SFEG and the original LDC.

State #	OP_j	${}_kf_{os2}$	Load level	${}_k\Phi_o$
7	30	0.010 008 91	−30	24
6	25	0.0	−25	24
5	20	0.089 999 19	−20	24
4	15	0.005 383 8	−15	24
3	10	0.0	−10	24
2	5	0.0	−5	24
1	0	0.894 608 1	0	24
			0	24
			5	24
			10	24
			15	10
			20	10

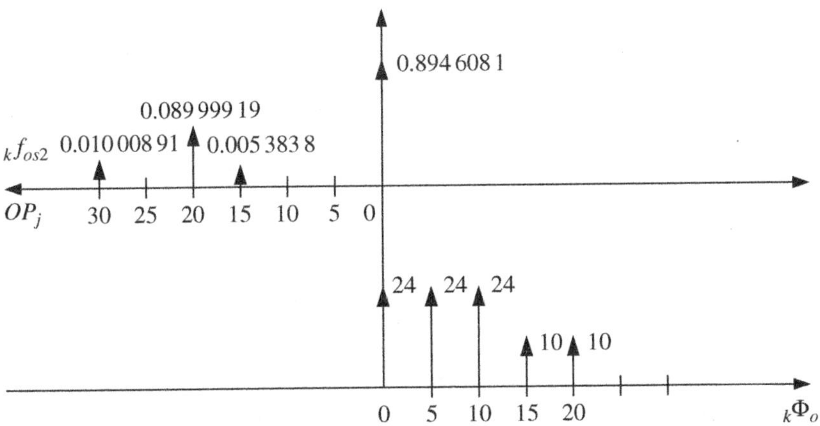

Figure 13.10 Outage capacity PDF of the SFEG and LDC.

$$E_2 = EENS_1 - EENS_2$$
$$= 34.271\ 944 - 12.672\ 138 = 21.599\ 806\ \text{MWh}$$

Probabilistic Production Cost

$$PC_2 = 4 \times 0.9 \times 2.454\ 032\ 4 + 3 \times 21.599\ 806$$
$$= 73.633\ 934\ 64\ \$$$

Table 13.9 Convolution integral processing.

State #	1	4	5	7	Total
Power level Load level	0	15	20	30	CMELDC($_k\Phi_2$)
0	21.475 944	0.129 211 2	2.159 980 6	0.240 213 84	24.0
5	21.475 944	0.129 211 2	2.159 980 6	0.240 213 84	24.0
10	21.475 944	0.129 211 2	2.159 980 6	0.240 213 84	24.0
15	8.946 081 0	0.129 211 2	2.159 980 6	0.240 213 84	11.475 486 6
20	8.946 081 0	0.129 211 2	2.159 980 6	0.240 213 84	11.475 486 6
25		0.129 211 2	2.159 980 6	0.240 213 84	2.529 405 64
30		0.053 838 0	2.159 980 6	0.240 213 84	2.454 032 42
35		0.053 838 0	0.899 991 9	0.240 213 84	1.194 043 7(LOLE2)
40			0.899 991 9	0.240 213 84	1.140 205 7
45				0.100 089 1	0.100 089 1
50				0.100 089 1	0.100 089 1

Additionally, the CMELDC ($_k\Phi_2$) obtained when loading generators 1 and 2 are shown in Figure 13.11.

13.5.2 Case Study: Modified Roy Billinton Test System

The program has been applied to the Modified Roy Billinton Test System (MRBTS) shown in Figure 13.12. in order to demonstrate the effectiveness of the proposed method. The capacities and forced outage rates of the generators and transmission lines are also given in Figure 13.12. All the generators in the base case study are assumed to be thermal units that have the operating cost functions shown in Table 13.10. Figure 13.13 presents the hourly chronological load variation curves of a standard day used to create the yearly load duration curves at the load points.

The six scenarios shown in Table 13.11 were studied. Case 0 assumes that the transmission system is an ideal system with the unlimited delivery capacity and zero forced outage rates. Case 1 is the base case with noncontingency. Cases 2, 3, and 4 include contingencies of lines. Case 5 assumes a new line addition from bus 2 to bus 3.

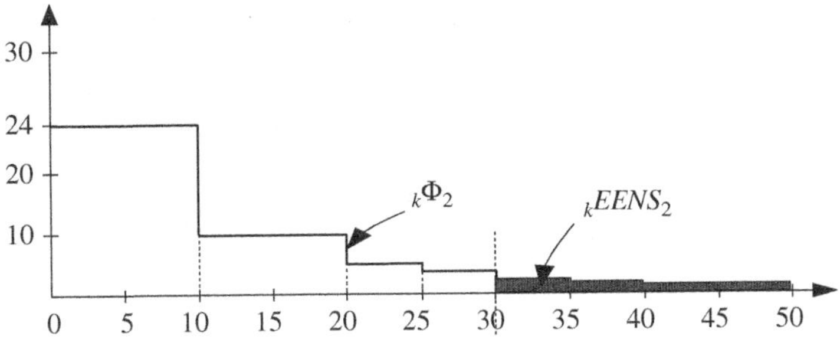

Figure 13.11 The CMELDC for loading generators 1 and 2.

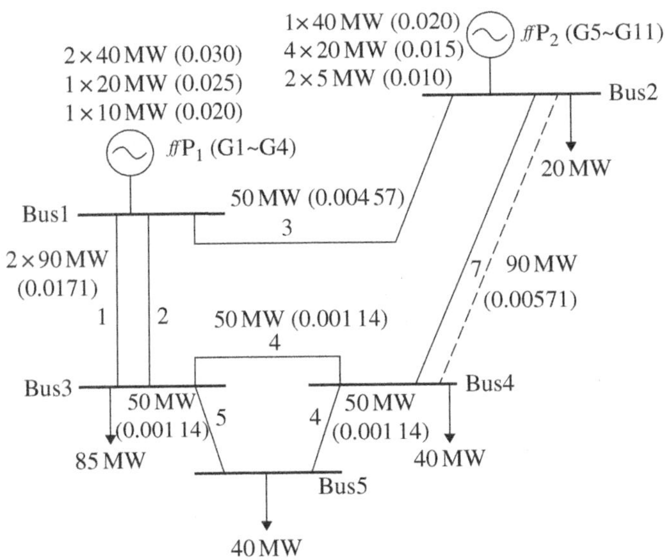

Figure 13.12 The 5-bus system used for the case study.

Figure 13.14 through Figure 13.19 show the variations in the nodal CMELDC (Φ_i,3) at bus 3 of base case (Case 1) after loading generators using the merit order list. Figure 13.8 is the given LDC (Load Duration Curve) because no generators have been loaded. The area under the curve (CMELDC) is the EENS (Expected Energy Not Served) after loading the indicated generators. The EENS and the maximum load (y-axis) of the nodal CMELDC at bus 3 decreases as generators are

Table 13.10 Operating cost functions ($/hr).

Bus #	Loading order	Capacity (MW)	Operating cost functions ($/hr)
2	G 5	40	11.41 + 0.5P
2	G 6 ~ G 9	20	5.71 + 0.5P
2	G 10 ~ G11	5	1.43 + 0.5P
1	G 1 ~ G2	40	90.18 + 12.0P
1	G3	20	77.63 + 12.25P
1	G4	10	68.49 + 12.5P

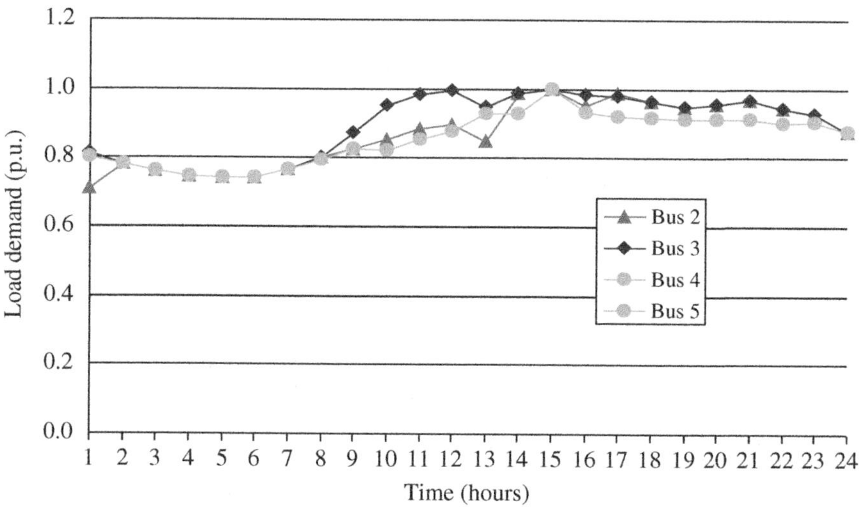

Figure 13.13 Hourly load variation curve of a standard day at the four load buses.

loaded using the merit order. The difference in the area between the previous CMELDC before loading the generator and post CMELDC after loading the generator is the nodal probabilistic production energy at bus 3 of the generator loaded at that time. The area under the final nodal CMELDC at bus 3 is the last EENS, which is the reliability level at bus 3. The y-axis intercept value of the final nodal CMELDC at bus 3 is the LOLE at bus 3.

Tables 13.12 and 13.13, respectively, show the nodal probabilistic production energies and probabilistic production costs at the load points obtained using the proposed method. This case is designated as the base case. Information on the

Table 13.11 System conditions of the cases.

Cases	System condition
Case 0	Ideal system with unlimited delivery capacity and no forced outages of lines
Case 1	Noncontingency (base case)
Case 2	T1-31 line contingency
Case 3	T1-21 line contingency
Case 4	T2-41 line contingency
Case 5	T2-31 new line addition

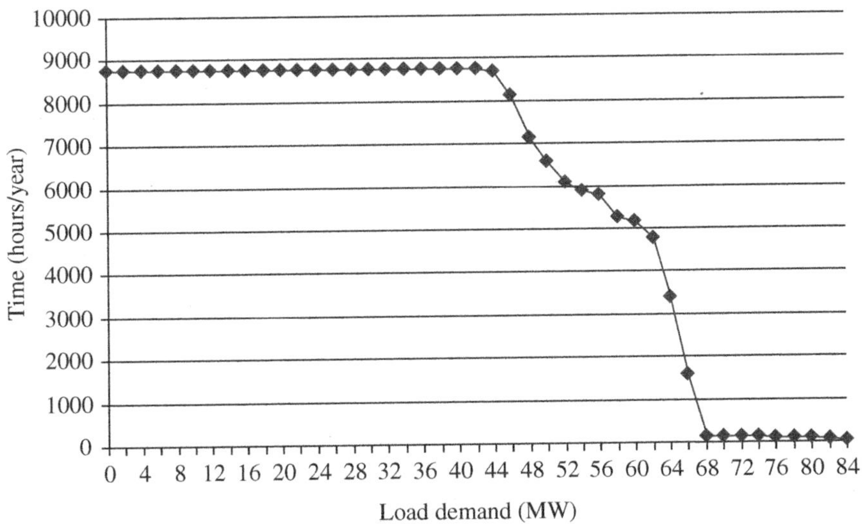

Figure 13.14 Nodal CMELDC (Φ5,3) at bus 3 after generator #5 is loaded (base case).

magnitudes of the total energy and cost of probabilistic production energy supplied from a generator to a load point can be determined from the results. Table 13.14 shows the nodal probabilistic marginal/average production cost per unit energy for the base case. The average unit production cost of the base case is 6.436 \$/MWh.

Figure 13.20 shows the nodal probabilistic production energy of each generator in Case 1 (Base case). The probabilistic production energy of generators allocated at each bus can be also obtained using the proposed method. Figure 13.21 shows the probabilistic production energy of generators at buses #2–#5. Information on

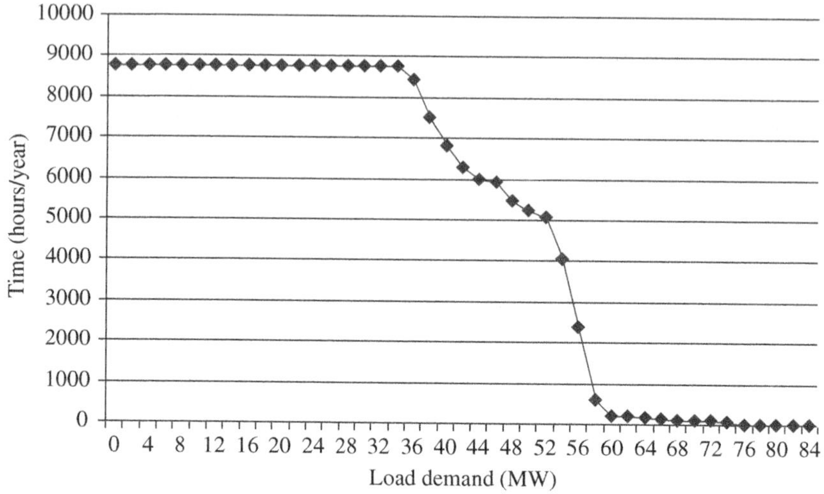

Figure 13.15 Nodal CMELDC (Φ6,3) at bus 3 after generator #6 is loaded (base case).

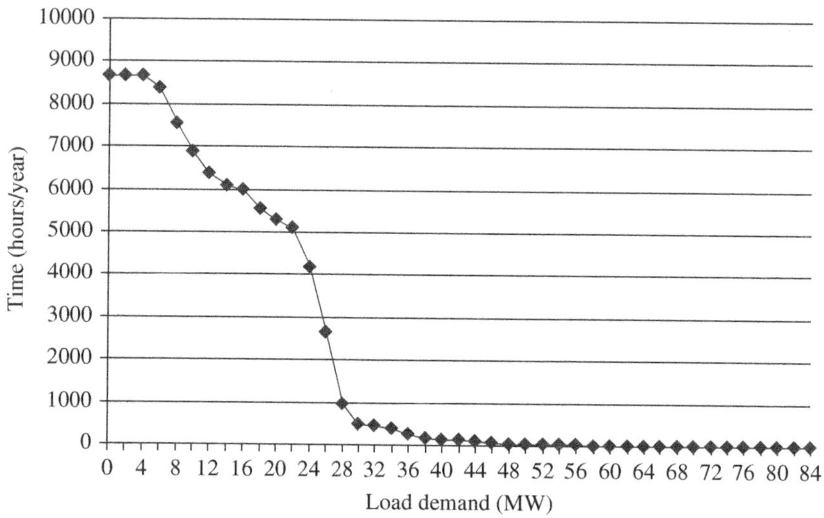

Figure 13.16 Nodal CMELDC (Φ10,3) at bus 3 after generator #10 is loaded (base case).

the nodal probabilistic production energy allocated from a generator to a load point should prove useful to an ISO or ITO.

Tables 13.15 and 13.16 show the probabilistic production energy and cost respectively under the assumption of perfect delivery (zero forced outage rate and

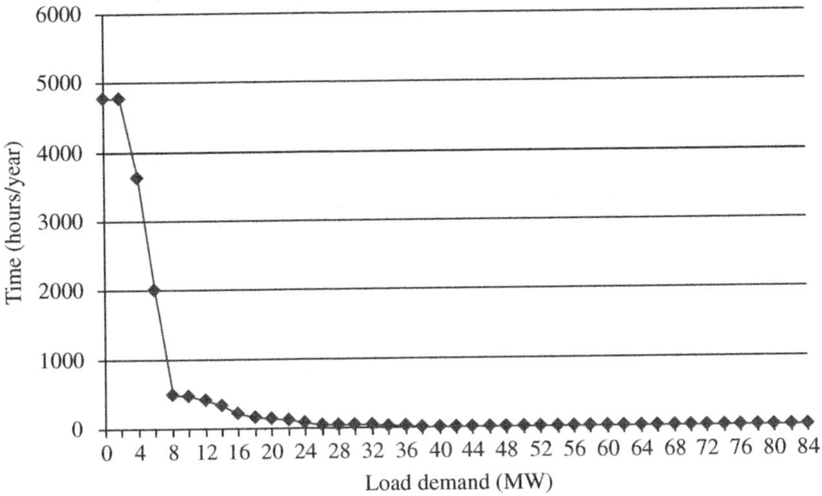

Figure 13.17 Nodal CMELDC (Φ1,3) at bus 3 after generator #1 is loaded (base case).

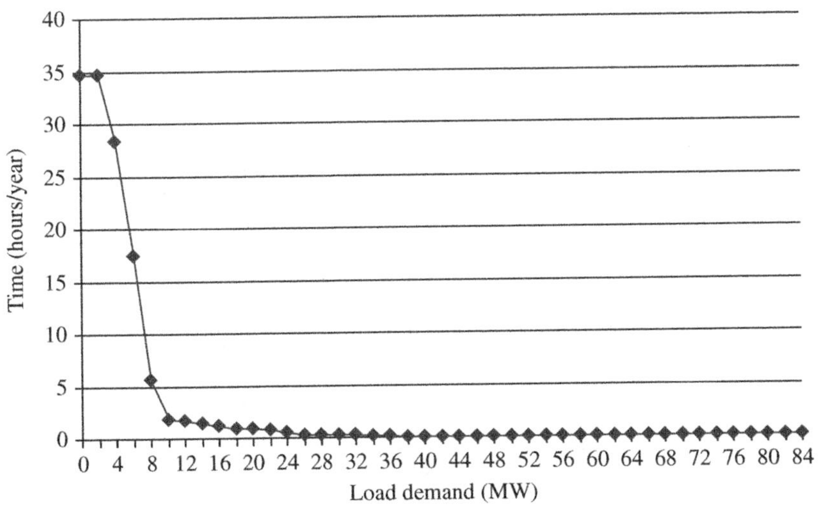

Figure 13.18 Nodal CMELDC (Φ3,3) at bus 3 after generator #3 is loaded (base case).

unlimited line capacity) of the transmission grid. This is designated as Case 0 in this chapter. The average unit production cost of Case 0 is 6.376 $/MWh and is lower than that of the base case (Case 1).

Table 13.12 Nodal probabilistic production energy (MWh/year) (base case).

Order	Bus 2	Bus 3	Bus 4	Bus 5	Total
G 5	37 122	157 771	74 245	74 245	343 383
G 6	18 635	79 201	37 271	37 271	172 378
G 7	18 578	78 959	37 154	37 154	171 846
G 8	18 588	79 005	37 173	37 173	171 940
G 9	18 510	78 678	37 017	37 017	171 222
G 10	4 993	21 378	9 943	9 943	46 256
G 11	4 868	21 117	9 803	9 803	45 592
G 1	26 052	122 310	52 052	52 052	252 466
G 2	4 736	24 807	4 920	4 920	39 384
G 3	189	965	269	269	1 692
G 4	24	161	26	26	237
Total	152 296	664 351	299 874	299 874	1 416 396

Table 13.13 Nodal probabilistic production cost ($/year) (base case).

Order	Bus 2	Bus 3	Bus 4	Bus 5	Total
G 5	116 514	176 838	135 075	135 075	563 502
G 6	58 587	88 870	67 905	67 905	283 266
G 7	58 558	88 749	67 847	67 847	283 000
G 8	58 553	88 761	67 845	67 845	283 004
G 9	58 493	88 577	67 746	67 746	282 562
G 10	14 072	23 075	17 357	17 357	71 861
G 11	13 458	22 830	17 082	17 082	70 452
G 1	957 621	2 218 704	1 357 524	1 357 524	5 891 373
G 2	240 283	715 190	187 062	187 062	1 329 596
G 3	8 624	22 814	8 924	8 924	49 286
G 4	1 344	4 333	1 041	1 041	7 758
Total	1 586 105	3 538 741	1 995 408	1 995 408	9 115 660

Simulated probabilistic production energy and cost values for the base case, Case 0, the cases of a line contingency and a line addition are compared in Tables 13.17 and 13.18, Case 2 is a case study for the line contingency between bus 1 and bus 3,

Table 13.14 Nodal probabilistic marginal production cost per unit energy ($/MWh) (base case).

Order	Bus 2	Bus 3	Bus 4	Bus 5	Average
G 5	3.139	1.121	1.819	1.819	1.641
G 6	3.144	1.122	1.822	1.822	1.643
G 7	3.152	1.124	1.826	1.826	1.647
G 8	3.150	1.123	1.825	1.825	1.646
G 9	3.160	1.126	1.830	1.830	1.650
G 10	2.818	1.079	1.746	1.746	1.554
G 11	2.764	1.081	1.743	1.743	1.545
G 1	36.758	18.140	26.080	26.080	23.335
G 2	50.736	28.831	38.017	38.017	33.760
G 3	45.626	23.642	33.160	33.160	29.125
G 4	55.845	26.988	39.934	39.934	32.771
Total	10.415	5.327	6.654	6.654	6.436

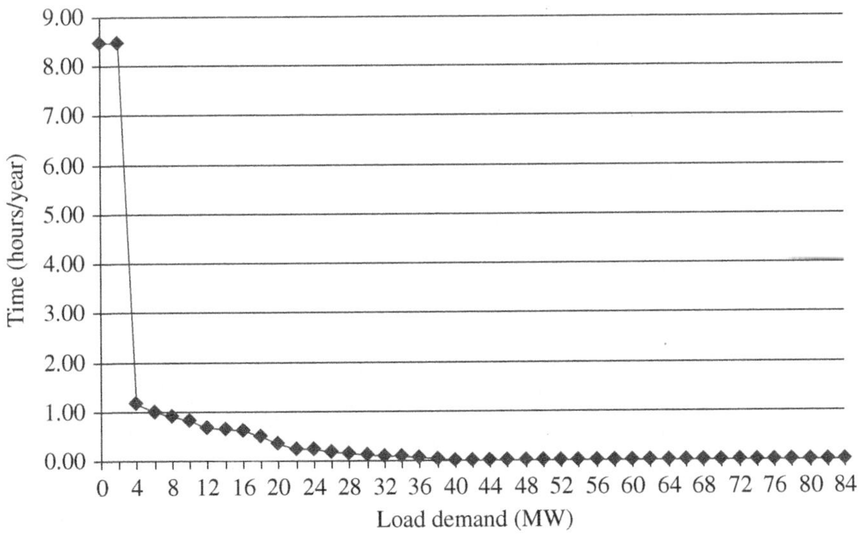

Figure 13.19 Nodal CMELDC (Φ4,3) at bus 3 after generator #4 is loaded (base case).

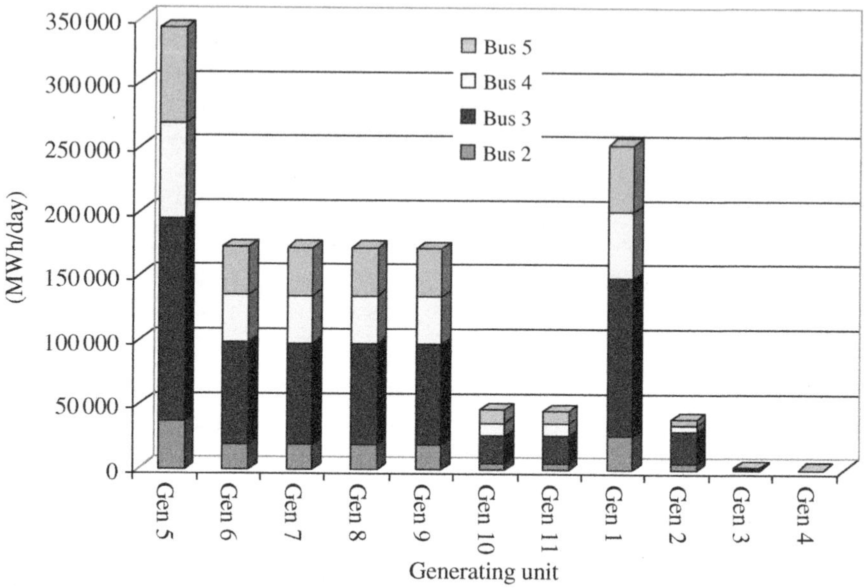

Figure 13.20 Nodal probabilistic production energy of the base case (base case).

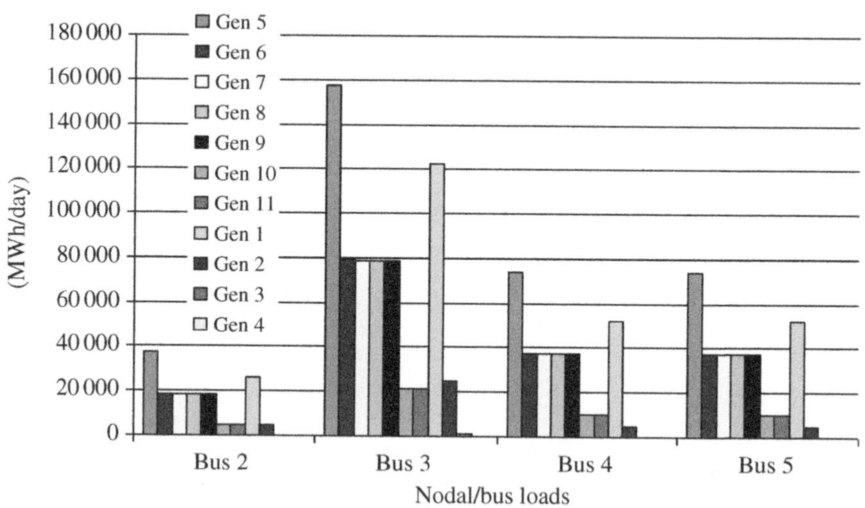

Figure 13.21 Probabilistic production energy of generators delivered to load points of the base case (base case).

Table 13.15 Probabilistic production energy under the assumption of a perfect grid (MWh/year) (Case 0).

Order	Bus 2	Bus 3	Bus 4	Bus 5	Total
G5	37 123	157 775	74 247	74 247	343 392
G6	18 656	79 290	37 313	37 313	172 572
G7	18 681	79 398	37 361	37 361	172 801
G8	18 693	79 451	37 383	37 383	172 911
G9	18 692	79 452	37 381	37 381	172 907
G10	5036	21 562	10 029	10 029	46 658
G11	4 910	21 299	9 888	9 888	45 986
G1	25 889	121 717	51 721	51 721	251 049
G2	4 491	23 731	4 403	4 403	37 027
G3	124	652	144	144	1 064
G4	9	47	10	10	76
Total	152 305	664 375	299 881	299 881	1 416 443

Table 13.16 Probabilistic production cost under the assumption of a perfect grid ($/year) (Case 0).

Order	Bus 2	Bus 3	Bus 4	Bus 5	Total
G5	116 514	176 840	135 076	135 076	563 506
G6	58 598	88 914	67 926	67 926	283 363
G7	58 610	88 968	67 950	67 950	283 478
G8	58 605	88 984	67 950	67 950	283 490
G9	58 584	88 964	67 928	67 928	283 404
G10	14 085	23 167	17 400	17 400	72 053
G11	13 466	22 921	17 124	17 124	70 635
G1	954 567	2 211 591	1 353 364	1 353 364	5 872 887
G2	232 982	699 964	175 728	175 728	1 284 401
G3	5 693	16 475	5 070	5 070	32 307
G4	399	1 186	353	353	2 291
Total	1 572 103	3 507 974	1 975 870	1 975 870	9 031 815

Table 13.17 Nodal probabilistic production energy for the cases (MWh/year).

Cases	Bus 2	Bus 3	Bus 4	Bus 5	Total
Case 0	152 305	664 375	299 881	299 881	1 416 443
Case 1	152 296	664 351	299 874	299 874	1 416 396
Case 2	151 934	662 687	299 170	299 170	1 412 964
Case 3	152 126	663 528	299 547	299 547	1 414 747
Case 4	150 685	660 433	298 738	298 738	1 408 597
Case 5	152 304	664 373	299 880	299 880	1 416 437

Table 13.18 Nodal probabilistic production costs for the cases ($/year).

Cases	Bus 2	Bus 3	Bus 4	Bus 5	Total
Case 0	1 572 103	3 507 974	1 975 870	1 975 870	9 031 815
Case 1	1 586 105	3 538 741	1 995 408	1 995 408	9 115 660
Case 2	1 588 919	3 525 505	1 994 627	1 994 627	9 103 679
Case 3	2 254 700	4 893 299	2 961 453	2 961 453	13 070 905
Case 4	3 489 414	7 813 030	4 625 140	4 625 140	20 552 724
Case 5	1 574 578	3 512 778	1 979 545	1 979 545	9 046 446

Case 3 is a case study for the line contingency between bus 1 and bus 2, Case 4 is a case study for the line contingency between bus 2 and bus 4, and Case 5 is a case study for the line addition between bus 2 and bus 3. Table 13.19 shows the differences in the costs of other cases from those of Case 0. The differences come from the delivery limitations and unavailability of the transmission system. The cost differences represent the congestion cost of the grid for each case. The delivery limitations and unavailability of the lines are included in the congestion cost and it can be designated as the probabilistic congestion cost. The base case study indicates that the model is a good delivery system because the production cost of the present system is similar to that of an assumed ideal system with a perfect grid and the congestion cost is relatively small. A one-circuit line contingency between bus 1 and bus 3 results in a large congestion cost. Case 5 (new line addition between bus 2 and bus 4) provides a decreased congestion cost (69 214 $/year).

Tables 13.20 and 13.21, respectively show the conventional nodal reliability indices (LOLE (Hrs/year), EENS (MWh/year) and ELC (MW/curtailment/year)) of the base case and the case (Case 0) of perfect grid delivery. As expected, the perfect

Table 13.19 Differences (probabilistic congestion cost) between the probabilistic production costs of Case 0 and those of the other cases ($/year) (reference is Case 0).

Cases	Bus 2	Bus 3	Bus 4	Bus 5	Total
Case 0	—	—	—	—	—
Case 1	14 002	30 767	19 538	19 538	83 845
Case 2	16 816	17 531	18 757	18 757	71 864
Case 3	682 597	1 385 325	985 583	985 583	4 039 090
Case 4	1 917 311	4 305 056	2 649 270	2 649 270	11 520 909
Case 5	2 475	4 804	3 675	3 675	14 631

Table 13.20 Reliability indices in the base case (Case 1).

Indices	Bus 2	Bus 3	Bus 4	Bus 5	Bulk
LOLE	4.595	8.461	2.217	2.217	4.174
EENS	11.516	32.557	9.941	9.941	63.955
ELC	2.506	3.848	4.484	4.484	15.322

Table 13.21 Reliability indices under the assumption of a perfect grid (Case 0).

Indices	Bus 2	Bus 3	Bus 4	Bus 5	Bulk
LOLE	0.651	0.994	0.634	0.634	0.796
EENS	1.943	9.383	2.793	2.793	16.913
ELC	2.986	9.443	4.406	4.406	21.240

delivery system (Case 0) gives higher reliability than the base case considering delivery capacity limitations and the grid unavailability.

Table 13.22 compares the probabilistic reliability indices of all the cases. As expected, the perfect delivery system (Case 0) gives a higher reliability level than the other cases considering delivery capacity limitations and grid unavailability. Case 4 is the most severe contingency case in regard to the total congestion and the average congestion cost per EENS is the highest in Case 3 (Table 13.23).

The costs (Tables 13.9 and 13.10) and the reliability EENS ratios are calculated in order to identify the relationship and effect between production and congestion costs and reliability (EENS). Tables 13.24 and 13.25 show the unit nodal

Table 13.22 Probabilistic reliability indices for the bulk system.

Cases	LOLE (hrs/year)	EENS (MWh/year)	ELC (MW/Curt./year)	Remark
Case 0	0.796	16.913	21.240	
Case 1	4.174	63.955	15.322	Base case
Case 2	57.966	3497.226	60.332	
Case 3	44.349	1712.070	38.604	
Case 4	569.222	7864.140	13.816	
Case 5	0.965	22.051	22.846	

Table 13.23 Bulk system unit probabilistic congestion cost per EENS ($/MWh).

Cases	Congestion cost ($/year)	EENS (MWh/year)	Reliability congestion cost ($/MWh)	Remark
Case 0	—	16.913	0	
Case 1	83 845	63.955	1311.0	Base case
Case 2	71 864	3497.226	20.6	
Case 3	4 039 090	1712.070	2359.2	
Case 4	11 520 909	7864.140	1465.0	
Case 5	14 631	22.051	663.5	

Table 13.24 Nodal probabilistic production costs per EENS ($/MWh).

Cases	Bus 2	Bus 3	Bus 4	Bus 5	Total
Case 0	809 111.2	373 864.9	707 436.4	707 436.4	2 597 848.9
Case 1	137 730.5	108 693.7	200 725.1	200 725.1	647 874.4
Case 2	4 260.2	2 078.0	2 794.2	2 794.2	11 926.7
Case 3	12 439.8	5 717.0	8 775.9	8 775.9	35 708.7
Case 4	2 150.9	1 977.6	4 037.3	4 037.3	12 203.2
Case 5	552 871.5	333 977.8	455 801.3	455 801.3	1 798 451.8

Table 13.25 Nodal probabilistic congestion cost per EENS ($/MWh).

Cases	Bus 2	Bus 3	Bus 4	Bus 5	Bulk
Case 0	—	—	—	—	—
Case 1	1215.87	945.02	1965.40	1965.40	1215.87
Case 2	45.09	10.33	26.28	26.28	45.09
Case 3	3766.07	1618.52	2920.67	2920.67	3766.07
Case 4	1181.86	1089.70	2312.57	2312.57	1181.86
Case 5	869.03	456.74	846.19	846.19	869.03

production cost and congestion cost per nodal EENS, respectively. Figures 13.22 and 13.23 show them graphically. In a higher reliability system, the unit production cost per EENS is larger and the unit congestion cost per EENS is smaller. Case 4 has the most serious production cost for transmission system congestion, but the largest unit congestion cost per reliability (EENS) goes to Case 3. Figures 13.22 and 13.23 indicate to the system planner or operator which line should be managed in view of the nodal production cost, congestion, and reliability.

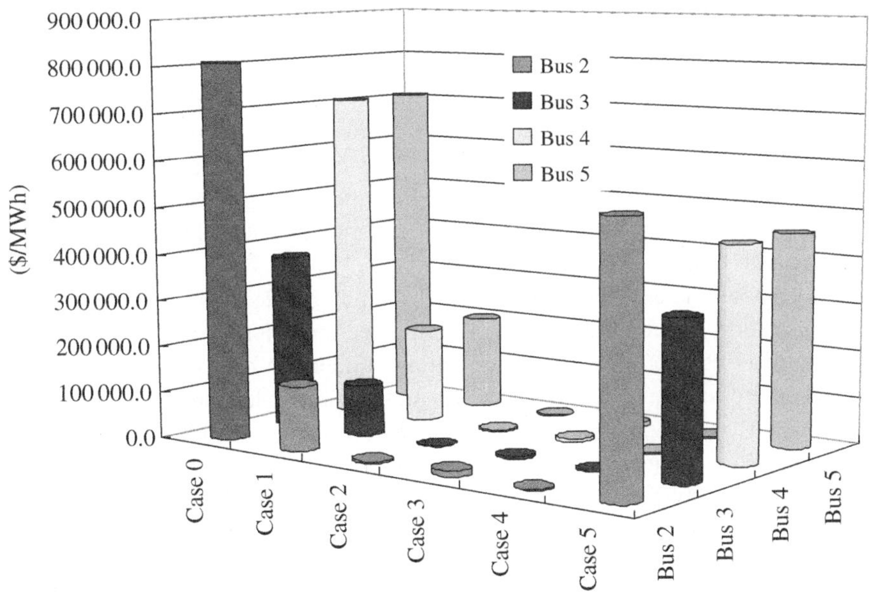

Figure 13.22 Nodal probabilistic production cost per EENS ($/MWh).

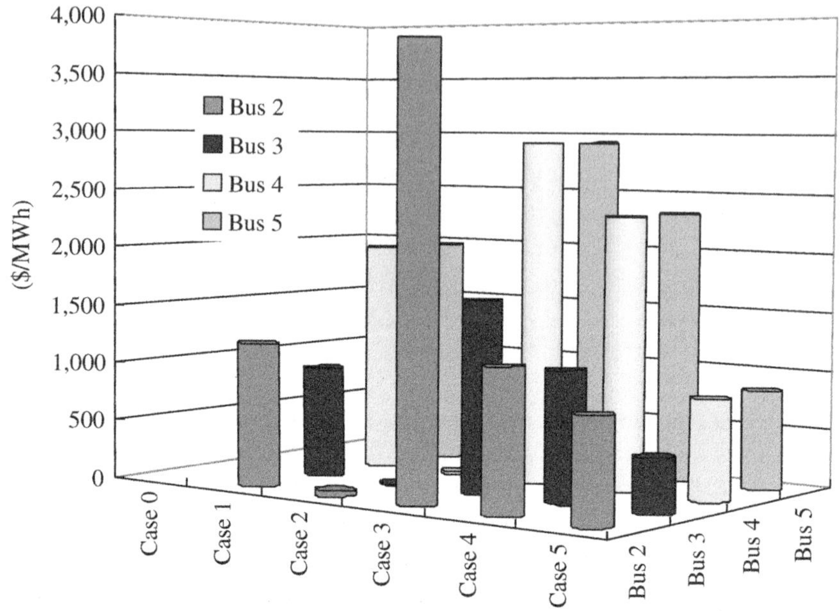

Figure 13.23 Nodal probabilistic congestion cost per EENS ($/MWh).

13.6 Conclusion

This chapter illustrates the concept and formulation of an extended effective load model for probabilistic production cost simulation and reliability evaluation at a load point in a composite power system. The extended effective load model includes generators as well as transmission line uncertainties. The proposed effective load model is obtained from convolution integral processing of the outage capacity PDF of the SFEG.

This proposed nodal effective load model should provide essential solutions for many problems based on nodal and decentralized operation and the control philosophy of electric power systems operations under the competitive electricity market environment. The CMELDC based on the effective load model at HLII will extend the various application research areas of nodal probabilistic production cost simulation, outage cost assessment, reliability evaluation, etc. As an example, the differences between the probabilistic production costs at HLI and HLII come from the uncertainties associated with failures in the transmission system facilities. This difference, therefore, indicates the price/value of the probabilistic congestion and reliability in viewpoints of operation of the transmission system. The price concept-based nodal-effective load model can provide essential tools for the development of methodologies for ISO and TRANSCO under future competitive and deregulated electricity markets. For the other example, the nodal ELDC within EENS area can be assessed from this model and this curve is useful for outage cost assessment using the analytical method. Additionally, the proposed SFEG model will be used for reliability evaluation of the multi-area interconnected systems because the forced outage rates of transmission lines of an assisting system have been considered in the model.

This chapter is the first step in the application of the new effective load model. Research on the development of the methodology using optimal AC or DC load flow for evaluating more accurate maximum arrival power and the application of Monte Carlo simulation for large-sized power systems will be conducted in the future.

14

Reliability Constraints

14.1 Deterministic Reliability Constraint Using Contingency Constraints

14.1.1 Introduction

Transmission expansion planning (TEP) with open access to the grid has become an important issue in the electric utility industry in recent years. The recent blackouts that have occurred in countries worldwide suggest that more reliable grid structures may be needed to establish successful deregulated electricity markets. These incidents call for the development of new tools that can address system uncertainties and significantly enhance the effectiveness of transmission planning. However, the basic objective of strengthening a transmission grid is relevant to most countries.

The TEP addresses the problem of augmenting an existing generation and transmission network to optimally serve a growing electric load while satisfying a set of economic and technical constraints. The problem is to minimize the cost of expansion subject to the constraints needed to meet an explicit reliability level. Various techniques, including branch and bound, sensitivity analysis, Bender decomposition, simulated annealing, genetic algorithms, tabu search, and Greedy Randomized Adaptive Search Procedure (GRASP), have been used to study the problem. Since it is difficult to obtain the optimal solution for a realistic system considering both generators and transmission lines simultaneously, TEP is usually performed after generation expansion planning. Typically, deterministic reliability criteria such as the $(N-1)$ or $(N-2)$ contingency criteria and load balance constraints are used in practice for transmission expansion planning because they are computationally tractable.

In a typical power system planning problem, adequacy or security standards may be used initially to select the reasonable plans from preliminary scenarios suggested from the strategic policy is called *the first macro stage*. More detailed technical analyses, which are mainly contingency analysis, fault analysis, and stability

Probabilistic Power System Expansion Planning with Renewable Energy Resources and Energy Storage Systems, First Edition. Jaeseok Choi and Kwang Y. Lee.

Published 2022 by John Wiley & Sons, Inc.

analysis, are applied to check the engineering feasibility of the plans. This is called *the second micro stage*. The conventional work procedure for power system planning is shown in Appendix 14.A. A deterministic reliability criterion, such as load-balancing constraints (adequacy), is often used in the first stage.

This chapter presents a methodology for choosing the best transmission expansion plan using an adequacy-based security criterion based on an $(N-\alpha)$ contingency criterion done by eliminating contingencies with probabilities lower than a prescribed probability limit. An $(N-\alpha)$ contingency is a single contingency with *α simultaneous component failures*. The objective function considers construction (investment) costs, operation/production costs and generator standby costs. An assessment of congestion costs, caused by the noncoherent characteristics of generators on a grid and conflicting situations occurring under deregulation, may be considered because the operation/production cost of generators is included in the objective function. However, congestion rents and the shadow prices of transmission are not considered because the focus of this chapter is on evaluating different contingency/security criteria for TEP rather than on evaluating market designs. The TEP problem in this chapter is modeled in an integer programming framework and determines the optimum mix of additions to a transmission network using a branch and bound method based on a network flow approach and the maximum flow-minimum cut-set theorem [29].

14.1.2 Transmission Expansion Planning Problem

A composite power system that includes generation and transmission facilities is shown in Figure 14.1. *TS* refers to the transmission system, *NG* is the number of generators, ${}_k\Phi_0$ is the load duration curve at the *kth* load bus, and *NL* is the number of load points (bus). Some generators such as solar or wind power may be sited at a load point. A composite power system is designated as HLII (hierarchical level II) in this chapter. For GEP in Part I, HLI is used to designate generation and load components only.

1) *The Objective Function* – The conventional definition of cost in TEP is to minimize the total construction costs plus the operating costs. This chapter proposes a new definition of the total cost (C^T) that includes the construction cost (*CC*) associated with investing in new transmission lines, the operation/production cost (*OC*), and the standby cost (*SC*) of generators as expressed in (14.1). In the case of a competitive electricity market environment, a grid owner, or ISO, or grid reliability committee will want to consider the problem of searching for a better plan from the perspective of a GENCO as well as the grid owners. This chapter considers production cost instead of the transmission line loss cost for the *OC*. A network model and a network flow method that

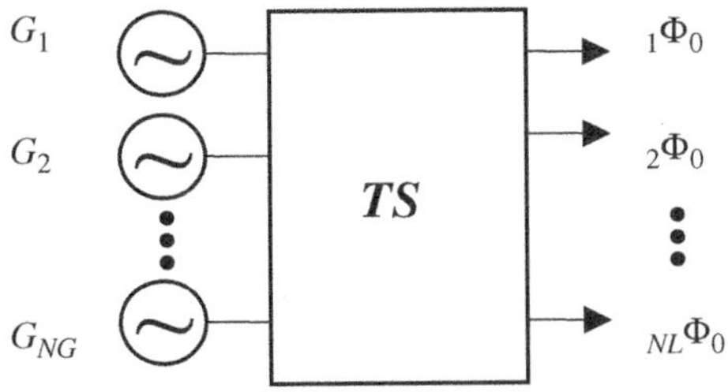

Figure 14.1 A composite power system with nodal load duration curves.

neglects line loss are used in this chapter. Consequently, the objective is to minimize the total cost given by:

$$C^T = CC + OC + SC \quad (\$) \tag{14.1}$$

i) *The Construction Cost*

The total cost of construction is the summation of the construction costs of new elements. In general, stricter reliability criteria are expected to increase the construction cost because additional transmission lines will be required in order to increase reliability. Therefore, the construction cost, CC, of new lines and transformers, can be formulated as:

$$CC = \sum_{(x,y)\in\sigma}\left[\sum_{i=1}^{m(x,y)} C_{(x,y)}^{(i)} U_{(x,y)}^{(i)}\right] \quad (\$) \tag{14.2}$$

where

ρ: the set of all transformers and transmission lines, which are represented as branches (segments connecting nodes) in a network model,

$m(x,y)$: the number of circuits (ckt) of new candidate branches (segments connecting nodes in a network) connecting buses x and y, which are represented as nodes (points to which branches are connected in a network),

$C_{(x,y)}^{(i)}$: the sum of construction costs of the 1-st ckt through the i-th ckt of the new transformers and lines connecting buses x and y. Specifically, $C_{(x,y)}^{(i)} = \sum_{j=1}^{i} \Delta C_{(x,y)}^{(j)}$,

$\Delta C_{(x,y)}^{(j)}$: the construction cost of the j-th ckt connecting buses x and y,

$U_{(x,y)}^{(i)}$: a decision (integer) variable associated with the ckt connecting buses x and y (1 if the 1-st ckt to i-th ckt lines are to be constructed, and 0 otherwise).

$$\sum_{i=1}^{m(x,y)} U_{(x,y)}^{(i)} = 1$$

$$P_{(x,y)}^{(i)} = \sum_{j=1}^{i} \Delta P_{(x,y)}^{(j)}$$

where

$P_{(x,y)}^{(i)}$: the sum of the capacities of the 1-st ckt through the i-th ckt connecting buses x and y,

$\Delta P_{(x,y)}^{(j)}$: the capacity of the j-th ckt connecting buses x and y

ii) *Strategic bidding-based operating (production) cost*

Operating cost based on locational marginal prices (LMPs) in a competitive electricity market could also be evaluated. In this chapter, however, it is assumed that generator offers are based on their true marginal costs because this chapter is focused not on LMPs but rather the effects of various types of contingency criteria. Consequently, a standard merit order method is used to evaluate the operating cost. Therefore, offer-based operating (production) cost is calculated as the energy produced times the true marginal operating cost for each generator. The operating cost is formulated as:

$$OC = \text{minimum}\left(\sum_{x=1}^{NG} MC_x\left(\sum_{k=1}^{NL}\left(\int_{AC_{x-1,k}}^{AC_{x,k}} {}_k\Phi_O(AC_{x,k}(p))dp\right)\right)\right) \quad (\$) \tag{14.3}$$

where

$AC_{x,\,k}$: the total arrival capacity at the k-th-load bus after generators #1 to # x are loaded,

p: the variable for real power (MW),

${}_k\Phi_0$: the load duration curve at load bus k,

MC_x: the marginal operating cost of generator$x$ (\$/MWh),

$$AC_{x,k} = \sum_{i=1} AP_{(i,k)}(p)$$

AP(i,k): maximum arrival of power delivered from generator i to load bus k (MW),

i: the merit order number of a generator

Figure 14.2 shows the deterministic production of energy by generator i at load bus k using the merit order method (8760 hours is used for T in the case studies).

iii) *Standby cost*

In some situations, a generator may be committed but may not produce electrical energy to balance the load in real time. This generator is not shut

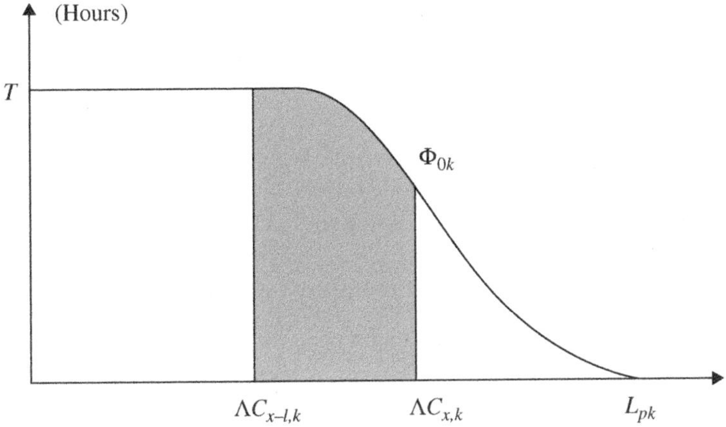

Figure 14.2 Deterministic operating/production energy of *x*-generator on load duration curve at the load bus *k*.

down but is operated on a standby (reserve) basis to maintain the reliability of the power system. A reasonable payment (cost) for standby by a generator should therefore be paid (called the standby cost in this chapter). Once again, if a stricter reliability criterion is required, the standby cost is expected to decrease because the grid will have more flexibility for delivering energy. The standby cost of generators, *SC*, is formulated as:

$$SC = \sum_{x=1}^{NG} MS_x Eg_x T \quad (\$) \tag{14.4}$$

where

MS_x: marginal standby cost of generator x (\$/MWh),

Eg_x: reserve-generating capacity of generator x (MW),

T: study period time (hours)

2) *Constraints*

i) *Minimum cut-set flow constraint*

In a general deterministic approach, meeting load requires that the total capacity of the branches (segments connecting nodes in a network) in a minimum cut-set should be greater than or equal to the total load (L), including the reserve rate. This is also referred to as the "bottleneck" capacity. Therefore, a no-shortage power supply constraint can be expressed as follows [28–31]:

$$P_c(S, T) \geq L \; (S \in S, T \in T) \tag{14.5}$$

where

$$L = \sum_{k=1}^{NL} L_{pk}(1 + BRR_k/100)$$

Here, $P_c(s,t)$ is the capacity of the minimum cut-set of the two subsets S and T containing source node s and terminal node t, respectively, when all nodes are separated by a minimum cut-set, L_{pk} is the peak load at load bus k, BRR_k is the bus reserve/marginal rate at load bus k, which is defined as $(AP_k - L_{pk}) \times 100/L_{pk}$, and AP_k and L_{pk} are the maximum arrival power and peak load, respectively at load bus k.

The demand constraint can be expressed as:

$$\sum_{(x,y)\in(X_k,\overline{X}_k)} \left[P_{(x,y)} = P^{(0)}_{(x,y)} + \sum_{i=1}^{m(x,y)} P^{(i)}_{(x,y)} U^{(i)}_{(x,y)} \right] \geq L \tag{14.6}$$

where k is the cut-set number ($k = 1, n$) and n is the number of cut-sets.

In a security-constraint approach to adequacy, the criterion of no-shortage of power supply requires that the total capacity of the branches in the minimum cut-set under $(N - \alpha)$ contingencies should be greater than or equal to the total load, L. This is also referred to as the bottleneck capacity. Therefore, the constraint of no-shortage power supply under $(N - \alpha)$ contingencies can be expressed by (14.7) [20]:

$$P_c(S, T)_{N-A} \geq L \ (S \in S, Tt \in T) \tag{14.7}$$

where

α: the depth of a contingency,

$P_c(s,t)_{N-\alpha}$: the capacity of the minimum cut-set of two subsets (S and T) containing source node s and terminal node t, respectively, when all nodes are separated by a minimum cut-set under $(N - \alpha)$ contingencies.

ii) *Security state probability constraint*

Generally, using a security criterion stricter than $(N - 3)$ contingencies requires a substantial amount of computation for a realistic system. Development of a modified or simplified method (for example, by considering outages of selected elements only) is necessary in order to make deep contingency criteria practical. In a contingency analysis, the actual occurrence of a deep contingency has a very low probability. For example, the occurrence of an $(N - 6)$ contingency in a power system with 100 generators each having a *FOR* (forced outage rate) of 0.01 and 500 lines each having a *FOR* of 0.001 is less than 10^{-9} when the outages are mutually independent. This type of rare contingency can be ignored for planning purposes. Based on this principle, the method used in this chapter is to eliminate all contingencies with probabilities lower than a given probability limit ε_p (Note that $\varepsilon_p = 10^{-9}$ was used in the case studies presented later in the chapter). The system probability ($\omega_{N-\alpha}$) under the $(N-\alpha)$ security criterion is formulated as:

$$\omega_{N-a} = \prod_{i \in \sigma N-1} (1-Q_i) \prod_{i \in \sigma N-1} Q_i \tag{14.8}$$

where

Q_i: the forced outage rate of element i (generators and lines),
$\rho_{N-\alpha}$: the set of branches not included with $(N-\alpha)$ contingency,
$\rho_{N-\alpha}$: the set of branches included with $(N-\alpha)$ contingency

Therefore, the system security probability ($\Omega_{N-\alpha}$) for the $N-\alpha$ security criterion is formulated as

$$\Omega_{N-a} = \sum_{i \in \beta(N-a)} \omega_{(N-a)} \tag{14.9}$$

where

β: the set of security systems existing within the probability limit under $(N-\alpha)$ contingencies.

14.1.3 Maximum Flow Under Contingency Analysis for Security Constraint

In order to consider the security constraints from the viewpoint of adequacy in this study, it is necessary to evaluate whether or not the power flows under $(N-\alpha)$ contingencies satisfy the peak load. The power flow in the network of each arc (transmission line) $(i, j) \in \rho$ is limited with a non-negative real number P_{ij} that is called the transmission line capacity (ratings) from node i to node j. The maximum static power flow, $F_{m\,N-\alpha}$ from s (source) to t (terminal) in branch set, $\rho_{N-\alpha}$ under $(N-\alpha)$ contingencies can be formulated as a linear optimization problem in (14.10). Note that the summation of flows at all nodes except the source and terminal nodes in (14.10) is always zero by Kirchhoff's first law:

$$\text{Maximize } F_{N-a} = \sum_{j \in \Lambda s} x_{sj}$$

$$\sum X_{ij} = \begin{cases} -K, & i = s \\ 0, & i \neq s, t \\ K, & i = t \end{cases} \tag{14.10}$$

$$0 \leq x_{ij} \leq P_{ij} \ (i,j) \in \rho_{(N-\alpha)}$$

where

$F_{N-\alpha}$: the flow under $(N-\alpha)$ contingencies,
x_{ij}: the real power flow on the branch between nodes i and j,
Λ_s: a set of nodes connected with a source (or terminal) node,
$\rho_{N-\alpha}$: a set of branches/lines under $N-\alpha$ contingency,
K: the total load ($= L$),
P_{ij}: a branch (generator and line) capacity between nodes i and j

The maximum power flow ($F_{mN-\alpha}$) using branch set, $\rho_{N-\alpha}$ under $(N-\alpha)$ contingencies can be calculated by an LP. This is a solution to the shortest path problem to determine the bottleneck capacity under $(N-\alpha)$ contingencies. It can be expressed as follows using the maximum flow minimum cut-set theorem:

$$P_c(S,T)_{N-\alpha} = F_{m\,N-\alpha} \tag{14.11}$$

where $F_{mN-\alpha}$ is the maximum power flow for branch set $\rho_{N-\alpha}$ under $(N-\alpha)$ contingencies.

14.1.4 Alternative Types of Contingency Criteria

This chapter proposes various new kinds of contingency criteria. The $(N-\alpha)$ criterion can be categorized in terms of the outage number, where α describes the depth of the contingency. In other words, the "$N-1$" criterion considers contingencies in which one element fails out of "N" elements (generators, lines, transformers, and switching gear, etc.).

In this chapter, the three types of contingency criteria, defined in Table 14.1, are proposed. First, the "$N-n$" criterion considers all possible $N-1$, $N-2$,..., $N-n$ contingencies with depths 1, 2,..., and n, respectively, with a maximum of "n" elements failing. Second, the "$N-nGmT$" criterion considers all possible $N-1$, $N-2$,..., $N-n$ contingencies for generators as well as all possible $N-1$, $N-2$,..., $N-m$ contingencies for transmission lines up to a maximum (n,m) elements failing. Third, the "$N-nG-mT$" criterion considers all possible $N-1$, $N-2$,..., $N-n$ contingencies for generators combined with all possible $N-1$, $N-2$,..., $N-m$ contingencies for transmission lines up to a maximum of $(n+m)$ elements failing.

Table 14.1 Alternative definitions of $(N - \alpha)$ contingencies.

	Contingency number configuration			
Contingency stem type	**Generator**	**Line**	**Maximum**	**Remarks**
$N - n$	n	N	n	$(= N - n\text{G}n\text{T})$
$N - n\text{G}m\text{T}$	n	M	Maximum (n,m)	
$N - n\text{G} - m\text{T}$	n	M	$n + m$	

Where G and T present generators and transmission lines.

14.1.5 Solution Algorithm

The proposed integer programming problem for TEP could be solved by any one of implicit enumeration, branch and bound, or the Gomory cutting plane methods [38]. Since the branch and bound method has significant merit for problems with many constraints, it is used in this chapter.

The proposed solution algorithm uses the following steps:

1) Check the "need" for transmission expansion for the system and the "possibility" of meeting load using the candidate lines. These can be checked for a given reliability criterion by considering the system with no candidate lines and with all candidate lines, respectively.
2) Set $j = 1$ (initial system), $jopt = 0$, $jmax = 0$, $C_{opt}^{T} = \infty$ and $ENNOD_j = 0$, where *jopt* is a parameter of the optimal solution in a solution graph for the branch and bound; *jmax* is the maximum number of branches that should be searched in a solution graph; *ENNOD* is a parameter that indicates whether a branch can be terminated in a solution graph (If it is 1, the branch is bounded).
3) If $ENNOD_j = 1$, the #*j* system is an end node at which the branch operation of branch and bound is finished (*Bound*) in a solution graph. In this situation, there is no need to consider any of the other graphs following the #*j* system. Go to 13.
4) Calculate the minimum cut-set using the maximum flow method for #*j* system (#*j* solution in the solution graph).
5) Select a #*i* branch/line of the candidate branches/lines set (S_j) in the minimum cut-set and add it to the #*j* system. In what follows, the new system is named the #*ji* system and it is called a *Branch* in a solution graph.
6) If the #*ji* system has already been considered in the solution graph, go to 14.
7) If $\omega_{N-\alpha} \leq \varepsilon$, go to 14.

8) Calculate the total cost $C_{ji}^T = C_j^T + CC_{ji} + OC_{ji} + SC_{ji}$ for the $\#ji$ system and evaluate the security level of the system.
9) If $C_{ji}^T < C_{jopt}^T$, the current system ($\#ji$) with a cost of C_{ji}^T may be optimal. If not, go to 12.
10) Set $jmax = jmax + 1$.
11) Check the security criterion; If $P_c(s,t)_{N-\alpha} > L_p$, set $C_{opt}^T = C_{ji}^T$, and $F_{moptN-\alpha} = P_c(s,t)_{N-\alpha}$, $jopt = jmax$, and go to 13.
12) Set $C^T_{j\,max} = C_{ji}^T$, $ENNOD_{jmax} = 1$, and go to 14.
13) Add this $\#jmax(\#ji)$ solution to the solution graph.
14) If all the candidate branches/lines in the cut-set S_j have been considered, go to 15. Otherwise, set $i = i + 1$ and go to 5.
15) If $j = jmax$, continue to the next step. Otherwise, set $j = j + 1$ and go to 4.
16) For $j = jmax$, the solution graph has been constructed fully and the optimal solution $jopt$ has the lowest cost C_{jopt}^T, and it also satisfies the required security criterion in step 11.

14.1.6 Case Studies

1) *Case Study I: 5 Bus System* – The method presented in this section was applied to a 5-bus sample system, shown in Figure 14.3, with 1530 MW of generation capacity and 900 MW of peak load. This initial system satisfies at least the ($N-3$) contingency criterion. After 10 years, it is forecasted that loads will increase from the initial level of 300 in all bus to 550 MW, 600 and 650 MW

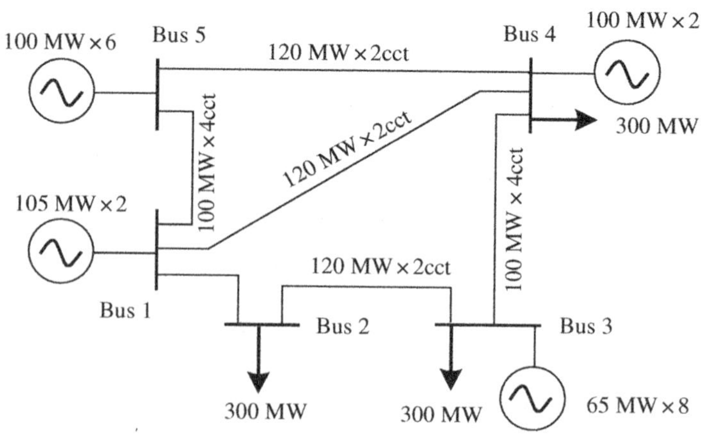

Figure 14.3 5-bus sample system (present year).

at load buses 2, 3, and 4, respectively. It is assumed that all generators are assumed to increase the existing generating capacity by 50%. The TRANSCOs should respond to this decision by generators and determine how to strengthen the grid in order to deliver the electrical energy reliably to meet the increased load. The TRANSCOs identify the candidate elements for constructing new lines shown in Table 14.2, where *GN*, *TL*, and *LD* represent the generators, transmission lines, and loads, respectively; *SB* and *EB* are the start and end buses of a line; $\Delta P^{(0)}_{(x,y)}$ and $\Delta C^{(0)}_{(x,y)}$ are the capacity and cost of the existing line that connects nodes *x* and *y*. In this study, three candidate lines are considered, appearing $m(x,y) = 2$ in (14.2) and (14.6). In Table 14.2, the parentheses for $\Delta P^{(j)}_{(x,y)}$ and $\Delta C^{(j)}_{(x,y)}$ are omitted for convenience. The cost unit, M$ in this table stands for a million dollars. Table 14.3 shows the forced outage rates of the generators and transmission lines. The "NN" column is the ckt of elements (generators or lines). Figure 14.4 shows the inverted load duration curves at the load buses for the higher forecast loads. Therefore, the grid expansion problem is "what is the optimal/best line choice (solution) with the minimum total cost under a security criterion constraint for the increased load?" Initially, the

Table 14.2 New generation capacity, future load and capacity, and cost data of system candidate lines.

NL	SB	EB	ID	NN	ΔP^0_{xy}	ΔP^1_{xy}	ΔP^2_{xy}	ΔP^3_{xy}	ΔC^1_{xy}	ΔC^2_{xy}	ΔC^3_{xy}
1	0	1	GN	3	105	—	—	—	—	—	—
2	0	3	GN	12	65	—	—	—	—	—	—
3	0	4	GN	3	100	—	—	—	—	—	—
4	0	5	GN	9	100	—	—	—	—	—	—
5	1	2	TL	4	100	100	100	100	45	45	45
6	2	3	TL	2	120	120	120	120	40	40	40
7	3	4	TL	4	100	100	100	100	55	55	55
8	4	5	TL	2	120	120	120	120	50	50	50
9	1	4	TL	2	120	120	120	120	60	60	60
10	1	5	TL	4	100	100	100	100	40	40	40
11	2	6	LD	1	550	—	—	—	—	—	—
12	3	6	LD	1	600	—	—	—	—	—	—
13	4	6	LD	1	650	—	—	—	—	—	—

$\Delta P(*)$: (MW) and $\Delta C(*)$: (M$)); (SB #0 and EB #6 represent the source and terminal nodes, respectively.

Table 14.3 Operating and standby marginal costs of generators and FOR.

	Operating marginal cost ($/MWh)	Standby marginal cost ($/MWh)	FOR
G1	15	10	0.015
G3	31	10	0.005
G4	30	10	0.015
G5	10	10	0.005
Lines	–	–	0.004 57

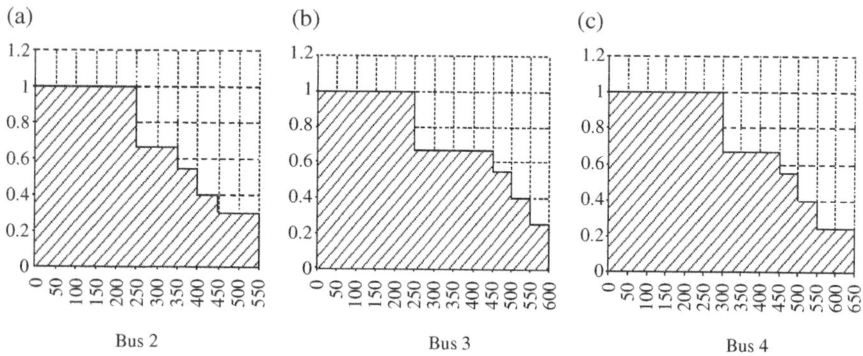

Figure 14.4 Inverted load duration curves (Φ_0) at the load buses.

Bus Reserve Rate in (14.5) is set to zero, $BRR_k = 0$, for cases 1–15 defined in Table 14.7, and the effects of setting $BRR_k > 0$ are evaluated in cases 16–18 defined in Table 14.9.

Figures 14.5 and 14.6 show the optimal systems for the $(N-1)$ and $(N-2)$ criteria, respectively, where the dotted line and bold number represent the new construction of generators and lines. Configurations of the power dispatched, energy delivered and operating cost for the optimal system using the $(N-1)$ criterion are shown in Tables 14.4–14.6, respectively.

From these tables, it is interesting to note that although the marginal cost of the generator at bus 5 is the lowest, part of this generator's capacity should be operated on standby and the standby cost (22.776 M$/year) should be paid to the generator's owner. The reason for this is that the delivery capacity of the grid is still limited for generator #5 even though a new line is constructed in the optimal plan.

Table 14.7 shows the components of cost for the optimal expansion plans for different contingency criteria. Case 0 corresponds to an $(N-0)$ criterion that meets only the constraint to balance the load. The results show that the number

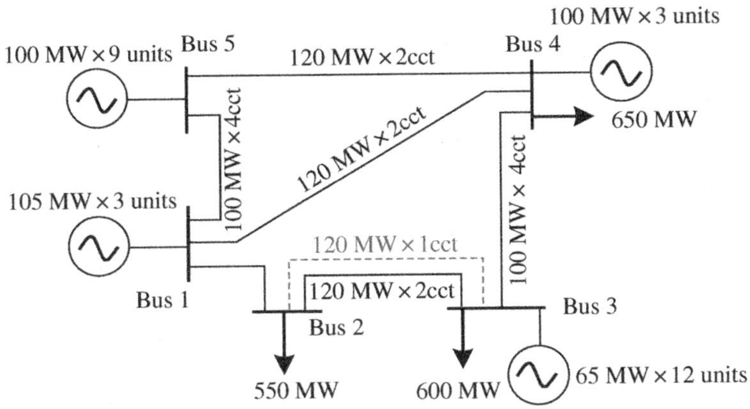

Figure 14.5 Optimal system under (*N* − 1) contingency security criterion (Case 1).

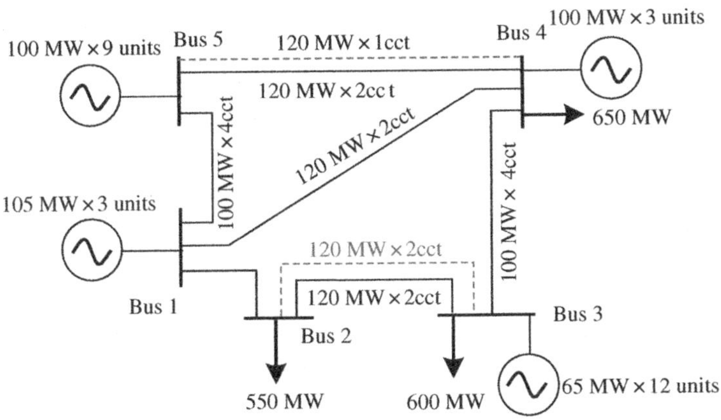

Figure 14.6 Optimal system under (*N* − 2) contingency security criterion (Case 2).

Table 14.4 Configuration of power delivered from generators to loads for the new system of Case 1 ((*N* − 1) contingency): (MW).

Plant # (cap.)	L2	L3	L4	Total	Standby
G1 (315 MW)	0	0	315	315	0
G3 (780 MW)	150	300	95	545	235
G4 (300 MW)	0	300	0	300	0
G5 (900 MW)	400	0	240	640	260
Total	550	600	650	1800	495

Table 14.5 Configuration of energy delivered from generators to loads for the new system of Case 1(($N-1$) contingency): (GWh/year).

Plant # (cap.)	L2	L3	L4	Total	Standby
G1 (315 MW)	0	0	1566.3	1566.3	0
G3 (780 MW)	350.4	1285.1	208.1	1843.5	2058.6
G4 (300 MW)	0	2336.3	0	2336.3	0
G5 (900 MW)	2752.4	0	2102.4	4854.8	2277.6
Total	3102.8	3621.4	3876.7	10 600.9	4336.2

Table 14.6 Configuration of operating cost from generators to load for the new system of Case 1(($N-1$) contingency): (M$/year).

Plant # (cap.)	L2	L3	L4	Total	Standby
G1 (315 MW)	0	0	23.494	23.494	0
G3 (780 MW)	10.862	39.838	6.450	57.150	20.586
G4 (300 MW)	0	70.089	0	70.089	0
G5 (900 MW)	27.524	0	21.024	48.548	22.776
Total	38.386	10.993	50.968	199.281	43.362

of new construction elements increases and the total cost is higher, as the contingency/security criterion gets stronger. For example, it is interesting to see that Cases 4, 5, and 6 have higher costs than Cases 1–3, respectively. Cases 7–10 present results using the more complex criteria of ($N-n\text{G}-m\text{T}$) contingencies. Cases 11–15 present results for ($N-n\text{G}m\text{T}$) contingencies. Testing ($N-4$) contingencies for the 5-bus system failed because the candidate lines cannot support a feasible solution.

The security probabilities for all contingency criteria were evaluated and the results are shown in the last column of Table 14.7. The highest security probability (0.999 95) was obtained for Case 6 (($N-3\text{G}-3T$) contingencies). Although the ($N-3\text{G}-3T$) criterion has the highest security probability, this does not mean that it is necessarily the most reliable. The expansion plans for Case 10 (($N-2\text{G}-3T$) contingencies) and Case 6 (($N-3\text{G}-3\text{T}$) contingencies) are identical and they are both equally reliable. The security probability is lower for Case 10 because fewer contingency events are considered compared to Case 6.

Table 14.7 Optimal expansion plans and costs using alternative security criteria (5-bus system).

Case (security criterion)	Construction of new lines	*CC* (M$/yr)	*OC* (M$/yr)	*SC* (M$/yr)	Total cost (M$/yr)	Security probability
Case 0 ($N-0$)	—	0	199.281	43.362	242.643	—
Case 1 (N–1)	T_{2-3}^1	40.0	199.281	43.362	282.643	0.967 64
Case 2 (N–2)	T_{2-3}^1, T_{2-3}^2 and T_{4-5}^1	130.0	171.456	43.362	344.818	0.996 90
Case 3 (N–3)	T_{1-2}^1, T_{2-3}^1, T_{2-3}^2, T_{4-5}^1, and T_{1-5}^1	215.0	166.397	43.362	424.759	0.999 76
Case 4 ($N-1G-1T$)	T_{1-2}^1	45.0	199.281	43.362	287.643	0.980 58
Case 5 ($N-2G-2T$)	T_{1-2}^1, T_{2-3}^1, T_{4-5}^1 and T_{4-5}^2	185.0	159.134	43.362	387.496	0.998 92
Case 6 ($N-3G-3T$)	T_{1-2}^1, T_{1-2}^2, T_{1-2}^3, T_{4-5}^1, T_{1-5}^1 and T_{1-5}^2	265.0	165.324	43.362	473.686	0.999 95
Case 7 ($N-2G-1T$)	T_{1-2}^1 and T_{4-5}^1	95.0	171.456	43.362	309.818	0.995 31
Case 8 ($N-1G-2T$)	T_{1-2}^1, T_{2-3}^1 and T_{4-5}^1	135.0	171.456	43.362	349.818	0.983 80
Case 9 ($N-3G-2T$)	T_{2-3}^1, T_{2-3}^2, T_{4-5}^1, T_{4-5}^2 and T_{4-5}^3	230.0	160.402	43.362	433.764	0.999 80
Case 10 ($N-2G-3T$)	T_{1-2}^1, T_{1-2}^2, T_{1-2}^3, T_{4-5}^1, T_{1-5}^1 and T_{1-5}^2	265.0	165.324	43.362	473.686	0.999 06

(Continued)

Table 14.7 (Continued)

Case (security criterion)	Construction of new lines	*CC* (M$/yr)	*OC* (M$/yr)	*SC* (M$/yr)	Total cost (M$/yr)	Security probability
Case 11 ($N-0G1T$)	$T_{2\text{-}3}^{1}$	40.0	199.281	43.362	282.643	0.967 64
Case 12 ($N-2G1T$)	$T_{1\text{-}2}^{1}$	45.0	199.281	43.362	287.643	0.994 56
Case 13 ($N-1G2T$)	$T_{2\text{-}3}^{1}$, $T_{2\text{-}3}^{2}$ and $T_{4\text{-}5}^{1}$	130.0	171.456	43.362	344.818	0.983 15
Case 14 ($N-3G2T$)	$T_{1\text{-}2}^{1}$, $T_{2\text{-}3}^{2}$ and $T_{4\text{-}5}^{1}$	135.0	171.456	43.362	349.818	0.999 70
Case 15 ($N-2G3T$)	$T_{1\text{-}2}^{1}$, $T_{2\text{-}3}^{1}$, $T_{2\text{-}3}^{2}$, $T_{4\text{-}5}^{1}$, and $T_{1\text{-}5}^{1}$	215.0	166.397	43.362	424.759	0.998 96

Where, *CC*: construction cost, *OC*: operating cost, *SC*: standby cost.

Table 14.8 Alternative security criteria ranked by total cost (5-bus system).

Order	Case number	Contingency criterion	Total cost (M$)
1	Case 0	$N-0$	242.643
2	Cases 1 and 11	$N-1$ & $N-0$G1T	282.643
3	Cases 4 and 12	$N-1$G $-$ 1T & $N-2$G1T	287.643
4	Case 7	$N-2$G $-$ 1T	309.818
5	Cases 2 and 13	$N-2$ & $N-1$G2T	344.818
6	Cases 8 and 14	$N-1$G $-$ 2T & $N-3$G2T	349.818
7	Case 5	$N-2$G $-$ 2T	387.496
8	Cases 3 and 15	$N-3$ & $N-2$G3T	424.759
9	Case 9	$N-3$G $-$ 3T	433.764
10	Cases 6 and 10	$N-2$G $-$ 3T	473.686

Table 14.8 ranks the security criteria according to the total cost. The following inequalities can be determined from the results, where $TC_{(N-n)}$ is the total cost for $(N-n)$ contingencies, and show, as expected, that the deeper contingency criteria require higher total costs for transmission expansion planning.

$$TC_{(N-n)} \leq TC_{(N-nG-mT)} \leq TC_{(N-(n+m))}$$

Figure 14.7 shows how the components of the total cost vary for the different security criteria. Cases 11–15 are omitted for convenience because the costs are identical to Cases 1, 4, 2, 8, and 3, respectively. It is interesting to note how much the operating costs decrease with the stronger contingency criteria. For example, while the operating cost is 199.281 M$/year for $(N-1)$ contingencies, it is 171.456 and 166.397 M$/year for $(N-2)$ and $(N-3)$ contingencies, respectively, because congestion costs are lower when the grid is more flexible. In general, as the security criterion gets stronger, the investment cost increases substantially while the operating cost decreases slightly. These characteristics do not always hold because of the noncoherency of the composite power system. Security criteria using composite contingencies of the $(N-n\text{G}-n\text{T})$ type generally require higher total costs than the simpler criteria using $(N-n)$ contingencies. However, the ranking of the security criteria in Figure 14.7.is based on the total cost and this is not necessarily the same as ranking the level of reliability.

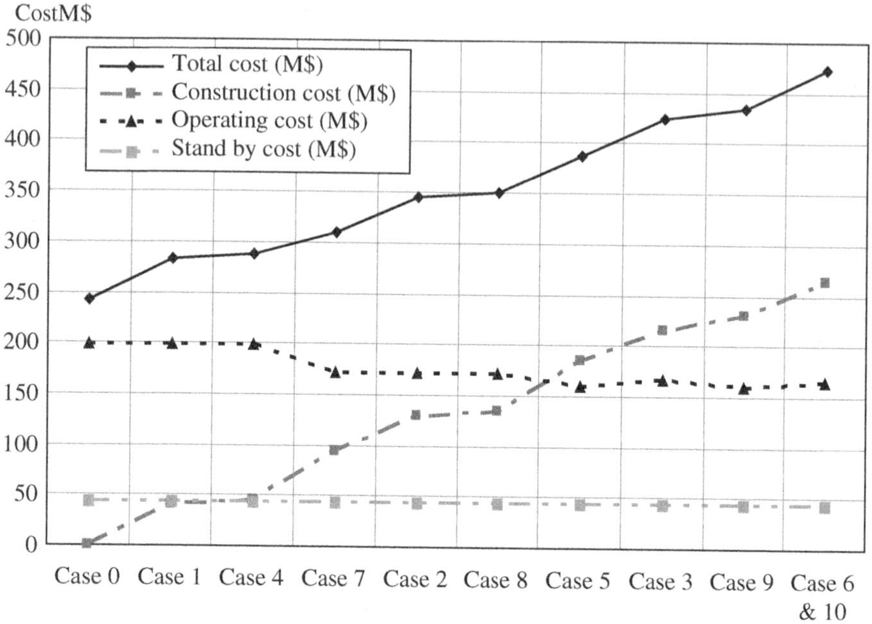

Figure 14.7 Costs variation due to security criterion types (cases).

This chapter evaluates a number of modified contingency/security criteria as alternatives to the standard criterion of $(N-\alpha)$ contingencies. The basic objective is to provide additional intermediate criteria between $(N-2)$ and $(N-3)$ contingencies, for example, when the former is considered too weak and the latter too strong. In practice, the step from $(N-2)$ to $(N-3)$ contingencies may be too ambitious because meeting an $(N-3)$ criterion is too expensive and computationally demanding for real systems. One alternative criterion uses the Bus Reserve Rate index (BRR_k), defined in (14.5) and below in (14.12), to represent situations in which customers at a particular load bus ask for additional reliability. The deterministic reliability (reserve) constraint for a specified load bus increases when $BRR_k > 0$.

$$P_c(s,t)_{N-a} \geq \sum_{k=1}^{NL} L_{pk}(1 + BRR_k/100) \tag{14.12}$$

where, BRR_k is $(AP_k - L_{pk}) \times 100/L_{pk}$ and AP_k and L_{pk} are the maximum arrival power and peak load, respectively at load bus k.

Table 14.9 shows the results for three case studies using the modified security criterion. In Cases 16–18, the basic criterion is $(N-1)$ contingencies, and in addition, $BRR_k = 20$ at load buses 2, 3, and 4, respectively. Comparing the

Table 14.9 Optimal expansion plans by combining an $(N-1)$ contingency criterion with $BRR_k = 20$ at buses 2, 3, and 4.

Case (security criterion)	Construction of new lines	CC (M $/yr)	OC (M$/yr)	SC (M $/yr)	Total cost (M$/yr)	Bus reserve rate (%)
Case 16 $(N-1)$	$T_{1-2}{}^1$ and $T_{2-3}{}^1$	85.0	199.281	33.726	318.007	$BRR_2 = 20$
Case 17 $(N-1)$	$T_{2-3}{}^1$ and $T_{4-5}{}^1$	90.0	171.361	32.850	294.711	$BRR_3 = 20$
Case 18 $(N-1)$	$T_{1-2}{}^1$ and $T_{1-5}{}^1$	85.0	194.222	31.974	311.196	$BRR_4 = 20$

Where $BRR_k = (AP_k - L_{pk}) \times 100/L_{pk}$.

results in Table 14.9 with Case 1 in Table 14.7 shows that more candidate lines are connected directly or indirectly with load bus k when $BRR_K > 0$. Therefore, the modified constraint in (14.12) with $BRR_K > 0$ can be used as an alternative criterion to give greater reliability than the simple criterion of $(N-1)$ contingencies.

2) *Case Study II: 21 Bus System* – The proposed reliability criteria were also tested on the 21-bus system shown in Figure 14.8 that is part of the grid in the south-eastern region (Youngnam) of Korea. After specifying a forecast of the future system load, the alterative criteria were applied and the results compared. The marginal operating costs, the Forced Outage Rates, and the capacity and construction costs are shown in Tables 14.A.2, 14.A.3 and 14.A.4, respectively, of Appendix 14.A. Figure 14.9 shows the inverted load duration curves for the four largest loads.

Using the criterion with $(N-1)$ contingencies in Case II-1, Figure 14.10 shows the new system with the dotted lines presenting the three new lines that were required to meet reliability and the higher forecasted load. The optimal expansion plan has a total cost of 1254.81 M$ and the new transmission lines are $T_{13-17}{}^1$, $T_{13-15}{}^1$ and $T_{9-11}{}^1$. Table 14.10 shows the configuration of power delivered from generators to loads for the new system in Figure 14.10.

The criterion with $(N-2)$ contingencies was applied to the same initial system shown in Figure 14.8 (Case II-2), and Figure 14.11 shows the new system. The optimal expansion plan in Case II-2 has a total cost of 1979.81 M$, and the new transmission lines are $TF_{6-21}{}^1$, $TF_{16-17}{}^1$, $TF_{1-2}{}^1$, $T_{13-15}{}^1$, $T_{13-15}{}^2$, $T_{9-10}{}^1$, $T_{9-11}{}^1$ and $T_{9-11}{}^2$. Table 14.11 shows the configuration of power delivered from generators to loads in Case II-2. Comparing Cases II-1 and II-2 in Tables 14.10 and 14.11,

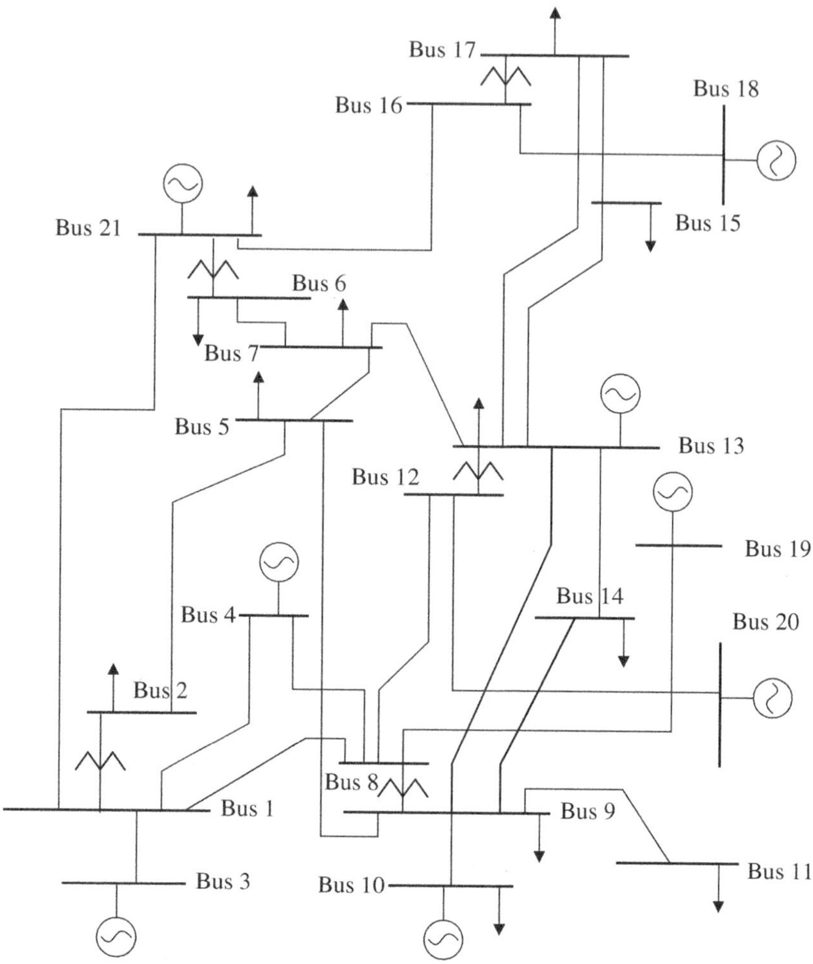

Figure 14.8 21-bus model system.

the pattern of deliveries from G3 and the levels of generation for G10, G18, and G20 are substantially different. Therefore, the distributions of both generation and reserve capacity change when the reliability criterion changes even though the total generation and the total reserve capacity remain unchanged.

Table 14.12 shows a summary of the components of the total cost of the optimal expansion plans using different security criteria. It is interesting to note that the results are identical for $(N-2G1T)$ and $(N-1G-1T)$ contingencies and for $(N-1G2T)$ and $(N-2)$ (which could also be written as $(N-2G2T)$)

contingencies. This suggests that transmission lines may be more important than generators for maintaining reliability in this particular system. It also implies that the initial system in Figure 14.8 has a transmission system that is relatively weak given the amount and location of installed generation capacity. Evaluating the relative importance of system reliability between adding new transmission lines or new generators to a given system is an interesting topic for future research.

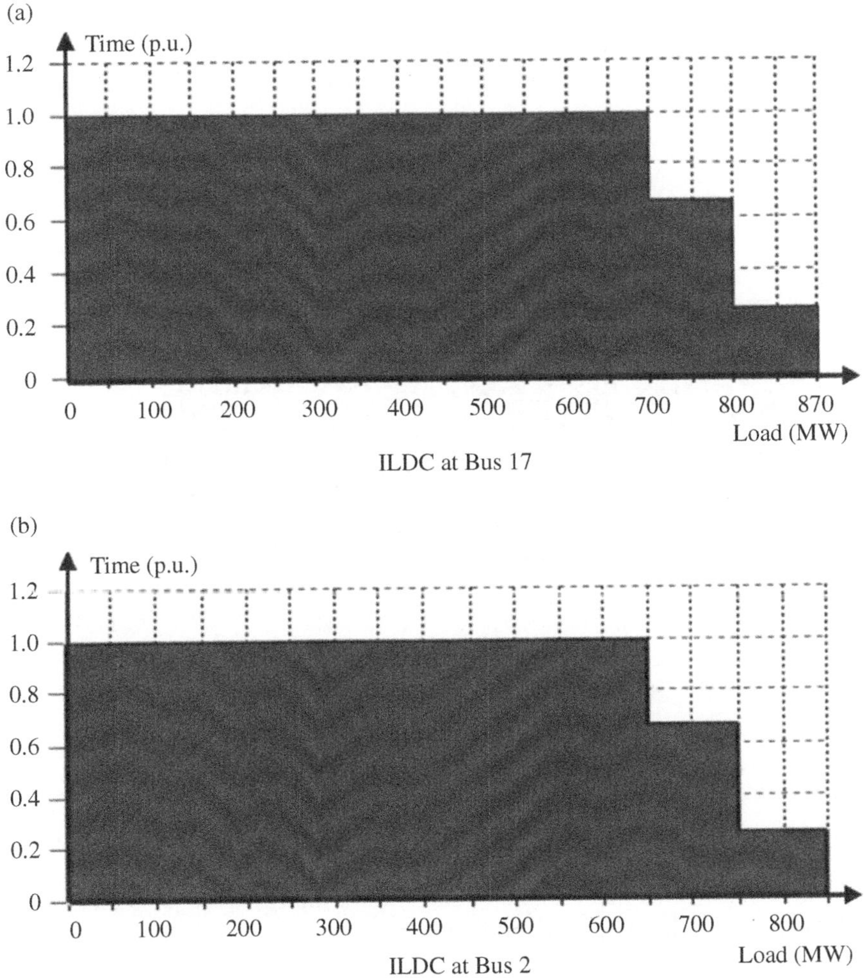

Figure 14.9 Inverted load duration curves at the buses with the four largest loads.

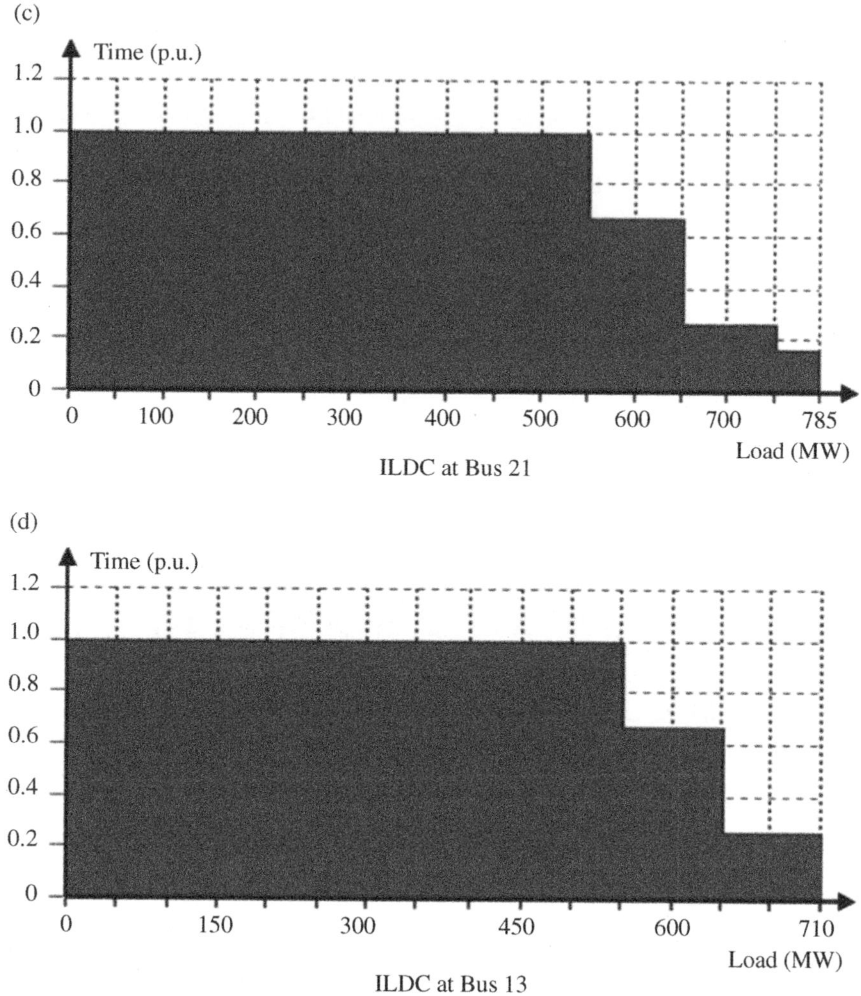

Figure 14.9 (Continued)

Table 14.13 summarizes the results by ranking Cases II-0 through II-7 by the total cost. This ranking is consistent with the results for the 5-Bus system in Table 14.8 and shows that the following inequalities hold. The table shows also computation time for the various criteria. The computation time is introduced in Table 14.13. It shows that the computation time is increased exponentially more as the applied contingency criterion is stronger.

$$TC_{(N-n)} \leq TC_{(N-nG-mT)} \leq TC_{(N-(n+m))}$$

Table 14.10 Configuration of power delivered from generators to loads for the optimal system in Case II-1 (MW).

	L21	L6	L2	L9	L10	L11	L14	L13	L15	L17	L7	L5	Total	Res.
G3	0	0	0	50	0	350	0	0	450	0	0	0	850	0
G21	785	115	0	0	0	0	0	0	0	0	0	0	900	0
G4	0	600	0	250	0	0	0	0	0	0	0	0	850	0
G10	0	0	720	0	17	0	0	0	0	0	0	0	737	163
G20	0	35	120	75	0	0	0	0	0	0	80	70	380	820
G18	0	0	10	220	0	0	190	0	0	70	210	0	700	150
G13	0	0	0	0	0	200	0	560	0	0	0	00	760	0
G19	0	0	0	0	0	0	0	150	0	800	0	0	950	0
Sum	785	750	850	595	17	550	190	710	450	870	290	70	6127	1133

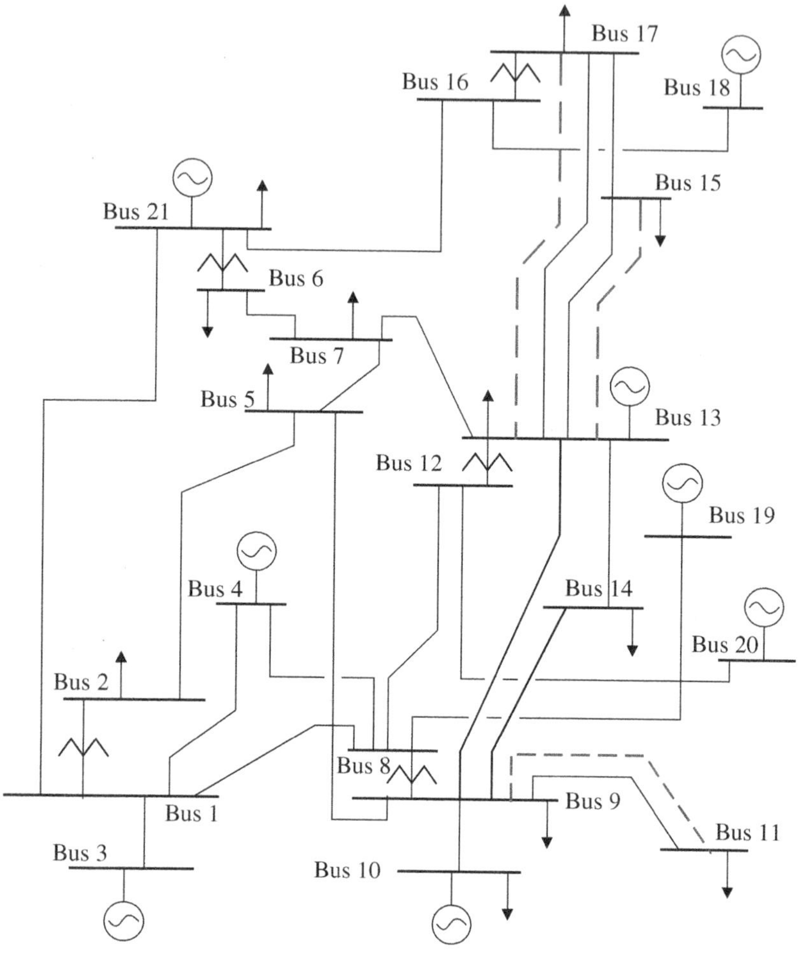

Figure 14.10 Optimal system using an $(N-1)$ criterion (Case II-1).

Table 14.14 shows the results from four case studies using $(N-1)$ contingencies and setting BRR_K in (14.12) to 20 and 30% at the two main load buses, Bus 6 and Bus 17. The results show that the optimal expansion plans have more candidate lines connected directly or indirectly to the load bus with $BRR_K > 0$ in order to satisfy the higher reliability requirements of that load. These results also confirm that setting $BRR_K > 0$ for a load is a practical way to augment the reliability of a conventional $(N-1)$ criterion using a larger, more realistic system.

Table 14.11 Configuration of power delive-ed from generators to load for the new system of Case II-2 ($N-2$ Contingency): (MW).

	L21	L6	L2	L9	L10	L11	L14	L13	L15	L17	L7	L5	Total	Res.
G3	0	0	370	0	0	40	0	0	440	0	0	0	850	0
G21	785	115	0	0	0	0	0	0	0	0	0	0	900	0
G4	0	600	0	250	0	0	0	0	0	0	0	0	850	0
G10	0	0	480	0	17	0	0	0	0	0	0	0	497	403
G20	0	35	0	125	0	110	190	0	10	70	290	70	900	300
G18	0	0	0	220	0	200	0	0	0	0	0	0	420	430
G13	0	0	0	0	0	200	0	560	0	0	0	00	760	0
G19	0	0	0	0	0	0	0	150	0	800	0	0	950	0
Sum	785	750	850	595	17	550	190	710	450	870	290	70	6127	1133

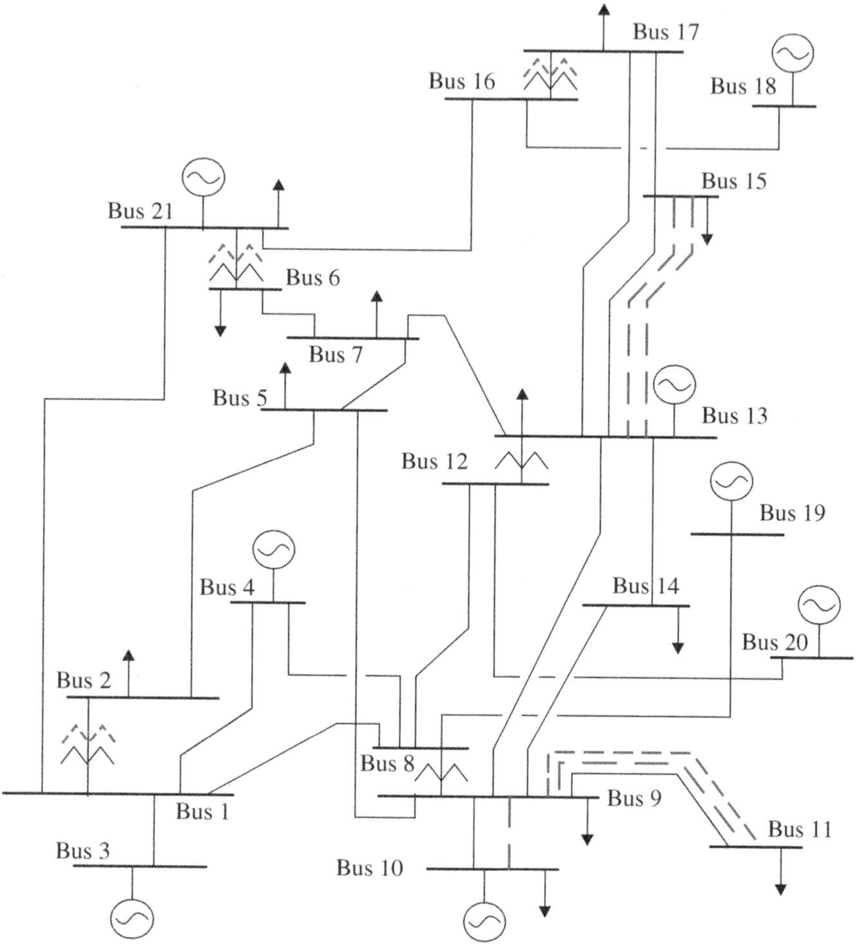

Figure 14.11 Optimal system by the $N-2$ security criterion approach (Case II-2).

14.1.7 Conclusion

This chapter presents a methodology for choosing the best transmission expansion plan using alternative security criteria based on different specifications of $(N - \alpha)$ contingencies. The proposed method minimizes the total cost of the investment, operation, and standby costs (reserve-generating capacity) of the system. It is challenging to consider these three costs simultaneously when searching for an optimal expansion plan. Specifically, the standby cost model proposed in this chapter implicitly puts a value on reliability using a deterministic approach. This is similar to the practice of including outage costs using a probabilistic approach. The optimal locations and capacities of new transmission lines can be determined using the proposed method.

Table 14.12 Optimal expansion plans and costs using alternative security criteria (21-bus system).

Cases	Construction of new lines	CC (M$)	OC (M$)	SC (M$)	Total cost (M$)	Remarks (security probability)
Case II-0 ($N-0$)	—	0	1260.131	297.752	1557.883	—
Case II-1 ($N-1$)	T_{13-17}^{1}, T_{13-15}^{1}, and T_{9-11}^{1}	197.0	1260.058	297.752	1754.810	0.981 39
Case II-2 ($N-2$)	TF_{6-21}^{1}, TF_{16-17}^{1}, TF_{1-2}^{1}, T_{13-15}^{1}, T_{13-15}^{2}, T_{9-10}^{1}, T_{9-11}^{1}, and T_{9-11}^{2}	744.0	1240.929	297.752	2282.682	0.995 64
Case II-3 ($N-1G-1T$)	T_{13-17}^{1}, T_{13-15}^{1}, T_{9-10}^{1}, and T_{9-11}^{1}	278.0	1243.620	297.752	1819.372	0.992 40
Case II-4 ($N-2G-1T$)	T_{21-1}^{1}, T_{13-17}^{1}, T_{13-17}^{2}, T_{13-15}^{1}, T_{8-19}^{1}, T_{12-20}^{1}, T_{9-10}^{1}, T_{9-10}^{2}, and T_{9-11}^{1}	549.0	1236.379	297.752	2083.132	0.993 43
Case II-5 ($N-0G1T$)	T_{13-17}^{1}, T_{13-15}^{1}, and T_{9-11}^{1}	197.0	1260.058	297.752	1754.810	0.981 39
Case II-6 ($N-2G1T$)	T_{13-17}^{1}, T_{13-15}^{1}, T_{9-10}^{1}, and T_{9-11}^{1}	278.0	1243.620	297.752	1819.372	0.992 40
Case II-7 ($N-1G2T$)	TF_{6-21}^{1}, TF_{16-17}^{1}, TF_{1-2}^{1}, T_{13-15}^{1}, T_{13-15}^{2}, T_{9-10}^{1}, T_{9-11}^{1}, and T_{9-11}^{2}	744.0	1240.929	297.752	2282.682	0.978 21

Where *CC*: construction cost, *OC*: operating cost, *SC*: standby cost.

Table 14.13 Alternative security criteria ranked by total cost and computational time for the criteria (21-bus system).

Order	Case number	Contingency type	Total cost (M$)	Computation time (sec)
1	Case II-0	$N-0$	1557.883	2
2	Case II-1 and II-5	$N-1$ and $N-0\text{G}1\text{T}$	1754.810	5
3	Case II-3 and II-6	$N-1\text{G}-1\text{T}$ and $N-2\text{G}1\text{T}$	1819.372	18
4	Case II-4	$N-2\text{G}-1\text{T}$	2083.132	25
5	Case II-2 and II-7	$N-2$ and $N-1\text{G}2\text{T}$	2282.682	3781

Pentium M, 1.4 GHz.

Table 14.14 Optimal expansion plans by combining an (N – 1) contingency criterion with BRR_k = 20 and 30 at buses 6 and 17.

Cases (security type)	Construction of new lines	CC (M$/yr)	OC (M$/yr)	SC (M$/yr)	Total cost (M$/yr)	BRR at buses 6 & 17
Case II-8 ($N-1$)	$T_{5-7}{}^{1}$, $T_{13-17}{}^{1}$, $T_{13-15}{}^{1}$, and $T_{9-11}{}^{1}$	274.0	1257.809	281.984	1813.793	$BRR_6 = 20\%$
Case II-9 ($N-1$)	$T_{16-17}{}^{1}$, $T_{13-15}{}^{1}$, and $T_{9-11}{}^{1}$	252.0	1260.654	252.025	1764.679	$BRR_{17} = 20\%$
Case II-10 ($N-1$)	$T_{6-21}{}^{1}$, $T_{13-15}{}^{1}$, $T_{13-15}{}^{2}$, and $T_{9-11}{}^{1}$	343.0	1258.065	255.704	1856.769	$BRR_6 = 30\%$
Case II-11 ($N-1$)	$T_{16-17}{}^{1}$, $T_{13-15}{}^{1}$, $T_{9-10}{}^{1}$, and $T_{9-11}{}^{1}$	333.0	1244.215	229.162	1806.377	$BRR_{17} = 30\%$

Where $BRR_k = (AP_k - L_{pk}) \times 100/L_{pk}$.

The proposed procedure represents the first macro-evaluation stage in the process of selecting a transmission expansion plan that would be followed by additional dynamic analyses. The chapter presents a practical approach to planning that should serve as a useful guide for decision makers in selecting a reasonable expansion plan prior to checking system stability, quality, and dynamics in more detail. In this chapter, modified security criteria of the form $N - n\text{G}m\text{T}$ and $N - n\text{G} - m\text{T}$ are proposed as intermediate steps between the typical $N - \text{c}$ and $N - \text{c} - 1$ criteria (for $c = m + n - 1$). The implicit assumption is that the cost and computational burden of the $N - c - 1$ criterion are too great, but the security provided by the $N - c$ criterion is too low. The various case studies for a 5-bus and a 21-bus system show that substantially different expansion plans are optimal for the same level of system load when different security criteria are used. In general, a stronger security criterion is associated with a higher investment cost, but at the same time, the operating cost may be slightly lower due to less congestion on the grid.

The chapter concludes that the modified security criteria can be used effectively for transmission expansion planning in both regulated and deregulated electricity markets. Since grid operators are often asked to adopt a stronger security criterion, such as a criterion covering ($N-2$) contingencies rather than the conventional ($N-1$) contingencies. If the ($N-2$) criterion requires too much investment, for example, it is useful to consider intermediate security criteria between the ($N-1$) criterion and the ($N-2$) criterion. This chapter proposes a more flexible security criterion, such as a criterion covering ($N-1G-1T$) contingencies, as a sensible way to find a transmission expansion plan that is more reliable than a plan based on the ($N-1$) criterion and less expensive than a plan based on the ($N-2$) criterion.

The objective of future research will be to extend the methodology in this chapter to consider probabilistic operating/production costs, nodal reliability criteria, stronger security criteria, simultaneous generation, and transmission expansion and AC load flow.

Appendix

A.1) *Traditional work procedure for Power System planning* – A typical work procedure for power system planning is given in Figure 14.A.1.

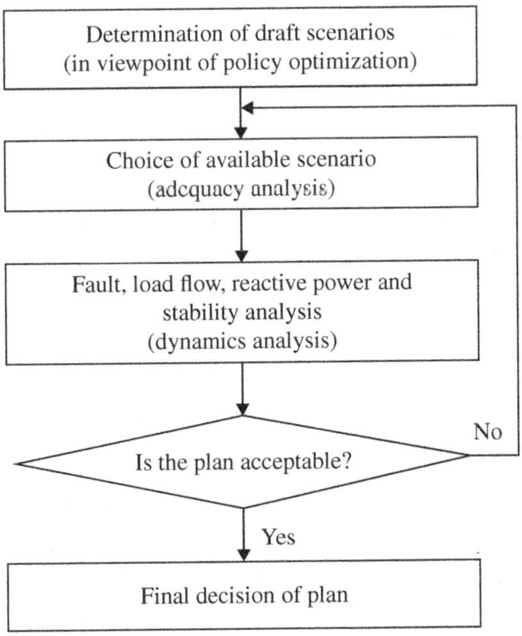

Figure 14.A.1 Traditional work procedure for power system planning.

A.2) *Network Modeling of Power System* – Generators, substations, and load points have limited capacities. It is difficult to check for a shortage of power supply in the system because these elements are presented as nodes in a system model. Network modeling of the system makes it convenient to check a shortage of power supply because the network elements mentioned above are presented as branches with capacity limitations. Aspects of a shortage of power supply according to a bottleneck are given in Table 14.A.1.

Table 14.A.1 Various aspects of power supply bottle neck.

Fm=L≤G	No shortage of supply
Fm=G<L	Shortage of power generation
Fm<L≤G	Shortage of transmission delivery capacity
Fm<G<L	Shortages of power generation and transmission delivery capacity

where

F_m: maximum flow of the network,
G: total power generation,
L: total system load

A.3) *Minimum Cut-Set: Example* – Cut-set with minimum flows of cut-sets of elements (lines and transformers) are divided as a source node set including a node with a binding generator is named a *source node* and terminal node set including a node with a binding load is referred to as *terminal* (*t* or *sink*) node is called the minimum cut-set. As it is, it is the same with set according to the bottleneck of the maximum flow. Figure 14.A.2 presents the network model of Figure 14.A.3 and cut-sets. The minimum cut-set is $\{G1, TL_{1-2}, L_{p2}\}$ and the capacity of minimum cut-set is 50 MW.

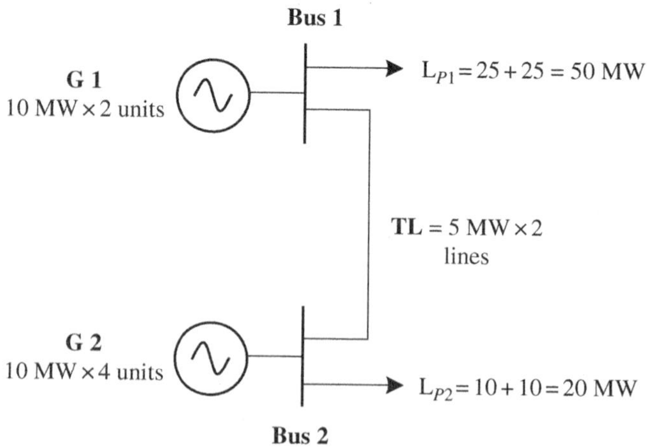

Figure 14.A.2 Power system.

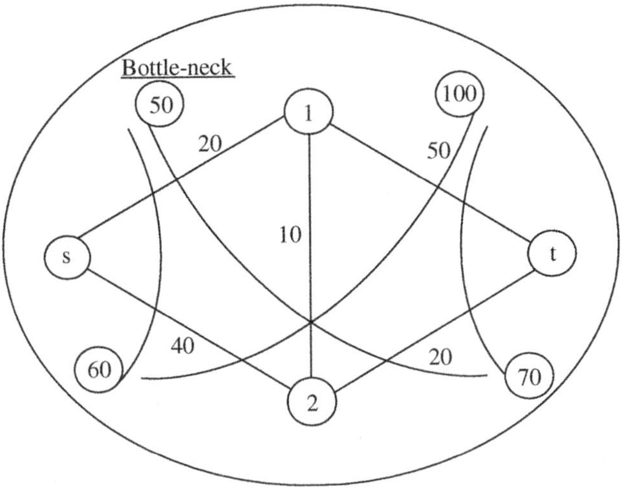

Figure 14.A.3 Network model, cut-sets and minimum cut-set of Figure 14.A.2 system.

A.4) Input Data for Case Study II

Table 14.A.2 Operating marginal cost and standby marginal cost of generators.

	Operating marginal cost ($/MWh)	Standby marginal cost ($/MWh)
G3	32	30
G21	28	30
G4	28	30
G10	25	30
G20	34	30
G18	32	30
G13	30	30
G19	22	30

Table 14.A.3 Forced outage rates of generators, transformers, and lines.

NL	FOR	NL	FOR	NL	FOR
1	0.012	13	0.0012	25	0.0014
2	0.015	14	0.0015	26	0.0020
3	0.010	15	0.0015	27	0.0018

(Continued)

Table 14.A.3 (Continued)

NL	FOR	NL	FOR	NL	FOR
4	0.015	16	0.0012	28	0.0022
5	0.010	17	0.0015	29	0.0020
6	0.012	18	0.0014	30	0.0025
7	0.0125	19	0.0015	31	0.0012
8	0.0155	20	0.0016	32	0.0015
9	0.0015	21	0.0018	33	0.0011
10	0.0020	22	0.0012	34	0.0011
11	0.0015	23	0.0012	35	0.0021
12	0.0012	24	0.0012	36	0.0022

14.2 Deterministic Reliability Constraints

14.2.1 Introduction

Transmission expansion planning (TEP) with open access to the transmission system has become an important issue in the electric energy industry in recent years. Electric market access has moved the industry from conventional monopolistic electricity markets to competitive markets. In a competitive market, the price of the delivered energy and the quality of energy supply including voltage quality and reliability of service are the main factors for business success. A key factor in today's competitive environment is the orientation toward customer needs and willingness to pay for quality. The characteristics are (i) Nodal, (ii) Decentralized, and (iii) More detail. The one of nodal concepts includes that a specified load demand may require a special reliability level. The grid should be able to service a function to be satisfied with the individual reliability-level requirement.

The TEP addresses the problem of broadening and strengthening an existing transmission network to optimally serve a growing electricity market while satisfying a set of economic and technical constraints. Normally, TEP problem can be classified into two models which are heuristic and mathematical optimization models. First, the heuristic models describe all the planning scenarios such as techniques, economic (investment, operation cost), reliability, quality, etc., to search for the best optimal solution. Various techniques including genetic algorithms (GAs), risk analysis (RA), game theory, simulated annealing (SA), expert system, fuzzy set theory, GRASP, tabu search, etc., have been used to study this problem. Secondly, the mathematical optimization models find an optimum solution by computing through a set of mathematical formulations which its variation parameters also describe as a set of techniques, economic (investment, operation cost, reliability cost, ...), reliability, quality, etc. Several methods have been used, like linear programming (LP), dynamic programming (DP), nonlinear programming, mixed integer programming, etc. The conventional TEP problem is to minimize the total cost subject to a reliability-level constraint.

This section proposes an alternative method for choosing the best TEP. The objective function is to minimize total cost for constructing new transmission lines which are an investment cost, subject to the nodal delivery marginal rate criterion which considers deterministic reliability level required at a specified load point. The nodal reliability criterion is one of the characteristics in competitive electricity market environment. The conventional branch and bound and network flow methods are used to search for the optimum mix of transmission network expansion. The proposed method also includes generation additions to determine optimum mix of generation and transmission facilities required to meet the composite system reliability criterion. It models the TEP problem as an integer programming problem and solves for the optimum mix of transmission network expansion using a conventional branch and bound method that utilizes a network flow approach and the maximum flow-minimum cut-set theorem [28].

14.2.2 Transmission System Expansion Planning Problem

1) *The Objective Function* – The conventional TEP problem is to minimize the total cost (C^T) which is construction cost associated with investing in new generators and transmission lines as

$$\text{minimize } C^T = \sum_{(x,y)\in B} \left[\sum_{i=1}^{m(x,y)} C^i_{(x,y)} U^i_{(x,y)} \right] \tag{14.13}$$

where

ρ: the set of all branches (generators and transmission lines)

$m(x,y)$: the number of new candidate branches connecting nodes x and y

$C^i_{(x,y)}$: sum of the construction costs of the new generators and lines 1st through i-th that connect buses x and y,

$$C^i_{(x,y)} = \sum_{j=1}^{i} \Delta C^j_{(x,y)}$$

$\Delta C^j_{(x,y)}$: construction cost of the new j-th generator or line connecting nodes x and y

$U^i_{(x,y)}$: the decision variable associated with the generator or line (1 if from 1-st to i-th generators or lines are to be constructed, and 0 otherwise):

$$\sum_{i=1}^{m(x,y)} U^i_{(x,y)} = 1$$

$$U^i_{(x,y)} = \begin{bmatrix} 1 & P_{(x,y)} = P^0_{(x,y)} + P^i_{(x,y)} \\ 0 & P_{(x,y)} \neq P^0_{(x,y)} + P^i_{(x,y)} \end{bmatrix} \tag{14.14}$$

$$P^i_{(x,y)} = \sum_{j=1}^{i} \Delta P^j_{(x,y)}$$

where

$P^i_{(x,y)}$: sum of the capacities of new branches (new generators or new transmission lines) between nodes x and y

$\Delta P^j_{(x,y)}$: capacity of the *j-th* element of the candidate branches connecting nodes x and y

$P^0_{(x,y)}$: capacity of the existing generators and lines that connect nodes x and y.

2) *Constraints* – The basic reliability criteria normally considered in a TEP problem can be categorized as two types of constraints. One is a deterministic reliability criterion and the other is the probabilistic reliability criterion.

In a deterministic approach, no shortage of power supply requires that the total capacity of the branches involved in the minimum cut-set should be greater than or equal to the system peak load demand, L_p. This is also referred to as the bottleneck capacity. Therefore, a no-shortage power supply constraint can be expressed by

$$P_c(S, T) \geq L_p \quad (s \in S, \ t \in T) \tag{14.15}$$

where, $P_c(s,t)$ is the capacity of the minimum cut-set of two subsets, S and T, containing source node, s and terminal node, t respectively, when all nodes are separated by a minimum cut-set. And, total load of system can be formulated using BMR_k as

$$L_p = \sum_{k=1}^{NL} L_{pk}(1 + BMR_k/100) \tag{14.16}$$

where BMR_k is bus/nodal reserve/marginal rate at load bus k, which is defined as $(AP_k - L_{pk}) \times 100/L_{pk}$, where L_{pk} is the peak load at load bus k. AP_k and L_{pk} are the maximum arrival power and peak load, respectively, at load bus k and NL is number of load buses. The demand constraint (14.15) can be expressed by (14.17) with c being the cut-set number ($c = 1,\ldots, n$), where n is number of cut-set.

$$\sum_{(x,y)\in(S_c,T_c)} \left[P_{(x,y)} = P^0_{(x,y)} + \sum_{i=1}^{m(x,y)} P^i_{(x,y)} U^i_{(x,y)} \right] \geq L_p \tag{14.17}$$

14.2.3 Maximum Flow Under Contingency Analysis for Security Constraint

In order to consider the security constraints from the p of adequacy in this study, it is necessary to evaluate whether or not the power flows. The power flow in the network of each arc (transmission line) $(i, j) \in \rho$ is associated with a non-negative real number P_{ij} that is called the transmission line capacity from node i to node j. The maximum static power flow, F_m from s (source) to t (terminal) in branch set, ρ can be formulated as a linear optimization problem in (14.18). This problem can be solved by LP. Note that the summation of flows at all nodes except the source and terminal nodes in (14.18) is always zero by the Kirchhoff's first law [35, 37].

$$\text{Maximize} \quad F_{N-a} = \sum_{j\in} x_{sj}$$
$$\sum(-x_{ij} + x_{ij}) = \begin{cases} -K, & i = s \\ 0 \quad , & i \neq s, t \\ K \quad , & i = t \end{cases} \tag{14.18}$$
$$0 \leq x_{ij} \leq P_{ij} \qquad (i,j) \in r_{(N-a)}$$

where

F: network flow
x_{ij}: power flow of branch between i and j nodes
Λ_s: set of nodes connected with source (or terminal) node
ρ: branch/lines set
K: total load (= L)
P_{ij}: branch (generator and line) capacity between nodes i and j

The maximum power flow (F_m) using branch set, ρ can be calculated by LP or the shortest path problem to determine the bottleneck capacity. It can be expressed as follows in (14.19) using the maximum flow minimum cut-set theorem, where F_m is the maximum power flow for branch set, ρ.

$$P_c(S, T) = F_m \tag{14.19}$$

14.2.4 Solution Algorithm

The objective in the proposed method using conventional branch and bound method is to minimize the total construction cost subject to a specified BMR_k reliability criterion.

The proposed solution algorithm uses the following steps:

1) Check the "need" for transmission expansion for the system and the "possibility" of meeting load using the candidate lines. Need and possibility can be checked for a given reliability criterion by considering the system with no candidate lines and with all candidate lines, respectively.

2) Set $j = 1$ (initial system), $jopt = 0$, $jmax = 0$, $C^T_{opt} = \infty$, and $ENNOD_j = 0$, where $jopt$ is a parameter of the optimal solution in a solution graph for the branch and bound; $jmax$ is the maximum number of branches that should be searched in a solution graph; $ENNOD$ is a parameter that indicates whether a branch can be terminated in a solution graph (if it is 1, the branch is bounded).
3) If $ENNOD_j = 1$, the #j system is an end node at which the branch operation of branch and bound is finished (*Bound*) in a solution graph. In this situation, there is no need to consider any of the other graphs following the #j system. Go to 13.
4) Calculate the minimum cut-set using the maximum flow method for #j system (#j solution in the solution graph).
5) Select a #i branch/line of the candidate branches/lines set (S_j) in the minimum cut-set and add it to the #j system. In what follows, the new system is named the #ji system and it is called a *Branch* in a solution graph.
6) If the #ji system has already been considered in the solution graph, go to 13.
7) Calculate the total cost $C^T_{ji} = C^T_j + CC_{ji}$ for the #ji system and evaluate the security level of the system.
8) If $C^T_{ji} < C^T_{jopt}$, the current system (#ji) with a cost of C^T_{ji} may be optimal. If not, go to 11.
9) Set $jmax = jmax + 1$.
10) Check the BMR_k criterion; If $P_c(s,t) > \sum L_{pk}\,(1 + BMR_k/100)$, set $C^T_{opt} = C^T_{ji}$, and $F_{mopt} = P_c(s,t)$, $jopt = jmax$, and go to 12.
11) Set $C^T{}_{jmax} = C^T_{ji}$, $ENNOD_{jmax} = 1$, and go to 13.
12) Add this #$jmax$(#ji) solution to the solution graph.
13) If all the candidate branches/lines in the cut-set S_j have been considered, go to 15. Otherwise, set $i = i + 1$ and go to 5.
14) If $j = jmax$, continue to the next step. Otherwise, set $j = j + 1$ and go to 4.
15) For $j = jmax$, the solution graph has been constructed fully and the optimal solution $jopt$ has the lowest cost C^T_{jopt}, and it also satisfies the required security criterion in step 10.

14.2.5 Case Studies

The proposed method was tested on the 21-bus model system shown in Figure 14.12. This is a part of the south-east area (Youngnam) in Korea. This expansion planning is a static problem, which is considered for a target year and can be described as follows: "What is the optimum TSEP to minimize total cost and satisfy the required nodal reliability criterion, BMR_k for the future forecast load?"

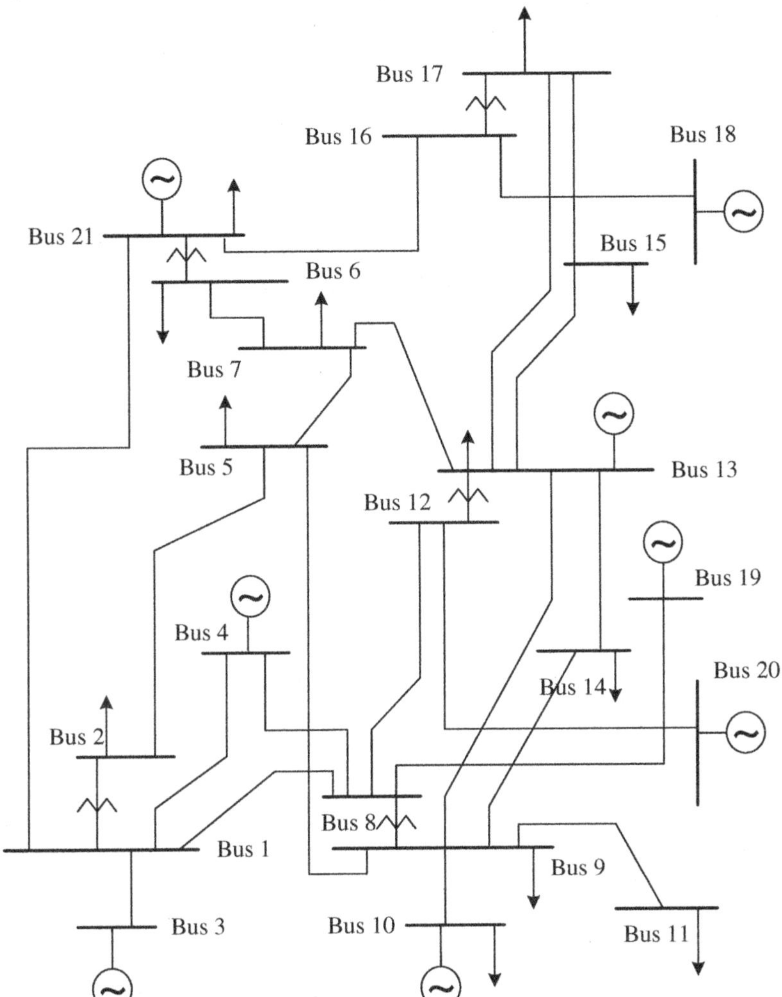

Figure 14.12 The 21-bus model system.

Table 14.15 shows the system data with *GN, TF, TL*, and *LD* representing generators, transformers, transmission lines, and loads, respectively. *SB* and *EB* are start (from) and end (to) buses of the line, respectively. $\Delta P_{(x,y)}^{(0)}$ and $\Delta C_{(x,y)}^{(0)}$ are, respectively, the capacities and costs of the existing lines that connect nodes x and y. In this study, four candidate generators and lines are considered as it is, which means $m(x,y) = 4$ in (14.13) and (14.17). In Table 14.15, parentheses in $\Delta P_{(x,y)}^{(j)}$ and $\Delta C_{(x,y)}^{(j)}$ are omitted for convenience. The cost unit, M$ in this table stands for a million dollars. Figure 14.13 shows the inverted load duration curves (ILDC) at the buses with the three largest loads.

Table 14.15 System capacity and cost data $P(*)$: (MW) and $C(*)$: (M$).

NL	SB	EB	ID	ΔP^0_{xy}	ΔP^1_{xy}	ΔP^2_{xy}	ΔP^3_{xy}	ΔP^4_{xy}	ΔC^0_{xy}	ΔC^1_{xy}	ΔC^2_{xy}	ΔC^2_{xy}	ΔC^4_{xy}
1	0	3	GN	850	0	0	0	0	0	0	0	0	0
2	0	21	GN	900	0	0	0	0	0	0	0	0	0
3	0	4	GN	850	0	0	0	0	0	0	0	0	0
4	0	10	GN	900	0	0	0	0	0	0	0	0	0
5	0	20	GN	1200	0	0	0	0	0	0	0	0	0
6	0	18	GN	850	0	0	0	0	0	0	0	0	0
7	0	13	GN	760	0	0	0	0	0	0	0	0	0
8	0	19	GN	950	0	0	0	0	0	0	0	0	0
9	6	21	TF	1020	510	510	0	0	0	132	132	0	0
10	16	17	TF	1020	510	510	0	0	0	124	124	0	0
11	12	13	TF	1020	510	510	0	0	0	123	130	0	0
12	8	9	TF	800	800	0	0	0	0	155	0	0	0
13	1	2	TF	800	800	0	0	0	0	151	0	0	0
14	21	1	TL	500	500	500	0	0	0	29	29	0	0
15	2	5	TL	220	220	0	0	0	0	54	0	0	0
16	1	4	TL	300	300	0	0	0	0	73	0	0	0
17	1	8	TL	400	400	0	0	0	0	70	0	0	0
18	1	3	TL	1000	250	250	250	250	0	20	20	20	20
19	4	8	TL	300	300	0	0	0	0	63	0	0	0
20	5	9	TL	220	220	0	0	0	0	82	0	0	0
21	5	7	TL	220	220	0	0	0	0	77	0	0	0
22	7	6	TL	220	220	0	0	0	0	85	0	0	0
23	21	16	TL	1000	250	250	250	250	0	30	30	30	30
24	7	13	TL	220	220	0	0	0	0	88	0	0	0
25	13	17	TL	220	220	0	0	0	0	69	0	0	0
26	13	15	TL	220	220	0	0	0	0	83	0	0	0
27	16	18	TL	1320	330	330	330	330	0	32	32	32	32
28	9	13	TL	220	220	0	0	0	0	71	0	0	0
29	9	14	TL	220	220	0	0	0	0	65	0	0	0
30	8	19	TL	620	620	0	0	0	0	64	0	0	0
31	12	20	TL	1240	310	310	310	310	0	28	28	28	28
32	12	8	TL	400	400	0	0	0	0	62	0	0	0
33	9	10	TL	240	240	0	0	0	0	81	0	0	0
34	9	11	TL	340	340	0	0	0	0	45	0	0	0
35	15	17	TL	220	220	0	0	0	0	80	0	0	0

Table 14.15 (Continued)

NL	SB	EB	ID	ΔP_{xy}^0	ΔP_{xy}^1	ΔP_{xy}^2	ΔP_{xy}^3	ΔP_{xy}^4	ΔC_{xy}^0	ΔC_{xy}^1	ΔC_{xy}^2	ΔC_{xy}^2	ΔC_{xy}^4
36	13	14	TL	220	220	0	0	0	0	80	0	0	0
37	21	22	LD	785	0	0	0	0	0	0	0	0	0
38	6	22	LD	750	0	0	0	0	0	0	0	0	0
39	2	22	LD	850	0	0	0	0	0	0	0	0	0
40	9	22	LD	595	0	0	0	0	0	0	0	0	0
41	10	22	LD	17	0	0	0	0	0	0	0	0	0
42	11	22	LD	550	0	0	0	0	0	0	0	0	0
43	14	22	LD	190	0	0	0	0	0	0	0	0	0
44	13	22	LD	710	0	0	0	0	0	0	0	0	0
45	15	22	LD	450	0	0	0	0	0	0	0	0	0
46	17	22	LD	870	0	0	0	0	0	0	0	0	0
47	7	22	LD	290	0	0	0	0	0	0	0	0	0
48	5	22	LD	70	0	0	0	0	0	0	0	0	0

#0 and #22 represent source and terminal nodes, respectively.

The proposed method can solve a composite power system expansion planning (CPSEP) well but this chapter focuses only on the TSEP problem. Case studies have been worked for the three largest loads in convenience. Figure 14.14 is an optimal solution of TSEP which considered the deterministic nodal reliability criterion, $BMR_{L2} = 20\%$ at load bus #2 only. It required constructing the new transmission lines T^1_{21-1}, T^1_{1-8}, T^1_{8-19}, T^1_{9-11}, and T^1_{15-17} with the total cost of 288 M\$. Otherwise, Figure 14.15 is an optimal solution of TEP which considered both deterministic nodal reliability criterion, $BMR_{L17} = 20\%$. Five new transmission lines T^1_{21-1}, T^1_{1-8}, T^1_{13-15}, T^1_{8-19}, and T^1_{9-11} were required to adding the existing system with a total cost of 291 M\$. The total cost of cases (case 4, case 5, and case 6) which has to be satisfied with *BMR* criteria at specified two or three load points is smaller than the sum of the costs of cases obtained by individual $BMR_k = 20\%$ criterion and also they are larger than each cost. For example, total cost of case 4 with $BMR_{L2} = 20\%$ (case 1) and $BMR_{L17} = 20\%$ (case 2) criteria is 311 M\$. The costs of the cases are 288 and 291 M\$, respectively. As it is, the cost of case 4 is smaller than the sum (579 M \$) of the two cases. And, the cost of case 4 is larger than that of case 1 or case 2. The characteristics are also applicable for case 7 which requires a specified *BMR* simultaneously at three loads.

The construction cost is a budget for investment power system elements, which are transformers and transmission lines to satisfy the required nodal and deterministic reliability criterion (BMR_k) (Figure 14.16, Table 14.16). Tables 14.17,

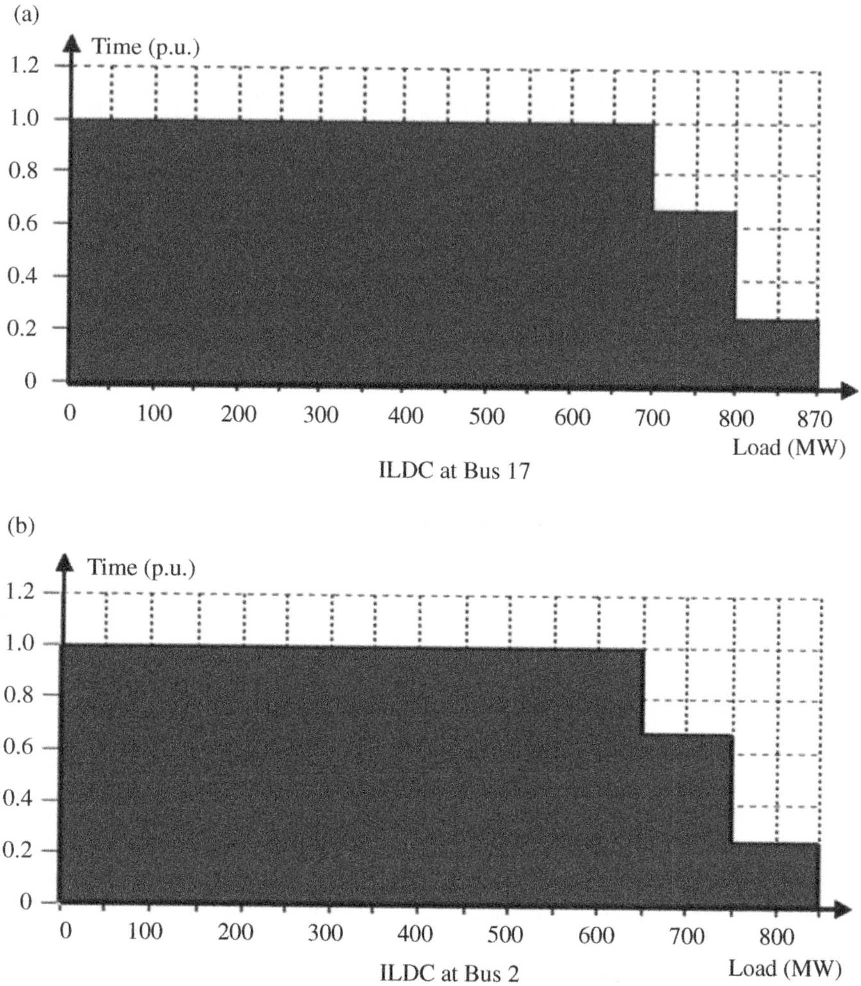

Figure 14.13 Inverted load duration curves (ILDCs) at the buses with the three largest loads.

14.18 and 14.19 present the characteristics of the optimal TEP results while $BMRL_{17}$, BMR_{L2} and $BMRL_{21}$, respectively, were changed.

Tables say that T^1_{21-1}, T^1_{1-4}, T^1_{9-10}, T^1_{9-11}, and T^1_{15-17} are very important lines. Especially, T^1_{21-1} was included in all cases. It means that T^1_{21-1} should be constructed first in order to satisfy the increased load demand immediately at the three load points (L17, L2, and L21). In viewpoint of the nodal concept

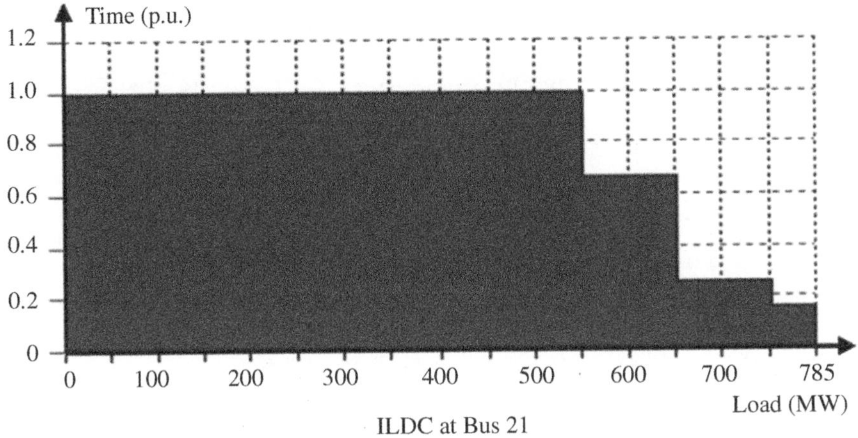

Figure 14.13 (Continued)

under competitive electricity market environments, *BMR* looks a more useful criterion than *SMR* (system marginal rate) which means bulk system reliability criterion. For practical system, different areas have different increased rates at demand load. The cost for the increased load demand at load point #12 is higher than that of load point #17, although load #17 is higher than load #12. Therefore, it can be considered that load point #12 is located in a more prosperous area than the other load points.

Figure 14.17 shows relative characteristics of the results graphically. Figure 14.17 answers to the question about "How much can a given investment budget cover the increased demand load at a load point?" For example, the investment budget of 400 M$ can cover the increased demand of 20% at load point #2, 40% at load point #17, and 60% at load point #21.

14.2.6 Conclusion

This section addresses TEP problem considering deterministic nodal delivery marginal rate criterion associated with construction cost, subject to the deterministic nodal reliability criterion which was described as nodal/bus delivery marginal rate (BMR_k). Optimal placements and the capacity of transformers as well as transmission lines can be determined using the proposed method. It presents a new and practical approach that should serve as a useful guide for the decision maker to select a reasonable expansion plan prior to checking system stability and dynamics in detail. The proposed method finds the optimal TEP considering a reliability

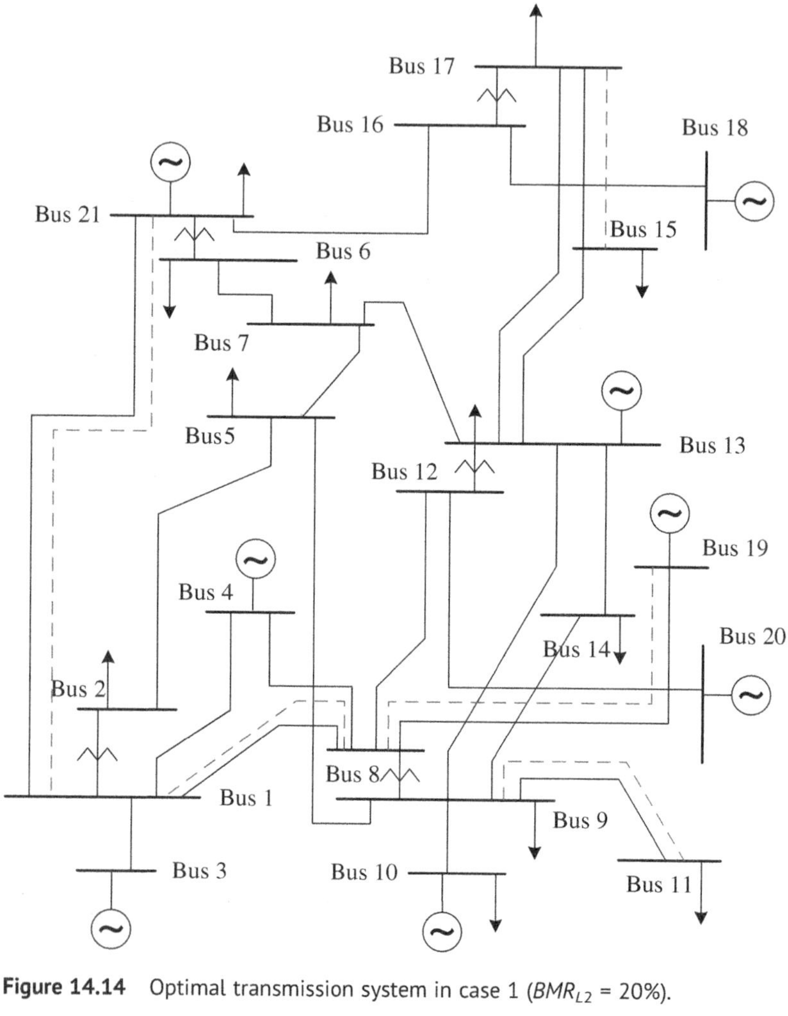

Figure 14.14 Optimal transmission system in case 1 (BMR_{L2} = 20%).

level required at a specified load point. It models the problem as an integer programming one. A proposed branch and bound algorithm which includes the network flow method and the maximum flow-minimum cut-set theorem is proposed to solve the problem. The 21-bus system analysis used to illustrate the method shows that quite different expansion plans can result from applying various *BMRs*. The chapter shows that the proposed method can be used to perform the TSEP that grid owner wants to include a specified nodal reliability level in macro viewpoint for the decision-making investment cost.

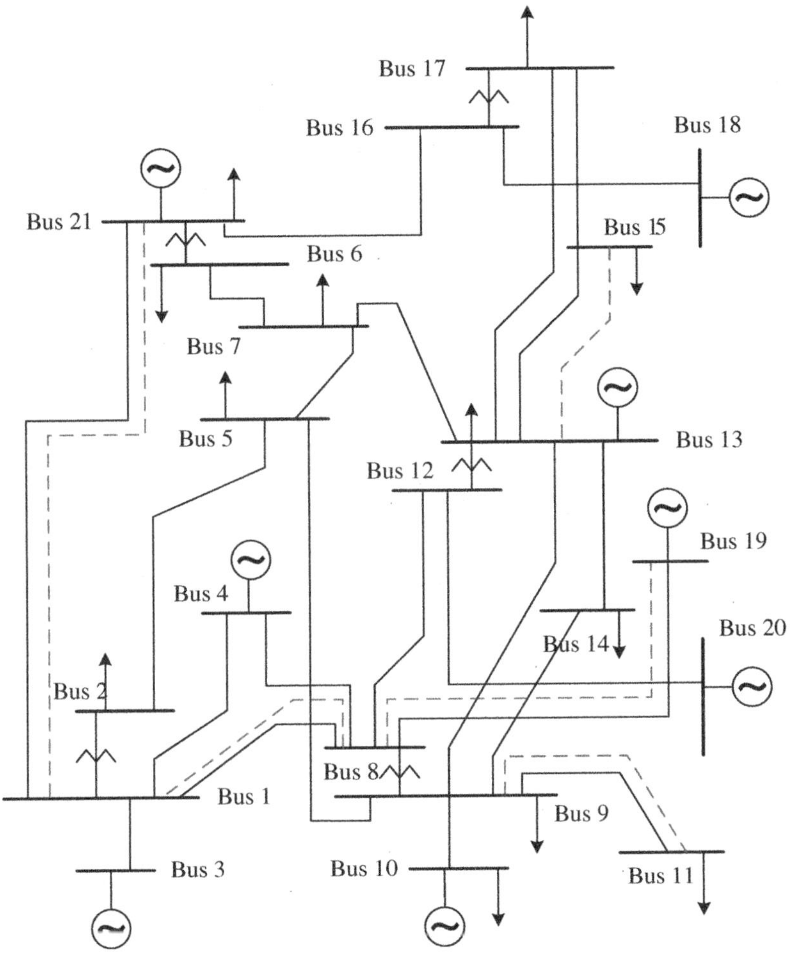

Figure 14.15 Optimal transmission system in case 2 (BMR_{L17} = 20%).

14.3 Probabilistic Reliability Constraints

14.3.1 Introduction

As an alternative to the deterministic reliability constraint approach in Section 14.2, this section presents the probabilistic reliability constraint approach. Deterministic reliability criteria such as an $N-1$ or $N-2$ contingency criteria and load balance constraints are used in most transmission system and composite power system expansion planning because of computation time

Table 14.16 Various optimal solutions for transmission system expansion planning due to variation of BMR_k.

Case	BMR (%)	Optimal solution	Const. cost (M$)
0	$BMR_{All} = 10$	T^1_{21-1}, T^2_{21-1}, T^1_{1-8}, T^1_{4-8}, T^1_{8-19}, T^1_{9-10}, T^1_{9-11}, and T^1_{15-17}	461
1	$BMR_{L2} = 20$	T^1_{21-1}, T^1_{1-8}, T^1_{8-19}, T^1_{9-11}, and T^1_{15-17}	288
2	$BMR_{L17} = 20$	T^1_{21-1}, T^1_{1-8}, T^1_{13-15}, T^1_{8-19}, and T^1_{9-11}	291
3	$BMR_{L21} = 20$	T^1_{21-1}, T^1_{1-4}, T^1_{9-11}, and T^1_{15-17}	227
4	$BMR_{L2} = 20$ and $BMR_{L17} = 20$	T^1_{21-1}, T^1_{1-4}, T^1_{13-15}, T^1_{9-10}, and T^1_{9-11}	311
5	$BMR_{L2} = 20$ and $BMR_{L21} = 20$	T^1_{21-1}, T^1_{1-4}, T^1_{9-10}, T^1_{9-11}, and T^1_{15-17}	308
6	$BMR_{L17} = 20$ and $BMR_{L21} = 20$	T^1_{21-1}, T^1_{1-4}, T^1_{13-15}, T^1_{9-10}, and T^1_{9-11}	311
7	$BMR_{L2} = 20$, $BMR_{L17} = 20$ and $BMR_{L21} = 20$	T^1_{21-1}, T^2_{21-1}, T^1_{1-4}, T^1_{13-15}, T^1_{8-19}, T^1_{9-10}, and T^1_{9-11}	404

Table 14.17 Optimal solution of transmission system expansion planning due to variation of BMR_{L2}.

Case	BMR (%)	Optimal solution	Const. cost (M$)
8	$BMR_{L2} = 10$	T^1_{21-1}, T^1_{1-4}, T^1_{9-11}, and T^1_{15-17}	227
9	$BMR_{L2} = 20$	T^1_{21-1}, T^1_{1-8}, T^1_{8-19}, T^1_{9-11}, and T^1_{15-17}	288
10	$BMR_{L2} = 30$	TF^1_{1-2}, T^1_{1-4}, T^1_{13-15}, T^1_{8-19}, and T^1_{9-11}	416
11	$BMR_{L2} = 40$	TF^1_{1-2}, T^1_{1-4}, T^1_{13-15}, T^1_{9-10}, and T^1_{9-11}	433
12	$BMR_{L2} = 50$	TF^1_{1-2}, T^1_{1-4}, T^1_{1-8}, T^1_{13-15}, T^1_{8-19}, and T^1_{9-11}	486
13	$BMR_{L2} = 60$	TF^1_{1-2}, T^1_{1-4}, T^1_{13-15}, T^1_{8-19}, T^1_{9-10}, and T^1_{9-11}	497
14	$BMR_{L2} = 70$	TS^1_{1-2}, T^1_{1-8}, T^1_{4-8}, T^1_{13-15}, T^1_{8-19}, T^1_{9-10}, and T^1_{9-11}	557
15	$BMR_{L2} = 80$	TS^1_{1-2}, T^1_{1-4}, T^1_{1-8}, T^1_{13-15}, T^1_{8-19}, T^1_{9-10}, and T^1_{9-11}	567

Table 14.18 Optimal solution of transmission system expansion planning due to variation of BMR_{L17}.

Case	*BMR* (%)	Optimal solution	Const. cost (M$)
16	$BMR_{L17} = 10$	T^1_{21-1}, T^1_{1-4}, T^1_{9-11}, and T^1_{15-17}	227
17	$BMR_{L17} = 20$	T^1_{21-1}, T^1_{1-8}, T^1_{13-15}, T^1_{8-19}, and T^1_{9-11}	291
18	$BMR_{L17} = 30$	T^1_{21-1}, T^1_{1-4}, T^1_{13-15}, T^1_{9-10}, and T^1_{9-11}	311
19	$BMR_{L17} = 40$	T^1_{21-1}, T^1_{1-4}, T^1_{13-15}, T^1_{9-10}, and T^1_{9-11}	311
20	$BMR_{L17} = 50$	T^1_{21-1}, T^1_{1-8}, T^1_{13-17}, T^1_{13-15}, T^1_{8-19}, T^1_{9-10}, and T^1_{9-11}	441
21	$BMR_{L17} = 60$	T^1_{21-1}, T^2_{21-1}, T^1_{1-4}, T^1_{13-17}, T^1_{13-15}, T^1_{8-19}, T^1_{9-10}, and T^1_{9-11}	473
22	$BMR_{L17} = 70$	TF^1_{16-17}, T^1_{21-1}, T^2_{21-1}, T^1_{1-8}, T^1_{4-8}, T^1_{13-15}, T^1_{8-19},T^1_{9-10}, and T^1_{9-11}	588
23	$BMR_{L17} = 80$	TF^1_{16-17}, T^1_{21-1}, T^2_{21-1}, T^1_{1-4}, T^1_{1-8},T^1_{13-15}, T^1_{8-19},T^1_{9-10}, and T^1_{9-11}	598

Table 14.19 Optimal solution of transmission system expansion planning due to variation of BMR_{L21}.

Case	*BMR* (%)	Optimal solution	Const. cost (M$)
24	$BMR_{L21} = 10$	T^1_{21-1}, T^1_{1-4}, T^1_{9-11}, and T^1_{15-17}	227
25	$BMR_{L21} = 20$	T^1_{21-1}, T^1_{1-4}, T^1_{9-11}, and T^1_{15-17}	227
26	$BMR_{L21} = 30$	T^1_{21-1}, T^1_{1-8}, T^1_{8-19}, T^1_{9-11}, and T^1_{15-17}	288
27	$BMR_{L21} = 40$	T^1_{21-1}, T^1_{1-4}, T^1_{9-10}, T^1_{9-11}, and T^1_{15-17}	308
28	$BMR_{L21} = 50$	T^1_{21-1}, T^1_{1-4}, T^1_{13-15}, T^1_{9-10}, and T^1_{9-11}	311
29	$BMR_{L21} = 60$	T^1_{21-1}, T^1_{1-8}, T^1_{13-15}, T^1_{8-19}, T^1_{9-10}, and T^1_{9-11}	372
30	$BMR_{L21} = 70$	T^1_{21-1}, T^2_{21-1}, T^1_{1-4}, T^1_{13-15}, T^1_{8-19}, T^1_{9-10}, and T^1_{9-11}	404
31	$BMR_{L21} = 80$	T^1_{21-1}, T^2_{21-1}, T^1_{1-8}, T^1_{4-8},T^1_{13-15}, T^1_{8-19}, T^1_{9-10}, and T^1_{9-11}	464

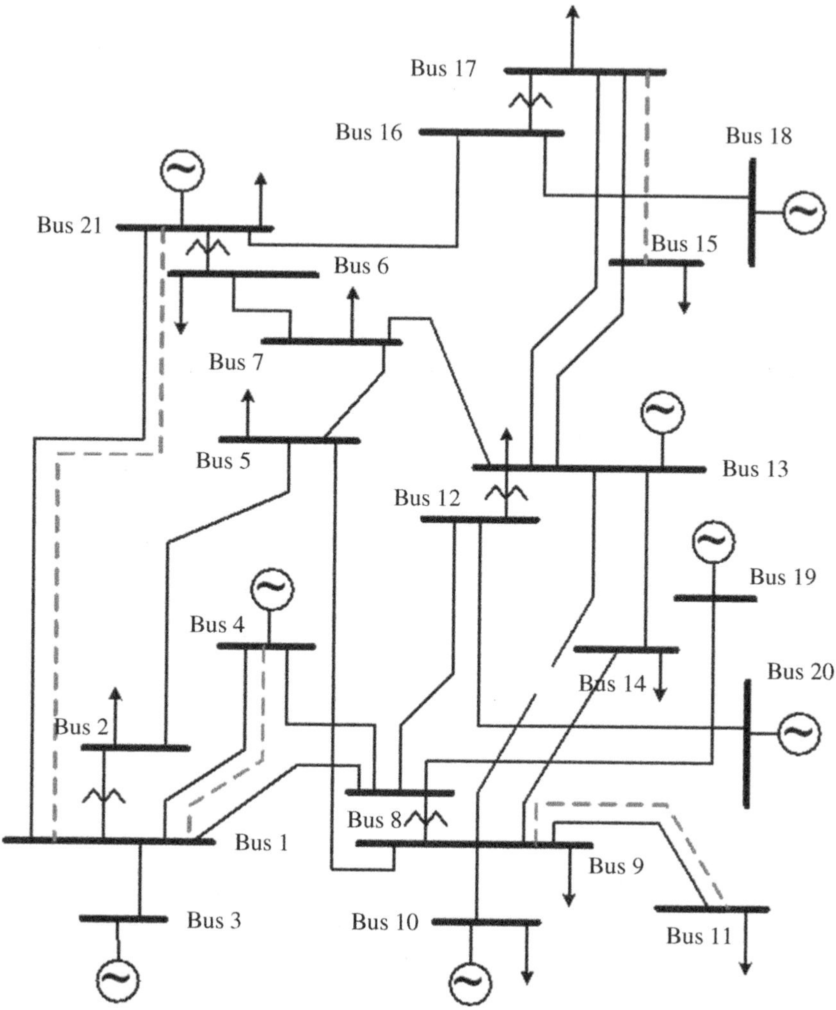

Figure 14.16 Optimal transmission system in case 2 (BMR_{L21} = 20%).

problems. The recent blackouts that have occurred in countries worldwide call for strengthening the grid structure in order to establish successful deregulated electricity markets. The incidents call for the development of tools that can address uncertainties and significantly enhance the ability to conduct effective transmission planning. Available Transfer Capability (ATC) is one good key parameter that indicates the ability of a power system to reliably increase the transferred power between two zones or two points. The NERC suggests the

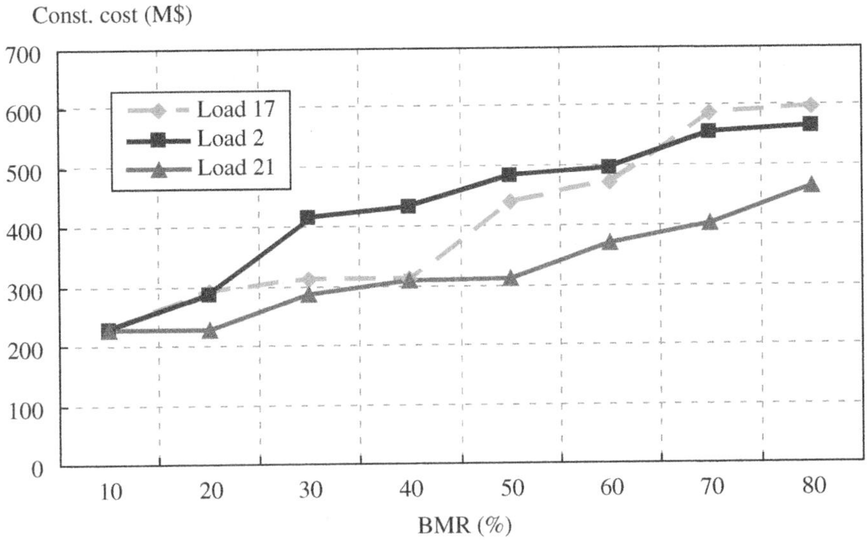

Figure 14.17 Variation of the cost due to change of BMR_{L17}, BMR_{L2}, and BMR_{L21}.

Transfer Reliability Criterion (TRC) based on the ATC for ISO operating and planning of a transmission system. The TRC is based on the $N-1$ contingency deterministic criteria concept. It has been used effectively for transmission system planning in regulated environments. A probabilistic Total Transfer Capability (TTC) methodology has been recently proposed.

Normally, the power system expansion planning problem is analyzed using a macro approach and then a micro approach considering the stability and dynamic characteristics of the new system. In a deregulated environment, electric utilities are expected to be winners in competition. Successful electricity market operation in such an environment depends on the transmission system and nodal (bus) reliability management. Deregulated electricity markets, therefore, call for nodal-based indices in system operation and planning. Nodal reliability indices together with related information can be used for the management and control of congestion and reliability by ISO and TRANSCO in deregulated markets.

This environment makes it important to assess and provide reasonable reliability criteria at the load points. In such an environment, probabilistic reliability indices become important parameters in transmission system expansion. In addition, in a competitive electricity market, there is more variability in the investment budget for construction and higher uncertainty in the transfer reliability of the transmission system. This is because profit maximization for the system owner is the major focus while, for a conventional power system, the primary function is to provide electrical energy to its customers economically and with an acceptable degree of continuity and quality. System planners and owners are therefore expected to evaluate the reliability and economic parameters with more detail in grid planning where the problem involves many uncertainties including those of the investment budget, reliability criterion, load forecast, and system characteristics, etc. It is a challenging task to develop an expansion plan that considers all these items in an effective and practical manner. Under such uncertain circumstances, methodologies that are based on fuzzy set theory and probabilistic approaches become attractive and useful to accomplish the task. The former is attractive because the experience and knowledge of experts and decision makers can be very helpful in dealing with subjective ambiguity in planning problems. The latter is also valuable for considering the objective uncertainties, such as the Forced Outage Rates (FOR) of power system elements.

This section proposes a new method for choosing the best expansion plan for a transmission system considering a transmission system probabilistic reliability criterion ($_{R}LOLE$). Two probabilistic reliability criteria are used as constraints. One is a transmission system reliability criterion constraint, Required Loss of Load Expectation of Transmission System ($_{R}LOLE_{TS}$) and the other is a bus/nodal reliability criterion constraint, Required Loss of Load Expectation at Bus/Node ($_{R}LOLE_{Bus}$). The proposed method minimizes the investment budget for constructing new transmission lines subject to probabilistic reliability criteria, which considers the uncertainties of transmission system elements. It models the TEP problem as an integer programming problem and solves for the optimum mix of transmission network expansion using a probabilistic branch and bound method that utilizes a network flow approach and the maximum flow-minimum cut-set theorem [39].

14.3.2 Transmission System Expansion Planning Problem

A composite power system that includes generation and transmission facilities is shown in Figure 14.18, where TS refers to the transmission system, NG is the number of generators, $_{k}\Phi_{0}$ is the inverted load duration curve at load point k, and NL is the number of load points. In this section, a composite power system is designated

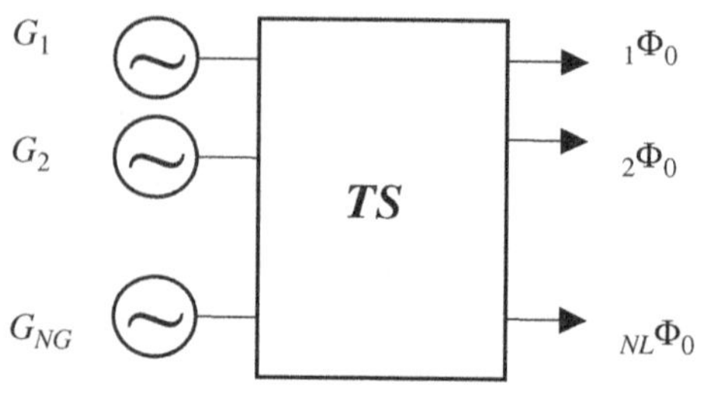

Figure 14.18 A composite power system including the transmission system.

as Hierarchical level II (HLII) and Hierarchical level I (HLI) is used to designate generation and load components only [44]. It is assumed that the generation system and transmission system plans are separated and the construction of new generators is determined independently by GENCOs.

1) *The Objective Function* – The conventional TEP problem is to minimize the total construction cost C^T associated with investing in new transmission lines as

$$\text{minimize } C^T = \sum_{(x,y)\in\rho} \left[\sum_{i=1}^{m(x,y)} C_{(x,y)}^{(i)} U_{(x,y)}^{(i)} \right] \tag{14.20}$$

where
ρ: the set of all branches (transmission lines)
$m(x,y)$: the number of new candidate branches connecting nodes x and y
$C_{(x,y)}^{(i)}$: sum of the construction costs of the new lines 1st through i-th that connect buses x and y,

$$C_{(x,y)}^{(i)} = \sum_{j=1}^{i} \Delta C_{(x,y)}^{(j)}$$

with $\Delta C_{(x,y)}^{(j)}$: construction cost of the new j-th line connecting nodes x and y,
$U_{(x,y)}^{(i)}$: the decision variable associated with the line (1 if from 1-st to i-th lines are to be constructed, and 0 otherwise):

$$\text{with } U_{(x,y)}^{(i)} \sum_{i=1}^{m(x,y)} U_{(x,y)}^{(i)} = 1$$

$$U_{(x,y)}^{(i)} = \begin{cases} 1 & P_{(x,y)} = P_{(x,y)}^{(0)} + P_{(x,y)}^{(i)} \\ 0 & P_{(x,y)} \neq P_{(x,y)}^{(0)} + P_{(x,y)}^{(i)} \end{cases} \tag{14.21}$$

$$P_{(x,y)}^{(i)} = \sum_{j=1}^{i} \Delta P_{(x,y)}^{(j)}$$

where
$P_{(x,y)}^{(i)}$: sum of the capacities of new branches (new transmission lines) between nodes x and y,
$\Delta P_{(x,y)}^{(j)}$: capacity of the j-th element of the candidate branches connecting nodes x and y,
$P_{(x,y)}^{(0)}$: capacity of the existing lines that connect nodes x and y.

2) *Constraints* – The basic reliability criteria normally considered in a composite power system planning problem can be categorized as two types of constraints. One is a deterministic reliability criterion and the other is the probabilistic reliability criterion.

In a deterministic approach, no shortage of power supply requires that the total capacity of the branches involved in the minimum cut-set should be greater than or equal to the system peak load demand, L_p. This is also referred to as the bottle-neck capacity. Therefore, a no-shortage power supply constraint can be expressed by

$$P_C(X,\overline{X}) \geq L_P\ (\mathrm{S} \in X, \mathrm{T} \in \overline{X}) \tag{14.22}$$

where $P_C(X,\overline{X})$ is the capacity of the minimum cut-set of two subsets, X and $\overline{X}$, containing source nodes s and terminal nodes t, respectively, when all nodes are separated by a minimum cut-set.

The demand constraint (14.22) can be expressed by (14.23) with k being the cut-set number ($k = 1,\ldots, n$), where n is the number of cut-set:

$$\sum_{(x,y)\in\left(X_k,\overline{X}_k\right)} \left[P_{(x,y)} = P_{(x,y)}^{(0)} + \sum_{i=1}^{m(x,y)} P_{(x,y)}^{(i)} U_{(x,y)}^{(i)}\right] \geq L \tag{14.23}$$

In the probabilistic approach, the probabilistic reliability criterion index, Loss of Load Expectation (*LOLE*), can be used as

$$LOLE_{TS}\left(P_{(x,y)}^{(i)}, \Phi\right) \leq {}_{R}LOLE \tag{14.24}$$

where ${}_{R}LOLE$ is the required transmission reliability criterion for the new system. The ${}_{R}LOLE$ is ${}_{R}LOLE_{TS}$ for the transmission system reliability criterion case and it is ${}_{R}LOLE_{Bus}$ for the bus/nodal reliability criterion case, Φ is a function of the load duration curve. A detailed discussion of Φ and *LOLE* is presented in the next section.

14.3.3 Composite Power System Reliability Evaluation

The following is a brief introduction to the methodology used to determine the transmission system reliability indices and the bus/nodal reliability indices. The methodology is based on the composite power system effective load model developed before

1) *Reliability Evaluation at HLI* – Reliability indices of $LOLE_{HLI}$ (loss of load expectation) and $EENS_{HLI}$ (expected energy not served) at HLI considering only the generation system are calculated using the effective load duration curve (ELDC), ${}_{HLI}\Phi(x)$ of HLI as following:

$$LOLE_{HLI} = {}_{HLI}\Phi(x)|_{x=IC} \text{ hours/yr} \tag{14.25}$$

$$EENS_{HLI} = \int_{IC}^{IC+Lp} {}_{HLI}\Phi(x)dx \text{ MWh/yr} \tag{14.26}$$

where
IC: total installed generating capacity (MW),
L_p: system peak load (MW)

and,

$$\begin{aligned} {}_{HLI}\Phi_i(x_e) &= {}_{HLI}\Phi_{i-1}(x_e) \otimes {}_{HLI}f_{oi}(x_{oi}) \\ &= \int_{HLI} \Phi_{i-1}(x_e - x_{oi})_{HLI}f_{oi}(x_{oi})dx \end{aligned} \tag{14.27}$$

where
$\otimes$: operator meaning convolution integral,
${}_{HLI}\Phi_0(x_e - x_{oi}) = {}_{HLI}\Phi(x_L)$,
${}_{HLI}f_{oi}(x_{oi})$: the probability distribution function of outage capacity of generator #i

2) *Reliability Evaluation at HL II (Composite Power System)* – The reliability indices at HLII can be classified as load point indices and bulk system indices depending on the object of the evaluation. The reliability indices can be evaluated using a Composite power system Equivalent Load Duration Curve (CMELDC) of HLII based on the composite power system effective load model in Figure 14.19. *CG, CT*, and q and q_l in Figure 14.19 are the capacities and forced outage rates of the generators and transmission lines, respectively.
3) *Reliability Indices at the Load Points (Buses)* – The load point reliability indices, $LOLE_k$ and $EENS_k$, can be calculated using (14.28) and (14.29) with the CMELDC, ${}_k\Phi_{NG}(x)$

$$LOLE_k = {}_k\Phi_{NG}(x)|_{x=AP_k} \text{ hours/yr)} \tag{14.28}$$

$$EENS_k = \int_{AP_k}^{AP_k+Lp_k} {}_k\Phi_{NG}(x)dx \text{ MWh/yr} \tag{14.29}$$

where
L_{pk}: peak load at load point k (MW)
AP_k: maximum arrival power at load point k (MW)

$$\begin{aligned} {}_k\Phi_i(x_e) &= {}_k\Phi_o(x_e) \otimes {}_kf_{osi}(x_{oi}) \\ &= \int_k \Phi_o(x_e - x_{oi})_k f_{osi}(x_{oi})dx_{oi} \end{aligned} \quad \text{MWh/yr} \tag{14.30}$$

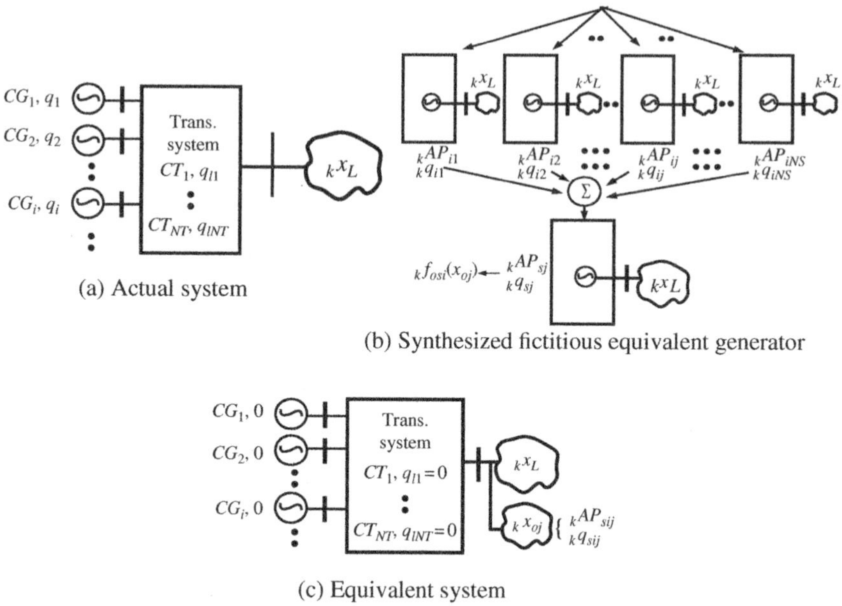

Figure 14.19 Composite power system effective load model at HLII.

where

$\otimes$: the operator representing the convolution integral

${}_k\Phi_0$ = original load duration curve at load point #k

${}_kf_{osi}$: outage capacity *pdf* of the synthesized fictitious generator created by generators *1* to *i*, at load point #*k*.

4) *Reliability Indices of the Bulk System* – While the $EENS_{HLII}$ of a bulk system is equal to the summation of the $EENS_k$ at the load points as shown in (14.31), the *LOLE* of a bulk system is entirely different from the summation of the $LOLE_k$ at the load points. The expected load curtailed at HLII (ELC_{HLII}) of bulk system is equal to the summation of ELC_k at the load points. The $LOLE_{HLII}$ of the bulk system can be calculated using (14.33):

$$EENS_{HLII} = \sum_{k=1}^{NL} EENS_k \quad \text{MWh/yr} \tag{14.31}$$

$$ELC_{HLII} = \sum_{k=1}^{NL} ELC_k \ \text{MW/CUR.yr} \tag{14.32}$$

$$LOLE_{HLII} = EENS_{HLII}/ELC_{HLII} \ \text{hours/yr} \tag{14.33}$$

$$EIR_k = 1 - EENS_k/DENG_k \ \text{PU} \tag{14.34}$$

where
NL: number of load points,
$ELC_k = EENS_k/LOLE_k$,
$DENG_k$: demand energy at bus $\#_k$

5) *Reliability Evaluation of Transmission System* – The reliability indices of a transmission system can be expressed as the difference between the HLII and HLI reliability indices as

$$EENS_{TS} = EENS_{HLII} \quad - EENS_{HLI} \text{ MWh/yr} \tag{14.35}$$

$$LOLE_{TS} = LOLE_{HLII} \quad - LOLE_{HLI} \text{ hrs/yr} \tag{14.36}$$

14.3.4 Solution Algorithm

The objective in the conventional branch and bound method is to minimize the total construction cost subject to a specified reliability criterion. The proposed probabilistic branch and bound-based method minimizes the total cost subject to the required probabilistic transmission system reliability criteria, $_RLOLE_{TS}$ or/and $_RLOLE_{Bus}$.

The solution algorithm for the proposed approach follows:

1) Check the need for transmission expansion for the system and its possibility using the candidate lines. Need and possibility can be checked respectively by the reliability evaluation for systems considering no candidate lines and considering all candidate lines.
2) Set $j = 1$ (initial system), $jopt = 0$, $jmax = 0$, $C^T_{opt} = \infty$ and $ENNOD_j = 0$.
3) If $ENNOD_j = 1$, the #j system is an end node at which the branch operation of a branch and bound is finished (bound) in the solution graph used to obtain the optimal solution, and there is no need to consider any of the other graphs following this system. Go to 13.
4) Calculate the minimum cut-set using the maximum flow method for system j (solution j in the solution graph).
5) Select a #i branch/line of the candidate branches/lines set (S_j) involved in the minimum cut-set and add to the #j system. In what follows, the new system is named the system ji.
6) If the system ji is already considered in the solution graph, go to 13.
7) Calculate the total cost $C^T_{ji} = C^T_j + C\left(P^{(i)}_{(x,y)}\right)$ for the system ji and evaluate the transmission system reliability index, $LOLE_{TSji}$ of the system.
8) If $C^T_{ji} < C^T_{jopt}$, the current system (ji) with a cost of C^T_{ji} can be optimal. If not, go to 11.
9) Set $jmax = jmax + 1$.

10) If $LOLE_{TSji}$ (or $LOLE_{Busji}$) < $_{R}LOLE_{TS}$, set $C^T_{opt} = C^T_{ji}$, and $_{R}LOLE_{opt} = LOLE_{ji}$, $jopt = jmax$, and go to 12.
11) Set $C^T_{j\,max} = C^T_{ji}$, $ENNOD_{jmax} = 1$, and go to 13.
12) Add the solution $jmax(ji)$ to the solution graph.
13) If all the candidate branches/lines in the cut-set S_j have been considered, go to 14. Otherwise, set $I = i + 1$ and go to 5.
14) If $j = jmax$, continue the next step. Otherwise, set $j = j + 1$ and go to 4.
15) For $j = jmax$, the solution graph has been constructed fully and the optimal solution *jopt* with C^T_{jopt} being the lowest cost and satisfying the required reliability criteria is obtained in 10.

14.3.5 Case Study

The proposed method was tested on the 21-bus model system shown in Figure 14.20. This is a part of the south-east area (Youngnam) in Korea. Considering a future forecast system load, the deterministic reliability criterion and the proposed probabilistic reliability approaches were applied and compared in a series of case studies. The probabilistic approach considers the probabilistic reliability criterion without the deterministic constraints (demand balance).

Table 14.20 shows the system data with *GN*, *TF*, *TL*, and *LD* representing generators, transformers, transmission lines, and loads, respectively; *SB* and *EB* are start and end buses of the line, respectively; $\Delta P^{(0)}_{(x,y)}$ and $\Delta C^{(0)}_{(x,y)}$ are, respectively, the capacities and costs of the existing lines that connect nodes x and y. In this study, four candidate generators and lines are considered as it is, which means $m(x,y) = 4$ in (14.20) and (14.23). In Table 14.20, parentheses in $\Delta P^{(j)}_{(x,y)}$ and $\Delta C^{(j)}_{(x,y)}$ are omitted for convenience. The cost unit, M$ in this table stands for a million dollars. Table 14.21 shows the forced outage rate of the generators and transmission lines. Figure 14.21 shows the inverted load duration curves at the buses with the four largest loads.

In the first case study, the required probabilistic transmission system reliability criterion, $_{R}LOLE_{TS} = 50$ hrs/yr is assumed for base case. The new optimal system is shown in Figure 14.22 with dotted lines presenting new lines. An optimal system which has the construction cost of 209 M$ and the new construction elements of $T_{9-10}{}^1$, $T_{9-11}{}^1$, and $T_{13-15}{}^1$ is obtained as the optimal solution of probabilistic approach proposed in this section. The actual reliability level, $LOLE_{TS}$ of the optimal system was evaluated as 45.47 hrs/yr and this level is satisfied with a required probabilistic reliability criterion level (constraint), $_{R}LOLE_{TS} = 50$ hrs/yr. It is interesting to note that this system is the same as that produced using the deterministic approach with a bus reserve rate at k load point $BRR_k = 0\%$ for all load points. AP_k

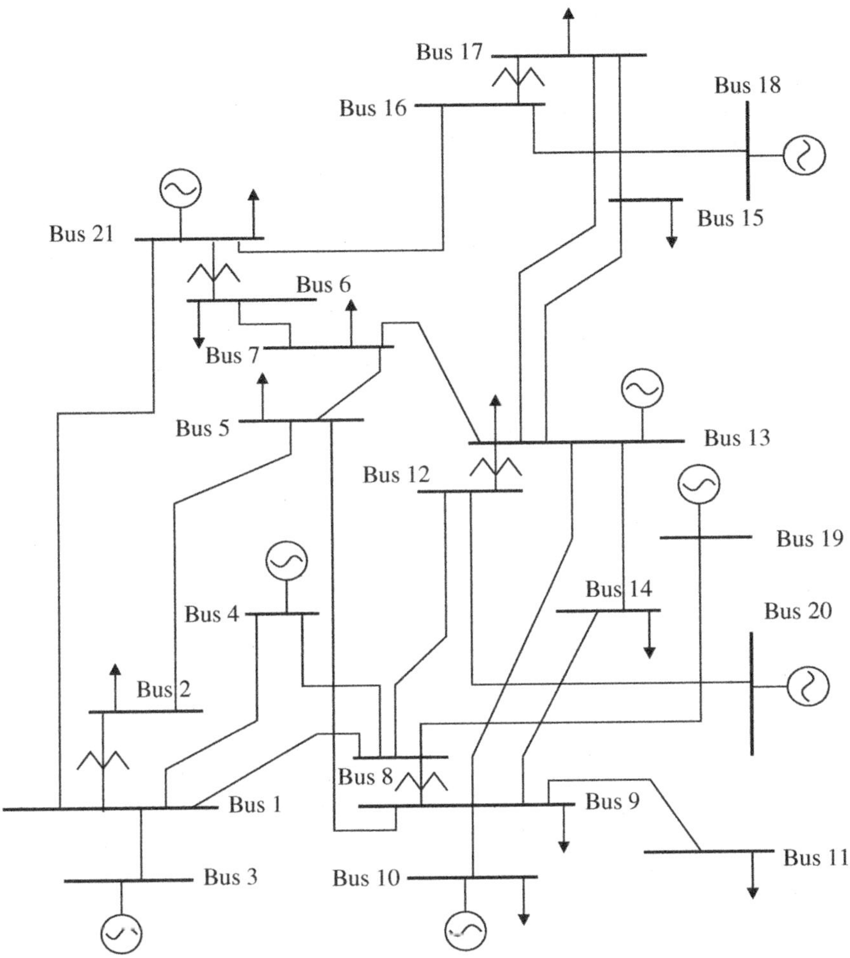

Figure 14.20 The 21-Bus model system.

and LP_k are the maximum arrival power and peak load, respectively, at the k load point. A deterministic bus/nodal reliability criterion, BRR_k is defined as:

$$BRR_k = (AP_k - Lp_k) \times 100/Lp_k$$

Table 14.22 shows the reliability indices at the load buses in the case of $_RLOLE_{TS} = 50$ hrs/yr. This table shows that the nodal reliability indices can have different values from the transmission system reliability level, $LOLE_{TS} = 45.67$ hrs/yr. But, this table shows that the *LOLE* indices' values at buses #2, #13, #17 exceed the system reliability criteria ($_RLOLE_{TS} = 50$ hrs/yr).

Table 14.20 System capacity and cost data $P(*)$: (MW) and $C(*)$: (M$).

NL	SB	EB	ID	ΔP^0_{xy}	ΔP^1_{xy}	ΔP^2_{xy}	ΔP^3_{xy}	ΔP^4_{xy}	ΔC^0_{xy}	ΔC^1_{xy}	ΔC^2_{xy}	ΔC^2_{xy}	ΔC^4_{xy}
1	0	3	GN	850	0	0	0	0	0	0	0	0	0
2	0	21	GN	900	0	0	0	0	0	0	0	0	0
3	0	4	GN	850	0	0	0	0	0	0	0	0	0
4	0	10	GN	900	0	0	0	0	0	0	0	0	0
5	0	20	GN	1200	0	0	0	0	0	0	0	0	0
6	0	18	GN	850	0	0	0	0	0	0	0	0	0
7	0	13	GN	760	0	0	0	0	0	0	0	0	0
8	0	19	GN	950	0	0	0	0	0	0	0	0	0
9	6	21	TF	1020	510	510	0	0	0	132	132	0	0
10	16	17	TF	1020	510	510	0	0	0	124	124	0	0
11	12	13	TF	1020	510	510	0	0	0	123	130	0	0
12	8	9	TF	800	800	0	0	0	0	155	0	0	0
13	1	2	TF	800	800	0	0	0	0	151	0	0	0
14	21	1	TL	500	500	500	0	0	0	29	29	0	0
15	2	5	TL	220	220	0	0	0	0	54	0	0	0
16	1	4	TL	300	300	0	0	0	0	73	0	0	0
17	1	8	TL	400	400	0	0	0	0	70	0	0	0
18	1	3	TL	1000	250	250	250	250	0	20	20	20	20
19	4	8	TL	300	300	0	0	0	0	63	0	0	0
20	5	9	TL	220	220	0	0	0	0	82	0	0	0
21	5	7	TL	220	220	0	0	0	0	77	0	0	0
22	7	6	TL	220	220	0	0	0	0	85	0	0	0
23	21	16	TL	1000	250	250	250	250	0	30	30	30	30
24	7	13	TL	220	220	0	0	0	0	88	0	0	0
25	13	17	TL	220	220	0	0	0	0	69	0	0	0
26	13	15	TL	220	220	0	0	0	0	83	0	0	0
27	16	18	TL	1320	330	330	330	330	0	32	32	32	32
28	9	13	TL	220	220	0	0	0	0	71	0	0	0
29	9	14	TL	220	220	0	0	0	0	65	0	0	0
30	8	19	TL	620	620	0	0	0	0	64	0	0	0
31	12	20	TL	1240	310	310	310	310	0	28	28	28	28
32	12	8	TL	400	400	0	0	0	0	62	0	0	0
33	9	10	TL	240	240	0	0	0	0	81	0	0	0

Table 14.20 (Continued)

NL	SB	EB	ID	ΔP_{xy}^{0}	ΔP_{xy}^{1}	ΔP_{xy}^{2}	ΔP_{xy}^{3}	ΔP_{xy}^{4}	ΔC_{xy}^{0}	ΔC_{xy}^{1}	ΔC_{xy}^{2}	ΔC_{xy}^{2}	ΔC_{xy}^{4}
34	9	11	TL	340	340	0	0	0	0	45	0	0	0
35	15	17	TL	220	220	0	0	0	0	80	0	0	0
36	13	14	TL	220	220	0	0	0	0	80	0	0	0
37	21	22	LD	785	0	0	0	0	0	0	0	0	0
38	6	22	LD	750	0	0	0	0	0	0	0	0	0
39	2	22	LD	850	0	0	0	0	0	0	0	0	0
40	9	22	LD	595	0	0	0	0	0	0	0	0	0
41	10	22	LD	17	0	0	0	0	0	0	0	0	0
42	11	22	LD	550	0	0	0	0	0	0	0	0	0
43	14	22	LD	190	0	0	0	0	0	0	0	0	0
44	13	22	LD	710	0	0	0	0	0	0	0	0	0
45	15	22	LD	450	0	0	0	0	0	0	0	0	0
46	17	22	LD	870	0	0	0	0	0	0	0	0	0
47	7	22	LD	290	0	0	0	0	0	0	0	0	0
48	5	22	LD	70	0	0	0	0	0	0	0	0	0

#0 and #6 represent source and terminal nodes, respectively.

Table 14.21 Forced outage rates of generators and lines.

NL	FOR	NL	FOR	NL	FOR
1	0.012	13	0.0012	25	0.0014
2	0.015	14	0.0015	26	0.0020
3	0.010	15	0.0015	27	0.0018
4	0.015	16	0.0012	28	0.0022
5	0.010	17	0.0015	29	0.0020
6	0.012	18	0.0014	30	0.0025
7	0.0125	19	0.0015	31	0.0012
8	0.0155	20	0.0016	32	0.0015
9	0.0015	21	0.0018	33	0.0011
10	0.0020	22	0.0012	34	0.0011
11	0.0015	23	0.0012	35	0.0021
12	0.0012	24	0.0012	36	0.0022

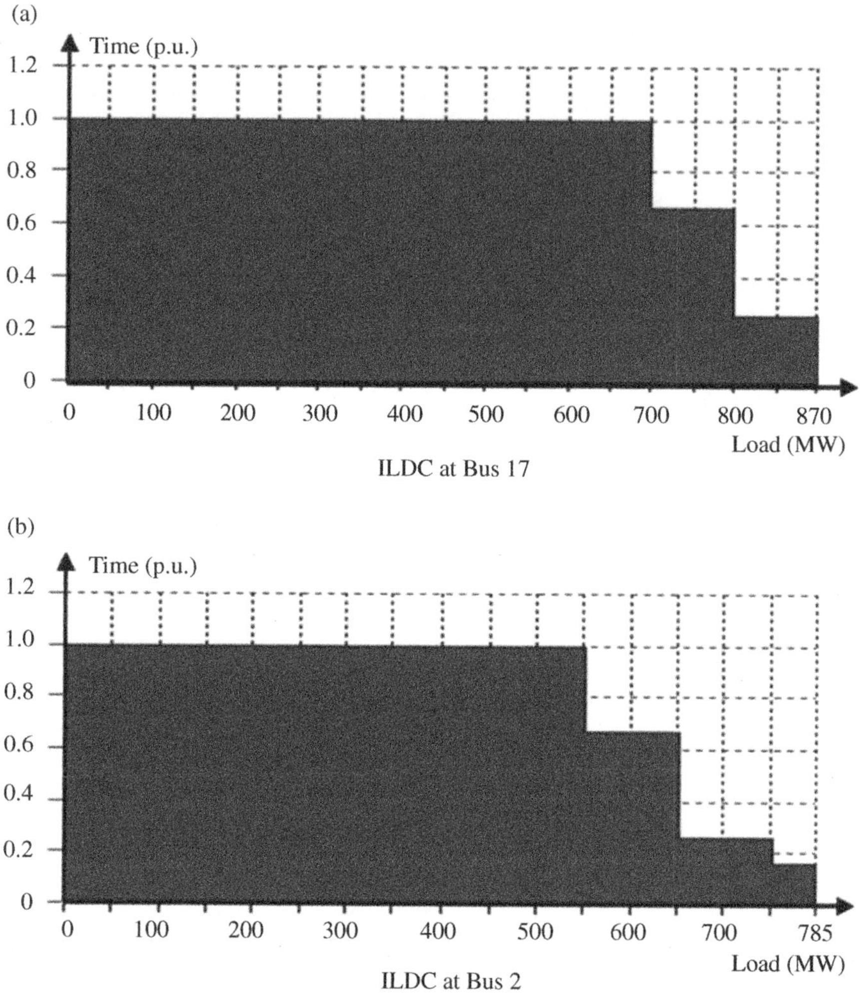

Figure 14.21 Inverted load duration curves at the buses with the four largest loads.

Figure 14.23 shows another new optimal system using the probabilistic approach with a $_RLOLE_{TS} = 25$ hrs/yr for the reliability criterion. This case utilizes a more severe reliability criterion than previous plans. The optimal solution has a construction cost of 511 M$ and the new construction elements of $TF_{1\text{-}2}{}^1$, $T_{5\text{-}9}{}^1$, $T_{13\text{-}17}{}^1$, $T_{13\text{-}15}{}^1$, $T_{9\text{-}10}{}^1$, and $T_{9\text{-}11}{}^1$. The evaluated reliability level, $LOLE_{TS}$ of the optimal system is 24.42 hrs/yr and this level satisfies the required probabilistic reliability criterion, $_RLOLE_{TS} = 25$ hrs/yr. The new plan suggests a strong grid system with a high cost.

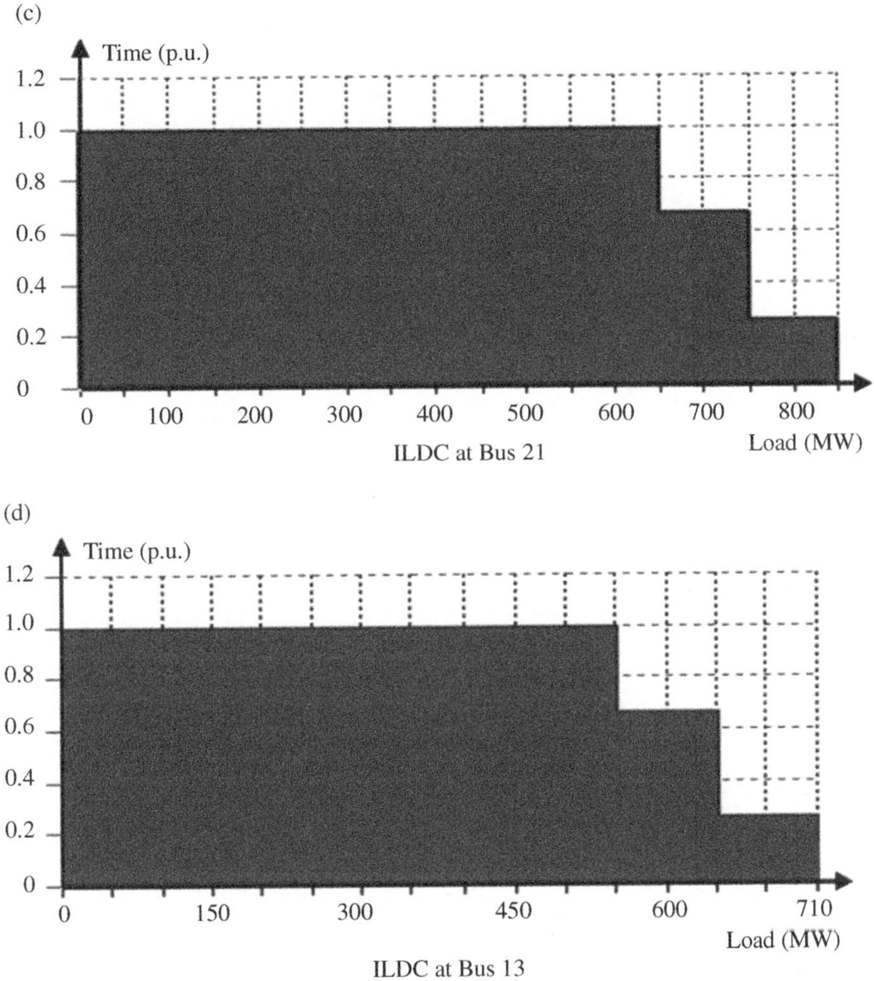

Figure 14.21 (Continued)

The optimal expansion plans that resulted from changing the transmission system reliability criteria, ${}_{R}LOLE_{TS}$, are given in Table 14.23. The tabulated results indicate that as ${}_{R}LOLE_{TS}$ increases, the total construction cost decreases, and reliability indices, $LOLE_{TS}$ and $EENS_{TS}$, of the new optimal transmission system increase.

In the second case study, the transmission system expansion plan using the bus/nodal reliability criteria, ${}_{R}LOLE_{Bus}$ instead of the system reliability criteria, ${}_{R}LOLE_{TS}$ in (14.24), was simulated so that all bus reliability levels of the new system should satisfy this bus reliability criteria, ${}_{R}LOLE_{Bus} = 50$ hrs/yr.

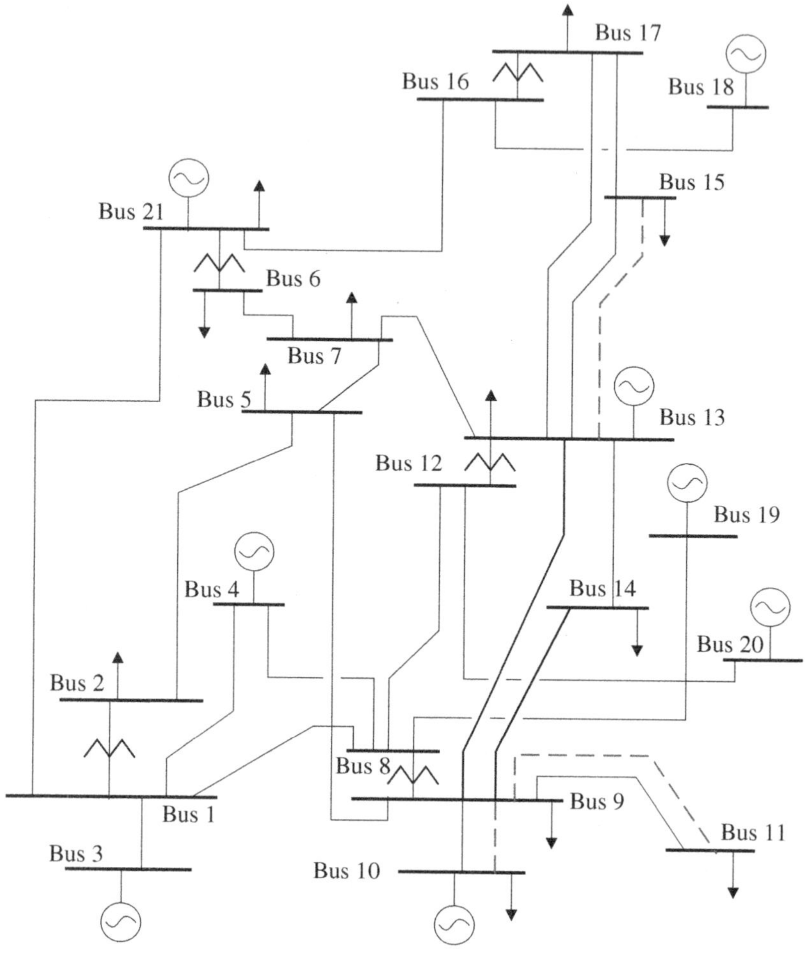

Figure 14.22 Optimal system by the probabilistic approach (${}_{R}LOLE_{TS}$ = 50 hrs/yr).

The new optimal system in the second case is shown in Figure 14.24. Table 14.24 shows the reliability indices at the load buses in the case of ${}_{R}LOLE_{Bus}$ = 50 hrs/yr. A maximum $LOLE_{Bus}$ in the new optimal system is the $LOLE_{Bus\#2}$ = 41.73 hrs/yr. This maximum value satisfies the required probabilistic bus reliability criterion, ${}_{R}LOLE_{Bus}$ = 50 hrs/yr. The optimal solution has a construction cost of 348 M$ and the new construction elements of $TF_{1-2}{}^{1}$, $T_{13-17}{}^{1}$, $T_{13-15}{}^{1}$, and $T_{9-11}{}^{1}$. This second plan using bus/nodal reliability criterion suggests a stronger grid system than the first case optimal plan using the system reliability criteria. The second plan has a construction cost of 139 M$ more than that of the former plan. The additional transmission

Table 14.22 Reliability indices at the load buses in the case of $_{R}LOLE_{TS}$ = 50 hrs/yr.

Load bus number	$LOLE_{Bus}$ (hrs/yr)	$EENS_{Bus}$ (MWh/yr)	EIR_{Bus} (PU)	Remark ($_{R}LOLE_{TS}$)
21	45.22	8486	0.998 504	
6	44.81	9307	0.998 344	
2	55.77	12 073	0.998 142	Over
9	38.20	6496	0.998 534	
10	0.0	0	1.000 000	
11	37.35	5926	0.998 468	
14	14.25	1016	0.999 159	
13	63.06	10 378	0.998 118	Over
15	45.52	4968	0.998 340	
17	65.57	12 770	0.998 134	Over
7	16.73	1,564	0.999 046	
5	7.12	356	0.999 299	

elements are required in order to decrease the *LOLE* indices of buses 2, 13, and 17, which have reliability levels in excess of the approved bus reliability criterion.

The optimal expansion plans that resulted from changing bus reliability criteria, $_{R}LOLE_{Bus}$, are given in Table 14.25. The tabulated results indicate that as $_{R}LOLE_{Bus}$ increases, the total construction cost decreases, and reliability indices, $LOLE_{TS}$ and $EENS_{TS}$, of the new optimal transmission system increase.

The third case study has a market characteristic that the grid owner has to construct the new grid to supply electrical energy with the bus reliability level, $_{R}LOLE_{Bus17}$ = 50 hrs/yr for the customers of bus 17 (this load point is the third largest city (Daegu) in Korea) and with a system reliability level, $_{R}LOLE_{TS}$ = 50 hrs/yr, for the other customers.

Figure 14.25 shows the new optimal system produced in this case study. The optimal solution has the construction cost of 278 M$ and the new construction elements of $T_{13-17}{}^{1}$, $T_{13-15}{}^{1}$, $T_{9-10}{}^{1}$, and $T_{9-11}{}^{1}$. Table 14.26 shows the bus reliability indices of the optimal plan. The system reliability level, $LOLE_{TS}$ and the bus 17 reliability level of the optimal system are 39.68 and 49.49 hrs/yr, respectively. These levels satisfy the two required probabilistic reliability criteria, $_{R}LOLE_{TS}$ = 50 hrs/yr for the system reliability criterion and $_{R}LOLE_{Bus17}$ = 50 hrs/yr for the bus reliability criterion at the bus 17. The new third plan using the specified bus and system reliability criteria creates a stronger grid system with higher cost than the first case plan using the system reliability criterion only. It is interesting to note that the

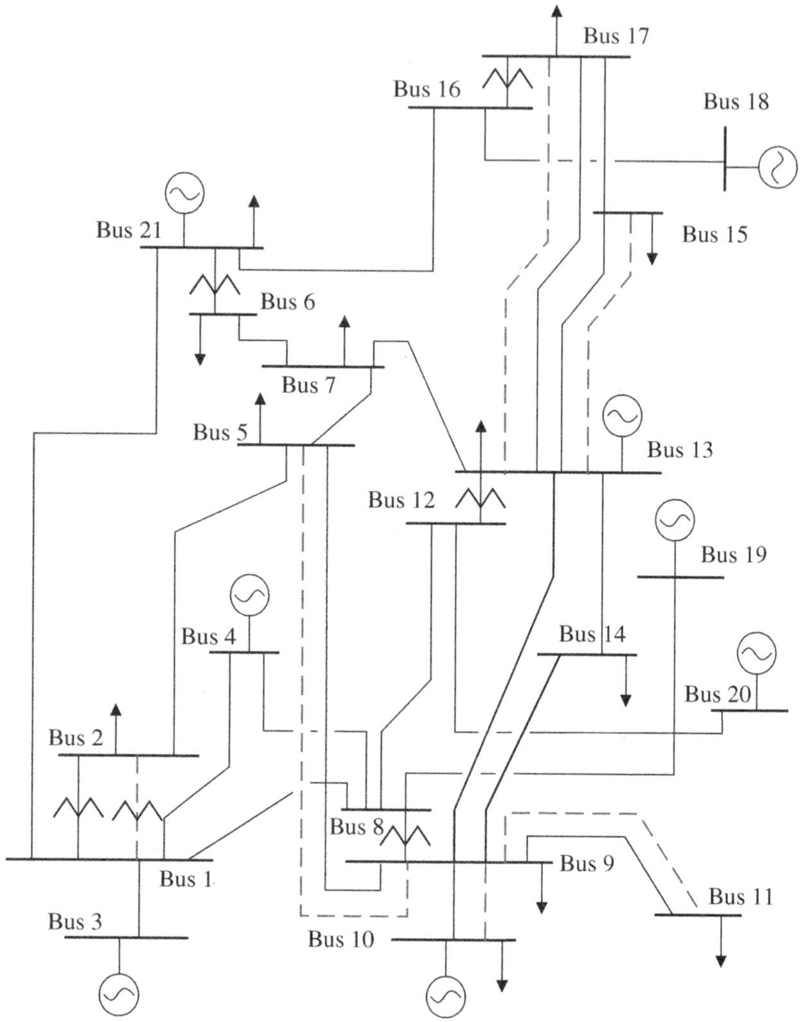

Figure 14.23 Optimal system by the probabilistic approach (${}_{R}LOLE_{TS}$ = 25 hrs/yr).

optimal plan involves a new line between bus 17 and bus 13 in order to satisfy the bus 17 reliability criterion.

The optimal expansion plans resulting from changing the bus 17 reliability criterion, ${}_{R}LOLE_{BUS17}$, are given in Table 14.26. The tabulated results indicate that as the ${}_{R}LOLE_{BUS17}$ increases, the total construction cost decreases, and the actual reliability indices, the specified $LOLE_{BUS17}$, as well as $LOLE_{TS}$ and $EENS_{TS}$ of the new optimal transmission system increase. The same new plans are obtained for

Table 14.23 Optimal expansion plans due to changing the system reliability criterion, ${}_RLOLE_{TS}$.

${}_RLOLE_{TS}$ (hrs/yr)	Construction of new lines	$Cost_T$ (M$)	$LOLE_{TS}$ (hrs/yr)	$EENS_{TS}$ (MWh/yr)	Remarks
25	$TF_{1-2}{}^1$, $T_{5-9}{}^1$, $T_{13-17}{}^1$, $T_{13-15}{}^1$, $T_{9-10}{}^1$, and $T_{9-11}{}^1$	511	24.42	29 942	
30	$TF_{1-2}{}^1$, $T_{13-17}{}^1$, $T_{13-15}{}^1$, $T_{9-10}{}^1$, and $T_{9-11}{}^1$	429	26.54	31 645	
35	$TF_{1-2}{}^1$, $T_{13-17}{}^1$, $T_{13-15}{}^1$, and $T_{9-11}{}^1$	348	31.64	39 841	
40, 45	$T_{13-17}{}^1$, $T_{13-15}{}^1$, $T_{9-10}{}^1$, and $T_{9-11}{}^1$	278	39.68	69 580	
50, 55	$T_{13-15}{}^1$,$T_{9-10}{}^1$, and $T_{9-11}{}^1$	209	45.47	73 339	Same with BRR = 0%
60, 65, 70	$T_{21-1}{}^1$, $T_{13-15}{}^1$, and $T_{9-11}{}^1$	157	55.10	83 654	
75,..., 100	$T_{21-1}{}^1$, $T_{9-11}{}^1$, and $T_{15-17}{}^1$	154	73.52	105 454	

Where *BRR*: Bus Reserve Rate in the deterministic approach.

Table 14.24 Reliability indices at the load buses in the case of ${}_RLOLE_{BUS}$ = 50 hrs/yr.

Load bus number	$LOLE_{Bus}$ (hrs/yr)	$EENS_{Bus}$ (MWh/yr)	EIR_{Bus} (PU)	Remark ($LOLE_{Bus}$)
21	31.45	4375	0.999 229	
6	39.08	5517	0.999 019	
2	41.73	6784	0.998 956	Maximum
9	30.76	3534	0.999 203	
10	0.00	0	1.000 000	
11	29.82	3257	0.999 158	
14	9.75	492	0.999 593	
13	39.36	5653	0.998 975	Satisfied
15	27.64	2429	0.999 188	
17	36.57	7050	0.998 970	Satisfied
7	10.45	626	0.999 618	
5	2.50	125	0.999 754	

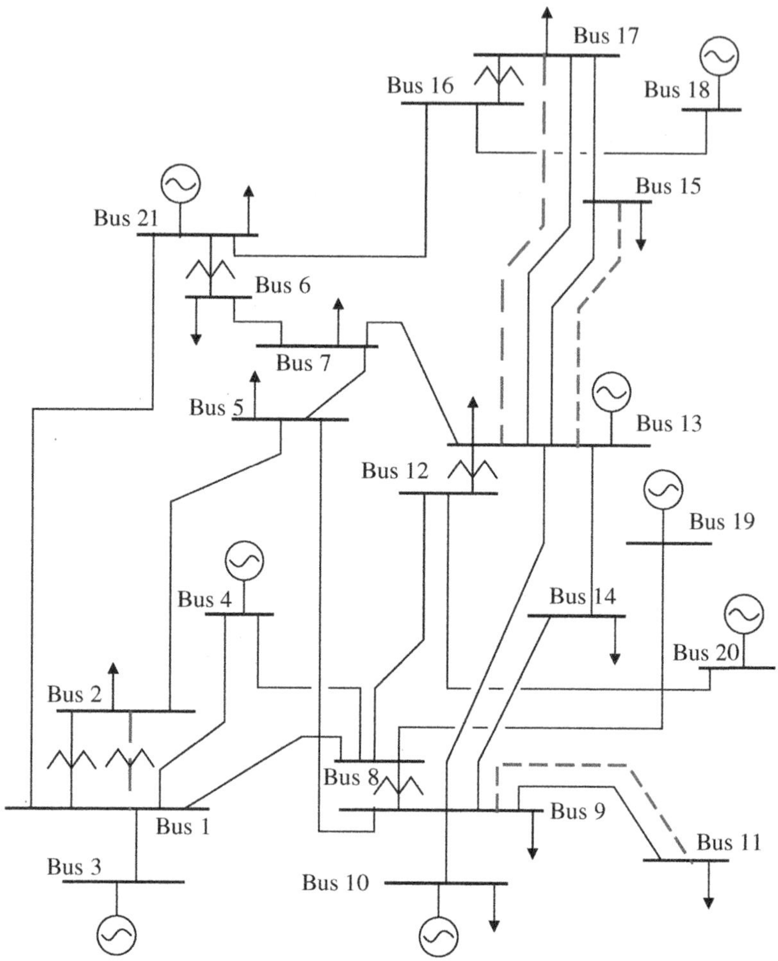

Figure 14.24 Optimal system by the probabilistic approach (${}_{R}LOLE_{BUS}$ = 50 hrs/yr).

${}_{R}LOLE_{Bus17}$ = 50 hrs/yr and over. This comes from the fact that the system reliability constraint is more dominant than the bus reliability constraint and the determination of the optimal plan is dependent on the system reliability constraint rather than the bus reliability constraint in the cases of 50 hrs/yr and over (Table 14.27).

In summary, Figures 14.22–14.25 in three cases show that very different expansion plans can result by using different types and magnitudes for the probabilistic reliability criterion. The case studies suggest that bus/nodal probabilistic reliability criteria result in a stronger grid than that produced using system reliability criteria.

Table 14.25 Optimal expansion plans due to changing the bus reliability criterion, ${}_RLOLE_{BUS}$.

${}_RLOLE_{Bus}$ (hrs/yr)	Construction of new lines	$Cost_T$ (M$)	$LOLE_{TS}$ (hrs/yr)	$EENS_{TS}$ (MWh/yr)	Remarks
35	$TF_{1-2}{}^1$, $T_{5-9}{}^1$, $T_{13-17}{}^1$, $T_{13-15}{}^1$, $T_{9-10}{}^1$, and $T_{9-11}{}^1$	511	24.42	29 942	
40	$TF_{1-2}{}^1$, $T_{13-17}{}^1$, $T_{13-15}{}^1$, $T_{9-10}{}^1$, and $T_{9-11}{}^1$	429	26.54	31 645	
45, 50	$TF_{1-2}{}^1$, $T_{13-17}{}^1$, $T_{13-15}{}^1$, and $T_{9-11}{}^1$	348	31.64	39 841	
55, 60, 65	$T_{13-17}{}^1$, $T_{13-15}{}^1$, $T_{9-10}{}^1$, and $T_{9-11}{}^1$	278	39.68	69 580	
70	$T_{13-15}{}^1$,$T_{9-10}{}^1$, and $T_{9-11}{}^1$	209	45.47	73 339	Same with $BRR = 0\%$
75, ..., 100	$T_{21-1}{}^1$, $T_{13-15}{}^1$, and $T_{9-11}{}^1$	157	55.10	83 654	

Table 14.26 Reliability indices at the load buses in the case of two constraints, ${}_RLOLE_{TS}$ = 50 hrs/yr and ${}_RLOLE_{bus17}$ = 50 hrs/yr.

Load bus number	$LOLE_{Bus}$ (hrs/yr)	$EENS_{Bus}$ (MWh/yr)	EIR_{Bus} (PU)	Remark (${}_RLOLE_{TS}$)
21	38.19	8094	0.998 574	
6	44.78	9308	0.998 344	
2	50.08	11 220	0.998 273	
9	34.52	6496	0.998 578	
10	0.0	0	1.000 000	
11	37.28	5918	0.998 470	
14	14.22	1015	0.999 161	
13	49.26	9523	0.998 274	
15	36.92	4523	0.998 489	
17	49.49	11 755	0.998 282	Satisfied
7	16.73	1564	0.999 046	
5	7.11	356	0.999 300	

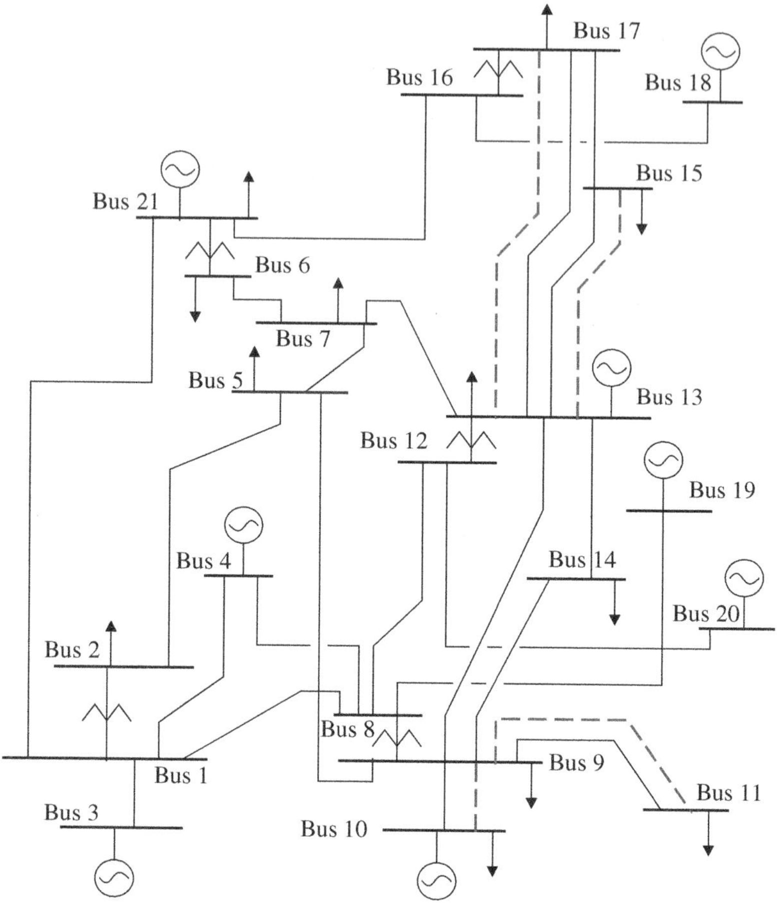

Figure 14.25 Optimal system by the probabilistic approach (${}_{R}LOLE_{TS}$ = 50 hrs/yr and ${}_{R}LOLE_{Bus17}$ = 50 hrs/yr).

Table 14.27 Optimal expansion plans due to changing the reliability criterion, ${}_{R}LOLE_{BUS\#17}$ and setting on ${}_{R}LOLE_{TS}$ = 50 hrs/yr.

${}_{R}LOLE_{BUS\#17}$ (hrs/yr)	Construction of new lines	$Cost_T$ (M$)	$LOLE_{BUS\#17}$ (hrs/yr)	$LOLE_{TS}$ (hrs/yr)	$EENS_{TS}$ (MWh/yr)
35	TF_{1-2}^{1}, T_{13-17}^{1}, T_{13-15}^{1}, T_{9-10}^{1}, and T_{9-11}^{1}	429	32.99	26.54	31 645
40, 45	TF_{1-2}^{1}, T_{13-17}^{1}, T_{13-15}^{1}, and ${}_{9-11}^{1}$	348	36.57	31.64	39 841
50, 55, 60, and 65	T_{13-17}^{1}, T_{13-15}^{1}, T_{9-10}^{1}, and T_{9-11}^{1}	278	49.49	39.68	69 580

14.3.6 Conclusion

This section addresses TEP using a probabilistic reliability criterion. The proposed procedure is a first step in preparing a TEP employing probabilistic reliability assessment methods to ensure the reliability of the electric power grid. Optimal locations and capacities of transmission lines can be determined using the proposed method. The chapter presents a new and practical approach that should serve as a useful guide for the decision maker in selecting a reasonable expansion plan prior to checking system stability and dynamics in detail. The proposed method finds the optimal transmission system expansion plan considering uncertainties associated with the forced outage rates of the grid elements (transformers and lines). It models the problem as a probabilistic integer programming one and considers problem uncertainties through probabilistic modeling. A probabilistic branch and bound algorithm, which includes the network flow method, and the maximum flow-minimum cut-set theorem is used to solve the problem. The case studies show that quite different planning alternatives can be determined from the use of the deterministic and probabilistic reliability approaches and different reliability criteria. The case studies on the 21-bus test system suggest that bus/nodal probabilistic reliability criteria result in a stronger grid than a grid produced using system reliability criteria. The chapter shows that the proposed method can be used to perform TEP considering different individual bus reliability criteria. The proposed approach can therefore accommodate customers' requirements in a competitive electricity market environment.

14.4 Outage Cost Constraints

14.4.1 Introduction

In general, an objective power system expansion planning problem can be classified into two main methods. One method is to minimize the total cost (investment cost) subject to the reliability-level constraints which are the deterministic reliability criteria, $N-1$ or $N-2$ contingency criterion and load balance constraints, and the probabilistic reliability criteria constraints to be bulk system/nodal (bus). Another method is also the minimization of the total cost which is investment cost and outage cost (reliability cost) subject to reliability criterion-level constraints or not subject to it.

This section proposes an alternative method for choosing the best TEP. The objective function is to minimize the total cost for constructing new transmission lines which are an investment cost and the annual outage cost, subject to probabilistic transmission system reliability criterion ($LOLE_{TS}$) which considers the uncertainties of transmission system elements. The annual outage cost of a pure transmission system can be obtained by a yearly transmission system-based outage energy *EENS* (MWh/year) times Interruption Energy Assessment Rate (IEAR) ($(k/Wh)) or Value of Loss of Load (VOLL) ($/k/Wh). The conventional branch and bound and network flow methods are used to search for the optimum mix of transmission network expansion. The proposed method also includes generation additions to determine the optimum mix of generation and transmission facilities required to meet the composite system probabilistic reliability criterion. It models the TEP problem as an integer programming problem. It solves for the optimum mix of transmission network expansion using a probabilistic branch and bound method that utilizes a network flow approach and the maximum flow-minimum cut-set theorem [30].

14.4.2 The Objective Function

The conventional composite power system expansion planning problem is to minimize the total cost (C^T) which is construction cost associated with investing in new generators and transmission lines and annual outage cost (*OTC*) which is yearly system *EENS* (MWh/year) times *IEAR* ($ (k/Wh) as

$$\text{minimize } C^T = \sum_{(x,y)\in B} \left[\sum_{i=1}^{m(x,y)} C^i_{(x,y)} U^i_{(x,y)} + OTC^i_{(x,y)} \right] \tag{14.37}$$

where

ρ: the set of all branches (generators and transmission lines),

$m(x,y)$: the number of new candidate branches connecting nodes x and y,

$C^i_{(x,y)}$: sum of the construction costs of the new generators and lines 1st through i-th that connect buses x and y, and

$$C^i_{(x,y)} = \sum_{j=1}^{i} \Delta C^j_{(x,y)},$$

$\Delta C^j_{(x,y)}$: annualized construction cost of the new *j-th* generator or line connecting nodes x and y (M$/year),

$OTC^i_{(x,y)}$: annual outage cost of the construction of the new generators and lines 1st through i-th that connect buses x and y,

$U^i_{(x,y)}$: the decision variable associated with the generator or line (1 if from 1-st to i-th generators or lines are to be constructed, and 0 otherwise),

$$\sum_{i=1}^{m(x,y)} U^i_{(x,y)} = 1$$

$$U^i_{(x,y)} = \begin{cases} 1 & P_{(x,y)} = P^0_{(x,y)} + P^i_{(x,y)} \\ 0 & P_{(x,y)} \neq P^0_{(x,y)} + P^i_{(x,y)} \end{cases}$$
$$P^i_{(x,y)} = \sum_{j=1}^{i} \Delta P^j_{(x,y)} \tag{14.38}$$

where

$P^i_{(x,y)}$: sum of the capacities of new branches (new generators or new transmission lines) between nodes x and y,

$\Delta P^j_{(x,y)}$: capacity of the j-th element of the candidate branches connecting nodes x and y,

$P^0_{(x,y)}$: capacity of the existing generators and lines that connect nodes x and y.

14.4.3 Constraints

The basic reliability criteria normally considered in a composite power system planning problem can be categorized into two types of constraints. One is a deterministic reliability criterion and the other is the probabilistic reliability criterion.

In a deterministic approach, no shortage of power supply requires that the total capacity of the branches involved in the minimum cut-set should be greater than or equal to the system peak load demand, L_p. This is also referred to as the bottleneck capacity. Therefore, a no-shortage power supply constraint can be expressed by

$$P_c(S, T) \geq L_p \quad (s \in S,\ \ t \in T) \tag{14.39}$$

where, $P_c(S, T)$ is the capacity of the minimum cut-set of two subsets, S and T, containing source nodes s and terminal nodes t, respectively, when all nodes are separated by a minimum cut-set.

The demand constraint (14.39) can be expressed by (14.40) with k being the cut-set number ($k = 1,\dots, n$), where n is the number of cut-set.

$$\sum_{(x,y)\in(S_k,T_k)} \left[P_{(x,y)} = P^0_{(x,y)} + \sum_{i=1}^{m(x,y)} P^i_{(x,y)} U^i_{(x,y)} \right] \geq L_p \tag{14.40}$$

In the probabilistic approach, the probabilistic reliability criterion index, *LOLE* (Loss of Load Expectation), can be used as

$$LOLE_{TS}\left(P^i_{(x,y)}, \Phi\right) \leq {}_R LOLE_{TS} \tag{14.41}$$

Where ${}_{R}LOLE_{TS}$ is the required transmission reliability criterion for the new system, and Φ is a function of the load duration curve. A detailed discussion of Φ and *LOLE* is presented in Section 14.3.

14.4.4 Outage Cost Assessment of Transmission System

Reliability evaluation at hierarchical level II (HLII) – The reliability indices at HLII can be classified as load point indices and bulk system indices depending on the object of the evaluation. The reliability indices can be evaluated using a CMELDC of HLII based on the composite power system effective load model. Formulating the Synthesized Fictitious Equivalent Generator (SFEG) mathematically is the core of the proposed model and the definitions of the components of the model follow.

Nomenclature

pdf: abbreviation of probability distribution function
NG: number of generators
NT: number of transmission lines
CG_i: capacity of generator # i ($i = 1,\ldots, NG$)
q_{gi}: forced outage rate of generator # i
CT_i: capacity of transmission line # i ($i = 1,\ldots, NT$)
q_{ti}: forced outage rate of transmission line # i
${}_{k}AP_{ij}$: maximum arrival power for system state j at load point #k considering the operation of generators from #1st to # ith
${}_{k}q_{ij}$: state probability for system state j at load point #k considering the operation of generators from #1st to # ith
${}_{k}AP_{sij}$: operating power of SFEG at load point #k considering the operation of generators from #1st to # ith
${}_{k}q_{sij}$: operating state probability of SFEG at load point #k considering the operation of generators from #*1*st to # ith
${}_{k}f_{osi}$: outage capacity *pdf* of SFEG at load point #k considering the operation of generators from #*1*st to # ith
${}_{k}x_{e}$: random variable of the effective load at load point #k
${}_{k}x_{L}$: random variable of the original load at load point #k
${}_{k}x_{oj}$: random variable of the probabilistic load caused by the forced outage of generators and transmission lines at load point #k
NS: total number of system states (contingency states)
${}_{k}\overline{AP}_{i}$: the largest value of the maximum arrival powers (${}_{k}AP_{sij}$) (= supremum (${}_{k}AP_{sij}$))

$_k\Phi_i$: nodal equivalent load duration curve at load point #*k* considering the operation of generators from #*1*st to # *i*th

$_k\Phi_0$: original load duration curve at load point #*k*

$\otimes$: the operator meaning convolution integral

T: study period (8760 hours were used in here)

Reliability indices at the load points (buses): Figure 14.26 presents the main components of the nodal equivalent system and the nodal-effective load for HLII developed recently. Figure 14.26a is the original power system. Let's consider the operation of generators from #1st to # *i*th. It is possible to calculate the maximum arrival power ($_kAP_{ij}$) at the load point and the state probabilities ($_kq_{ij}$) for system

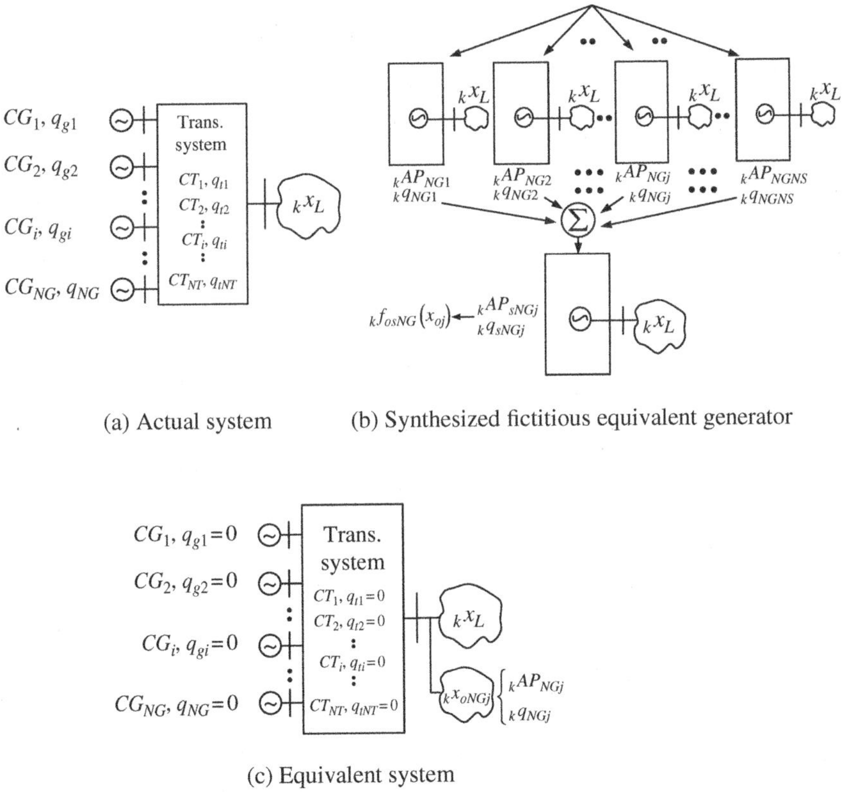

Figure 14.26 Composite power system-effective load model at HLII.

state j using optimal power flow analysis with the maximum arrival power being the objective function. Then the same maximum arrival power for different states may be obtained. The probabilities of the states with the same maximum arrival power ($_kAP_{ij}$) can be cumulated. The cumulated state probability and the same maximum arrival power were designated as $_kq_{sij}$ and $_kAP_{sij}$, respectively in here. The pdf composed from $_kq_{sij}$ and $_kAP_{sij}$ is equivalent to it of a supply source unit with the outage state probability, $_kq_{sij}$ and operating power, $_kAP_{sij}$ with multi-operating states at load point #k. This is designated as SFEG and the capacity of the SFEG is the largest value of the maximum arrival powers ($_kAP_{sij}$). Figure 14.26b shows the SFEG at load point #k and $_kf_{osNG}$ is the outage capacity pdf. This generator is referred to here as SFEG or nodal SFEG. The pdf of the SFEG can be extended to multi-states (partial failures), although the pdf in this example uses two states (on or off). Therefore, the nodal-effective load for the HLII criterion is defined by the summation of the original load and the probabilistic load caused by the forced outage of generators and transmission lines. This can be formulated as in (14.42). Figure 14.26c shows the nodal-effective load model for HLII criterion. Finally, when the effective load is composed as a load duration curve, it is called a CMELDC of HLII.

$$_kx_e = {_kx_L} + \sum_{j=1}^{NS} {_kx_{oj}} \tag{14.42}$$

Therefore, the load point reliability indices, $LOLE_k$ and $EENS_k$, can be calculated using (14.43) and (14.44) with the CMELDC, $_k\Phi_{NG}(x)$

$$LOLE_k = {_k\Phi_{NG}(x)}|_{x = AP_k} \text{ hr/year} \tag{14.43}$$

$$EENS_k = \int_{AP_k}^{AP_k + Lp_k} {_k}\Phi_{NG}(x)dx \text{ MWh/year} \tag{14.44}$$

where

L_{pk}: peak load at load point k (MW),
AP_k: maximum arrival power at load point k (MW).

$$\begin{aligned} {_k\Phi_i(x_e)} &= {_k\Phi_o(x_e)} \otimes {_kf_{osi}(x_{oi})} \\ &= \int_k \Phi_o(x_e - x_{oi})_k f_{osi}(x_{oi})dx_{oi} \end{aligned}$$

Reliability indices of the bulk system: While the $EENS_{HLII}$ of a bulk system is equal to the summation of the $EENS_k$ at the load points as shown in (14.45), the $LOLE$ of a bulk system is entirely different from the summation of the $LOLE_k$ at the load points. The ELC_{HLII} (expected load curtailed) of the bulk system is equal to the summation of ELC_k at the load points. The $LOLE_{HLII}$ of the bulk system can be calculated using (14.48).

$$EENS_{HLII} = \sum_{k=1}^{NL} EENS_k \text{ MWh/year} \tag{14.45}$$

$$ELC_{HLII} = \sum_{k=1}^{NL} ELC_k \text{ MW/CUR.year} \tag{14.46}$$

$$LOLE_{HLII} = EENS_{HLII}/ELC_{HLII} \text{ hrs/yr} \tag{14.47}$$

$$EIR_k = 1 - EENS_k/DENG_k \text{ PU} \tag{14.48}$$

where

NL: number of load points,
$ELC_k = EENS_k/LOLE_k$,
$DENG_k$: demand energy at bus #k

14.4.5 Reliability Evaluation of Transmission System

The reliability indices of a transmission system can be expressed as the difference between the HLII and HLI reliability indices as

$$EENS_{TS} = EENS_{HLII} - EENS_{HLI} \text{ MWh/year} \tag{14.49}$$

$$EENS_{TS} = EENS_{HLII} - EENS_{HLI} \text{ hrs/yr} \tag{14.50}$$

14.4.6 Outage Cost Assessment

The annual outage cost assessment can be formulated by taking annual $EENS^i_{(x,y)}$ (expected energy not served) of the construction element of the new generators and lines 1-st through i-th that connect buses x and y multiplied with Interrupted Energy Assessment Rate (IEAR) system, that is sometimes called Value of Loss Load ($VOLL$) as

$$OTC^i_{(x,y)} = IEAR \times \sum_{k=1}^{NL} EENS^i_{k(x,y)} \text{ Million \$} \tag{14.51}$$

where

$IEAR$: interrupted energy assessment rate (\$ (k/Wh]))

14.4.7 Solution Algorithm

The objective of the conventional branch and bound method is to minimize the total construction cost subject to a specified reliability criterion. The proposed probabilistic branch and bound-based method minimizes the total cost subject to the required probabilistic transmission system reliability criteria, $_RLOLE_{TS}$ or/and $_RLOLE_{BUS}$, The solution algorithm for the proposed approach follows:

1) Check the need for transmission expansion for the system and its possibility of using the candidate lines. Need and possibility can be checked respectively by the reliability evaluation for systems considering no candidate lines and considering all candidate lines.
2) Set $j = 1$ (initial system), $jopt = 0$, $jmax = 0$, $C^T_{opt} = \infty$ and $ENNOD_j = 0$.
3) If $ENNOD_j = 1$, the #j system is an end node at which the branch operation of a branch and bound is finished (bound) in the solution graph used to obtain the optimal solution, and there is no need to consider any of the other graphs following this system. Go to 13.
4) Calculate the minimum cut-set using the maximum flow method for system j (solution j in the solution graph).
5) Select a #i branch/line of the candidate branches/lines set (S_j) involved in the minimum cut-set and add to the #j system. In what follows, the new system is named the system ji.
6) If the system ji is already considered in the solution graph, go to 13.
7) Probabilistic reliability evaluation indices of system ji, ($LOLE_{ji}$, $EENS_{ji}$), and calculate an outage cost of the system ji ($OTC_{ji} = IEAR_{sys}$ x $EENS_{ji}$).
8) Calculate the total cost $C^T_{ji} = C^T_j + C\left(P^{(i)}_{(x,y)}\right) + OTC\left(P^{(i)}_{(x,y)}\right)$ for the system ji and evaluate the transmission system reliability index, $LOLE_{TSji}$ of the system.
9) If $C^T_{ji} < C^T_{jopt}$, the current system (ji) with a cost of C^T_{ji} can be optimal. If not, go to 11.
10) Set $jmax = jmax + 1$.
11) If $LOLE_{TSji}$ (or $LOLE_{BUSji}$) $< {_R}LOLE_{TS}$, set $C^T_{opt} = C^T_{ji}$, and $_RLOLE_{opt} = LOLE_{ji}$, $jopt = jmax$, and go to 12.
12) Set $C^T{}_{j\,max} = C^T_{ji}$, $ENNOD_{jmax} = 1$, and go to 13.
13) Add the solution $jmax(ji)$ to the solution graph.
14) If all the candidate branches/lines in the cut-set S_j have been considered, go to 14. Otherwise, set $i = i + 1$ and go to 5.
15) If $j = jmax$, continue the next step. Otherwise, set $j = j + 1$ and go to 4.
16) For $j = jmax$, the solution graph has been constructed fully and the optimal solution $jopt$ with C^T_{jopt} being the lowest cost and satisfies the required reliability criteria is obtained in 10.

14.4.8 Case Study

The proposed method was tested on the 21-bus model system shown in Figure 14.27. This is a part of the south-east area (Youngnam) in Korea. This expansion planning is a static problem, which is considered for a target year and can be described as follows. "What is the optimum TEP to minimize total cost and satisfy the required probabilistic transmission system reliability criterion, $LOLE_{TS}$, for the future forecast load?"

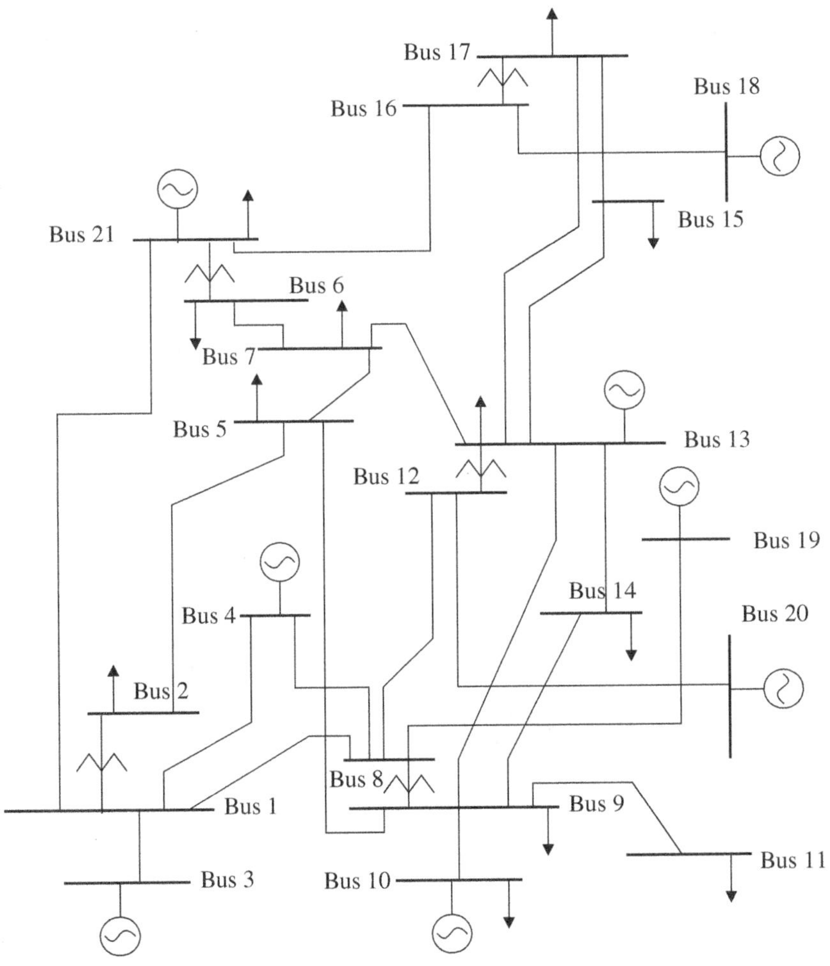

Figure 14.27 The 21-Bus model system.

Table 14.28 shows the system data with *GN*, *TF*, *TL*, and *LD* representing generators, transformers, transmission lines, and loads, respectively; *SB* and *EB* are start and end buses of the line, respectively; $\Delta P^{(0)}_{(x,y)}$ and $\Delta C^{(0)}_{(x,y)}$ are, respectively, the capacities and costs of the existing lines that connect nodes x and y. In this study, four candidate generators and lines are considered, which means $m(x,y) = 4$ in (14.1) and (14.4). In Table 14.28, parentheses in $\Delta P^{(j)}_{(x,y)}$ and $\Delta C^{(j)}_{(x,y)}$ are omitted for convenience. The cost unit, M$ in this table stands for a million dollars. Table 14.29 shows the forced outage rate of the generators and ransmission lines. Figure 14.28 shows the inverted load duration curves at the buses with the four largest loads. In this case study, the construction costs are assumed on annualized basement considering the annual expenditure rate.

Table 14.28 System capacity and cost data.

NL	SB	EB	ID	ΔP^0_{xy}	ΔP^1_{xy}	ΔP^2_{xy}	ΔP^3_{xy}	ΔP^4_{xy}	ΔC^0_{xy}	ΔC^1_{xy}	ΔC^2_{xy}	ΔC^3_{xy}	ΔC^4_{xy}
1	0	3	GN	850	0	0	0	0	0	0	0	0	0
2	0	21	GN	900	0	0	0	0	0	0	0	0	0
3	0	4	GN	850	0	0	0	0	0	0	0	0	0
4	0	10	GN	900	0	0	0	0	0	0	0	0	0
5	0	20	GN	1200	0	0	0	0	0	0	0	0	0
6	0	18	GN	850	0	0	0	0	0	0	0	0	0
7	0	13	GN	760	0	0	0	0	0	0	0	0	0
8	0	19	GN	950	0	0	0	0	0	0	0	0	0
9	6	21	TF	1020	510	510	0	0	0	132	132	0	0
10	16	17	TF	1020	510	510	0	0	0	124	124	0	0
11	12	13	TF	1020	510	510	0	0	0	123	130	0	0
12	8	9	TF	800	800	0	0	0	0	155	0	0	0
13	1	2	TF	800	800	0	0	0	0	151	0	0	0
14	21	1	TL	500	500	500	0	0	0	29	29	0	0
15	2	5	TL	220	220	0	0	0	0	54	0	0	0
16	1	4	TL	300	300	0	0	0	0	73	0	0	0
17	1	8	TL	400	400	0	0	0	0	70	0	0	0
18	1	3	TL	1000	250	250	250	250	0	20	20	20	20

Table 14.28 (Continued)

NL	SB	EB	ID	ΔP^0_{xy}	ΔP^1_{xy}	ΔP^2_{xy}	ΔP^3_{xy}	ΔP^4_{xy}	ΔC^0_{xy}	ΔC^1_{xy}	ΔC^2_{xy}	ΔC^3_{xy}	ΔC^4_{xy}
19	4	8	TL	300	300	0	0	0	0	63	0	0	0
20	5	9	TL	220	220	0	0	0	0	82	0	0	0
21	5	7	TL	220	220	0	0	0	0	77	0	0	0
22	7	6	TL	220	220	0	0	0	0	85	0	0	0
23	21	16	TL	1000	250	250	250	250	0	30	30	30	30
24	7	13	TL	220	220	0	0	0	0	88	0	0	0
25	13	17	TL	220	220	0	0	0	0	69	0	0	0
26	13	15	TL	220	220	0	0	0	0	83	0	0	0
27	16	18	TL	1320	330	330	330	330	0	32	32	32	32
28	9	13	TL	220	220	0	0	0	0	71	0	0	0
29	9	14	TL	220	220	0	0	0	0	65	0	0	0
30	8	19	TL	620	620	0	0	0	0	64	0	0	0
31	12	20	TL	1240	310	310	310	310	0	28	28	28	28
32	12	8	TL	400	400	0	0	0	0	62	0	0	0
33	9	10	TL	240	240	0	0	0	0	81	0	0	0
34	9	11	TL	340	340	0	0	0	0	45	0	0	0
35	15	17	TL	220	220	0	0	0	0	80	0	0	0
36	13	14	TL	220	220	0	0	0	0	80	0	0	0
37	21	22	LD	785	0	0	0	0	0	0	0	0	0
38	6	22	LD	750	0	0	0	0	0	0	0	0	0
39	2	22	LD	850	0	0	0	0	0	0	0	0	0
40	9	22	LD	595	0	0	0	0	0	0	0	0	0
41	10	22	LD	17	0	0	0	0	0	0	0	0	0
42	11	22	LD	550	0	0	0	0	0	0	0	0	0
43	14	22	LD	190	0	0	0	0	0	0	0	0	0
44	13	22	LD	710	0	0	0	0	0	0	0	0	0
45	15	22	LD	450	0	0	0	0	0	0	0	0	0
46	17	22	LD	870	0	0	0	0	0	0	0	0	0
47	7	22	LD	290	0	0	0	0	0	0	0	0	0
48	5	22	LD	70	0	0	0	0	0	0	0	0	0

Units: $\Delta P(*)$: (MW); $\Delta C(*)$: (Million Dollars/Year); 0 and 22 represent source and terminal nodes, respectively.

Table 14.29 Forced outage rates of generators and lines.

NL	FOR	NL	FOR	NL	FOR
1	0.012	13	0.0012	25	0.0014
2	0.015	14	0.0015	26	0.0020
3	0.010	15	0.0015	27	0.0018
4	0.015	16	0.0012	28	0.0022
5	0.010	17	0.0015	29	0.0020
6	0.012	18	0.0014	30	0.0025
7	0.0125	19	0.0015	31	0.0012
8	0.0155	20	0.0016	32	0.0015
9	0.0015	21	0.0018	33	0.0011
10	0.0020	22	0.0012	34	0.0011
11	0.0015	23	0.0012	35	0.0021
12	0.0012	24	0.0012	36	0.0022

The proposed method can be solved a composite power system expansion planning (CPSEP) well but this chapter focused on only TEP problem. The required probabilistic transmission system reliability ($LOLE_{TS}$) was 50 h/year. Figure 14.28 is an optimal solution of TEP which considered only probabilistic transmission reliability criterion, $LOLE_{TS}$ = 50 h/year, without considering annual outage cost ($IEAR = 0$ \$(k/Wh)). It required constructing the new transmission lines T^1_{13-15}, T^1_{9-10}, T^1_{9-11} with a total cost of 209 million dollars per year (M\$/year). Otherwise, Figure 14.29 is an optimal solution of TEP which considered both the probabilistic reliability criterion, $LOLE_{TS}$ = 50 h/year and annual outage cost [$IEAR = 5$ \$(k/Wh]. A transformer and two transmission lines TF^1_{1-2}, T^1_{13-15}, T^1_{9-11} were required in adding the existing system, with a total cost of 499.7 M\$ which is the summation of the construction cost of 279 M\$ and an annual outage cost of 220.7 M\$.

(This is the same as not considering outage cost model (model I))

The construction cost is a budget for investment power system elements which are transformers and transmission lines to satisfy the required probabilistic transmission system reliability ($LOLE_{TS}$). The annual outage cost is multiplied to annual *EENS* (MWh/year) while a new element is added to the actual power system and *IEAR* (k/Wh) together. The total cost is the summation of a construction cost and an annual outage cost. The optimal TEP may be sensitive at the places where there are big load demands as shown in Figures 14.29 and 14.30. Bus 2 is one of the big load demands in this system. Table 14.30 and Figure 14.31 present characteristics of the optimal TEP results while *IEAR* was

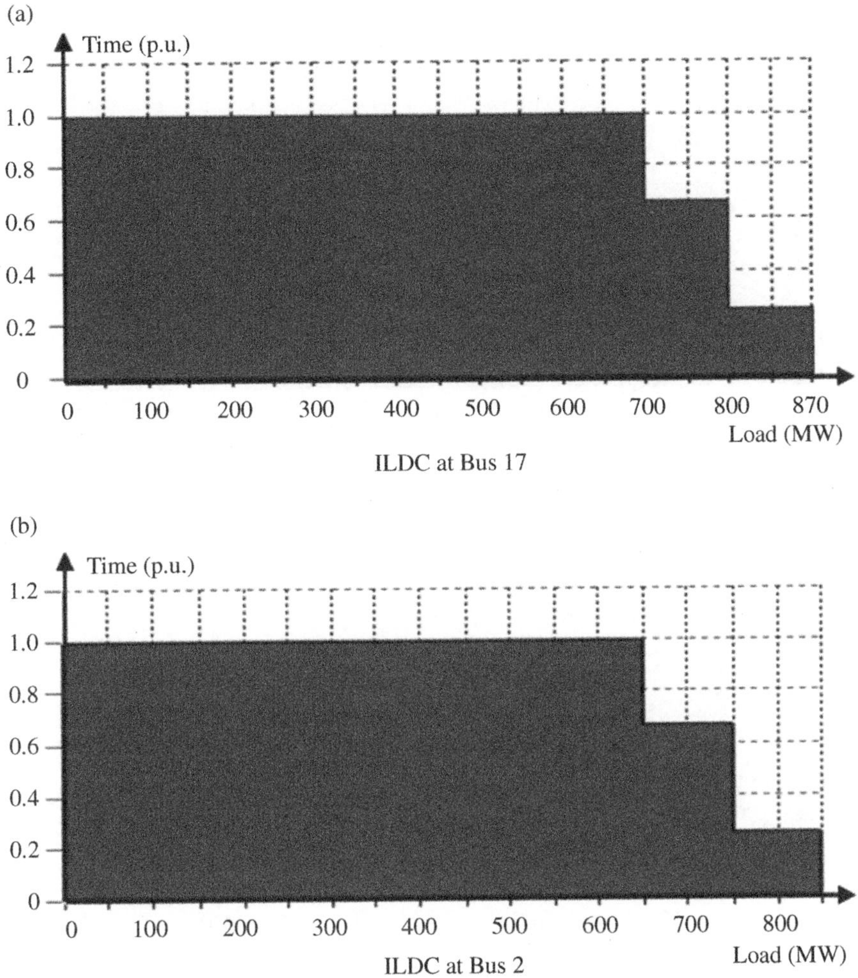

Figure 14.28 Inverted load duration curves (ILDCs) at the buses with the four largest loads.

increasing. The general characteristic of the construction cost curve is increasing and saturated while *IEAR* is still increasing because the probabilistic transmission system reliability is satisfied for the new system.

14.4.9 Conclusion

This chapter addresses TEP problem considering annual outage cost assessment (*OTC*) associated with construction cost (*C*), subject to the probabilistic transmission system reliability criterion ($LOLE_{TS}$). Optimal placements and the capacity of

(c)

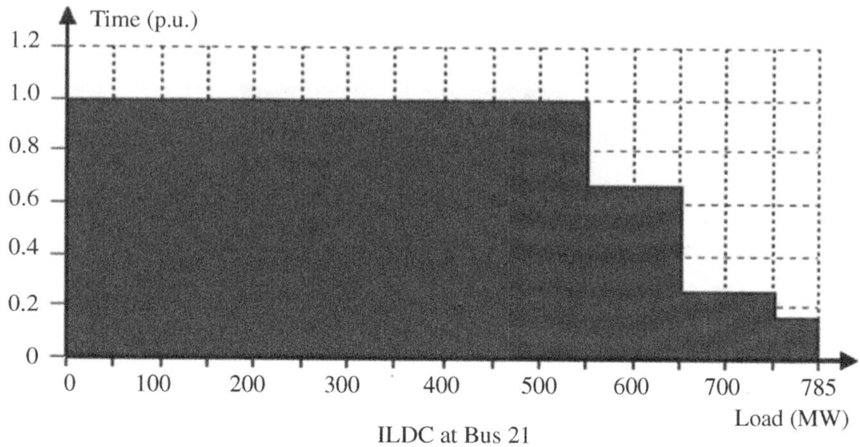

(d)

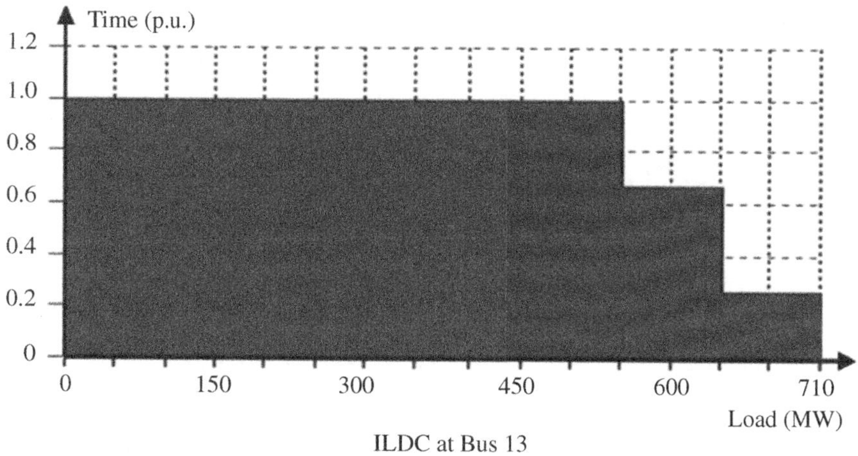

Figure 14.28 (Continued)

transformers as well as transmission lines can be determined using the proposed method. It presents a new and practical approach that should serve as a useful guide for the decision maker to select a reasonable expansion plan prior to checking system stability and dynamics in detail. The proposed method finds the optimal CPSEP considering uncertainties associated with the forced outage rates of generators, transformers, and transmission lines. It models the problem as a probabilistic integer programming one and considers problem uncertainties through probabilistic modeling. A proposed probabilistic branch and bound algorithm,

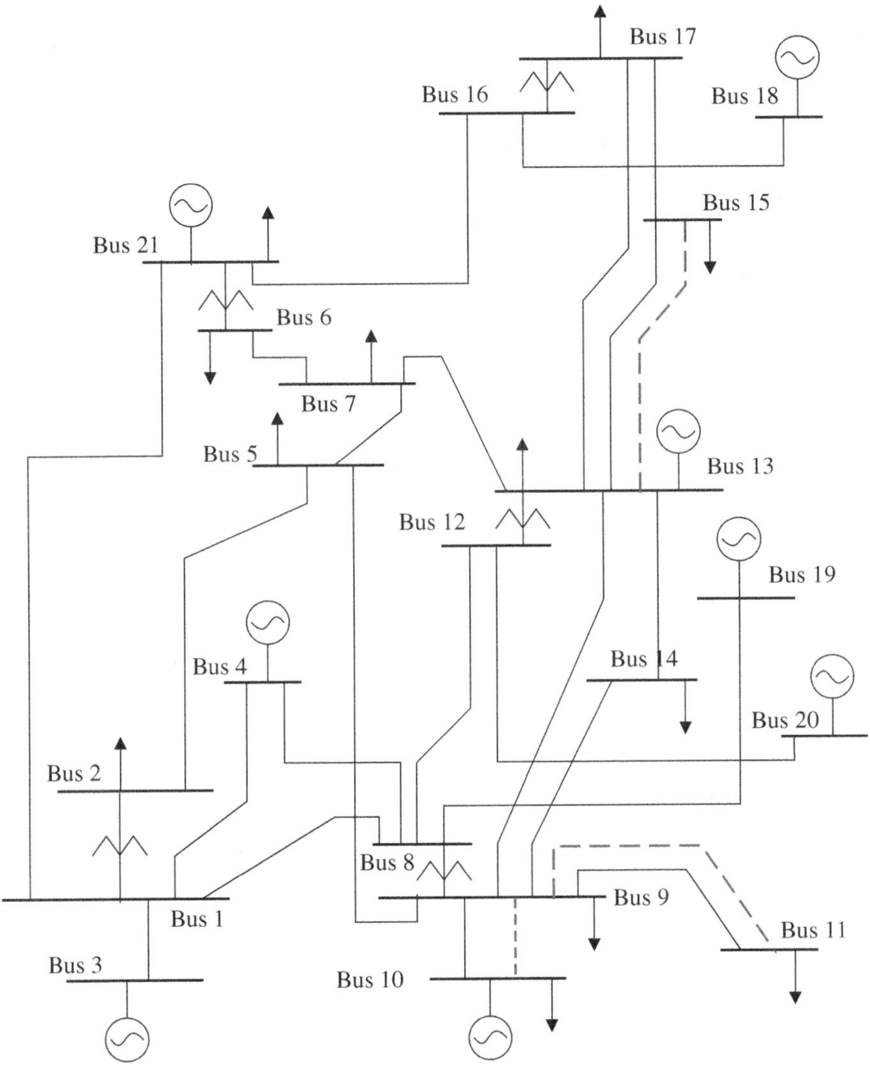

Figure 14.29 Optimal transmission system in the base case (${}_{R}LOLE_{TS}$ = 50 h/year, *IEAR* = 0 \$(k/Wh)).

which includes the network flow method, and the maximum flow-minimum cut-set theorem, is proposed to solve the problem. The 21-bus system analysis used to illustrate the method show that quite different expansion plans can result from applying various *IEAR*. The chapter shows that the proposed method can be used to perform the CPSEP that includes the residual uncertainties existing in practical systems.

Table 14.30 Various optimal solutions of transmission system expansion planning due to increase *IEAR* \$(k/Wh), ($_{R}LOLE_{TS}$ = 50 h/year).

Cases	*IEAR* (\$(k/Wh'))	Optimal solution	Const. cost (Million \$)	Out. cost (Million \$)	Total cost (Million \$)
Base case	0.0	T^1_{13-15}, T^1_{9-10}, and T^1_{9-11}	209	0.0	209
1	0.1	T^1_{13-15}, T^1_{9-10}, and T^1_{9-11}	209	7.3	216.3
2	0.5	T^1_{13-15}, T^1_{9-10}, and T^1_{9-11}	209	36.7	245.7
3	1.0	TF^1_{1-2}, T^1_{13-15}, and T^1_{9-11}	279	44.1	323.1
4	5.0	TF^1_{1-2}, T^1_{13-15}, and T^1_{9-11}	279	220.7	499.7
5	10.0	TF^1_{1-2}, T^1_{13-15}, and T^1_{9-11}	279	441.4	720.4
6	15.0	TF^1_{1-2}, T^1_{13-15}, and T^1_{9-11}	279	662.1	941.1

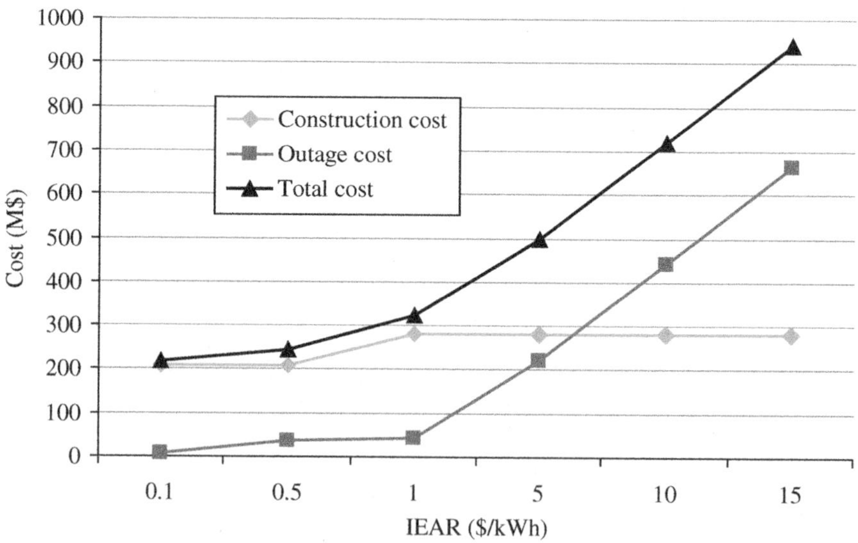

Figure 14.30 Optimal transmission system in Case 4 [$_{R}LOLE_{TS}$ = 50 h/year, *IEAR* = 5 \$(k/Wh).

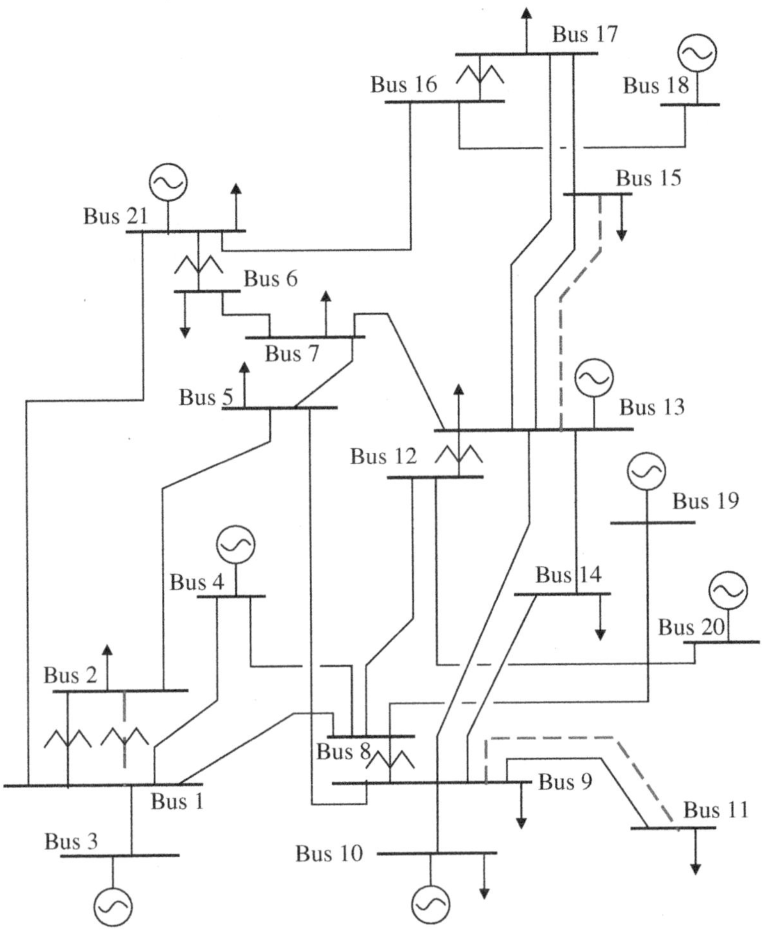

Figure 14.31 Variation of the investment cost, outage cost, and total cost due to increased IEAR (\$(k/Wh)). (${}_{R}LOLE_{TS}$ = 50 h/year).

14.5 Deterministic–Probabilistic (D–P) Criteria

This technique was developed by Bao [31] in 2007. The Deterministic-Probabilistic (D-P) concept is a deterministic framework incorporating a probabilistic criterion for TEP. According to the deterministic $N-1$ criterion, the system should be able to withstand the loss of any single element without any damage or load curtailment at all demand levels. An acceptable SI value is defined as the probabilistic criterion using a probabilistic approach. The D-P technique adds additional probabilistic risk information to the traditional accepted deterministic $N-1$ criterion that is useful when making system reinforcement decisions. This technique D-P approach combines the probabilistic criterion in a deterministic framework that incorporates both the deterministic and probabilistic perspectives. The D-P approach is driven by the deterministic $N-1$ criterion and influenced by the probabilistic criterion (Pc). The D-P technique adds additional probabilistic risk information to the traditional accepted deterministic $N-1$ criterion that is useful when making system reinforcement decisions [31].

15

Fuzzy Decision Making for TEP

15.1 Introduction

Transmission expansion planning (TEP) addresses the problem of expanding and strengthening an existing transmission network to optimally serve a growing electricity market while satisfying a set of economic and technical constraints. The problem is to minimize the cost subject to a reliability constraint. Various techniques including branch and bound, sensitivity analysis, Bender decomposition, simulated annealing, genetic algorithms, tabu search, and GRASP were used to study the problem. Transmission system expansion planning is performed following generation expansion planning (GEP) because it is very difficult to obtain the optimal solution of composite power systems considering generators and transmission lines simultaneously.

The transmission system expansion planning problem is analyzed from the viewpoint of a macro approach and then a micro approach considering the stability and dynamic characteristic of the new system. In the regulated monopolistic electricity market, global and bulk system criteria and indices are needed recently. However, in a deregulated environment, electric utilities are expected to be winners of competitions. And, successful electricity market operation in such an environment depends on the transmission system and nodal (bus) reliability management. This is because deregulated electricity markets call for nodal and decentralized operation and planning mode-based indices. Nodal reliability indices together with related information can be used for the management and control of congestion and reliability by ISO and TRANSCO in deregulated markets. This environment makes it more important to assess and construct reasonable reliability criterion at load points. In such an environment, the reliability of the delivery marginal rate of a transmission system becomes a more important parameter in transmission planning. In addition, in competitive electricity markets, there exists more ambiguity of the investment budget for construction cost and higher uncertainty of delivery marginal rate of the transmission system. This is

Probabilistic Power System Expansion Planning with Renewable Energy Resources and Energy Storage Systems, First Edition. Jaeseok Choi and Kwang Y. Lee.

Published 2022 by John Wiley & Sons, Inc.

because of the profit maximization of the system owner being the major focus while the primary function of a conventional power system is to provide electrical energy to its customers economically and with an acceptable degree of continuity and quality.

Furthermore, system planners and owners are expected to evaluate the reliability and economic parameters in grid planning when the problem involves many uncertainties including those of the investment budget, reliability criterion, load forecasting, and system characteristics, etc. It is a challenging task to develop an expansion plan that considers all these items in an effective and practical manner. When the database available for evaluating reliability indices and the investment budget for constructing new equipment is limited in size, it becomes difficult, if not impossible, to use general probabilistic methods to solve the problem. Under such circumstances, an approach that is based on fuzzy-set theory becomes attractive and useful to accomplish the task. The experience and knowledge of experts and decision makers can be very helpful in dealing with subjective uncertainties and ambiguity in planning problems. Such experience can be reflected in the membership functions of fuzzy objective functions and constraints. In addition, the fuzzy-set theory can conveniently handle multi-objective optimization problems. Accordingly, a fuzzy optimal decision policy can be formulated as the maximization of the satisfaction level of a decision maker.

This chapter proposes a new approach that is designed to facilitate sensitivity analysis and trend evaluation in view of a macro approach prior to the micro-level checking (stability and dynamics, etc.) in TEP. It uses fuzzy-set theory to model the uncertainties and ambiguities associated with construction cost and delivery marginal rate through appropriate membership functions. A new fuzzy branch and bound algorithm, which includes the use of network flow theory and the maximum flow-minimum cut set theorem, is proposed. It is designed to obtain the optimal solution with the highest satisfaction level of the decision maker. The effectiveness of the proposed approach is demonstrated by testing it on a 21-bus test system, which is a part of south–east area system of Korean Electric Power Company (KEPCO) system. Following are used assumptions:

a) A network flow method considering active power is used.
b) The network flow method is acceptable for the long-term planning problem. The focus is on model development of practical algorithm considering the expert experiences of decision makers. Instead of AC or DC load flow, maximum flow is used in the proposed algorithm.
c) A set of draft plans/scenarios is used as candidate plans.
d) The problem is limited to a static expansion-planning problem for a single-stage or horizon-year.

In the following discussion, a node with a binding generator is named a source node and a node with a binding load is referred to as a terminal (t or sink) node. And, transmission lines are represented as elements and referred to as branches [34, 35].

15.2 Fuzzy Transmission Expansion Planning Problem

The objective in the conventional crisp expansion planning is to minimize total construction cost CT for investing in new transmission lines as

$$\text{minimize } C^T = \sum_{(x,y)\in\rho}\left[\sum_{i=1}^{m(x,y)} C^{(i)}_{(x,y)} U^{(i)}_{(x,y)}\right] \tag{15.1}$$

where

ρ: the set of all branches (transmission lines),
$m(x,y)$: the number of new candidate branches connecting nodes x and y,
$C^{(i)}_{(x,y)}$: sum of construction costs of the new lines 1st through the ith that connect buses x and y,

$C^{(i)}_{(x,y)} = \sum_{j=1}^{i} \Delta C^{(j)}_{(x,y)}$

with $\Delta C^{(j)}_{(x,y)}$: construction cost of the new jth line connecting nodes x and y,
$U^{(i)}_{(x,y)}$: the decision variable concerning the line (*1* if from the *1st* to the ith, the lines are to be constructed and 0 otherwise),

with $\sum_{i=1}^{m(x,y)} U^{(i)}_{(x,y)} = 1$

$$U^{(i)}_{(x,y)} = \begin{cases} 1 & P_{(x,y)} = P^{(0)}_{(x,y)} + P^{(i)}_{(x,y)} \\ 0 & P_{(x,y)} \neq P^{(0)}_{(x,y)} + P^{(i)}_{(x,y)} \end{cases}$$

$$P^{(i)}_{(x,y)} = \sum_{j=1}^{i} \Delta P^{(j)}_{(x,y)}$$

with $P^{(i)}_{(x,y)}$: sum of capacities of branches (transmission lines) between nodes x and y,
$\Delta P^{(j)}_{(x,y)}$: capacity of the jth element of candidate branches connecting nodes x and y,
$P^{(0)}_{(x,y)}$: capacity of existing lines that connect nodes x and y.

Although the solution investment budget for construction in the crisp problem is optimal, it is not necessarily practical. The future investment budget cannot be guaranteed clearly. The secured investment budget for construction is usually over or below the optimal solution cost of (15.1). This ambiguity gives the decision maker some flexibility (freedom of decision). As it is, the decision maker might not want to actually minimize the total construction cost. Rather, one might want to reach some aspiration levels which might not even be definable crisply. The objective function considering the ambiguity can be expressed based on fuzzy set theory and is called a fuzzy goal function. The fuzzy goal function considering the decision maker's aspiration level of investment budget for the construction cost Z_{C*} can be expressed by (15.2). Here, "$\lesssim$" denotes the fuzzified version of "$\leq$" and means essentially smaller than or equal to

$$C^T\left(C_{(x,y)}\right) \lesssim Z_C^* \tag{15.2}$$

No shortage of power supply requires that the total capacity of branches involved in the minimum cut-set should be greater than or equal to the system load demand L. This is also referred to as the bottleneck capacity (or the maximum flow of the network) in Table 15.A.1 of the Appendix. Therefore, a no-shortage power supply constraint can be expressed by

$$P_C\left(X,\overline{X}\right) \geq L \quad \left(s \in X, t \in \overline{X}\right) \tag{15.3}$$

where, $P_C\left(X,\overline{X}\right)$ is the capacity of minimum cut-set of two subsets, X and $\overline{X}$, containing source nodes s and terminal nodes t respectively when all nodes are separated by a minimum cut-set.

The demand constraint (15.3) can be expressed by (15.4) with k being the cut-set number ($k = 1,\ldots,n$), where n is the number of cut-set.

$$\sum_{(x,y)\in\left(X_k,\overline{X}_k\right)} \left[P_{(x,y)} = P_{(x,y)}^{(0)} + \sum_{i=1}^{m(x,y)} P_{(x,y)}^{(i)} U_{(x,y)}^{(i)}\right] \geq L \tag{15.4}$$

A fuzzified version of "$\geq$" or "$\gtrsim$" means essentially greater than or equal to. A constraint with "$\gtrsim$" sign is not meant to be "$\geq$" in a strict mathematical sense but smaller violations might well be acceptable for the long-term grid planning problem. The conventional constraint without flexibility calls for the load to be strictly satisfied by the supply power. In order to give flexibility (freedom of decision) to decision makers, maximization of a reliability objective function defined by *DMR* (delivery marginal rate) as in (15.5) can be used. Maximization of the reliability function is considered here because decision makers always want to maximize the reliability level of the system.

$$\text{maximize } DMR(P_{(x,y)}) = \left[\sum_{(x,y)\in(X_k,\overline{X}_k)} P_{(x,y)} - L \right] \times 100/L \tag{15.5}$$

The second fuzzy goal function considering the decision maker's aspiration level of reliability criterion, $Z_R{}^*$, can be expressed by

$$DMR\left(P_{(x,y)}\right) \gtrsim Z_R^* \tag{15.6}$$

15.3 Equivalent Crisp Integer Programming and Branch and Bound Method

To solve the fuzzy multi-objective TEP problem (15.2)–(15.6), an equivalent crisp integer programming problem (15.7) is constructed as described in Appendix (15.A.5)–(15.A.8). If the decision maker wants to have a "crisp" decision proposal, it seems appropriate to suggest to him the dividend satisfaction level which has the highest degree of membership in the fuzzy set "decision." This decision in a fuzzy environment has formally been defined by Bellman and Zadeh in 1970 [49]. The problem here is to maximize the satisfaction level λ of the decision maker subject to fuzzy construction cost/budget and delivery marginal rate constraints. Membership functions that reflect uncertainties and ambiguities associated with construction cost and delivery marginal rate are used:

$$\begin{aligned} &\text{maximize } \lambda \\ &\text{s.t.} \quad \lambda \le \mu_C\left\{ P_{(x,y)}^{(i)} \right\} \\ &\qquad \lambda \le \mu_R\left\{ P_{(x,y)}^{(i)} \right\} \\ &\qquad \lambda \ge 0 \end{aligned} \tag{15.7}$$

where

$\mu_C(\cdot)$: membership function of the set for construction costs,
$\mu_R(\cdot)$: membership function of the set for delivery marginal rates,
$P_{(x,y)}^{(i)}$: sum of capacities of branches (transmission lines) between nodes x and y (MW), $P_{(x,y)}^{(i)} = \sum_{j=1}^{i} \Delta P_{(x,y)}^{(j)}$
$\Delta P_{(x,y)}^{(i)}$: capacity of the jth element of candidate branches connecting nodes x and y,
$P_{(x,y)}^{(0)}$: capacity of existing lines that connect nodes x and y.

The branch and bound algorithm has merits in the case of complex optimization problems of integer programming with a large number of constraints. A fuzzy branch and bound algorithm, which uses the network flow theory and the maximum flow-minimum cut set theorem is proposed to find the optimal solution with the highest satisfaction level.

15.4 Membership Functions

In this chapter, it is assumed that the threshold values of membership functions are determined from the results of simulating a conventional planning problem with crisp membership functions. Here, these membership functions are considered to be exponential in nature. The membership function of fuzzy construction cost is

$$\mu_C\left\{P_{(x,y)}^{(i)}\right\} = \begin{cases} 1 & \Delta C(\cdot) \le 0 \\ e^{-W_C \Delta C\left\{P_{(x,y)}^{(i)}\right\}} & \Delta C(\cdot) > 0 \end{cases} \tag{15.8}$$

where

$$\Delta C(\cdot) = \left\{C\left(P_{(x,y)}^{(i)}\right) - Z_C^*\right\}/Z_C^*,$$

$C\left(P_{(x,y)}^{(i)}\right)$: construction cost (M$) of $P_{(x,y)}^{(i)}$,
Z_C^*: aspiration level of budget for construction cost,
W_C: a weighting factor associated with the membership function of budget for the construction cost (a certainty grade),
$P_{(x,y)}^{(i)}$: capacity of branch i between nodes x and y (MW).

The membership function of the fuzzy delivery marginal rate is expressed as follows:

$$\mu_R\left\{P_{(x,y)}^{(i)}\right\} = \begin{cases} 1 & \Delta R(\cdot) \ge 0 \\ e^{W_R \Delta R\left\{P_{(x,y)}^{(i)}\right\}} & \Delta R(\cdot) < 0 \end{cases} \tag{15.9}$$

where

$$\Delta R(\cdot) = \left\{DMR\left(P_{(x,)}^{(i)}\right) - Z_R^*\right\}/Z_R^*,$$

$DMR\left(P_{(x,y)}^{(i)}\right)$: delivery marginal rate (p.u.) of $P_{(x,y)}^{(i)}$,
Z_R^*: aspiration level for delivery marginal rate,
W_R: weighting factor of the membership function for delivery marginal rate (a permissible grade of obligation).

15.5 Solution Algorithm

The objective of the conventional crisp branch and bound method is to minimize the total construction cost subject to a specified reliability criterion. The proposed fuzzy branch and bound method maximizes the satisfaction level λ of the decision maker and produces the optimal solution considering both the construction cost and the delivery marginal rate.

The solution algorithm for the proposed approach follows:

1) Check the need for transmission expansion for the system and its possibility of using the candidate lines. Need and possibility can be checked respectively by evaluating the maximum flow of minimum cut-set considering no candidate lines and considering all candidate lines.
2) Set $j = 1$ (initial system), $jopt = 0$, $j_{max} = 0$, $\lambda_{opt} = 0$ and ENNOD$j = 0$.
3) If ENNOD$j = 1$, the #j system is an end node at which branch operation of the branch and bound is finished (bound) in solution graph used to obtain the optimal solution, and there would be no need to consider any of the other graphs following this system. Go to step 15.
4) Calculate the minimum cut-set using the maximum flow method for #j system (#j solution in the solution graph).
5) Select a #i branch/line of the candidate branches/lines set (Sj) involved in the minimum cut-set and add to the #j system. In what follows, the new system is named the #ji system.
6) If the #ji system is already considered in the solution graph, go to step 14.
7) Calculate the total cost $CTji = CTj + C(P(i)(x,y))$ for the #ji system and evaluate the cost membership grade μ_{Cji}.
8) Calculate the minimum cut-set capacity $P_{Cji}(X, \overline{X})$ using the maximum flow/minimum cut-set theorem and evaluate the delivery marginal rate (reliability) membership grade μ_{Rji}.
9) Calculate λji = Minimum $\{\mu_{Cji}, \mu_{Rji}\}$.
10) If $\mu_{Cji} < \lambda_{opt}$ (the current system (#ji) with a cost of $CTji$ is optimal and the lambdas of the following systems will be lower than the current optimal solution (λ_{opt})), go to step 14.
11) Set $j_{max} = j_{max} + 1$.
12) If $\lambda ji > \lambda_{opt}$, set $\lambda_{opt} = \lambda ji$, $jopt = j_{max}$, and go to step 14.
13) If $\mu_{cji} < \mu_{Rji}$, set $\lambda j_{max} = \lambda ji$, ENNOD$j_{max} = 1$, and go to step 15.
14) Add this #j_{max}(#ji) solution to the solution graph.
15) If all the candidate branches/lines in set Sj have been considered, go to step 16. Otherwise, set $i = i + 1$ and go to step 5.
16) If $j = j_{max}$, continue. Otherwise, set $j = j + 1$ and go to step 4.

17) For $j = j_{max}$, the solution graph has been constructed fully and the optimal solution *jopt* with λ_{opt} being the highest satisfaction level of the decision maker is obtained in step 12.

15.6 Testing

The proposed method was tested on the 21-bus system shown in Figure 15.1.

Considering a forecasted future system load, the crisp case C_1 (minimization of construction cost) and four fuzzy test cases of maximizing the satisfaction level of the decision maker (F_0, F_1, F_2, and F_3) were studied. The crisp case is simply a conventional cost minimization problem subject to the capacity of minimum cut-set satisfied at peak load as described in (15.3). Tables 15.1 and 15.2 describe

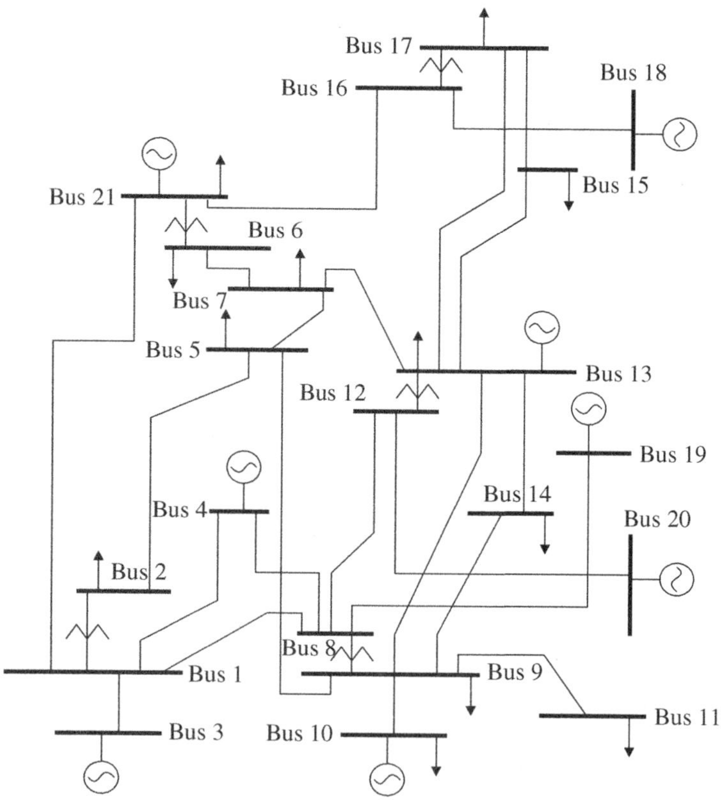

Figure 15.1 The 21-bus test system.

Table 15.1 Ambiguity and uncertainty of the budget guarantee and the ambiguity and obligation of reliability criteria.

Cases	Ambiguity and uncertainty of budget guarantee	Ambiguity and obligation of reliability criterion
F_0	Very sufficient and secure	Very low and strict
F_1	Not sufficient and very secure	Low and not strict
F_2	Sufficient and not secure	High and very strict
F_3	Not sufficient and somewhat secure	Medium and somewhat strict

Table 15.2 Test parameters.

Cases	Z_C^* (M$)	W_C	Z_R^* (%)	W_R
F_0	500 (VS)	10 (S)	5 (VL)	10 (S)
F_1	300 (NS)	20 (VS)	13 (L)	5 (NS)
F_2	370 (S)	5 (NS)	17 (H)	20 (VS)
F_3	315 (NS)	15 (SS)	15 (M)	15 (SS)

and give the parameters of the membership function for the fuzzy cases. Aspiration levels ($Z_C{}^*$ and $Z_R{}^*$) are input data, while the satisfaction level (λ) of decision maker is output data. The consideration of the fuzzy environments in the aspiration level is represented by three and four levels of ambiguity. These are: very sufficient, sufficient, and not sufficient according to budget aspiration level $Z_C{}^*$ and very low, low, medium, and high according to reliability aspiration level $Z_R{}^*$. Also, four levels of budget uncertainty and reliability obligation are assumed. These are: very secure, secure, somewhat secure, and not secure according to a certainty grade W_C, and very strict, strict, somewhat strict, and not strict according to an obligation grade W_R. Figures 15.2 and 15.3 show the membership functions of the cost and the delivery marginal rate, respectively.

Table 15.3 shows the system data with GN, TF, TL, and LD representing generators, transformers, transmission lines, and loads, respectively; SB and EB are start and end buses of the line, respectively; $P(x,y)(0)$ and $C(x,y)(0)$ are, respectively, the capacities and costs of existing lines that connect nodes x and y. In this study, four candidate lines are considered (or $m(x,y) = 4$) in (15.1) and (15.4). Parentheses in $P(x,y)(i)$ and $C(x,y)(j)$ are omitted for convenience in Table 15.3.

The problem is solved using the proposed approach and sensitivity analyses are performed. Using sensitivity analysis results, the decision maker can catch the

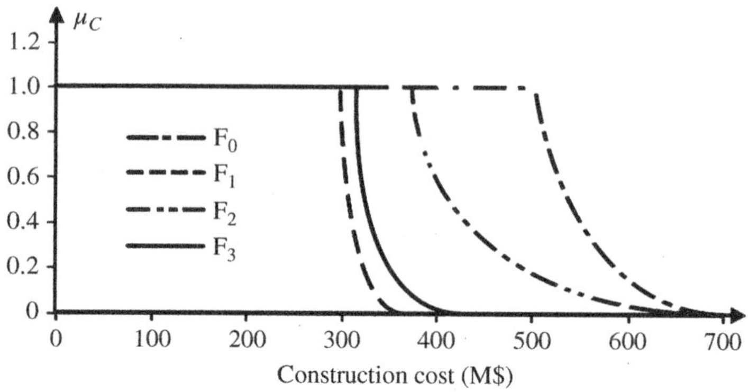

Figure 15.2 The membership function of the construction cost.

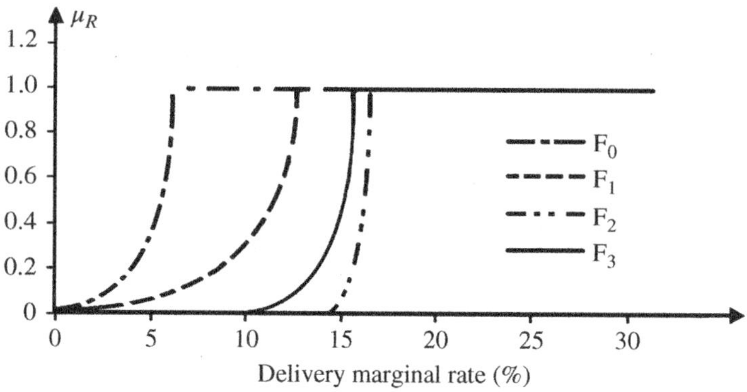

Figure 15.3 The membership function of the delivery marginal rate.

important information on what is the most effective range of the ambiguous budget guarantee and reliability criterion (how reasonable is the reliability criterion level) with uncertainty, to increase the satisfaction level efficiently.

15.6.1 Discussion of Results

Table 15.4 shows the optimal expansion plans for each of the test cases. It is observed that the results for the fuzzy case F_0 and the crisp C_1 are the same. This is because of the "very sufficient" budget for construction and the strict reliability obligations in F_0. The delivery marginal rates and satisfaction levels of the new system for the different test cases are shown in Table 15.5. These results indicate that the delivery marginal rates are different due to the consideration of the line

Table 15.3 System capacity and cost data $P(*)$: (MW) and $C(*)$: (M$).

NL	SB	EB	ID	ΔP^0_{xy}	ΔP^1_{xy}	ΔP^2_{xy}	ΔP^3_{xy}	ΔP^4_{xy}	ΔC^0_{xy}	ΔC^1_{xy}	ΔC^2_{xy}	ΔC^3_{xy}	ΔC^4_{xy}
1	0	3	GN	850	0	0	0	0	0	0	0	0	0
2	0	21	GN	900	0	0	0	0	0	0	0	0	0
3	0	4	GN	850	0	0	0	0	0	0	0	0	0
4	0	10	GN	900	0	0	0	0	0	0	0	0	0
5	0	20	GN	1200	0	0	0	0	0	0	0	0	0
6	0	18	GN	850	0	0	0	0	0	0	0	0	0
7	0	13	GN	760	0	0	0	0	0	0	0	0	0
8	0	19	GN	950	0	0	0	0	0	0	0	0	0
9	6	21	TF	1020	510	510	0	0	0	132	132	0	0
10	16	17	TF	1020	510	510	0	0	0	124	124	0	0
11	12	13	TF	1020	510	510	0	0	0	123	130	0	0
12	8	9	TF	800	800	0	0	0	0	155	0	0	0
13	1	2	TF	800	800	0	0	0	0	151	0	0	0
14	21	1	TL	500	500	500	0	0	0	29	29	0	0
15	2	5	TL	220	220	0	0	0	0	54	0	0	0
16	1	4	TL	300	300	0	0	0	0	73	0	0	0
17	1	8	TL	400	400	0	0	0	0	70	0	0	0
18	1	3	TL	1000	250	250	250	250	0	20	20	20	20
19	4	8	TL	300	300	0	0	0	0	63	0	0	0
20	5	9	TL	220	220	0	0	0	0	82	0	0	0
21	5	7	TL	220	220	0	0	0	0	77	0	0	0
22	7	6	TL	220	220	0	0	0	0	85	0	0	0
23	21	16	TL	1000	250	250	250	250	0	30	30	30	30
24	7	13	TL	220	220	0	0	0	0	88	0	0	0
25	13	17	TL	220	220	0	0	0	0	69	0	0	0
26	13	15	TL	220	220	0	0	0	0	83	0	0	0
27	16	18	TL	1320	330	330	330	330	0	32	32	32	32
28	9	13	TL	220	220	0	0	0	0	71	0	0	0
29	9	14	TL	220	220	0	0	0	0	65	0	0	0
30	8	19	TL	620	620	0	0	0	0	64	0	0	0
31	12	20	TL	1240	310	310	310	310	0	28	28	28	28
32	12	8	TL	400	400	0	0	0	0	62	0	0	0

(Continued)

Table 15.3 (Continued)

NL	SB	EB	ID	ΔP_{xy}^{0}	ΔP_{xy}^{1}	ΔP_{xy}^{2}	ΔP_{xy}^{3}	ΔP_{xy}^{4}	ΔC_{xy}^{0}	ΔC_{xy}^{1}	ΔC_{xy}^{2}	ΔC_{xy}^{3}	ΔC_{xy}^{4}
33	9	10	TL	240	240	0	0	0	0	81	0	0	0
34	9	11	TL	340	340	0	0	0	0	45	0	0	0
35	15	17	TL	220	220	0	0	0	0	80	0	0	0
36	13	14	TL	220	220	0	0	0	0	80	0	0	0
37	21	22	LD	785	0	0	0	0	0	0	0	0	0
38	6	22	LD	750	0	0	0	0	0	0	0	0	0
39	2	22	LD	850	0	0	0	0	0	0	0	0	0
40	9	22	LD	595	0	0	0	0	0	0	0	0	0
41	10	22	LD	17	0	0	0	0	0	0	0	0	0
42	11	22	LD	550	0	0	0	0	0	0	0	0	0
43	14	22	LD	190	0	0	0	0	0	0	0	0	0
44	13	22	LD	710	0	0	0	0	0	0	0	0	0
45	15	22	LD	450	0	0	0	0	0	0	0	0	0
46	17	22	LD	870	0	0	0	0	0	0	0	0	0
47	7	22	LD	290	0	0	0	0	0	0	0	0	0
48	5	22	LD	70	0	0	0	0	0	0	0	0	0

#0 and #22 mean source and terminal nodes, respectively.

Table 15.4 Optimal expansion plans for the test cases(*crisp case).

Cases	Solution	Total branch and bound	Cost (M$)
C_1*	T_{9-10}^{1}, T_{9-11}^{1}, T_{13-15}^{1}	119	209
F_0	T_{9-10}^{1}, T_{9-11}^{1}, T_{13-15}^{1}	113	209
F_1	T_{21-1}^{1}, T_{1-4}^{1}, T_{13-15}^{1}, T_{8-19}^{1}, T_{9-11}^{1}	272	294
F_2	T_{21-1}^{1}, T_{21-1}^{2}, T_{1-8}^{1}, T_{4-8}^{1}, T_{15-17}^{1}, T_{8-19}^{1}, T_{9-11}^{1}	1167	380
F_3	T_{21-1}^{1}, T_{21-1}^{2}, T_{1-4}^{1}, T_{15-17}^{1}, T_{8-19}^{1}, T_{9-11}^{1}	406	320

capacity constraints. The conventional supply reserve rates assuming unlimited capacity of the transmission system are, however, the same for all the test cases. A discussion of sensitivity analyses results follows.

Table 15.5 The delivery power marginal rates and satisfaction levels of the composite power system-results.

Cases	Supply reserve rate (%)	Delivery marginal rate (%)	Satisfaction level
F_1	18.492	14.738	0.916
F_2	18.492	17.186	0.874
F_3	18.492	15.554	0.788

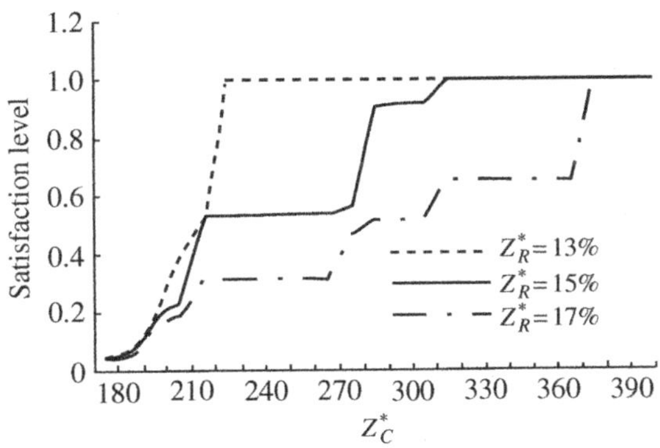

Figure 15.4 Sensitivity of the satisfaction level to changes in the aspiration levels of construction budget $Z_C{}^*$ for a given $Z_R{}^*$.

15.6.2 Solution Sensitivity to Reliability Criterion

Sensitivity analysis with respect to the reliability criterion aspiration $Z_R{}^*$, the aspiration construction cost/budget $Z_C{}^*$, and the corresponding weighting factors W_R and W_C for Case F_1 (the base case) were performed. Figure 15.4 indicates that higher aspiration levels of budget for construction cost $Z_C{}^*$ correspond to higher satisfaction levels of the decision maker. Conversely, higher aspiration levels of reliability as measured by $Z_R{}^*$ correspond to lower satisfaction levels. Figure 15.4 shows that the satisfaction level experiences a step behavior for changes in the aspiration level of construction cost/budget $Z_C{}^*$. Decision makers can catch the information for the most efficient incremental investment from the high gradient range of the step-type function of the satisfaction level. For example, within the budget guarantee range 220–270 M$ for case F_1, the investment budget is highly guaranteed as indicated by the membership function given in Figure 15.2. However, as shown in Figure 15.4, the satisfaction level of the decision maker cannot be increased efficiently because the satisfaction level is saturated in the range. As it is, it is meaningless for the decision maker to request increasing the budget within this range as it is not helpful to do so. On the other hand, within the budget guarantee range 270–290 M$, the satisfaction level of the decision maker can be increased very efficiently, although the investment budget guarantee is a small change, a request for an increase of the budget by the decision maker is quite meaningful because an increase in this range would have a sizable effect. This shows how a decision maker can make use of sensitivity graphs to identify the most effective budget in view of her/his satisfaction so she/he can suggest a reasonable budget of construction cost to the transmission owner.

Figure 15.5 indicates that for a given value of reliability criterion aspiration Z_R^*, lower satisfaction levels of decision-maker correspond to higher weighting factor W_R of the reliability criterion. Higher rates of decay of the satisfaction level correspond to higher aspiration levels Z_R^*. For a reliability criterion of 13%, the satisfaction level does not change for changes in the weighting factor (strict grade) of the reliability membership function. But for $Z_R^* = 17\%$, the satisfaction level is very sensitive to changes in the weighting factor of membership. This calls for careful handling of the weighting factors of membership functions when the aspiration level is high because the shape of the membership function is decided by its weighting factor.

Figure 15.6 shows that for a given value of reliability criteria aspiration Z_R^*, the satisfaction level of the decision maker decreases and eventually saturates as the

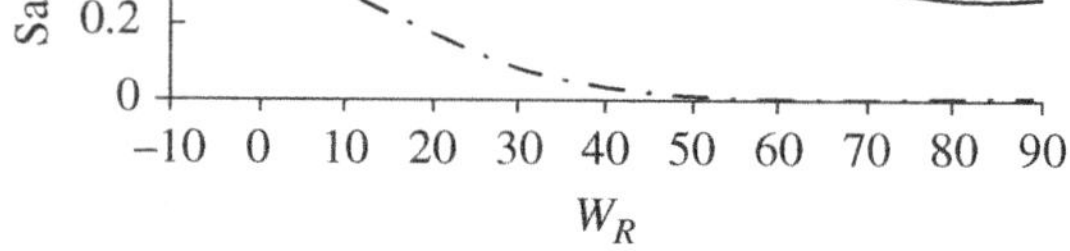

Figure 15.5 Variation of the satisfaction level of a decision maker due to changes in reliability criterion weighting factor W_R for a given Z_R^*.

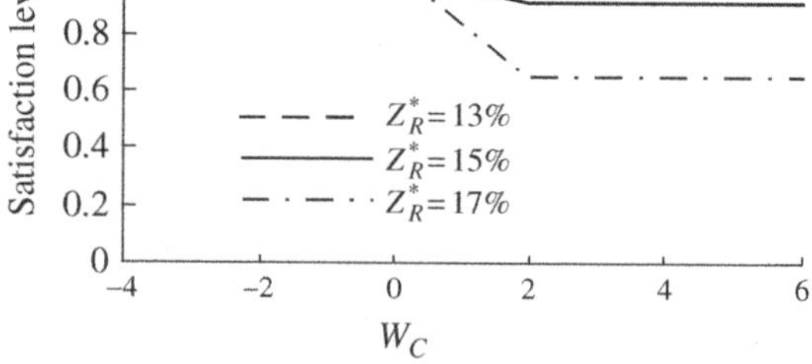

Figure 15.6 Variation of satisfaction level due to changes in construction cost weighting factor W_C for a given Z_R^*.

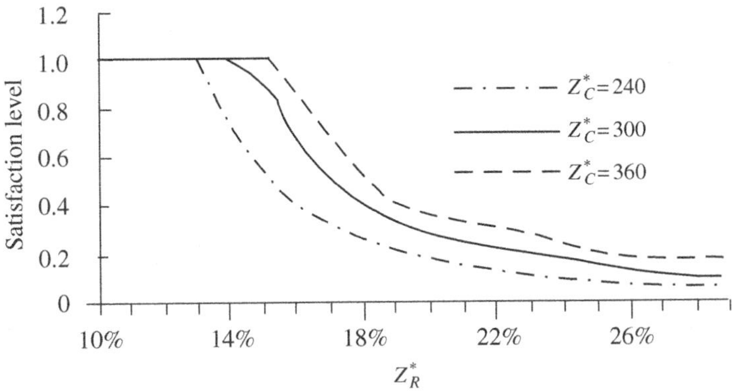

Figure 15.7 Changes of satisfaction level versus changes of the aspiration levels of the reliability criterion Z_R^* for a given Z_C^*.

weighting factor W_C of the construction cost increases. The rate of decay of the satisfaction level is higher for higher values of Z_R^*. The characteristics of this case are similar to those described in Figure 15.5.

15.6.3 Sensitivity to Budget for Construction Cost

For a given aspiration level of budget for construction cost Z_C^*, lower satisfaction levels of the decision maker correspond to higher aspiration levels of the reliability criterion Z_R^*. Conversely, higher aspiration levels of the construction budget Z_C^* correspond to higher satisfaction levels as shown in Figure 15.7.

Decision makers can catch the information for the most efficient incremental reliability criterion from the high gradient range of the linear-saturation-type function of the satisfaction level. For example, in the F_1 case, for a reliability criterion Z_R^* that is higher than 19%, the reliability criterion is set at a high value (1.0) as indicated by the membership function given in Figure 15.3. However, as shown in Figure 15.7, the satisfaction level of the decision maker cannot be increased efficiently because the satisfaction level is somewhat saturated in the range. As it is, it is meaningless for the decision maker to request a decrease of the reliability criterion within this range as it is not helpful to do so. On the other hand, within the reliability criterion range under 19%, the satisfaction level of the decision maker can be increased quite efficiently for a small change of the reliability criterion, a request by the decision maker to decreasing the criterion is quite meaningful because a decrease in this range would have a sizable effect. This shows how a decision maker can make use of sensitivity graphs to identify the most effective reliability criterion in view of one's satisfaction so one can suggest a reasonable reliability criterion to the transmission owner.

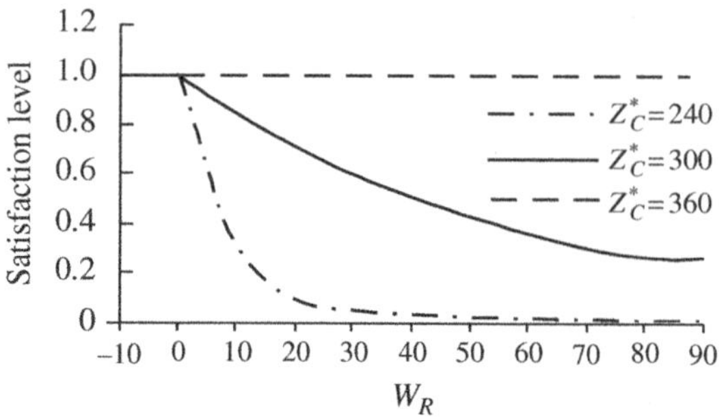

Figure 15.8 Variation of satisfaction level due to changes in reliability criterion weighting factor W_R for a given Z_C^*.

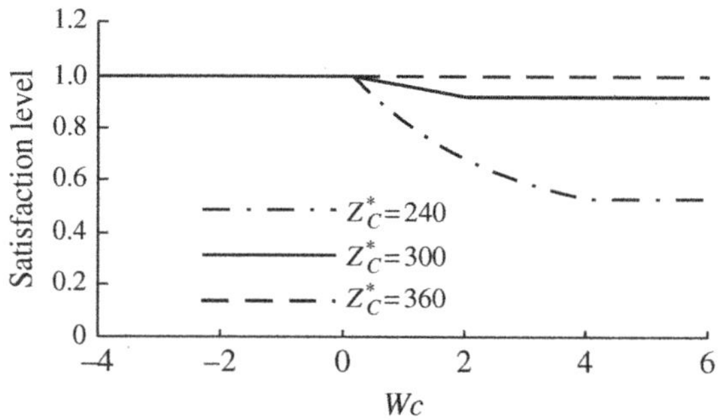

Figure 15.9 Variation of satisfaction level due to changes in W_C for a given Z_C^*.

Figure 15.8 reveals that increasing the weighting factor W_R of the reliability criterion and the construction cost Z_C^* correspond to lower satisfaction levels of the decision maker, which would eventually saturate. The rate of decay of the satisfaction level is higher for lower values of Z_C^*. The characteristics of this case are similar to those described in Figure 15.5.

Figure 15.9 indicates that as the weighting factor Wc of the construction cost increases, the satisfaction level of the decision maker decreases. On the other hand, for the same value of W_C, higher values of Z_C^* correspond to higher satisfaction levels of the decision maker. The characteristics of this case are similar to the case described in Figure 15.6.

15.7 Case Study

The case studies were demonstrated an effective proposed approach to the 63-bus radial style of near side practical power system. The total installation-generating units planning in the system are 12 666 MW. The generation mix is mainly 35.8%

hydro, 24.9% coal, 24.4% natural gas, and 14.9% (oil, IPP). The actual total load demands and forecasted load demands in long-term planning are 8015 and 9370 MW, respectively. The transmission system voltage levels are mainly 500, 220, and 110 kV.

Table 15.6 shows the system data with GN, TF, TL, and LD representing generators, transformers, transmission lines, and loads respectively; SB and EB are start and end buses of the line, respectively; $\Delta P(x,y)(0)$ and $\Delta C(x,y)(0)$ mean the capacity and cost, respectively, of constructed transmission lines. In this study, four candidate lines are considered. As it is $m(x,y) = 4$ is assumed for $\Delta P(x,y)(i)$ and $\Delta C(x,y)(i)$ in (15.1) and (15.4).

Tables 15.7 and 15.8 show the input data of the assumed membership functions environment for the cost/budget and the delivery marginal rate for two fuzzy cases based on the expert experiences of grid expansion planning. In these fuzzy environments, it was assumed that there are two levels of ambiguity as very sufficient (VS) and sufficient (S) for budget-guaranteed aspiration (Z_C), high (H) and medium (M) for reliability criteria aspiration (Z_R) of the grid owner asked for. It was also assumed that there are two levels of uncertainty as very secure (VS) and secure (S) according to certainty grade (W_C) of the budget guarantee and also very strict (VS) and strict (S) according to obligation grade (W_R) of the reliability criteria. Therefore, the configurations of the membership functions are shown in Figures 15.10 and 15.11.

Table 15.9 shows the optimal TEP for each of the test cases. It is observed that the results for the fuzzy case F_1 and the crisp case (conventional case) are the same. This is because of the "very sufficient" budget for construction associated with very secure weighting factors of construction cost and the high-reliability obligations associated with very strict delivery marginal rate of reliability criterion of membership functions. The delivery marginal rates and satisfaction levels of the new system for the different test cases are shown in Table 15.10. These results indicate that the delivery marginal rates are different due to consideration of the capacity constraints of transmission lines. The optimal configuration of TEP (case F_2) was obtained by using fuzzy modeling as shown in Figure 15.12 as a base case.

The delivery marginal rates and satisfaction levels of the power systems of the study cases are shown in Table 15.10. The delivery marginal rates are somewhat different because the capacity constraints of transmission lines are considered. The supply reserve rates for all the cases are the same since the rates are calculated independent of the transmission system. Moreover, the satisfaction levels are quite different with the highest being for case F_2. As a result, F_2 was chosen a base case for the transmission expansion plan and also for the sensitivity analysis on the new system.

Table 15.6 System capacity and cost data of 63-bus system [$P(^*)$: [MW] and $C(^*)$: [M$]].

NL	SB	EB	ID	ΔP_{xy}^0	ΔP_{xy}^1	ΔP_{xy}^2	ΔP_{xy}^3	ΔP_{xy}^4	ΔC_{xy}^0	ΔC_{xy}^1	ΔC_{xy}^2	ΔC_{xy}^3	ΔC_{xy}^4
1	0	1	GN	720	0	0	0	0	0	0	0	0	0
2	0	2	GN	690	0	0	0	0	0	0	0	0	0
:	:	:	:	:	:	:	:	:	:	:	:	:	:
18	41	43	TF	450	200	200	200	0	0	550	550	550	0
19	39	44	TF	550	250	250	250	0	0	650	650	650	0
:	:	:	:	:	:	:	:	:	:	:	:	:	:
39	1	41	TL	550	275	275	275	0	0	220	220	220	0
40	43	40	TL	550	275	275	275	0	0	397	397	397	0
41	4	40	TL	450	225	225	225	0	0	485	485	485	0
42	4	42	TL	552	276	276	276	0	0	395	395	395	0
43	42	36	TL	550	275	275	275	0	0	397	397	397	0
44	4	36	TL	550	275	275	275	275	0	397	397	397	397
45	36	39	TL	550	275	275	275	0	0	397	397	397	0
46	36	37	TL	750	250	250	250	0	0	970	970	970	0
:	:	:	:	:	:	:	:	:	:	:	:	:	:
123	41	64	LD	200	0	0	0	0	0	0	0	0	0
124	46	64	LD	600	0	0	0	0	0	0	0	0	0

#0 and #64 mean source and terminal nodes, respectively.

Table 15.7 The ambiguity and uncertainty of budget guarantee, the ambiguity and obligation of reliability criteria.

Cases	Ambiguity and uncertainty of budget guarantee	Ambiguity and obligation of reliability criteria
Case F_1	Very sufficient and very secure	High and strict
Case F_2	Sufficient and secure	Medium and very strict

Table 15.8 Conditions of the crisp and fuzzy cases.

Cases	Z_C (M$)	W_C	Z_R (%)	W_R
Case F_1	1900 (VS)	8 (VS)	45 (H)	6 (S)
Case F_2	1800 (S)	2 (S)	40 (M)	10 (VS)

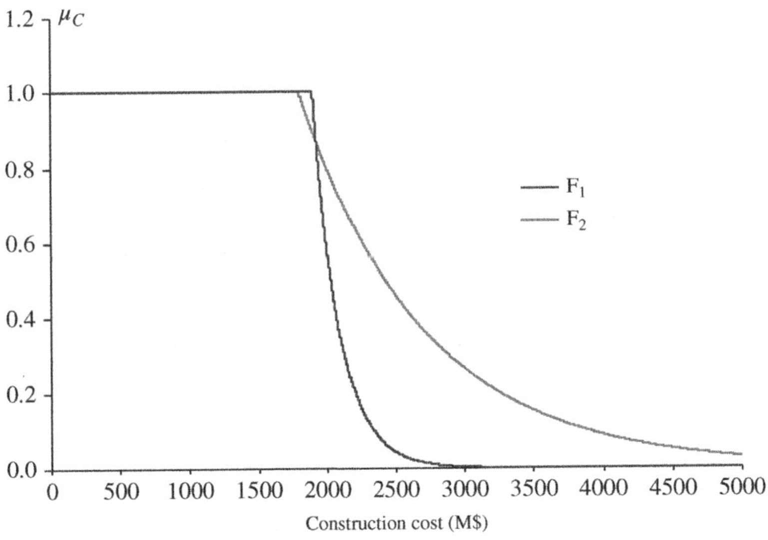

Figure 15.10 Membership function of the construction cost/budget.

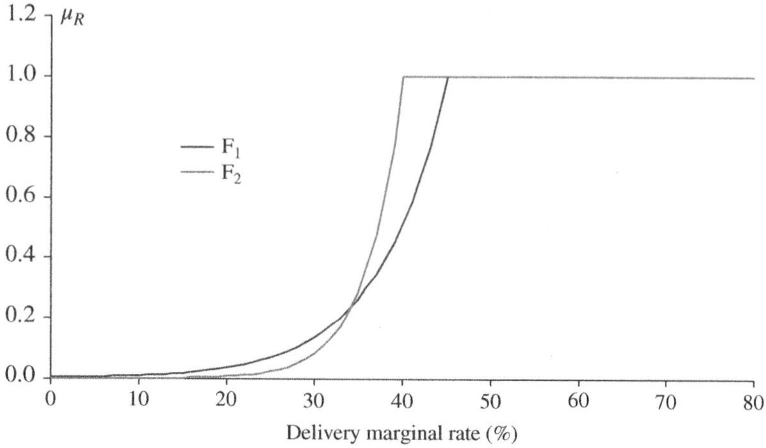

Figure 15.11 Membership function of the delivery marginal rate.

Table 15.9 Results of the crisp with total cost minimization and fuzzy cases with the maximization of the satisfaction level of decision maker.

Cases	Optimal solution	Total cost (M$)	Satisfaction level (λ)
Crisp case	TF^1_{42-46}, T^1_{4-35}, T^2_{4-35}, T^1_{35-30}, T^1_{24-20}	1 868	–
Case F_1	TF^1_{42-46}, T^1_{4-35}, T^2_{4-35}, T^1_{35-30}, T^1_{24-20}	1 868	0.261 54
Case F_2	TF^1_{42-46}, T^1_{4-35}, T^2_{4-35}, T^1_{35-30}, T^1_{24-20}, T^1_{14-15}	2 838	0.299 40

Table 15.10 The delivery power marginal rates and satisfaction levels of the power system for case studies.

Cases	Supply reserve rate (%)	Delivery marginal rate (%)	Satisfaction level (λ)
Case F_1	35.176	34.941	0.261 54
Case F_2	35.176	35.176	0.299 40

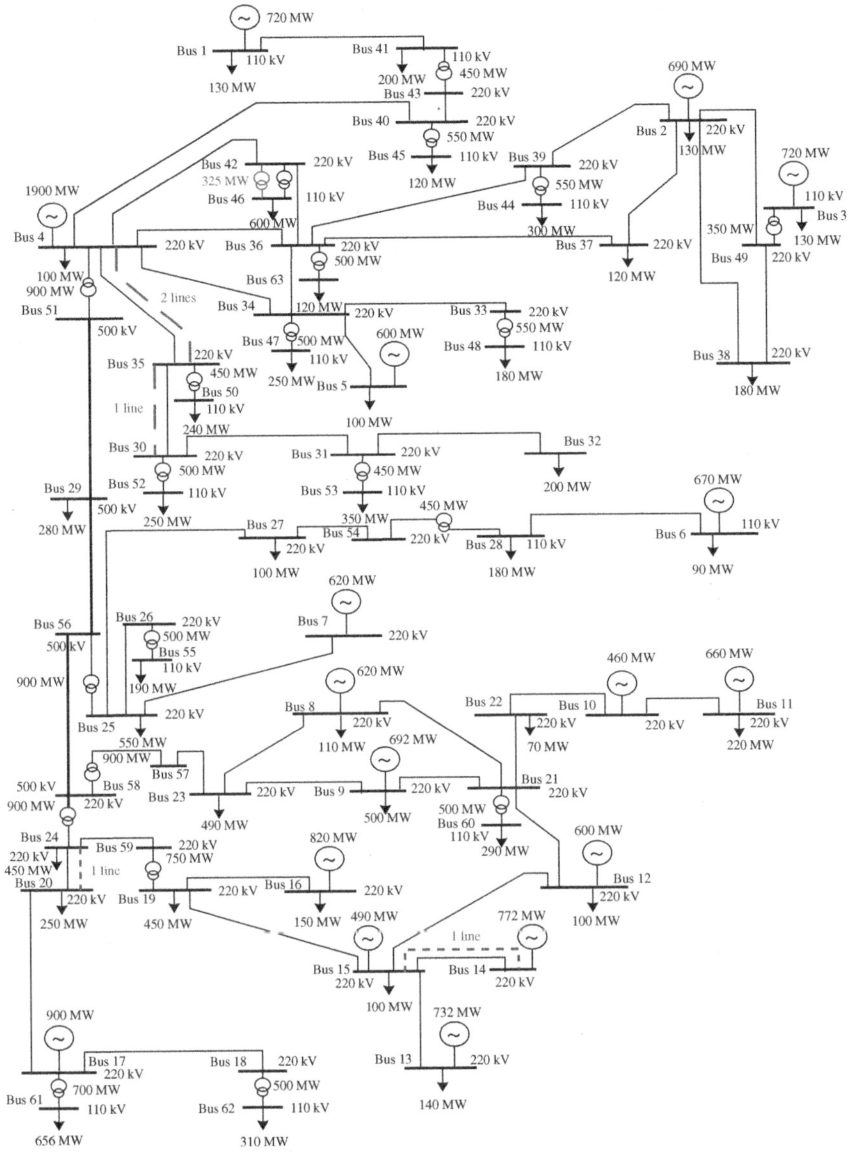

Figure 15.12 The configuration of the TEP of the fuzzy case F_2.

15.8 Conclusion

This chapter addresses the TEP problem when the database available for evaluating reliability indices and the investment budget for constructing new equipment is limited in size. It presents a new and practical approach that should serve as a useful guide for the decision maker to select a reasonable expansion plan prior to checking system stability and dynamics in detail. The proposed method finds the optimal transmission expansion plan (a reasonable and flexible plan that would not be significantly worsened by any assumed changes in the surrounding situations) in terms of the highest satisfaction level of decision maker considering uncertainties associated with the construction cost/budget (economics) and the delivery power marginal rate (reliability). It models the problem as a fuzzy integer programming (FIP) one and considers problem uncertainties through fuzzy modeling. A proposed fuzzy branch and bound algorithm, which includes the network flow method, and the maximum flow-minimum cut set theorem is proposed to solve the problem. The proposed method provides a valuable tool through sensitivity analysis and trend evaluation in the viewpoint of decision maker's satisfaction considering ambiguities involved in long-term expansion planning with a single stage. The new approach makes it possible for a decision maker to make very good use of her/his expert experiences when solving the problem. Testing of the proposed method on an existing 21-bus test system and an existing 63-bus test system including various sensitivity analyses shows that the proposed method is suitable for application to perform practical expansion planning of transmission systems.

15.A Appendix

15.A.1 Network Modeling of Power System

Generators, substations, and load points have limited capacities. It is difficult to check for a shortage of power supply in the system because these elements are presented as nodes in a system model. Network modeling of the system makes it convenient to check the shortage of power supply because the network elements mentioned above are presented as branches with capacity limitations. Aspects of a shortage of power supply according to a bottleneck are given in Table 15.A.1.

Table 15.A.1 Various aspects of power supply bottleneck.

$F_m = L \leq G$	**No shortage of supply**
$F_m = G < L$	Shortage of power generation
$F_m < L \leq G$	Shortage of transmission delivery capacity
$F_m < G < L$	Shortages of power generation and transmission delivery capacity

where

F_m: maximum flow of the network,
G: total power generation,
L: total system load

15.A.2 Definition

In 1970, Bellman and Zadeh [49] presented a classical model of a decision and suggested a model for decision making in a fuzzy environment. If the decision maker's desires are a "crisp" decision proposal, it seems appropriate to suggest to him the dividend which has the highest degree of membership in the fuzzy set "decision." As a step toward formulating the multi-objective TEP problem in a fuzzy environment later, let's define a decision in a fuzzy environment in the sense of Bellman and Zadeh.

The fuzzy decision set D resulting from q fuzzy goal sets $G_1, \ldots, G_q$ and p fuzzy constraint sets $C_1, \ldots, C_p$ results from their intersection.

$$D = \left(\bigcap_{i=1}^{q} G_i \right) \cap \left(\bigcap_{j=1}^{p} C_j \right) \tag{15.A.1}$$

Its membership function, μ_D resulting from fuzzy goals and constraints is defined by

$$\mu_D(x) = \min \left[\min_{i=1\sim p} \mu_{G_i}, \min_{j=1\sim q} \mu_{C_j} \right] \tag{15.A.2}$$

The fuzzy mathematical programming problem consists of finding the maximum of the fuzzy decision D

$$\mu_D(x^*) = \max \mu_D(x) \tag{15.A.3}$$

where

x: the optimal decision solution.

The vector (15.A.3) can be rewritten as follows:

$$\mu_D\left(x_1^*, x_2^*, \ldots, x_N^*\right) = \max_{x_1 \cdots x_N} \mu_D(x_1, x_2, \ldots, x_N) \tag{15.A.4}$$

15.A.3 Fuzzy Integer Programming (FIP)

The multi-objective-TEP can be formulated as an ordinary integer (0,1) problem as follows:

$$\left.\begin{array}{ll}\text{Minimize} & F(x)\\ s.t. & Ax \le b\\ & x \in \{0, 1\}\end{array}\right\} \tag{15.A.5}$$

where

x: decision vector,
F: coefficient matrix of the objective function $(q \times n)$,
A: coefficient matrix of the constraints $(p \times n)$,
b: constant vector of constraints (RHS) $(p \times 1)$,
p: number of constraints,
n: number of variables,
q: number of goal functions

Considering the fuzzy characteristics of its variables, the problem can be formulated as an FIP problem with aspiration level, Z^*, as follows [37, 38]:

$$\begin{array}{l} F(x)\, Z^* \text{ (fuzzy objective functions : } q)\\ Ax\, b \text{ (fuzzy constraints : } p)\\ x \in \{0, 1\} \text{ (0 or 1 variables : } n) \end{array} \tag{15.A.6}$$

From (15.A.1)–(15.A.4), the fuzzy mathematical programming problem consists of finding the maximum point of the membership functions. As it is, the fuzzy optimal decision policy based on Bellman–Zadeh's symmetrical decision model maximizes the satisfaction level of a decision maker. The optimal solution x^* for (15.A.6), therefore, can be obtained by (15.A.7) as described in.

$$\begin{aligned} &\max_{x \ge 0}\left[\min\left\{\min_{i=1,\cdots,q} \mu_i(\boldsymbol{F}(\boldsymbol{x})), \min_{i=1,\cdots,p} \mu_i(\boldsymbol{A}\boldsymbol{x})\right\}\right]\\ &\quad = \max_{x \ge 0}\left[\min_{i=1,\cdots,p+q} \mu_i(\boldsymbol{B}(\boldsymbol{x}))\right] \end{aligned} \tag{15.A.7}$$

where

$\mu_i(\bullet)$: the membership function of the ith fuzzy inequality constraint.

$$\boldsymbol{B}(\boldsymbol{x}) = \begin{bmatrix} \boldsymbol{F}(\boldsymbol{x}) \\ \boldsymbol{A}\boldsymbol{x} \end{bmatrix}$$

Introducing a parameter, λ (the satisfaction level of the decision maker), problem (15.A.7) can be expressed equivalently as a conventional optimization programming as follows [37, 38]:

$$\left.\begin{array}{l} \text{maximize} \quad \lambda \\ \text{s.t.} \quad \lambda \leq \mu_i(\boldsymbol{B}(\boldsymbol{x})) \\ \qquad \boldsymbol{x} \in \{\boldsymbol{0}, \boldsymbol{1}\} \\ \qquad \lambda \geq 0 \end{array}\right\} \tag{15.A.8}$$

The optimal solution can be obtained using a suitable optimization technique such as the branch and bound method. The optimal solution of the fuzzy set theory-based problem depends on the membership functions which reflect subjectively the experience of the decision maker. Because of the nonlinear nature of the original optimization problem, nonlinear membership functions reflect the subjective experience of the decision maker in more detail rather than linear membership functions used in the fuzzy problem mode.

16

Optimal Reliability Criteria for TEP

16.1 Introduction

The primary function of an electric power system is to provide electrical energy to its customers as economically as possible and with an acceptable degree of continuity and quality. Modern society, because of its pattern of social and working habits, has come to expect the supply to be continuously available on demand. But this is not physically possible in reality due to the random system failures which are generally outside the control of power system engineers. Bulk transmission system reliability is an important consideration in system planning and operation. In the past, electric utilities were continuously adding more facilities to their system in order to satisfy the growing customer energy requirements. Successful operation of power system under-regulated as well as deregulated electricity markets is very important. Additionally, power reliability and quality have been increasingly important in recent years due to the number of blackout events occurring throughout the world. Bulk transmission systems are planned to meet specified criteria in an attempt to provide consistently high reliability for utility customers. One very important requirement in the planning and operation of a bulk power system is maintaining the reliability of service to the loads. Planning engineers are interested in representing systems in as much detail as possible and in studying as many contingencies as possible, using accurate power flow algorithms. Therefore, reliability is an issue in future TEP.

16.2 Probabilistic Optimal Reliability Criterion

16.2.1 Introduction

A key factor in today's competitive environment is the orientation toward customer needs and willingness to pay for reliability and quality. Grid companies, ITO and ISO, under competitive environments are therefore becoming more

Probabilistic Power System Expansion Planning with Renewable Energy Resources and Energy Storage Systems, First Edition. Jaeseok Choi and Kwang Y. Lee.

Published 2022 by John Wiley & Sons, Inc.

interested in utilizing concepts of transmission system reliability evaluation and reliability criteria in optimal transmission system expansion planning. Nodal/bus reliability issues and questions of how to obtain the optimal reliability criteria considering the objective of minimizing investment for transmission rather than generation are increasing. In practice, a reliability criterion in terms of security as $(N-1)$ or $(N-2)$ contingency criterion is mainly used for transmission system expansion planning and operation.

As a methodology for deciding the optimal reliability criteria in TEP, a probabilistic reliability index, Loss of Load Expectation (LOLE), is used in this chapter. The optimal reliability criterion, $_{R}LOLE*$, for a transmission system is determined at the minimum cost point of the total cost curve, which is the sum of the utility cost associated with the construction and the customer outage cost associated with supply interruptions. Therefore, this approach is an extension of the conventional concept of optimal reliability criterion for GEP. The maximum flow-minimum cut set theorem is used to obtain the optimal solution to the objective of minimizing construction cost subject to probabilistic reliability constraints, capacity limitation and right-of-way constraints. The two curves, utility cost (reliability cost) and customer outage cost (reliability worth) are required in this methodology.

The first step is to create the utility cost (reliability cost) curve. In this study, a probabilistic reliability constraint-based TEP program (CmExpP.For) has been developed and is used to compose the utility cost curve. This educational program package uses Integer Programming (IP) with a probabilistic branch and bound algorithm, which includes minimization of the construction cost as the objective function under a probabilistic reliability criteria constraint. The second step is to create the customer outage cost (reliability worth) curve associated with the probabilistic reliability level of the transmission system. This second step is composed of two sub-steps. One sub-step is the reliability evaluation of the transmission system and the other sub-step is an assessment of the Interrupted Energy Assessment Rate (IEAR) or Value of Loss of Load (VOLL) by outage cost assessment. In this chapter, the CmExpP.For software is combined with an educational reliability evaluation program known as CmRel.For. This software is used in order to evaluate the optimal construction cost of an optimal expansion plan which is satisfied with a given reliability criterion and the system reliability indices, the expected energy not supplied (EENS), and LOLE. The educational CmRel. For program uses an enumeration method that is able to consider up to eighth order contingencies (4-deep for generation and 4-deep for transmission systems) and the maximum network flow method for optimal load flow. This program uses the effective load duration curve of the composite power system, a Composite power system Equivalent Load Duration Curve (CMELDC) developed earlier. The customer outage cost curve for the optimal scenarios/cases can be assessed by multiplying the EENS calculated from the optimal solution in the probabilistic composite power system expansion planning and the IEAR obtained from surveys or a macro approach. In this study, the IEAR is assumed as this study is focused on determining probabilistic optimal reliability criteria. The characteristics and effectiveness of this methodology are illustrated by a case study of a 21-bus test system [36].

16.2.2 Optimal Reliability Criterion Determination

The economics of alternate facilities play a major role in the decision-making process for TEP. The simplest approach, which can be used to relate economics with reliability, is to consider the investment cost only. The increase in reliability due to various alternatives is evaluated together with the investment cost associated with each scheme. A goal for future TEP should be to extend this adequacy comparison within the same hierarchical structure to include security and therefore to arrive at reliability-cost and reliability-worth evaluation. The extension of quantitative reliability analysis to the evaluation of service worth is deceptively simple but is fraught with potential misapplication. In addition, Figure 16.1 shows the Position of grid reliability evaluation in power systems. In order to understand the concept of reliability of the TEP, it is necessary to have some knowledge about the basic concept of reliability engineering, optimization theory, and operation and planning of the power system.

Figure 16.2 shows that utility cost will generally increase as customers are provided with higher reliability. On the other hand, customer outage costs associated with supply interruptions will decrease as the reliability increases. The total cost to society is the sum of these two individual costs. This total cost exhibits a minimum point at which an "optimal" or target level of reliability is achieved.

16.2.3 Optimal Composite Power System Expansion Planning

16.2.3.1 The Objective Function

The conventional composite power system expansion planning problem is to minimize the total construction cost CT associated with investing in new generators and transmission lines as expressed as

$$\text{minimize } C^T = \sum_{(x,y)\in\rho} \left[\sum_{i=1}^{m(x,y)} C_{(x,y)}^{(i)} U_{(x,y)}^{(i)} \right] \tag{16.1}$$

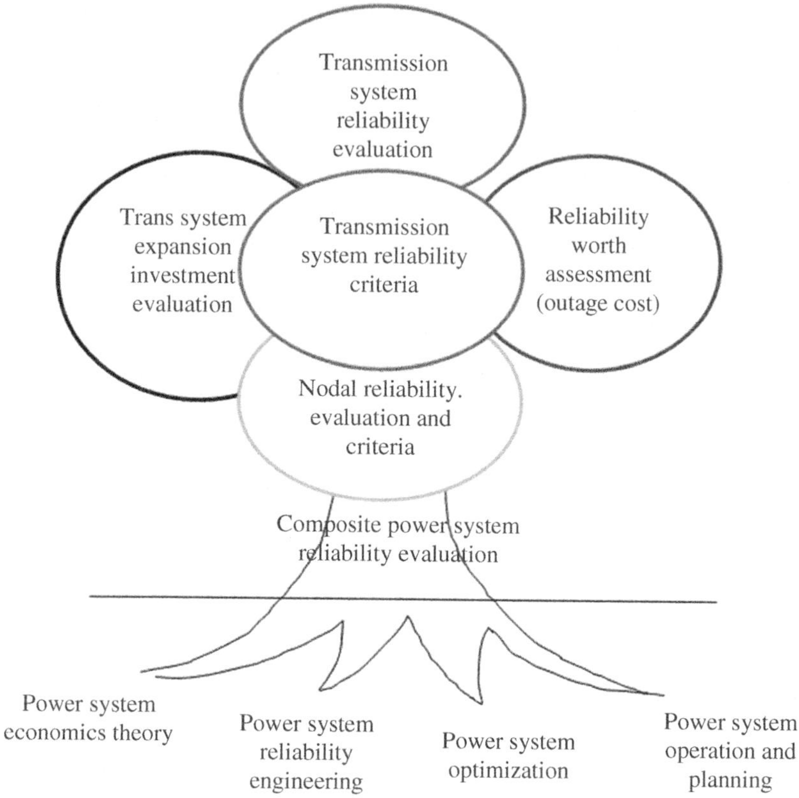

Figure 16.1 Position of grid reliability evaluation in power systems.

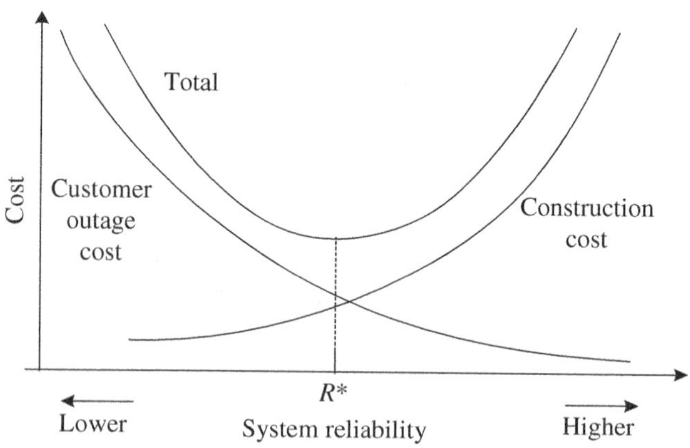

Figure 16.2 Customer, utility, and total cost as a function of system reliability.

where

ρ: the set of all branches (generators and transmission lines),
$m(x, y)$: the number of new candidate branches connecting nodes x and y,
$C_{(x,y)}^{(i)}$: sum of the construction costs of the new generators and lines 1st through ith that connect buses x and y,

$$C_{(x,y)}^{(i)} = \sum_{j=1}^{i} \Delta C_{(x,y)}^{(j)}$$

with

$\Delta C_{(x,y)}^{(j)}$: construction cost of the new jth generator or line connecting nodes x and y,
$U_{(x,y)}^{(i)}$: the decision variable associated with the generator or line (1 if from 1st to ith generators or lines are to be constructed, and 0 otherwise),
with

$$\sum_{i=1}^{m(x,y)} U_{(x,y)}^{(i)} = 1$$

$$U_{(x,y)}^{(i)} = \begin{cases} 1 & P_{(x,y)} = P_{(x,y)}^{(0)} + P_{(x,y)}^{(i)} \\ 0 & P_{(x,y)} \neq P_{(x,y)}^{(0)} + P_{(x,y)}^{(i)} \end{cases}$$

$$P_{(x,y)}^{(i)} = \sum_{j=1}^{i} \Delta P_{(x,y)}^{(j)}$$

with

$P_{(x,y)}^{(i)}$: sum of the capacities of the new branches (new generators or new transmission lines) between nodes x and y,
$\Delta P_{(x,y)}^{(j)}$: capacity of the jth element of the candidate branches connecting nodes x and y,
$P_{(x,y)}^{(0)}$: capacity of the existing generators and lines that connect nodes x and y.

16.2.3.2 Constraints

The problem considered here has two constraints which are deterministic, in addition to the probabilistic reliability criterion. First, no shortage of power supply requires that the total capacity of the branches involved in the minimum cut-set should be greater than or equal to the system peak load demand, L_p. This is also referred to as the bottleneck capacity. Therefore, a no-shortage power supply constraint can be expressed by

$$P_C(X,\overline{X}) \geq L_p \quad (s \in X, \ \in \overline{X}) \tag{16.2}$$

where $P_C(X,\overline{X})$ is the capacity of the minimum cut-set of two subsets, X and $\overline{X}$, containing source nodes s and terminal nodes t, respectively, when all nodes are separated by a minimum cut-set.

The demand constraint (16.2) can be expressed as following with k being the cut-set number ($k = 1, n$), where n is the number of cut-set:

$$\sum_{(x,y)\in(X_k,\overline{X}_k)} \left[P_{(x,y)} = P^{(0)}_{(x,y)} + \sum_{i=1}^{m(x,y)} P^{(i)}_{(x,y)} U^{(i)}_{(x,y)} \right] \geq L \tag{16.3}$$

Second, the probabilistic reliability criterion called Loss of Load Expectation (LOLE), which is discussed in detail in Section 5.2, can be used as in (16.4), where ${}_R LOLE$ is the required reliability criterion for the new system and Φ is a function of the load duration curve discussed in Section 5.2. As shown in (16.4), the investment for system expansion depends strongly on the reliability criterion, ${}_R LOLE$. The purpose of this study is to suggest and test a method that can be used to answer the question: "What is the optimal value of the ${}_R LOLE$" in order to establish a reasonable transmission system expansion plan:

$$LOLE\left(P^{(i)}_{(x,y)}, \Phi\right) \leq {}_R LOLE \tag{16.4}$$

16.2.4 Composite Power System Reliability Evaluation and Outage Cost Assessment

16.2.4.1 Reliability Evaluation at HLI

The reliability indices of $LOLE_{HLI}$ (Loss of load expectation) and $EENS_{HLI}$ (Expected energy not served) of the generation system only and the ELDC (Effective load duration curve), ${}_{HLI}\Phi(x)$ of HLI are calculated as following:

$$LOLE_{HLI} = {}_{HLI}\Phi(x)|_{x=IC} \quad (\text{h/yr}) \tag{16.5}$$

$$EENS_{HLI} = \int_{IC}^{IC+Lp} {}_{HLI}\Phi(x)\ dx \quad (\text{MWh/yr}) \tag{16.6}$$

where

IC total installed capacity of generators (MW),
L_p: peak load of the system (MW),

$$
\begin{aligned}
{}_{HLI}\Phi_i(x_e) &= {}_{HLI}\Phi_{i-1}(x_e) \otimes {}_{HLI}f_{oi}(x_{oi}) \\
&= \int_{HLI} \Phi_{i-1}(x_e - x_{oi})_{HLI} f_{oi}(x_{oi})\ dx \quad (\text{h/yr})
\end{aligned}
\tag{16.7}
$$

where

$\otimes$: operator meaning convolution integral,
${}_{HLI}\Phi_0(x_e - x_{oi}) = {}_{HLI}\Phi(x_L)$,
${}_{HLI}f_{oi}(x_{oi})$: the probability distribution function of outage capacity of generator #i

16.2.4.2 Reliability Evaluation at HLII (Composite Power System)

The reliability indices at HLII can be classified as load point indices and bulk system indices depending on the objective of the evaluation. The reliability indices can be evaluated using a CMELDC at HLII based on the composite power system effective load model in Figure 16.3, where CG, CT, q, and q_l are the capacities and forced outage rates of the generators and transmission lines, respectively.

i) *Reliability indices at the load points*
The load point reliability indices, $LOLE_k$ and $EENS_k$, can be calculated as following with the CMELDC, ${}_k\Phi_{NG}(x)$:

$$
LOLE_k = {}_k\Phi_{NG}(x)|_{x = AP_k}\ (\text{hr/yr}) \tag{16.8}
$$

$$
EENS_k = \int_{AP_k}^{AP_k + Lp_k} {}_k\Phi_{NG}(x)dx\ (\text{MWh/yr}) \tag{16.9}
$$

where

L_{pk}: peak load at load point k (MW),
AP_k: maximum arrival power at load point k (MW)

$$
\begin{aligned}
{}_k\Phi_i(x_e) &= {}_k\Phi_o(x_e) \otimes {}_k f_{osi}(x_{oi}) \\
&= \int_k \Phi_o(x_e - x_{oi})_k f_{osi}(x_{oi}) dx_{oi}\ (\text{hr/yr})
\end{aligned}
\tag{16.10}
$$

where

$\otimes$: the operator representing the convolution integral,
${}_k\Phi_0$: original load duration curve at load point #k,
${}_k f_{osi}$: outage capacity pdf of the synthesized fictitious generator created by generators from #*1* to #i, at load point #k.

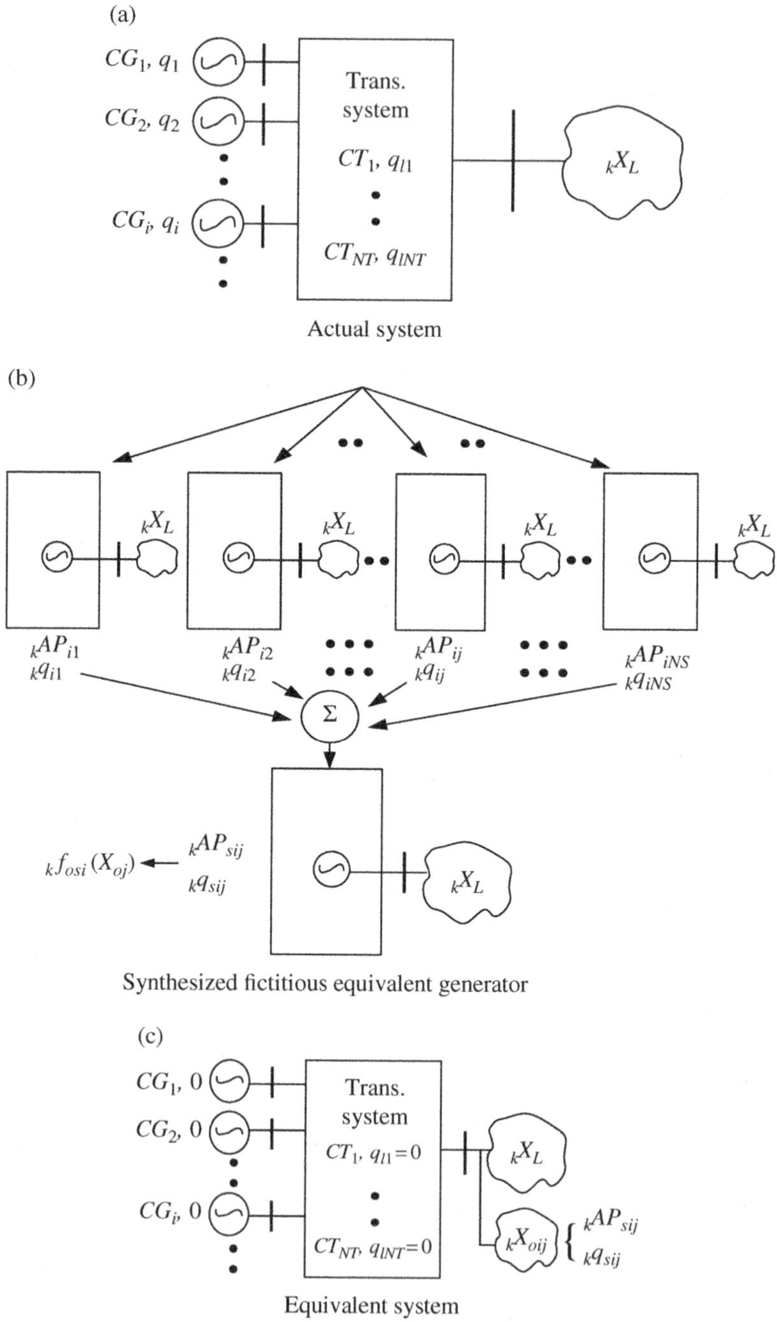

Figure 16.3 Composite power system effective load model at HLII.

ii) *Reliability indices of the bulk system*

While the $EENS_{HLII}$ of a bulk system is equal to the summation of the $EENS_k$ at the load points as shown below, the LOLE of a bulk system is entirely different. The expected load curtailed (ELC) of the bulk system, ELC_{HLII}, is equal to the summation of ELC_k at the load points, while the $LOLE_{HLII}$ of the bulk system can be calculated as following:

$$EENS_{HLII} = \sum_{k=1}^{NL} EENS_k(\mathrm{MWh/yr}) \tag{16.11}$$

$$ELC_{HLII} = \sum_{k=1}^{NL} ELC_k\,(\mathrm{MW/cur.yr}) \tag{16.12}$$

$$LOLE_{HLII} = EENS_{HLII}/ELC_{HLII}\,(\mathrm{h/yr}) \tag{16.13}$$

where

NL: number of load points

$ELC_k = EENS_k/LOLE_k$

iii) *Reliability Evaluation of the Transmission System*

The LOLE and EENS of the transmission system alone using the reliability indices at HLI and HLII are calculated as following:

$$LOLE_{TS} = LOLE_{HLII} - LOLE_{HLI}\,(\mathrm{hr/yr}) \tag{16.14}$$

$$EENS_{TS} = EENS_{HLII} - EENS_{HLI}\,(\mathrm{MWh/yr}) \tag{16.15}$$

iv) *Outage Cost Assessment*

Using the transmission system reliability index described in (16.15), the outage cost of the transmission system can be assessed as following, where IEAR is Interrupted Energy Assessment Rate which means average outage unit cost.

$$OC_{TS} = EENS_{TS} \times IEAR\,(\$) \tag{16.16}$$

16.2.4.3 Flow Chart of the Proposed Methodology for Optimal Reliability Criterion Determination in Transmission System Expansion Planning

A flow chart of the probabilistic optimal reliability criteria decision process is shown in Figure 16.4.

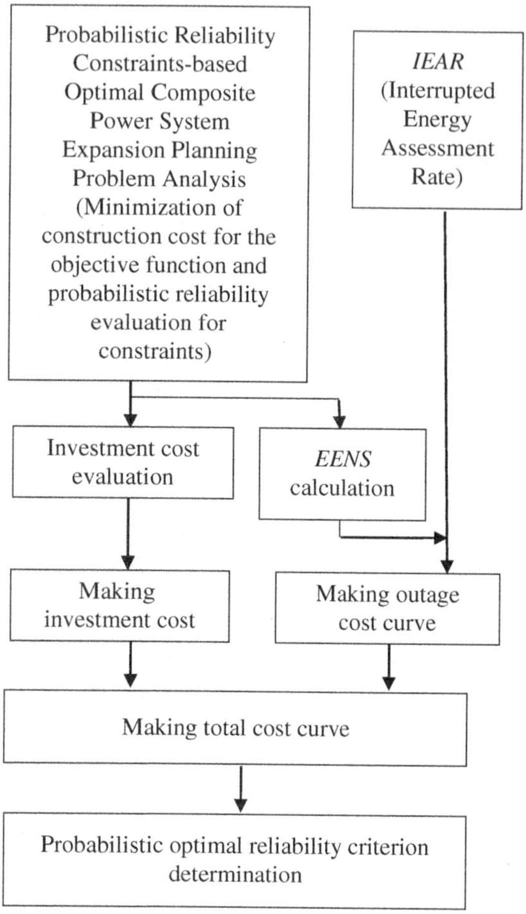

Figure 16.4 Flow chart of the probabilistic optimal reliability criterion determination process for transmission system expansion planning.

16.2.5 Case Study

The proposed method was tested on the 21-bus model system shown in Figure 16.5, which is a part of the south-east area (Youngnam) in Korea.

In this study, four candidate generators and lines are considered (or $m(x, y) = 4$) in (16.1) and (16.4). The capacities and costs of the generators and lines in the case study in [36] have been used also in this case study. Table 16.1 shows the forced outage rates of the branches (lines).

Table 16.2 shows the reliability indices for seven optimum cases obtained by changing the required reliability criterion. The cost and reliability indices of the optimal solutions were obtained from running the program CmExpP.For, where a deterministic bus/nodal reliability criterion, bus reserve rate BRR_k is defined in [23] as:

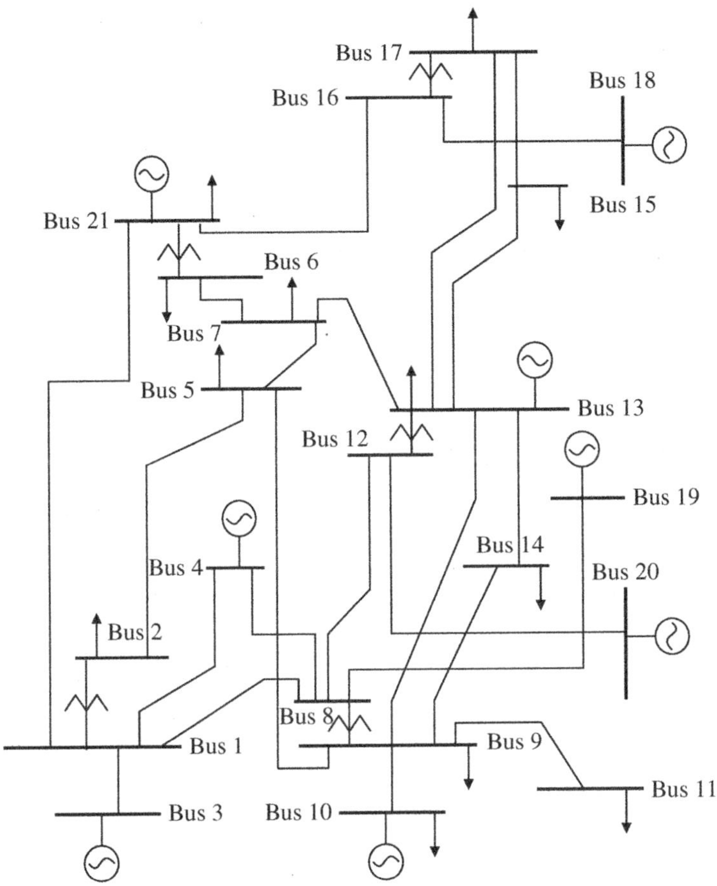

Figure 16.5 The 21-bus test system.

$$BRR_k = (AP_k - Lp_k) \times 100/L_{pk}$$

First, a base case study for the transmission system optimal reliability criterion determination has been tried. Table 16.3 shows the construction cost, customer outage, and total costs obtained assuming $IEAR$ = 5 \$/kWh. A $_RLOLE_{TS}*$ = 35 h/yr is determined as the optimal reliability criterion is determined at the lowest total cost. The single dotted line in Figure 16.6 shows the monotonic increasing characteristics of the construction cost due to changing the transmission system reliability criterion, $_{\mathrm{R}}\mathrm{LOLE}_{\mathrm{TS}}$. The monotonic decreasing characteristics of the customer outage cost are shown as the double dotted line in Figure 16.6. The customer outage cost curve indicates the reliability worth for each new system with minimum construction cost but it does not give the optimal reliability level.

Table 16.1 Forced outage rates of the branches (lines).

NL	FOR	NL	FOR	NL	FOR
1	0.012	13	0.0012	25	0.0014
2	0.015	14	0.0015	26	0.0020
3	0.010	15	0.0015	27	0.0018
4	0.015	16	0.0012	28	0.0022
5	0.010	17	0.0015	29	0.0020
6	0.012	18	0.0014	30	0.0025
7	0.0125	19	0.0015	31	0.0012
8	0.0155	20	0.0016	32	0.0015
9	0.0015	21	0.0018	33	0.0011
10	0.0020	22	0.0012	34	0.0011
11	0.0015	23	0.0012	35	0.0021
12	0.0012	24	0.0012	36	0.0022

Table 16.2 Optimal expansion plans due to changing of the reliability criterion, ${}_{R}LOLE_{TS}$.

${}_{R}LOLE_{TS}$ (h/yr)	Construction of new lines	$Cost_T$ (M$)	$LOLE_{TS}$ (h/yr)	$EENS_{TS}$ (MWh/yr)	Remarks
25	TF^1_{1-2}, T^1_{5-9}, T^1_{13-17}, T^1_{13-15}, T^1_{9-10}, and T^1_{9-11}	511	24.42	29 942	
30	TF^1_{1-2}, T^1_{13-17}, T^1_{13-15}, T^1_{9-10}, and T^1_{9-11}	429	26.54	31 645	
35	TF^1_{1-2}, T^1_{13-17}, T^1_{13-15}, and T^1_{9-11}	348	31.64	39 841	
40, 45	T^1_{13-17}, T^1_{13-15}, T^1_{9-10}, and T^1_{9-11}	278	39.68	69 580	
50, 55	T^1_{13-15}, T^1_{9-10}, and T^1_{9-11}	209	45.47	73 339	Same with BRR = 0%
60, 65, 70	T^1_{21-1}, T^1_{13-15}, and T^1_{9-11}	157	55.10	83 654	
75, 80, 100	T^1_{21-1}, T^1_{9-11}, and T^1_{15-17}	154	73.52	105 454	

Where *BRR*: bus reserve rate in the deterministic approach.

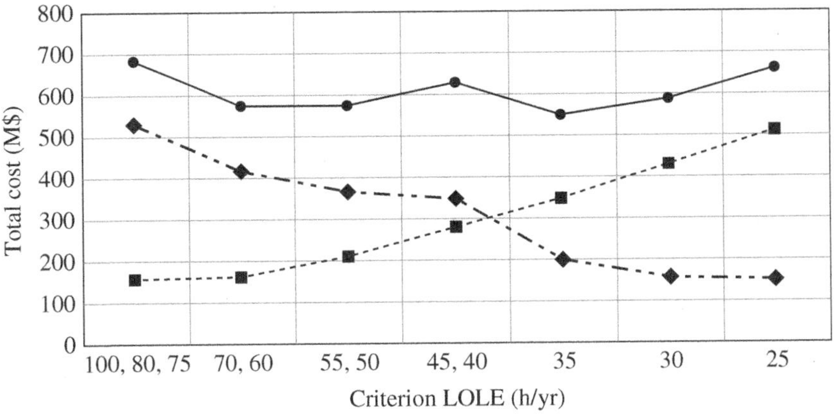

Figure 16.6 Curves of construction, customer outage, total costs, and optimal reliability level of the base case (solid line: total cost, single dotted line: construction cost, double dotted line: outage cost).

A solid line in Figure 16.6 shows the total cost as the sum of the construction cost and customer outage cost. It is noted that the total cost curve becomes a nonconvex function because of the noncoherent characteristic of the transmission system. The probabilistic optimal reliability criterion value (${}_RLOLE_{TS}*$) for transmission system expansion planning is found by the minimum point on this curve. The optimal ${}_RLOLE_{TS}*$ can be used by the transmission planner to determine the total budget required to construct the required transmission lines at the optimal reliability level. In this case study, the cost for transmission line construction is 348 M$ at the optimal reliability criterion of ${}_RLOLE_{TS}* = 35$ h/yr as shown in Table 16.3.

Secondly, two other case studies under assuming different IEAR values have been tried. Tables 16.4 and 16.5 show the customer outage cost assessments and optimal reliability criteria determination assuming that the IEARs of the system are 10 $/kWh (Case 1) and 2 $/kWh (Case 2), respectively. The construction costs and reliability indices are the same as in the previous case study, with IEAR = 5 $/kWh. Table 16.4 shows that the optimal reliability criterion moves from 35 h/yr in the case study with IEAR = 5 $/kWh to 30 h/yr at 10 $/kWh. Table 16.5 shows that the optimal reliability criterion moves from 35 h/yr in the case study with IEAR = 5 $/kWh to 70–60 h/yr range at 2 $/kWh. As expected, the optimal reliability level increases as the IEAR increases and the optimal reliability level decreases as the IEAR decreases. The result clearly shows that the reliability criterion for grid expansion planning in a developed society should be more stringent than that applied in expansion planning in under developed societies.

Table 16.3 Construction cost, reliability, and outage cost according to change of reliability criterion, ${}_RLOLE_{TS}$ (base case).

(*IEAR* = 5 $/kWh)						
Cases	${}_RLOLE_{TS}$	**Const. cost (M$)**	$EENS_{TS}$ **(MWh/yr)**	**Outage cost (M$)**	**Total cost (M$)**	**Remark**
1	100, 80, 75	154	105 454	527.27	681.27	
2	70, 60	157	83 654	418.27	575.27	
3	55, 50	209	73 339	366.695	575.695	
4	45, 40	278	69 580	347.9	625.9	
5	35	348	39 841	199.205	547.205	Optimal
6	30	429	31 645	158.225	587.225	
7	25	511	29 942	149.71	660.71	

Table 16.4 Construction cost, reliability, and outage cost in the case study of IEAR = 10 $/kWh: Case 1.

Cases	${}_RLOLE_{TS}$	**Const. cost (M$)**	$EENS_{TS}$ **(MWh/yr)**	**Outage cost (M$)**	**Total cost (M$)**	**Remark**
1	100, 80, 75	154	105 454	1 054.54	1 208.54	
2	70, 65, 60	157	83 654	836.54	993.54	
3	55, 50	209	73 339	733.39	942.39	
4	45, 40	278	69 580	695.8	973.8	
5	35	348	39 841	398.41	746.41	
6	30	429	31 645	316.45	745.45	Optimal
7	25	511	29 942	299.42	810.42	

Thirdly, a case study on the nodal/bus optimal reliability criteria determination was conducted. Table 16.6 shows the EENS at the load buses. The bus/nodal outage costs based on the EENS have been assessed under assuming IEAR = 5 $/kWh at all load bus. In this study, it is assumed that a bus/nodal construction cost charged to a load bus is allocated with the ratio of the bus demand energy as shown in the following equation, where BCC_k, BDE_k, TDE, and CC are the construction

Table 16.5 Construction cost, reliability, and outage cost in the case study of IEAR = 2 $/kWh: Case 2.

Cases	$_{R}LOLE_{TS}$	Const. cost (M$)	$EENS_{TS}$ (MWh/yr)	Outage cost (M$)	Total cost (M$)	Remark
1	100, 80, 75	154	105 454	210.908	364.908	
2	70, 65, 60	157	83 654	167.308	324.308	Optimal
3	55, 50	209	73 339	146.678	355.678	
4	45, 40	278	69 580	139.16	417.16	
5	35	348	39 841	79.682	427.682	
6	30	429	31 645	63.29	492.29	
7	25	511	29 942	59.884	570.884	

Table 16.6 Reliability indices and $EENS_{Bus}$ at the load buses.

Load bus number	$EENS_{Bus}$ (MWh/yr)						
	$_{R}LOLE_{TS}$ (h/yr) = 25	30	35	40, 45	50, 55	60, 65, 70	75,..., 100
21	3 254	3 400	4 375	8 093	8 486	9 468	11 914
6	4 287	4 484	5 517	9 307	9 307	10 840	14 667
2	5 157	5 440	6 784	11 219	12 073	13 921	17 895
9	2 549	2 652	3 534	6 304	6 496	7 370	9 505
11	2 527	2 640	3 257	5 917	5 926	7 135	7 192
14	362	362	492	1 014	1 016	1 145	1 405
13	4 173	4 506	5 653	9 522	10 378	11 683	15 057
15	1 914	2 008	2 429	4 522	4 968	5 667	6 852
17	5 134	5 556	7 050	11 754	12 770	14 351	18 744
7	484	489	626	1 563	1 564	1 697	1 843
5	102	102	125	356	356	379	380
Total	29 942	31 639	39 841	69 580	73 339	83 654	105 454

cost at bus k, the bus demand energy at bus k, the total demand energy, and the construction cost, respectively:

$$BCC_k = CC \times 100/TDE$$

$$TDE = \sum BDE_k$$

Table 16.7 shows a determination of the optimal reliability criterion at the buses. It is interesting to note that the optimal reliability criteria are the same, except bus 14, bus 7, and bus 5. The optimal reliability criteria at bus 14, bus 7, and bus 5, which have relatively lower load demand than the other buses, are lower than at the other buses. This comes from the reason that the allocated construction cost at a bus is more dominant than the outage cost at the bus.

16.2.6 Conclusion

This study introduces a new methodology for selecting an optimal probabilistic reliability criterion in the TEP in the perspective of the adequacy of power system energy delivery. The optimal reliability criterion, ${}_RLOLE^*$, for a transmission system, is located at the minimum cost point of the total cost curve, which is the sum of the utility cost associated with construction and the customer outage cost associated with supply interruptions. A case study using a 21-bus test system shows that an optimal probabilistic transmission system reliability criterion can be successfully determined using the probabilistic reliability constraints based on optimal expansion planning programs. The monotonic decreasing characteristic of the customer outage cost is obtained using the probabilistic reliability index, EENS. The bus optimal reliability criteria determination is demonstrated using a practical case study. The grid expansion planning using optimal bus probabilistic reliability criteria will become important in the competitive electricity market with an open power system grid.

16.3 Deterministic Reliability Criterion for Composite Power System Expansion Planning

16.3.1 Introduction

A methodology to decide the deterministic optimal reliability criteria for composite power system expansion planning is discussed in this section. The (N–n) contingency rule has been widely used for power system expansion planning. This is a criterion in the perspective of the redundancy. It is not based on coordination (site and power delivery, etc.) of generators and lines. In this section, a deterministic reliability criterion, Delivery Marginal Rate (DMR), considering

Load bus number	Cost name	Bus total cost (*BTC*) (M$/yr)							Bus optimal reliability criterion
		${}_{R}LOLE_{TS}$ (h/yr) = *25*	30	35	40, 45	50, 55	60, 65, 70	75,..., 100	${}_{R}LOLE^{*}_{Bus}$ (h/yr)
21	*BOC*	16.27	17.00	21.875	40.465	42.43	47.34	59.57	**35**
	BCC	64.51	54.16	43.93	35.09	26.38	19.82	19.44	
	BTC	80.78	71.16	65.805	75.555	68.81	67.16	79.01	
6	*BOC*	21.435	22.42	27.585	46.535	46.535	54.2	73.335	**35**
	BCC	63.90	53.64	43.52	34.76	26.13	19.63	19.26	
	BTC	85.335	76.06	71.105	81.295	72.665	73.83	92.595	
2	*BOC*	25.785	27.2	33.92	56.095	60.365	69.605	89.475	**35**
	BCC	73.86	62.00	50.30	40.18	30.21	22.69	22.26	
	BTC	99.645	89.2	84.22	96.275	90.575	92.295	112.735	
9	*BOC*	12.745	13.26	17.67	31.52	32.48	36.85	47.525	**35**
	BCC	50.38	42.30	34.31	27.41	20.61	15.48	15.18	
	BTC	63.125	55.56	51.98	58.93	53.09	52.33	62.705	
11	*BOC*	12.635	13.2	16.285	29.585	29.630	35.675	35.96	**35**
	BCC	43.98	36.92	29.95	23.93	17.99	13.51	13.26	
	BTC	56.615	50.12	46.235	53.515	47.62	49.185	48.22	
14	*BOC*	1.81	1.91	2.46	5.07	5.08	5.725	7.025	**60–70**
	BCC	13.75	11.54	9.36	7.48	5.62	4.22	4.14	
	BTC	15.56	13.45	11.82	12.55	10.70	9.945	11.145	

(*Continued*)

Table 16.7 (Continued)

		Bus total cost (*BTC*) (M$/yr)							
Load bus number	Cost name	${}_{R}LOLE_{TS}$ (h/yr) = *25*	30	35	40, 45	50, 55	60, 65, 70	75,..., 100	Bus optimal reliability criterion ${}_{R}LOLE*_{Bus}$ (h/yr)
13	*BOC*	20.865	22.53	28.265	47.61	51.89	58.415	75.285	35
	BCC	62.70	52.64	42.70	34.11	25.65	19.27	18.90	
	BTC	83.565	74.17	70.965	81.72	77.54	77.685	93.185	
15	*BOC*	9.570	10.04	12.145	22.61	24.84	28.335	34.26	35
	BCC	34.03	28.57	23.17	18.51	13.92	10.45	10.25	
	BTC	43.60	38.61	35.315	41.12	38.76	38.785	44.51	
17	*BOC*	25.67	27.78	35.25	58.77	63.85	71.755	93.72	35
	BCC	77.79	65.31	52.98	42.32	31.82	23.90	23.44	
	BTC	103.46	93.09	88.23	101.09	95.67	95.655	117.16	
7	*BOC*	2.42	2.445	3.13	7.815	7.82	8.485	9.265	60–70
	BCC	18.64	15.65	12.69	10.14	7.62	5.73	5.62	
	BTC	21.06	18.095	15.82	17.955	15.44	14.215	14.885	
5	*BOC*	0.51	0.51	0.625	1.78	1.78	1.895	1.9	75–100
	BCC	5.78	4.85	3.93	3.14	2.36	1.77	1.74	
	BTC	6.29	5.36	4.555	4.92	4.14	3.665	3.64	

Where *BOC*: bus/nodal outage cost, *BCC*: bus/nodal construction cost, *BTC*: bus/nodal total cost.

the coordination of generation system and transmission system is defined. Therefore, it depends on the location of generators and lines. The optimal reliability criterion, DMR*, for a composite power system, is determined at the minimum cost point of the total cost curve, which is the sum of the utility cost associated with the construction and the customer outage cost associated with supply interruptions. The two curves, utility cost (reliability cost) and customer outage cost (reliability worth), are required in this methodology.

The first step is creating a utility cost (reliability cost) curve. In this study, CmExpP.For, a composite power system expansion planning program, is used to compose the utility cost curve. The second step is to create a customer outage cost curve associated with the probabilistic reliability of the transmission system. This has two sub-steps. One sub-step is the reliability evaluation of the composite power system by CmExpP and the other sub-step is an assessment of the IEAR by outage cost assessment. The effective load duration curve of the composite power system (CMELDC) is used in order to obtain the reliability indices of composite power systems at the hierarchical level II (HLII). Therefore, the customer outage cost curve for the optimal scenarios/cases can be assessed by multiplying the EENS and IEAR given from surveys or a macro approach. In this study, outage cost assessment for IEAR is assumed as the scope of this study is focused on optimal reliability criteria determination.

The characteristics and effectiveness of this methodology are illustrated by a case study of the simplified Korea power system [37].

16.3.2 Optimal Reliability Criterion Determination

The reliability criterion considering generation and transmission systems simultaneously and economics of alternate facilities play a major role in the decision-making process for composite power system expansion planning in the viewpoint of the overall system coordination.

Figure 16.7 shows that utility cost will generally increase as customers are provided with higher reliability. On the other hand, customer outage costs associated with supply interruptions will decrease as the reliability increases. The total costs to society are the sum of these two individual costs. This total cost exhibits a minimum point at which an "optimal" or target level of reliability is achieved.

16.3.3 Optimal Composite Power System Expansion Planning

16.3.3.1 Composite Power System Expansion Planning Formulation in CmExpP.For

The static power system expansion problem can be stated as follows. Given demand patterns in a future target year, find a set of generators and transmission

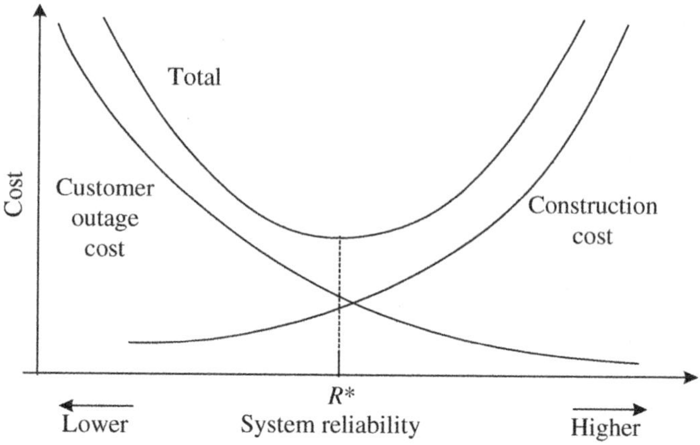

Figure 16.7 Customer, utility, and total cost as a function of system reliability.

line additions to minimize the total investment cost, subject to load constraints, reliability constraints, and right-of-way constraints. The composite power system expansion planning can be formulated as an Integer Programming (IP) problem as follows:

$$\text{minimize} \quad C^T = \sum_{(x,y)\in B} \left[\sum_{i=1}^{m(x,y)} C^i_{(x,y)} U^i_{(x,y)} \right] \tag{16.17}$$

where

C^T: the total construction cost of new equipment.

No power supply shortage requires that the total capacity of the branches involved in the minimum cut-set should be greater or equal to the total load of the system. This is also referred to as the bottleneck capacity. Therefore, a no-shortage power supply constraint can be expressed by

$$P_c(X,\overline{X}) \geq L \qquad (s \in X, t \in \overline{X}) \tag{16.18}$$

The demand constraint can be formulated by

$$\sum_{(x,y)\in(x_k,\overline{x}_k)} \left[P^{(0)}_{(x,y)} + \sum_{i=1}^{m(x,y)} P^{(i)}_{(x,y)} \; U^i_{(x,y)} \right] \geq L(1 + DMR/100) \tag{16.19}$$

$$C^{(i)}_{(x,y)} = \sum_{j=1}^{i} \Delta C^{(j)}_{(x,y)} \tag{16.20}$$

$$P^{(i)}_{(x,y)} = \sum_{j=1}^{i} \Delta P^{(j)}_{(x,y)} \tag{16.21}$$

$$\sum_{i=1}^{m(x,y)} U^{i}_{(x,y)} = 1 \tag{16.22}$$

$$U^{i}_{(x,y)} = \begin{cases} 1, & P_{(x,y)} = P^{(0)}_{(x,y)} + P^{(i)}_{(x,y)} \\ 0, & P_{(x,y)} \neq P^{(0)}_{(x,y)} + P^{(i)}_{(x,y)} \end{cases} \tag{16.23}$$

$$P_{(x,y)} = P^{(0)}_{(x,y)} + \sum_{i=1}^{m(x,y)} P^{(i)}_{(x,y)} U^{(i)}_{(x,y)} \tag{16.24}$$

where

L: total demand ($= \sum Lp_l$),
Lp_l: peak load at load point # l,
$P_{(x,y)}$: capacity of a transmission line or generator between node x and node y,
$\Delta C^{(j)}_{(x,y)}$: construction cost of new #j parallel element of branches between node x and node y,
$\Delta P^{(j)}_{(x,y)}$: capacity of new #j parallel element of branches between node x and node y,
K: cut-set subscript number ($= 1, 2, k, n$),
B: a set of all branches,
$m(x,y)$: the number of new branches between nodes x and y,
DMR: delivery marginal rate ($= \frac{\sum APl - L}{L}$),
AP_l: maximum arrival power at load bus #l

16.3.3.2 Flow Chart

The flow chart of the program is shown in Figure 16.8.

16.3.4 Composite Power System Reliability Evaluation

The indices of the composite power system (HLII) can be classified in terms of load point indices and bulk system indices according to the objective of the evaluation. The reliability indices can be evaluated from the CMELDC of HLII using the Synthesized Fictitious Equivalent Generator (SFEG) model shown in Figure 16.9. In this figure, ${}_kAP_{ij}$ and ${}_kq_{ij}$ are arrival power and state probability of contingency state #j at load point #k, respectively.

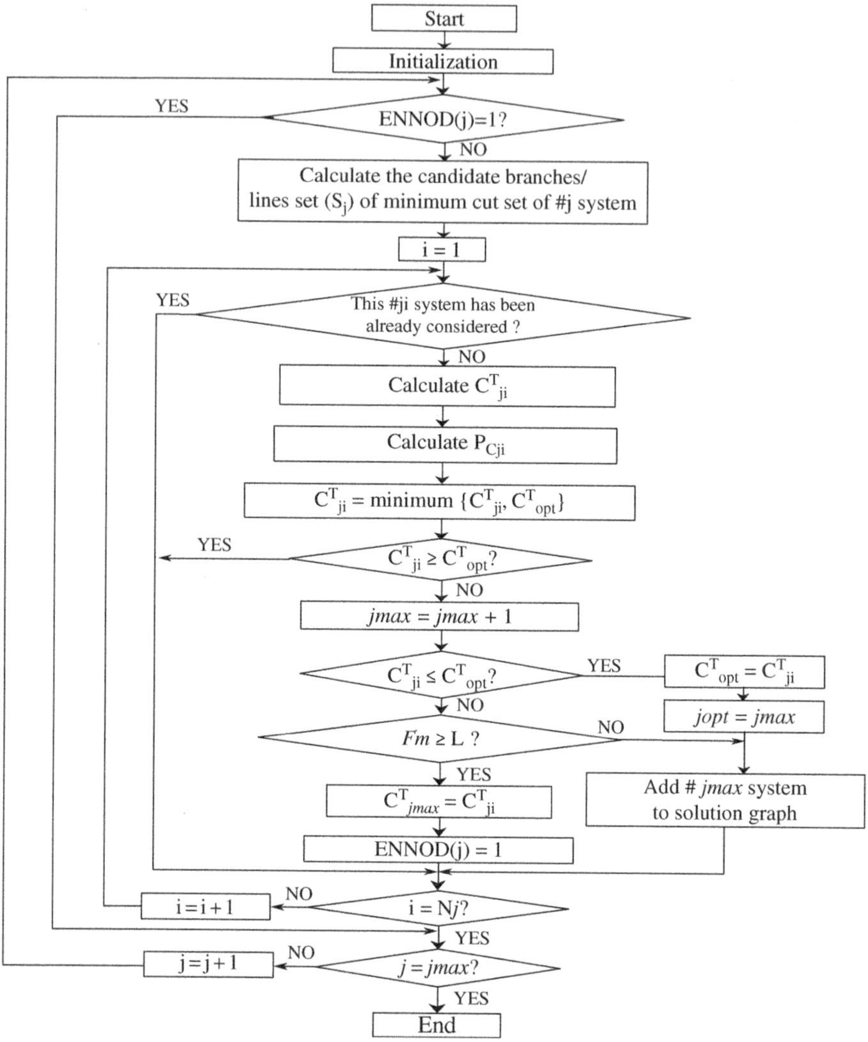

Figure 16.8 Flow chart of CmExpP program.

16.3.4.1 Reliability Indices at Load Points

The load point reliability indices, LOLEk and $EENS_k$, can be calculated as following with the nodal CMELDC, ${}_k\Phi_{NG}(x)$ of (16.27):

$$LOLE_k = {}_k\Phi_{NG}(x)\big|_{x=AP_k} \quad \text{(day)} \tag{16.25}$$

$$EENS_k = \int_{AP_k}^{AP_k + Lp_k} {}_k\Phi_{NG}(x)dx \quad (\text{MWh}) \tag{16.26}$$

where

AP_k: maximum arrival power at load point/bus #k,
L_{pk}: the peak load at load point/bus #k

$$\begin{aligned} {}_k\Phi_i(x_e) &= {}_k\Phi_o(x_e) \otimes {}_k f_{osi}(x_{oi}) \\ &= \int_k \Phi_o(x_e - x_{oi})_k f_{osi}(x_{oi}) dx_{oi} \end{aligned} \tag{16.27}$$

where

$\otimes$: the operator meaning convolution integral,
${}_k\Phi_0$: original load duration curve at load point #k,
${}_kf_{osi}$: outage capacity *pdf* of synthesis fictitious generator operated by generators from *#1* to *#i* at load point *#k*.

16.3.4.2 Reliability Indices of the Bulk System

The $EENS_{HLII}$ of the bulk system is equal to the summation of $EENS_k$ at load points as shown below. The LOLE of the bulk system is different from the summation of $LOLE_k$ at load points. The ELC_{HLII} of the bulk system is equal to the summation of the ELC_k at the load points, and therefore the $LOLE_{HLII}$ of the bulk system can be calculated as follows:

$$EENS_{HLII} = \sum_{k=1}^{NL} EENS_k (\text{MWh}) \tag{16.28}$$

$$ELC_{HLII} = \sum_{k=1}^{NL} ELC_k \, (\text{MW/cur.yr}) \tag{16.29}$$

$$LOLE_{HLII} = EENS_{HLII} / ELC_{HLII} (\text{h/yr}) \tag{16.30}$$

where

NL: number of load points
R: set of states of not supplied powers
$ELC_k = EENS_k / LOLE_k \quad (\text{MW/cur. yr})$

These conventional indices are calculated using the subprogram, CmRel.For version 3.2 included in CmExpP program. It can accommodate about 100 buses, 150 transmission lines, 8 deep contingencies, and a 10–9 cut off for the state probability of a contingency (outage).

16.3.5 DMR Evaluation using Maximum Flow Method

In order to consider the DMR criterion in the perspective of adequacy in this study, it should be evaluated whether the power flow satisfies the peak load or not. The power flow in the network of each arc (generator and transmission line) $(i,j) \in \rho$ is associated with its non-negative real number P_{ij} that is called the generator and transmission line capacity from node i to node j. The maximum static power flow, from s (source) to t (terminal) in branch set, ρ, of the composite power system can be formulated as a linear optimization problem by

$$Maximize \quad F = \sum_{j \in \Delta_i} x_{ij} \tag{16.31}$$

Subject to

$$\sum \left(-x_{ij} + x_{ij}\right) = \begin{cases} -K, & i = s \\ 0 \ , & i \neq s, t \\ K \ , & i = t \end{cases} \tag{16.32}$$

$$0 \leq x_{ij} \leq P_{ij} \qquad (i,j) \in \rho_N$$

where

Λ_i: set of nodes connected with i-node,
ρ: branch set,
K: total peak load ($= L$),
P_{ij}: branch (generator and line) capacity between nodes i and j

The maximum power flow, Fm, using branch set, ρ, can be calculated by LP (Linear Programming) which provides the bottleneck capacity, $P_c(S,T)$ of the given network. Therefore, the DMR criterion can be expressed as below, where the Fm is the maximum power flow obtained by using branch set of network and arrival power at load bus, $\sum AP_k$ is the same as Fm ($=P_c(S,T)$), and L_{pk} is peak load at load point k:

$$DMR = \frac{\sum AP^k - \sum Lp^k}{\sum Lp^k} \times 100[\%] \tag{16.33}$$

16.3.6 Flow Chart of Optimal Reliability Criterion Determination

A flow chart of optimal reliability criterion determination proposed in this study is shown in Figure 16.10.

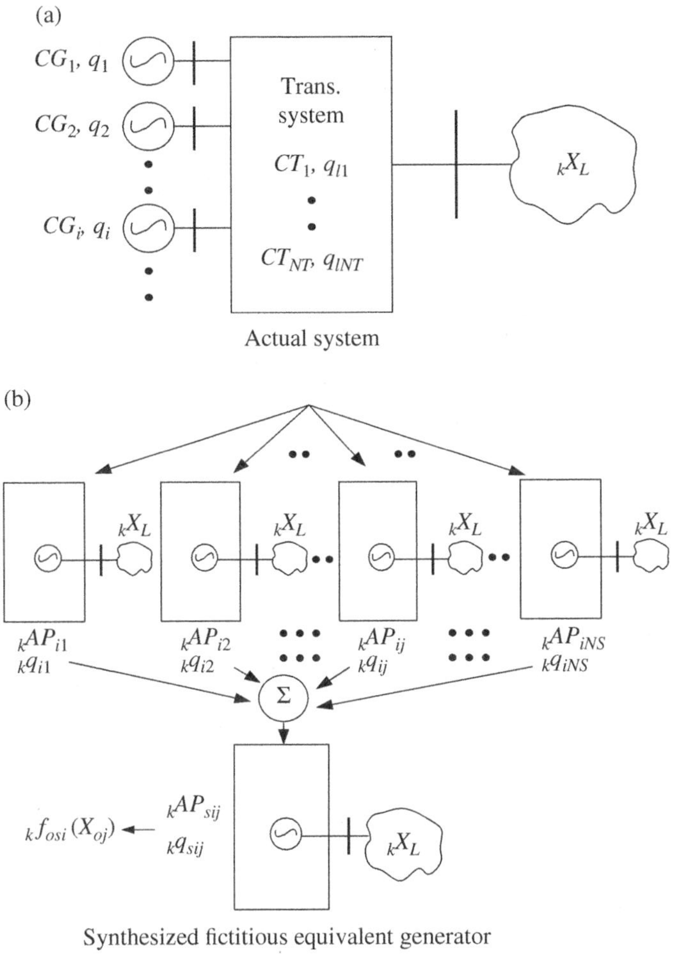

Figure 16.9 Synthesized fictitious equivalent generator model at HLII.

16.3.7 Case Study

16.3.7.1 Basic Input Data

The proposed method was tested on the simplified Korea power system shown in Figure 16.11. Table 16.8 shows the system input data with GN, TF, TL, and LD representing the generators, transformers, transmission lines, and loads, respectively and SB and EB are the start and end buses, respectively. T^0_{i-j} and $C(0)$ are the capacity and cost of constructing the original transmission lines or generators and T^k_{i-j} and $C(k)$, for $k = 1\text{-}4$ are, respectively, the capacity and cost of the

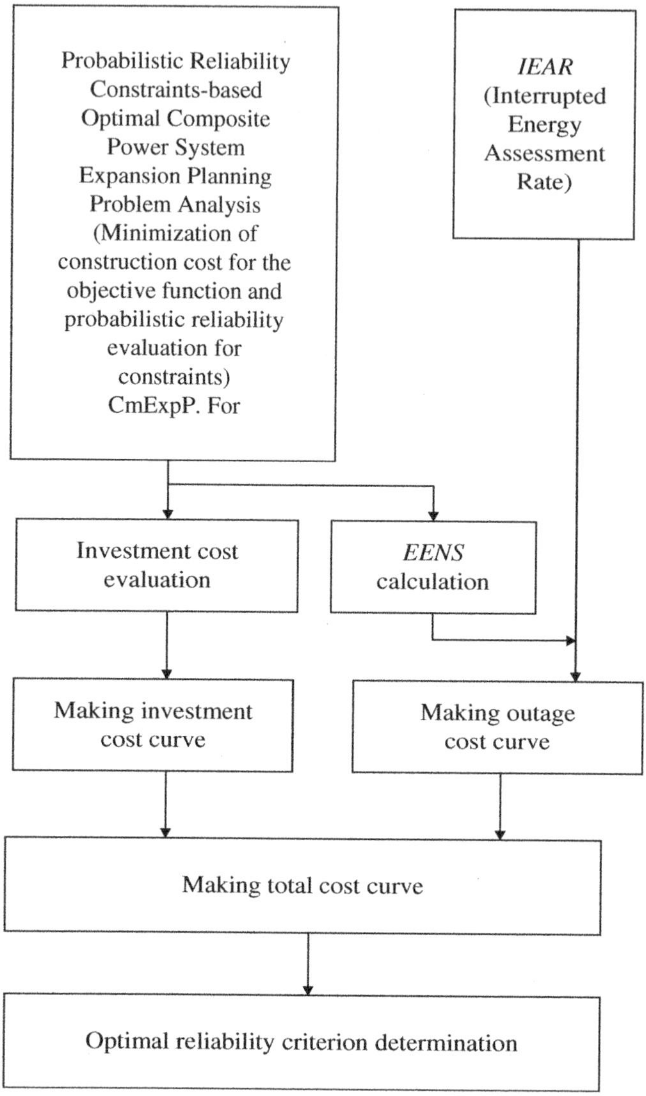

Figure 16.10 Flow chart of the optimal reliability criterion determination as proposed in this study.

candidate generators and lines, where subscripts i and j are numbers of starting and end buses. While the total construction costs for candidate lines are given, the costs for candidate generators are given as marginal cost (Million Won/MW), where US$1 is about 1500 (Won) on 30 November 2008.

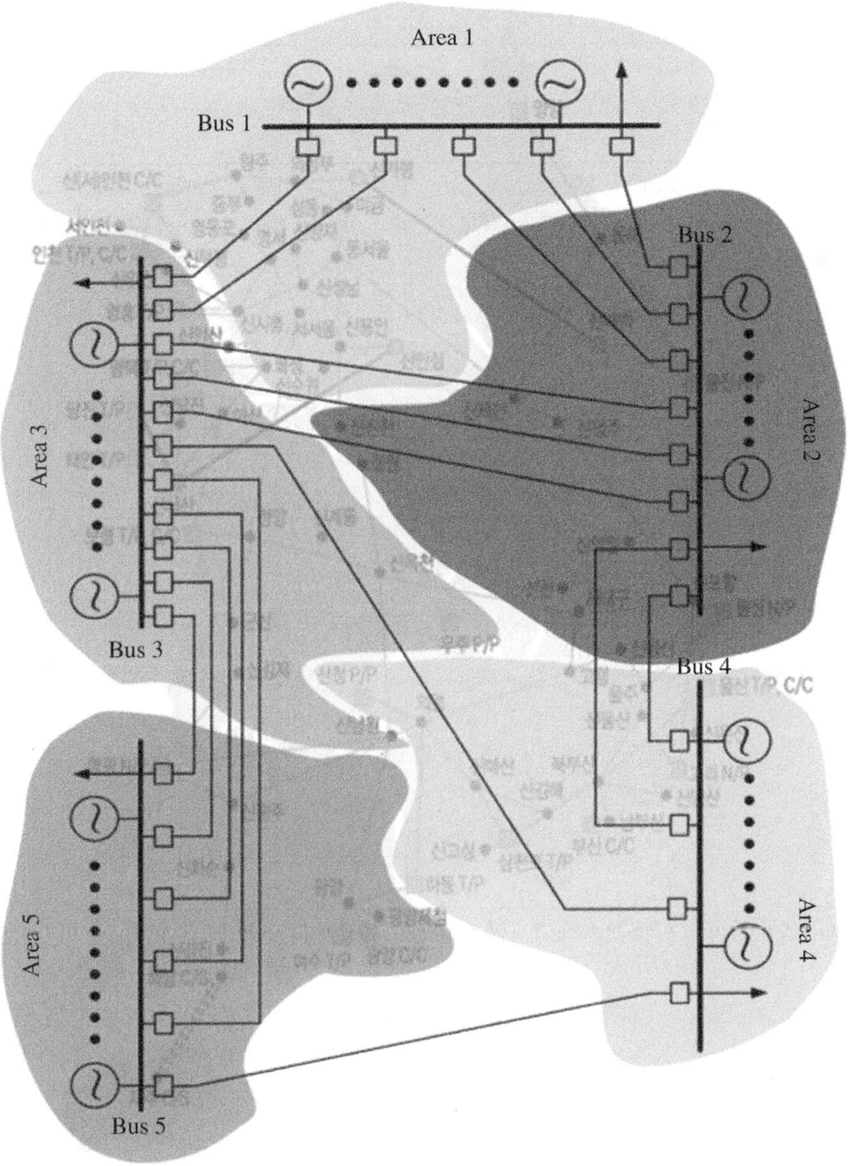

Figure 16.11 The simplified Korea power system – fiveAreas.

Table 16.8 System capacity and cost data.

				(T^k_{i-j}: [MW] and C^k_{i-j}: [MWon])									
NL	SB	EB	ID	T^0_{i-j}	T^1_{i-j}	T^2_{i-j}	T^3_{i-j}	T^4_{i-j}	C^0_{i-j}	C^1_{i-j}	C^2_{i-j}	C^3_{i-j}	C^4_{i-j}
1	0	1	GN	870	870	870	870	0	0	1 058	1 058	1 058	0
2	0	2	GN	1 400	1 400	1 400	1 400	0	0	1 400	1 400	1 400	0
3	0	3	GN	600	600	600	600	0	0	741	741	741	0
:	:	:	:	:	:	:	:	:	:	:	:	:	:
21	1	2	TL	7 290	7 290	7 290	7 290	0	0	201 856	201 856	201 856	0
22	1	3	TL	2 173	2 173	2 173	2 173	0	0	85 394	85 394	85 394	0
23	1	5	TL	1 986	1 986	1 986	1 986	0	0	54 847	54 847	54 847	0
24	2	4	TL	4 346	4 346	4 346	4 346	0	0	85 964	85 964	85 964	0
25	2	5	TL	7 290	7 290	7 290	7 290	0	0	184 524	184 524	184 524	0
:	:	:	:	:	:	:	:	:	:	:	:	:	:
39	2	6	LD	12 908	0	0	0	0	0	0	0	0	0
40	3	6	LD	7 769	0	0	0	0	0	0	0	0	0
41	4	6	LD	4 128	0	0	0	0	0	0	0	0	0
42	5	6	LD	11 449	0	0	0	0	0	0	0	0	0

#0 and #6 in SB and EB mean source and terminal nodes, respectively.

16.3.7.2 Results of Construction Costs of Cases

Table 16.9 shows the results of the construction cost according to increasing DMR. The results were obtained using the CmExpP program with minimizing construction cost (objective function), where $G_{i\text{-}jk}$ means the unit of #*k* generator at #*j* plant on #*i* area in simplified Korea power system. All the cases mean the best plans for pre-specified reliability criterion, DMR. Figure 16.12 shows the construction cost curve graphically.

16.3.7.3 Reliability Evaluation

Outage cost curve, and optimal DMR in case of IEAR = 20 000 (Won/kWh)*

Table 16.9 Optimal expansion planning of the system.

Cases	DMR (%)	Optimal solution of the new system (units)	Total cost (million won)
1	11	G2–31	250 800
2	11.5	G3–31	340 860
3	12	G4–51	666 900
4	12.5	G3–11, G4–51	855 000
5	13	G3–31, G4–51	1 007 760
6	13.5	G1–32, G3–31	1 230 060
7	14	G3–31, G3–32, G3–33, G5–31	1 389 375
8	14.5	G1–31, G3–31, G5–31	1 596 855
9	15	G3–11, G3–31, G3–32, G3–33	1 877 580
10	15.5	G1–41, G3–31, G3–32	2 015 520
11	16	G1–31, G1–41	2 223 000
12	16.5	G3–31, G3–32, G3–33, G4–51, G5–31, G5–32	2 423 070
13	17	G1–31, G3–31, G4–51, G5–21	2 630 550

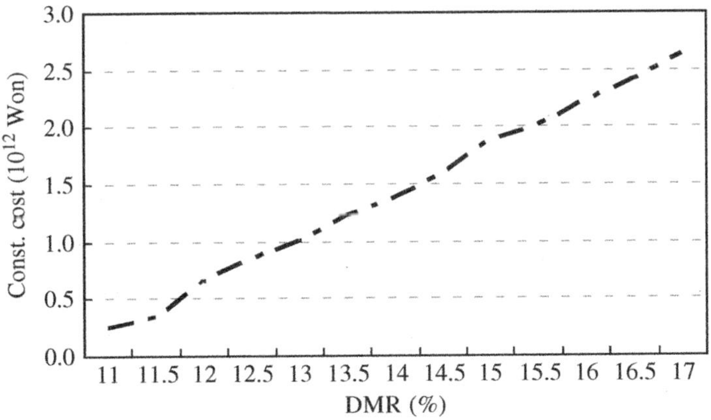

Figure 16.12 Construction cost curve.

The outage cost curve may be assessed using IEAR and EENS of the optimal plans of the cases resulting from the previous CmExpP. In this case study, it is assumed that IEAR is 20 000 Won/kWh. Figure 16.13 shows the total outage cost curve for each case/plan graphically.

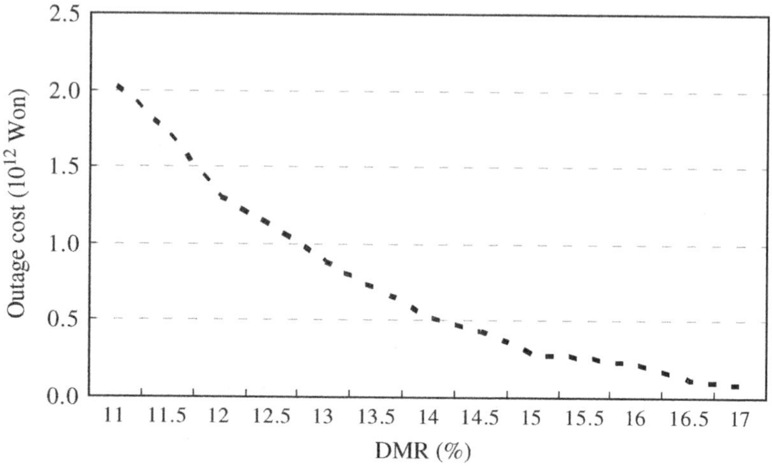

Figure 16.13 The outage cost curve.

Table 16.10 Construction cost, reliability, and outage cost at each case study.

Case num.	DMR (%)	Const. cost (MWon)	EENS (MWD/yr)	Outage cost (MWon)	Total cost (MWon)	Remark
1	11	250 800	4 228.65	2 029 752	2 280 552	
2	11.5	340 860	3 572.55	1 714 824	2 055 684	
3	12	666 900	2 696.30	1 294 224	1 961 124	
4	12.5	855 000	2 315.92	1 111 642	1 966 642	
5	13	1 007 760	1 849.95	887 976	1 895 736	Optimal
6	13.5	1 230 060	1 462.40	701 952	1 932 012	
7	14	1 389 375	1 084.25	520 440	1 909 815	
8	14.5	1 596 855	903.56	433 709	2 030 564	
9	15	1 877 580	608.16	291 917	2 169 497	
10	15.5	2 015 520	535.63	257 102	2 272 622	
11	16	2 223 000	436.56	209 549	2 432 549	
12	16.5	2 423 070	233.17	111 922	2 534 992	
13	17	2 630 550	176.20	84 576	2 715 126	

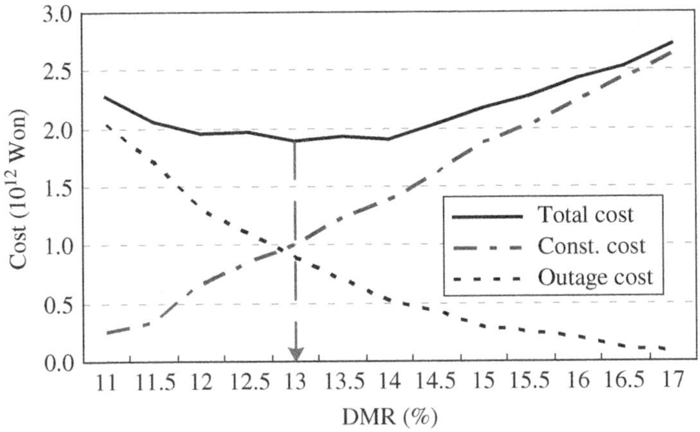

Figure 16.14 Curves of construction, customer outage, total costs, and optimal DMR∗ reliability level.

Table 16.10 shows the overall process that the optimal reliability criterion point (DMR∗) for the composite power system expansion planning of the system is determined at the minimum point of the total cost. In this study, the total budget is 1 895 736 Million Won at the reliability criterion of 13.0% as shown in the table. Figure 16.14 shows graphically that the optimal DMR* is determined at the minimum point of the total cost which is the sum of the construction cost and customer outage cost. The DMR* can be used by the generation and transmission planner to determine the total budget required to construct the required generators and transmission lines at optimal reliability level under their coordination.

16.3.8 Conclusion

This study introduces a new methodology for deciding the optimal reliability criterion for the composite power system expansion planning considering the generation system as well as the transmission system. A deterministic reliability index, delivery marginal rate (DMR) is used in this study. The optimal reliability criterion, DMR*, for a power system is located at the minimum cost point of the total cost curve, which is the sum of the utility cost associated with construction and the customer outage costs associated with supply interruptions. A case study using the five areas of simplified Korea power system shows that an optimal deterministic reliability criterion of the composite power system considering the coordination of grid owner and generation company can be determined successfully. The case study shows that the generator's site can be determined using the proposed method and tool in coordination. As expected, the optimal reliability level goes up as the IEAR increases. The procedure proposed in this chapter is able to provide a quantitative assessment of the relationship between the optimal DMR and IEAR.

17

Probabilistic Reliability-Based Expansion Planning with Wind Turbine Generators

17.1 Introduction

As a result of being environment-conscious, the utilization of renewable resources such as wind and solar energy to generate electricity has received considerable attention in recent years. Wind energy, in particular, has been fast-growing and is recognized as the most successful energy source among all the alternative energy sources. The location of wind turbine generators (WTGs) depends on the available wind speed conditions. Therefore, grid-constrained reliability evaluation has become very important for grid expansion planning and operation when WTGs are added to a power system. However, it is difficult to simulate the operation of WTG and evaluate its reliability contribution due to small effective load-carrying capability (ELCC) which means "how much can a generator cover load?"

The reason is large uncertainty of resource supply for WTG.

The uncertainties in the power system may be categorized into two kinds as like aleatory uncertainty and epistemic uncertainty. Former includes the forced outage rate of a generator, transmission lines, and main transformers, etc. The latter includes lack of information occurred in load forecasting, wind speed, and solar radiation, etc. Therefore, the latter occurs at renewable power plants while the former occurs at conventional power plants. Figure 17.1 shows differences in uncertainties of renewable energy resource power and conventional plants.

Composite power system reliability evaluation is complex and WTG cannot be adequately represented by simple two-state models. Since the two-state models create large errors in accuracy, multistate models are required for WTG. Relevant wind speed models are combined with WTG characteristics to create multistate WTG models. Research has been carried out to evaluate the reliability of generation systems (HLI; Hierarchical Level I) while including WTG by using multistate WTG models. As major WTG penetration levels are expected in the near future, the development of a methodology for evaluating the wind-integrated composite power system reliability has become an important and necessary task to accomplish.

Probabilistic Power System Expansion Planning with Renewable Energy Resources and Energy Storage Systems, First Edition. Jaeseok Choi and Kwang Y. Lee.

Published 2022 by John Wiley & Sons, Inc.

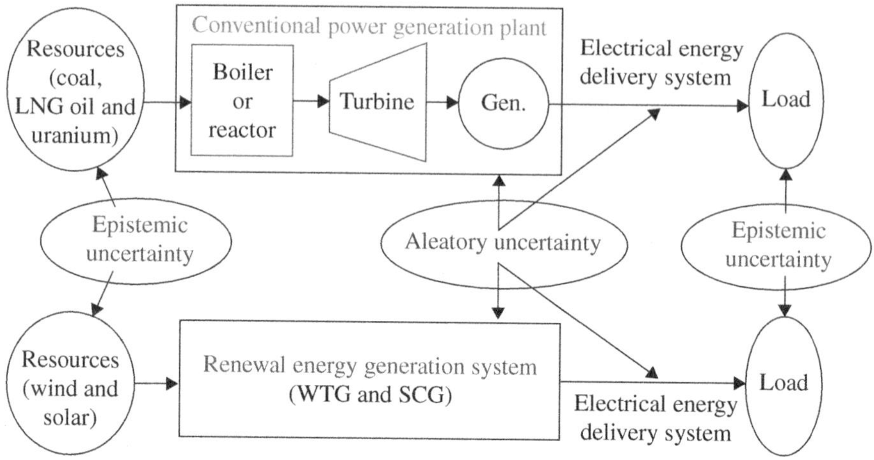

Figure 17.1 Differences of uncertainties of renewable energy resource power and conventional plants.

This chapter proposes a new methodology for grid-constrained probabilistic reliability evaluation of power system expansion including WTG. The reliability evaluation methodology based on the composite power system effective load model without WTG has been already developed in Part I. This methodology utilizes two-state models for both generators and transmission lines. But, this paper proposes a model extended from the two-state representation to multistate models for generators in order to consider WTGs. By using the proposed model, this chapter illustrates a grid expansion planning considering not only capacity limitations and uncertainties of transmission lines and generators, but also the uncertainty of power outputs of WTGs. The simulation methodology requires a nodal probabilistic model in order to implement the reliability evaluation of the composite power system including WTGs. This chapter describes the effectiveness of the proposed method through a case study with Jeju Island's power system in Korea.

17.2 The Multistate Operation Model of WTG

17.2.1 WTG Power Output Model

Figure 17.2 shows the relationship between the power output of a WTG and the wind speed, where V_{ci} is the cut-in speed (m/s), V_R the rated speed (m/s), V_{co} the cut-out speed (m/s), and P_R is the rated power (MW).

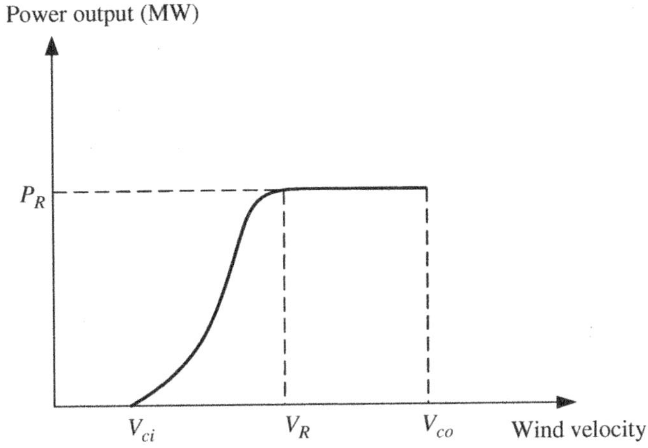

Figure 17.2 A typical power output curve for a WTG.

A mathematical model for the power output of a WTG is given by (17.1) [26, 30–33]. The power P_i generated by wind speed band SW_i is formulated in (17.1), where i is the wind speed index. The A, B, and C parameters are given by (17.A.1)–(17.A.3).

$$\begin{aligned} P_i &= 0, 0 \le SW_i < V_{ci} \\ &= P_R\left(A + B \times SW_i + C \times SW_i^2\right), V_{ci} \le SW_i < V_R \\ &= P_R, V_R \le SW_i \le V_{co} \\ &= 0, V_{co} < SW_i \end{aligned} \tag{17.1}$$

17.2.2 Wind Speed Model

Wind speeds vary in both time and space. It has been reported that the actual wind speed distribution is described by a Weibull probability distribution close to a normal distribution.

This chapter uses the normal probability distribution function (*pdf*) to model the wind speed in terms of the mean wind speed value μ and the standard deviation σ as shown in Figure 17.3. The negative wind speed in Figure 17.3 has no physical meaning and can be considered as zero wind speed.

17.2.3 The Multistate Model of WTG using Normal Probability Distribution Function

The power output curve of the WTG is combined with the wind speed model shown in Figure 17.4 to create the multistate WTG model. Each state has a pair

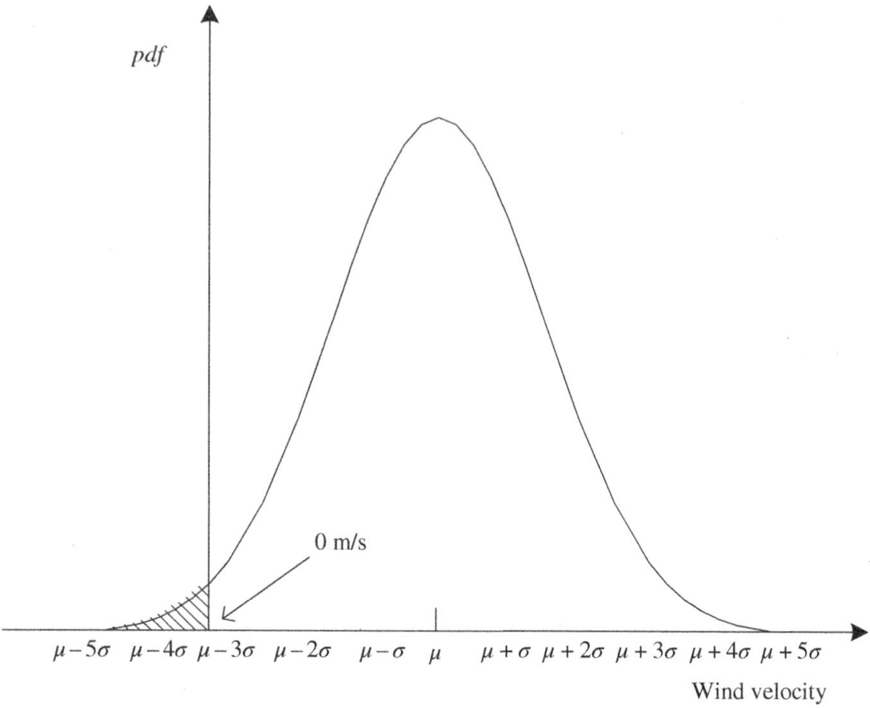

Figure 17.3 Wind speed model.

of associated parameters, namely the power (P_i) and probability (PB_i) of #i wind band, where PB_i is the probability of #*i* wind speed band (SW_i) as shown in Figure 17.4. Because this paper uses the normal probability distribution function (*pdf*) in terms of the mean wind speed value μ and the standard deviation σ, the probability (PB_i) corresponding to wind speed band (SW_i) can be calculated as

$$PB_i = \int_{CSW_{i-1}}^{CSW_i} \frac{1}{\sigma\sqrt{2\pi}} \exp\left(-\frac{(x-\mu)^2}{2\sigma^2}\right) dx \tag{17.2}$$

where

μ: mean of wind speed,
σ: standard deviation of wind speed,
$CSW_i = SW_1 + SW_2 + \cdots + SW_i$

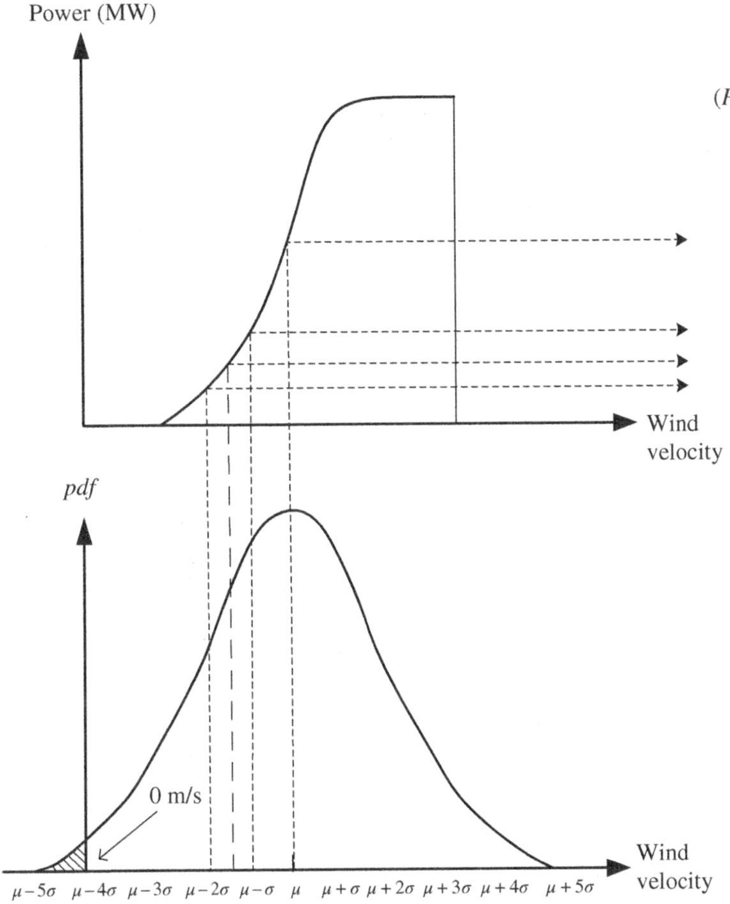

Figure 17.4 Components of a model describing the power output states of a WTG and the corresponding probabilities.

The operating model of a WTG is a multistate model which is described by a capacity outage probability density function. The number of multistates depends on the capacity of WTG. More states permit better accuracy. Even if the number increases over a multistate number, however, the accuracy is increased no more and is saturated at a state number. While the accuracy is saturated, computation time is exhausted more. Sensitivity analysis according to number variation, typically from 2 states to 100 states of multistates, should be evaluated first.

17.3 Reliability Evaluation of a Composite Power System with WTG

Composite power system reliability evaluation methodologies based on both enumeration and Monte Carlo methods have been developed. A composite power system reliability evaluation methodology based on the composite power system effective load model has also been developed. This methodology uses two-state models to describe the generators and transmission lines. The reliability indices for composite power systems (HLII; Hierarchical Level II) can be classified as load point indices and bulk system indices depending upon the objective of the evaluation. They can be evaluated using a Composite power system Equivalent Load Duration Curve (CMELDC) based on the composite power system effective load model in Figure 17.5. Parameters, CG, CT, g_o, and q_l in Figure 17.4 are the capacities and capacity outage density functions of the generators and the unavailability of the transmission lines, respectively. The model uses the capacity outage density functions of the WTG and considers them as multistate generators. The state model for transmission lines remains a two-state one.

The capacity outage pdf of the synthesized fictitious generator created by generators $I - i$ at load point k ($_k f_{osi}$) is also a multistate function. The convolution integral involving the original load duration curve at load point k ($_k\Phi_0$) and $_k f_{osi}$ is processed at HLII. The general multistate convolution integral calculation method for probabilistic reliability evaluation has been used extensively for generation expansion and can be calculated using the multistate recursive equation shown by

$$\begin{aligned} {}_k\Phi_i &= {}_k\Phi_{i-1} \otimes {}_k f_{osi} \\ &= \left(1 - \sum_{n=1}^{NS} {}_k q_{ni}\right) {}_k\Phi_{i-1}(x) + \sum_{n=1}^{NS} {}_k q_{ni} {}_k\Phi_{i-1}(x - {}_k C_{ni}) \end{aligned} \quad (17.3)$$

where,

$\otimes$: operator representing the convolution integral,
$_k\Phi_0$: original load duration curve at load point k,
$_k f_{osi}$: outage capacity pdf of the synthesized fictitious generator created by generators 1 to i, at load point k,
x: random variable of Φ,
NS: the multi-state number of the synthesized fictitious generator,
$_k C_{ni}$: outage capacity of state n of the synthesized fictitious generator created by the generator at load point k,

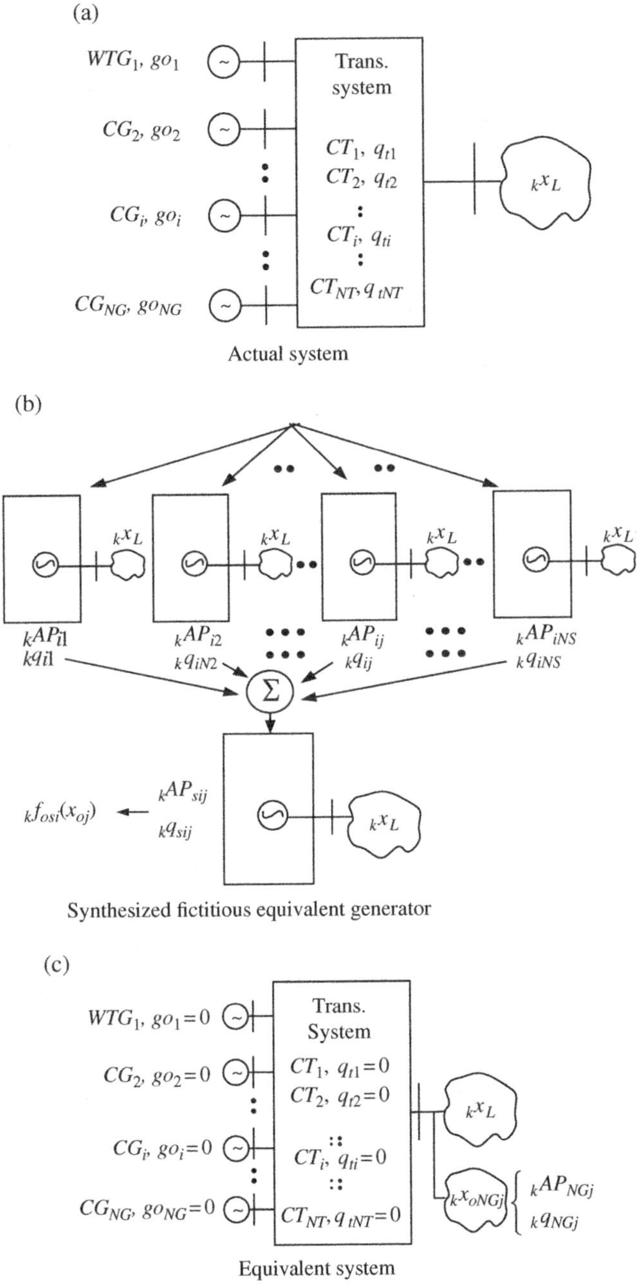

Figure 17.5 Composite power system effective load model at HLII.

${}_k q_{ni}$: the probability corresponds to the outage capacity of state k of the synthesized fictitious generator at load point k

17.3.1 Reliability Indices at Load Buses

The load point reliability indices, $LOLE_k$ and $EENS_k$, can be calculated using (17.4) and (17.5) using the CMELDC, ${}_k\Phi_{NG}(x)$, and EIR (Energy Index of Reliability) is calculated as (17.6), where $DENG_k$ is demand energy at load point #k.

$$LOLE_k = {}_k\Phi_{NG}(x)|_{x=AP_k}\ (\mathrm{hr/yr}) \tag{17.4}$$

$$EENS_k = \int_{AP_k}^{AP_k + Lp_k} {}_k\Phi_{NG}(x)dx\ (\mathrm{MWh/yr}) \tag{17.5}$$

$$EIR_k = 1 - EENS_k/DENG_k\ (\mathrm{p.u.}) \tag{17.6}$$

where

$DENG_k$: demand energy at bus k,
L_{pk}: peak load at load point k (MW),
AP_k: maximum arrival power at load point k (MW)

17.3.2 System Reliability Indices

The $EENS_{HLII}$ of a bulk system is equal to the summation of the $EENS_k$ at the load points as given by (17.7). The $LOLE$ of a bulk system, however, is entirely different from the summation of the $LOLE_k$ at the load points. The ELC_{HLII} (expected load curtailed) of a bulk system is equal to the summation of ELC_k at the load points as given by (17.8). An equivalent representative $LOLE_{HLII}$ of a bulk system can be obtained using (17.9), and EIR of the system is calculated as (17.10), where $DENG_{HLII}$ is sum of load point demand energy, $DENG_k$

$$EENS_{HLII} = \sum_{k=1}^{NL} EENS_k \quad (\mathrm{MWh/yr}) \tag{17.7}$$

$$ELC_{HLII} = \sum_{k=1}^{NL} ELC_k\ (\mathrm{MWh/yr}) \tag{17.8}$$

$$LOLE_{HLII} = EENS_{HLII}/ELC_{HLII}\ (\mathrm{hr/yr}) \tag{17.9}$$

$$EIR_{HLII} = 1 - EENS_{HLII}/DENG_{HLII}\ (\mathrm{p.u.}) \tag{17.10}$$

where

NL: number of load points,
ELC_k: expected load curtailed ($=EENS_k/\ LOLE_k$)

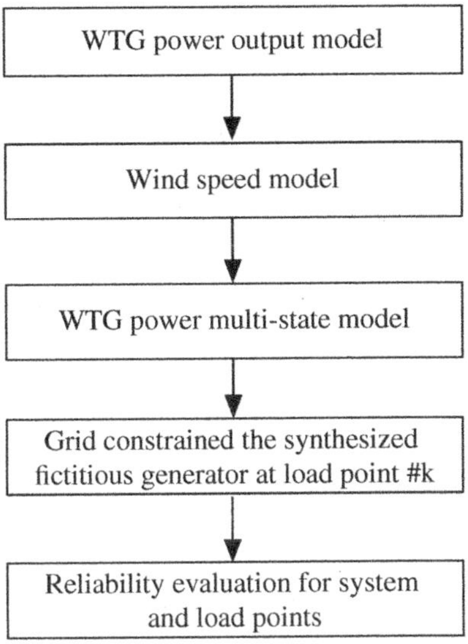

Figure 17.6 The flow chart for reliability evaluation of a power system involving WTGs.

Figure 17.6 shows a flow chart of the proposed method to evaluate the reliability of the grid including WTGs.

17.4 Case Study

The proposed method was implemented on the Jeju power system in 2012. Figure 17.7 shows the Jeju power system, Table 17.1 shows bus number matching bus name, and Table 17.2 shows the generation system in Jeju power system. The HVDC between the main land and Jeju power system can be modeled as an equivalent generator. Figure 17.8 shows the system's load variation curve-pattern in 2007. It is assumed that the load pattern in 2012 is the same as that of 2007. In 2012, the system's generation capacity and peak load will have reached 1140 and 681 MW, respectively. Table 17.3 shows the transmission system of Jeju power system. In Jeju Island, three wind farms are constructed at three different locations, namely Hangwon (HWN), Sungsan (SSN), and Hanlim (HLM). Data related to the three wind farms are given in Table 17.4. The parameters A, B, and C of the power generation function (17.1) of WTG, calculated using (17.A.1)–(17.A.3) for the three wind farms are presented in Table 17.5.

The outage capacity probability distribution function (OCPDF) considering a 5-state, a 7-state, and an 11-state models for the three wind farms is obtained as shown in Figures 17.9, 17.10, and 17.11, respectively. The number of multistates depends on the capacity of WTG. More states permit better accuracy. Even if the

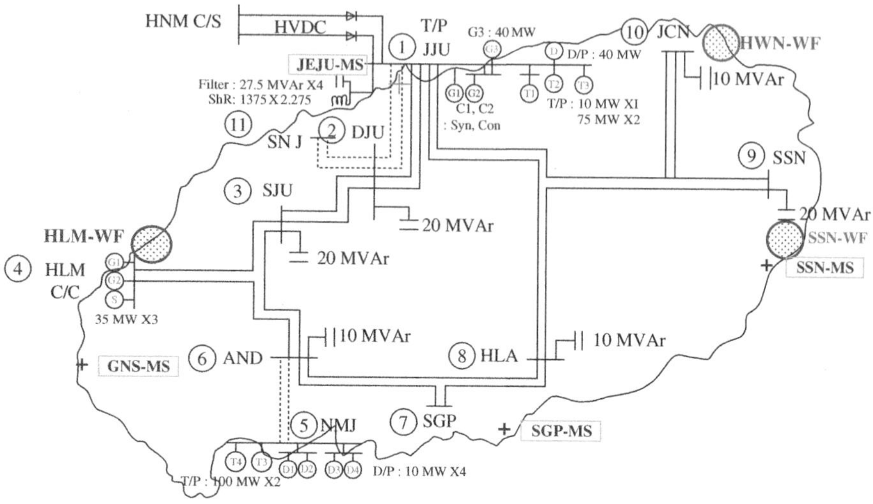

Figure 17.7 Power system of Jeju Island.

Table 17.1 Bus number matched bus name in Figure 17.7.

Bus name	JJM	DJU	SJU	HLM	NMJ	AND	SGP	HLA	SSN	JCN	SNJ
Bus number	1	2	3	4	5	6	7	8	9	10	11

number increases over a multistate number, however, the accuracy is increased no more and is saturated at a certain state number. While the accuracy is saturated, computation time is exhausted more. Sensitivity analysis according to number of states (from 2 states to 100 states) in Jeju power system at HLI was performed previously. The saturation curve of accuracy according to number of states of WTGs in Jeju power system at HLI is shown in Figure 17.A.1. In order to provide more stable accuracy, 11 states at HWN, 7 states at SSN, and 5 states at HLM are selected in this case study.

Tables 17.6 and 17.7 show the calculated LOLE, EENS, and EIR reliability indices for buses and system, respectively.

Table 17.8 shows comparison of reliability indices for five assumed scenarios of transmission plans in Jeju system. The result yields that the plan 2 is the best plan among the given five scenarios. The five scenarios (plans) are assumed entirely in order to evaluate the proposed method.

Table 17.2 Data of Jeju's power system (year 2012).

Bus name	Unit name	Type	Capacity (MW)	Unit and circuit	FOR
10	JCN1	WTG	50	1	–
9	SSN2	WTG	30	1	–
4	HLM3	WTG	20	1	–
1	HVDC	HVDC	150	2	0.028
5	NMJ3	T/P	100	2	0.012
1	JJU1	T/P	55	3	0.015
1	JJU2	T/P	75	2	0.012
4	HNM1	G/T	35	2	0.013
4	HNM1	S/T	35	1	0.013
1	JJU3	D/P	40	2	0.018
5	NMJ1	D/P	10	4	0.018
Total			1140	21	-

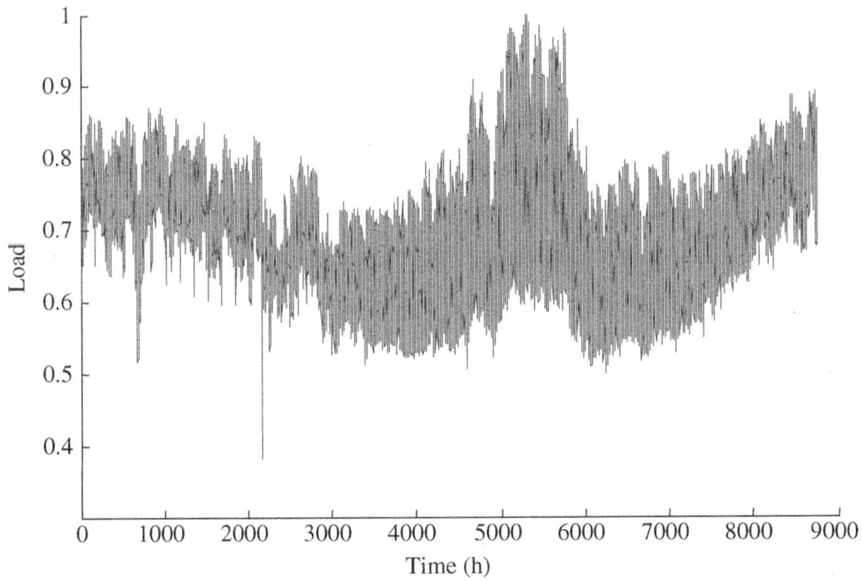

Figure 17.8 The load variation curve (pattern) of Jeju Island power system in 2007.

Table 17.3 Transmission system of Jeju power system.

Line #	SB	EB	Capacity (MW)	Type	Circuit	FOR	Remark
1	1	2	300	T/L	2	0.001 713	
2	1	2	200	C/L	2	0.001 000	
3	2	3	200	T/L	2	0.005 710	
4	3	4	200	T/L	1	0.001 142	
5	4	6	200	T/L	1	0.001 142	
6	3	6	200	T/L	1	0.001 142	
7	6	7	200	T/L	1	0.001 142	
8	5	6	226	C/L	2	0.001 000	
9	7	8	200	T/L	1	0.005 710	
10	6	8	200	T/L	1	0.001 142	
11	8	9	200	T/L	1	0.004 568	
12	8	1	200	T/L	1	0.001 142	
13	1	10	200	T/L	1	0.004 568	
14	10	9	200	T/L	1	0.001 142	
15	2	11	220	C/L	1	0.001 000	

Table 17.4 Data of HWN, SSN, and HLM wind farms.

Wind data			
Wind farm	**HWN-WF**	**SSN-WF**	**HLM-WF**
Peak speed	45 m/s	40 m/s	35 m/s
Mean wind speed	8.5 m/s	7.6 m/s	6.4 m/s
Standard deviation	7 m/s	6 m/s	5 m/s
WTG data			
WTG capacity	**50 MW**	**30 MW**	**20 MW**
Cut-in speed (V_{ci})	5 m/s	5 m/s	5 m/s
Rated speed (V_R)	16 m/s	15 m/s	14 m/s
Cut-out speed (V_{co})	25 m/s	25 m/s	25 m/s

Table 17.5 The *A*, *B*, and *C* parameters of WTGs.

	HWN-WF	SSN-WF	HLM-WF
A	0.1203	0.1111	0.0928
B	−0.06 $(m/s)^{-1}$	−0.063 $(m/s)^{-1}$	−0.0649 $(m/s)^{-1}$
C	0.0072 $(m/s)^{-2}$	0.0081 $(m/s)^{-2}$	0.0093 $(m/s)^{-2}$

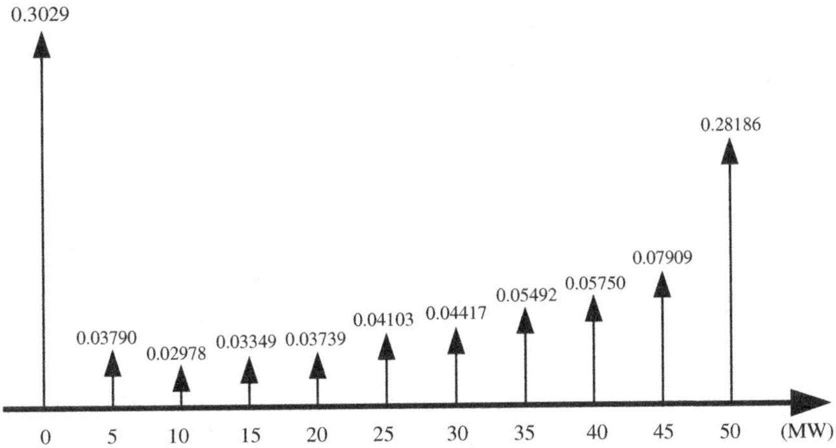

Figure 17.9 OCPDF of HWN wind farm (11-state model).

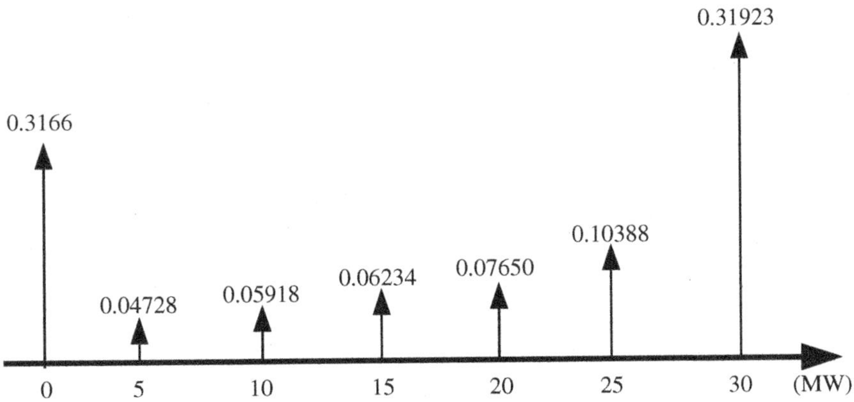

Figure 17.10 OCPDF of SSN wind farm (7-state model).

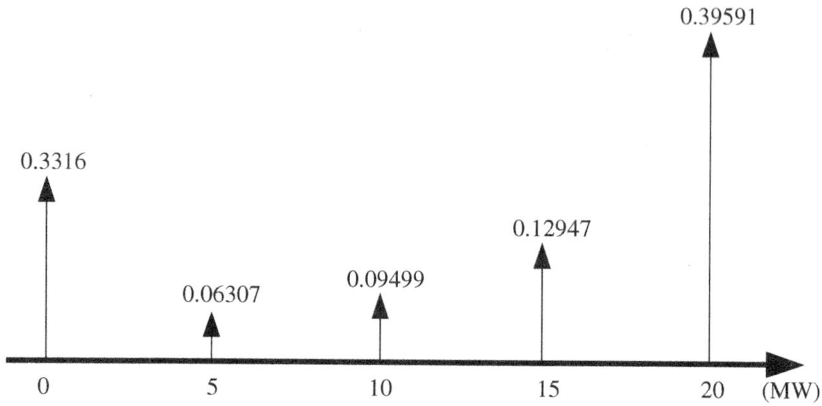

Figure 17.11 OCPDF of HLM wind farm (5-state model).

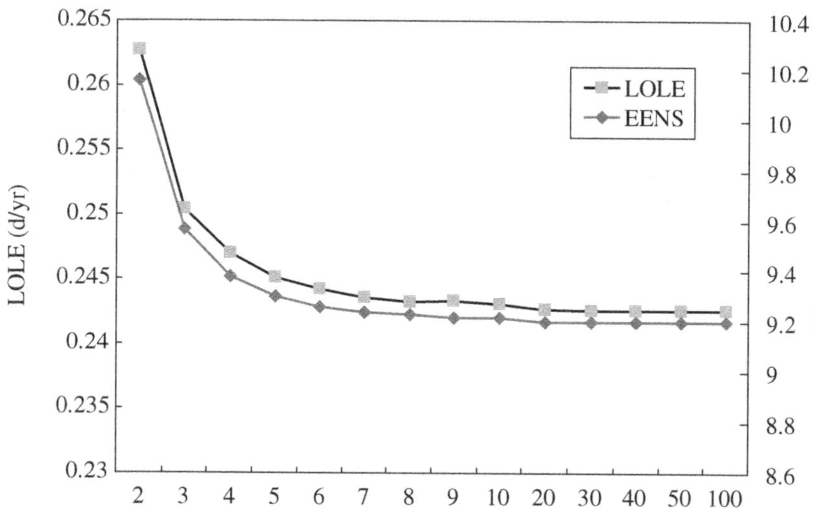

Figure 17.A.1 Saturation of accuracy according to number variation of multistates of WTGs in Jeju power system at HLI.

Table 17.6 Reliability indices at buses of Jeju system.

Load point #	Load name	L_p (MW)	LOLE (h/yr)	EENS (MWh/yr)	EIR (p.u.)
1 (bus 1)	JEJU	200	7.890 42	1 371.88	0.999 099
2 (bus 3)	SIJU	230	7.981 18	1 633.79	0.999 092
3 (bus 7)	SEGP	100	7.980 39	680.604	0.999 092
4 (bus 8)	HALA	50	7.980 19	340.301	0.999 092
5 (bus 9)	SUSN	30	7.980 88	213.103	0.999 092
6 (bus 4)	HALM	20	7.978 15	136.121	0.999 092
7 (bus 11)	SAJI	100	7.980 39	680.604	0.999 092

Table 17.7 System reliability indices of Jeju system.

	Grid not constrained	Grid-constrained case
LOLE (h/d)	1.26	7.955 98
EENS (MWh/day)	44.52	5 056.40
EIR (p.u.)	0.999 99	0.999 094

Table 17.8 Comparison of reliability indices for five transmission expansion plans of Jeju power system.

Plan # (scenario)	Addition item	LOLE (h/yr)	EENS (MWh/yr)	EIR (p.u.)
Plan 1	HVDC 150 × 2 circuit	7.802	4 919.73	0.999 111
Plan 2	T_{2-11} (200 × 1circuit)	0.112	62.10	0.999 989
Plan 3	T_{1-8} T_{8-7}	7.898	5 020.82	0.999 100
Plan 4	T_{5-6}	7.948	5 052.95	0.999 095
Plan 5	T_{1-2} T_{2-3} T_{3-4}	7.947	5 052.42	0.999 095

17.5 Conclusion

This paper presents a new method for grid expansion planning considering the probabilistic reliability of a composite power system with WTGs. The new model results from an extension of the composite power system effective load model including WTG. The proposed model utilizes a multistate operation model of WTG obtained by combining the wind speed model and the WTG's power output model. Test results on a Jeju power system demonstrate the effectiveness of the proposed method to perform grid expansion planning considering wind turbine generators. Only five scenarios are assumed in order to evaluate the proposed method in this chapter. From effective use of the proposed method in the case study, however, actual grid expansion planning problem involving WTGs can be evaluated for more scenarios.

It is expected that the proposed method can provide useful information in determining grid expansion planning and operation according to the location and capacity of WTGs in view of the composite power system reliability.

17.A Appendix

The parameters A, B, and C in (17.1) are given by (17.A.1)–(17.A.3).

$$A = \frac{1}{(V_{ci} - V_R)^2}\left[V_{ci}(V_{ci} + V_R) - 4(V_{ci}V_R)\left(\frac{V_{ci} + V_R}{2V_R}\right)^3\right] \tag{17.A.1}$$

$$B = \frac{1}{(V_{ci} - V_R)^2}\left[4(V_{ci} + V_R)\left(\frac{V_{ci} + V_R}{2V_R}\right)^3 - (3V_{ci} + V_R)\right] \tag{17.A.2}$$

$$C = \frac{1}{(V_{ci} - V_R)^2}\left[2 - 4\left(\frac{V_{ci} + V_R}{2V_R}\right)^3\right] \tag{17.A.3}$$

18

Probabilistic Reliability-Based HVDC Expansion Planning with Wind Turbine Generators

This chapter describes the probabilistic reliability-based grid expansion planning of high-voltage direct current (HVDC) in a power system including wind turbine generators. A new HVDC terminal point can be determined at the best reliability level. This chapter introduces two case studies. One is the NORDIC system and another is Jeju island-sized model power system in Korea. The case study considers the outage cost for HVDC expansion planning [37–44].

18.1 The Status of HVDC

HVDC transmission can be classified according to the three basic HVDC converter technologies in use:

- Line Commutated Converters (LCC)
- Voltage Source Converters (VSC)
- Capacitive Commutated Converters (CCC)

HVDC Scheme has two configurations:

- Monopole Configuration
- Bipole Configuration

The terms "monopole" and "bipole" refer to the use of one or two high-voltage DC polarities to interconnect the converters of an HVDC bulk power transmission scheme with the following:

Probabilistic Power System Expansion Planning with Renewable Energy Resources and Energy Storage Systems, First Edition. Jaeseok Choi and Kwang Y. Lee.

Published 2022 by John Wiley & Sons, Inc.

- The configuration with metallic return uses a low-voltage DC conductor to carry the returning DC.
- The bipole configuration can be regarded as being made up of two monopoles.

Various kinds of HVDC scheme configuration are presented in Figures 18.1–18.7. Figure 18.8 is a photo of the Jeju pilot HVDC in Jeju island.

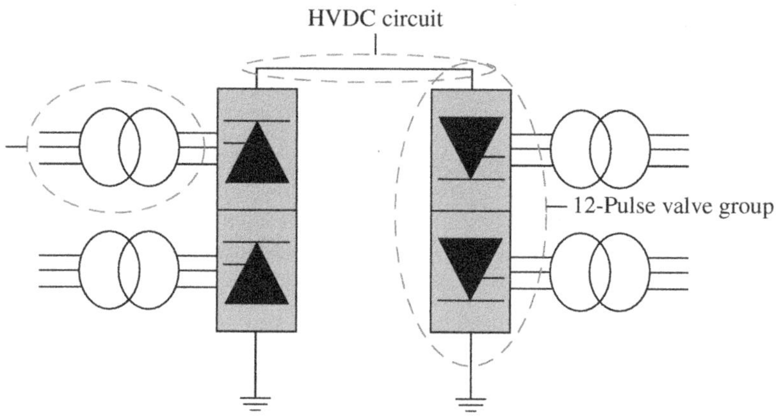

Figure 18.1 Monopole configuration with a ground return.

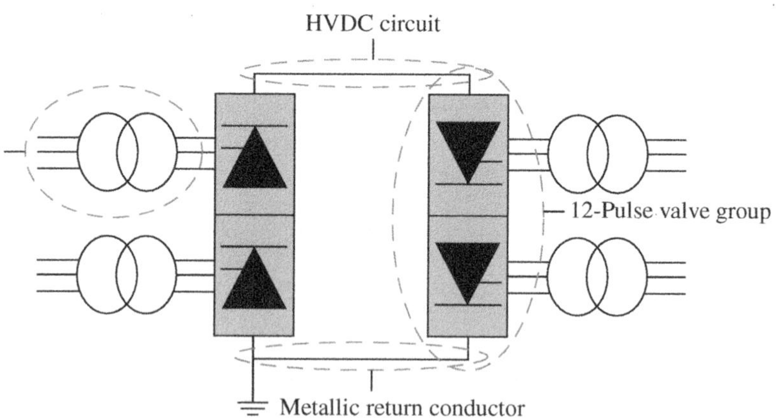

Figure 18.2 Monopole configuration with metallic conductor return.

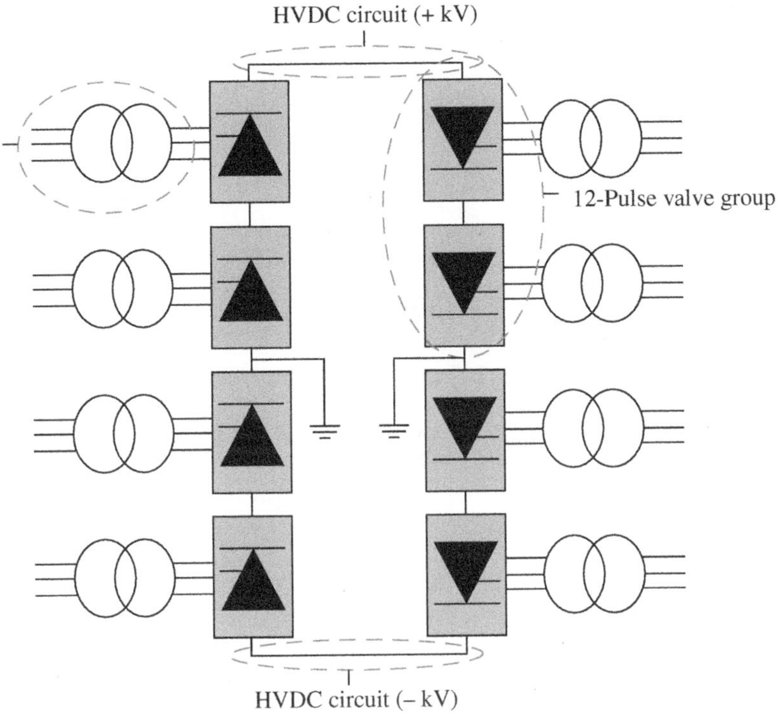

Figure 18.3 Bipole configuration with a ground return.

18.2 HVDC Technology for Energy Efficiency and Grid Reliability

Classic HVDC is an excellent technology for transmitting large amounts of electricity.

Example:

A 2000 km-long line rated at 800 kV and 6000 MW

- HVDC loses the electricity by about 5% but HVAC by about 10%
- HVDC needs just one power line versus three lines for an AC link of equivalent power and length

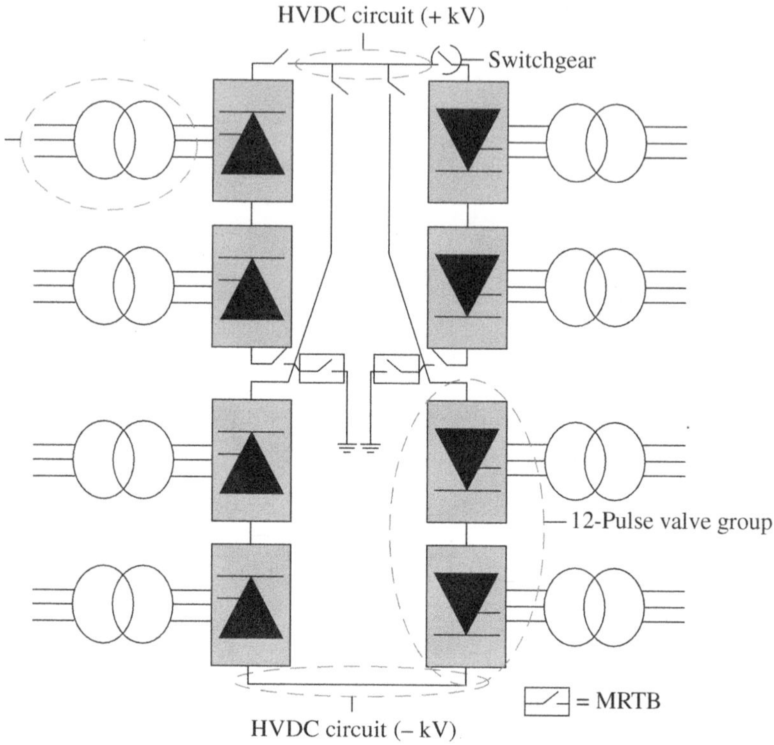

Figure 18.4 Bipole configuration with metallic conductor return.

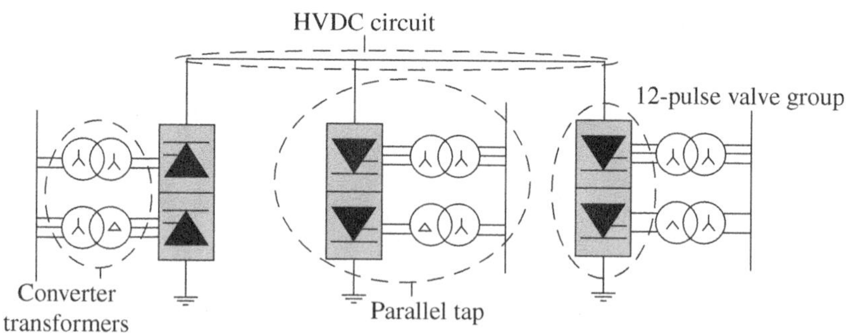

Figure 18.5 Simplified configuration of a parallel tap MT HVDC scheme.

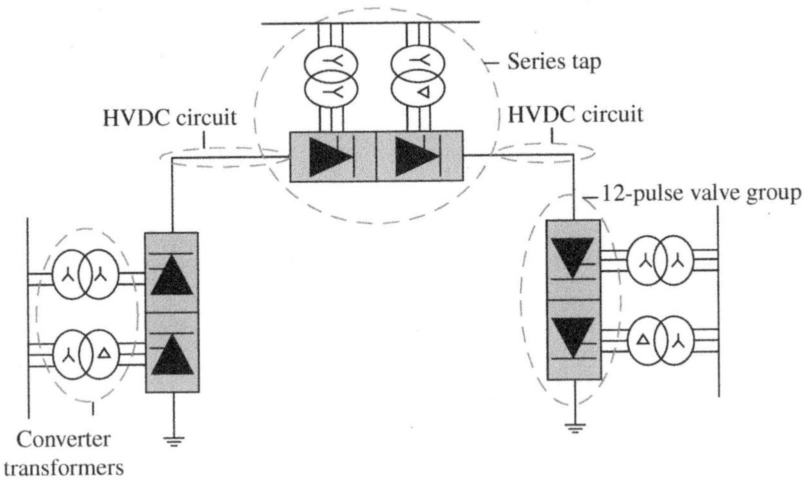

Figure 18.6 Simplified configuration of a series tap MT HVDC scheme.

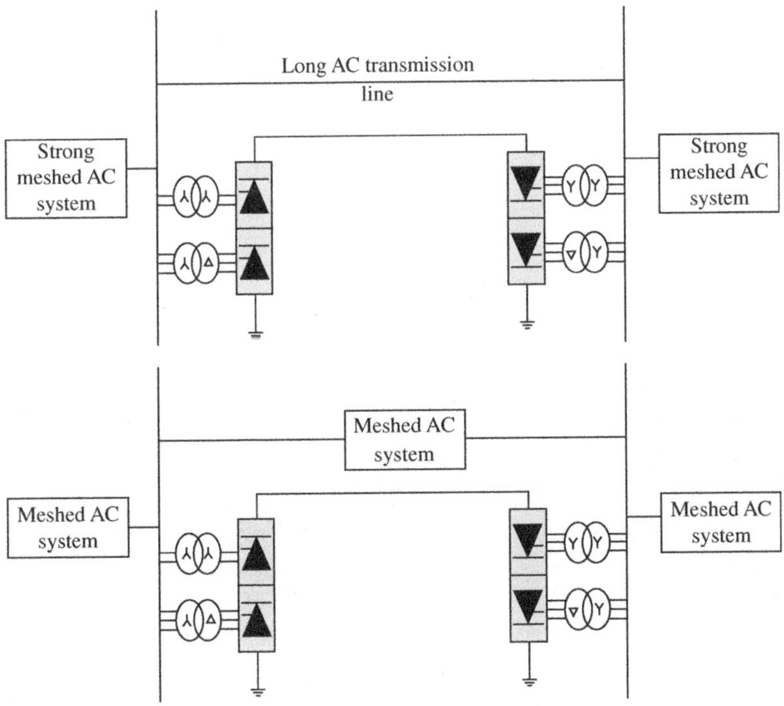

Figure 18.7 Embedded HVDC schemes.

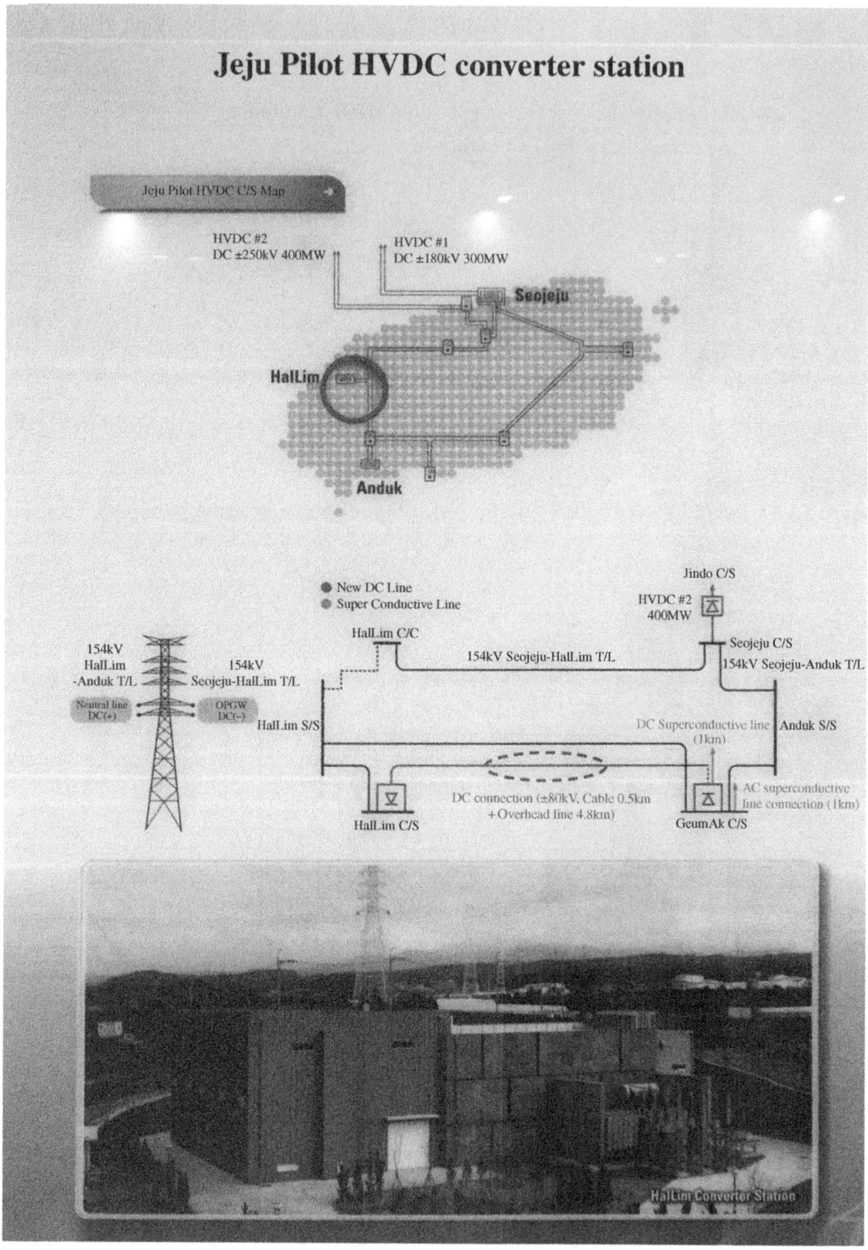

Figure 18.8 Jeju pilot HVDC converter station. *Source*: Photoed by Dr. Jaeseok Choi.

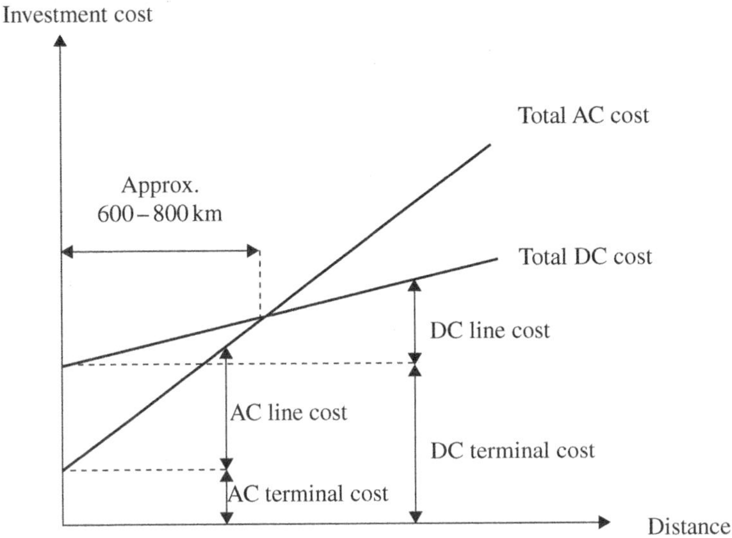

Figure 18.9 HVDC vs. HVAC cost.

- Increasing long-distance transmission efficiency, stabilizing the transmission network, and preventing the cascading outages
- The cost-benefit of DC over AC transmission is apparent over long distances, typically more than 600 km for overhead lines and more than 50 km for underground or underwater cables (Figure 18.9)
- HVDC technology is also used to connect asynchronous (incompatible) AC power networks, increasing the efficiency and stability of each.

18.3 HVDC Impacts on Transmission System Reliability

Figure 18.10 shows a reliability model of HVDC.

- The failure and repair rate of the converter transformer λ_{ct}, r_{ct}, λ_{cs}, and r_{cs}
- The failure rate and repair rate of the DC cable λ_{DCL} and r_{DCL}

18.4 Case Study

Two case studies are presented in this section: Nordic system and Jeju Island system.

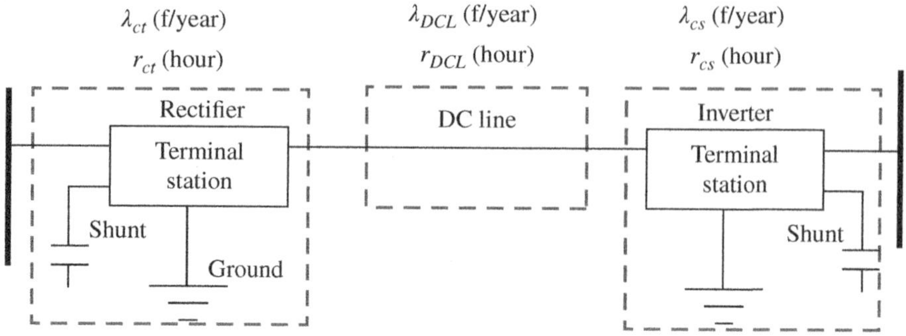

Figure 18.10 Reliability model of HVDC.

1) Nordic Power System

Figure 18.11 shows the Nordic system considered for a case study in this section, where 9 case studies are performed as defined in Table 18.1:

- Some HVDC cases will be proposed and added to the Nordic32.
- Checking how system reliability will be impacted.

Reliability Input Data:
The failure rate and repair time of components, respectively:

- The generating unit: 0.1691 (1/year) and 32.7 hour/failure
- The reactor (on 400 kV level): 0.354 (1/year) and 39.5 hour/failure
- The shunt capacitor: 0.005 12 (1/year) and 10 hour/failure
- The converter transformer: 0.0153 (1/year) and 1664 hour/failure
- The DC cable: 0.000 781 (1/year) and 68.8 hour/failure

The failure rate and repair time of the 400 kV transmission lines are tabulated in Table 18.2.

Case 1: An HVDC link between bus 4032 and 4043 (adding). The connected HVDC is a monopolar link between North area and Central area. The total system indices of the conventional AC system as a base case is given in Table 18.3 and the total system indices of Case 1 is summarized in Table 18.4. The reliability indices of all cases are tabulated in Table 18.5.

2) Jeju Island Power System

This section presents a case study for HVDC expansion planning in Jeju Island-sized Model Power System in Korea (Figure 18.12). Wind power construction planning is shown in Figure 18.13. The new HVDC construction is planned to connect the island to the main Korean peninsular as shown in Figure 18.14.

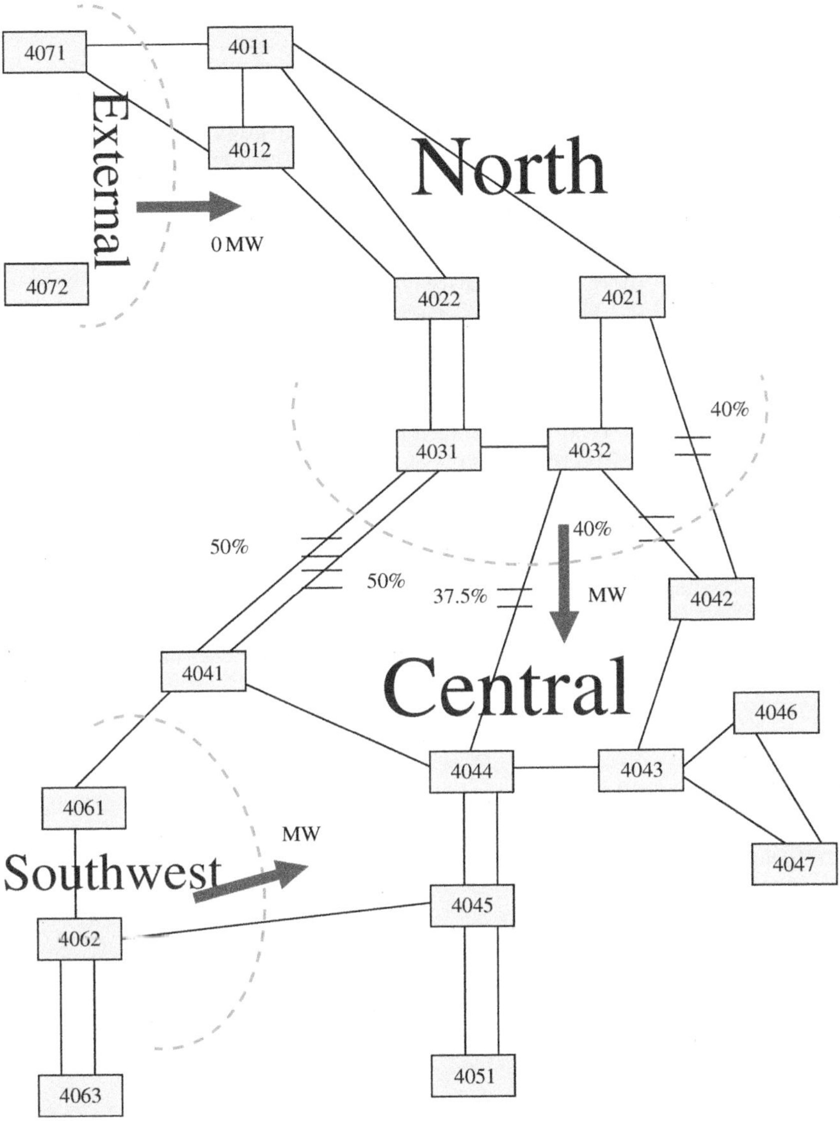

Figure 18.11 Case study: Nordic system. *Source*: Refs. [42–44] © John Wiley & Sons.

Table 18.1 Cases of Nordic HVDC case studies.

Cases	
1	An HVDC link between bus 4032 and 4043 (Adding)
2	An HVDC link between bus 4021 and 4042 (Adding)
3	An HVDC link between bus 4021 and 4042 (Replacing)
4	An HVDC link between bus 4031 and 4042 (Adding)
5	An HVDC link between bus 4032 and 4042 (Adding)
6	An HVDC link between bus 4032 and 4042 (Replacing)
7	An HVDC link between bus 4032 and 4044 (Adding)
8	An HVDC link between bus 4032 and 4044 (Replacing)
9	An HVDC link between bus 4032 and 4043 (Replacing) and HVDC link between 4021 and 4042 (Replacing)

Table 18.2 The failure rate and repair time of the 400 kV transmission lines.

Connected from Bus	Length (km)	Failure (1/year)	Repair time (hour)
4031 To bus 4041	770	3.92	38.89
4031 To bus 4041	770	3.92	38.89
4032 To bus 4044	810	44.13	40.91
4032 To bus 4042	580	2.95	29.29
4021 To bus 4042	1000	5.1	50.51
4042 To bus 4044	190	0.96	9.60
4042 To bus 4043	140	0.71	7.07
4022 To bus 4031	370	1.88	18.69
4022 To bus 4031	370	1.88	18.69
4021 To bus 4032	380	1.93	19.19
4012 To bus 4022	285	1.453	14.39
4011 To bus 4021	535	2.728	27.02
4011 To bus 4022	320	1.63	16.16
4011 To bus 4012	100	0.51	5.05
4071 To bus 4011	380	1.93	19.19
4071 To bus 4012	420	2.142	21.21
4071 To bus 4012	300	1.53	15.15

Table 18.2 (Continued)

Connected from Bus	Length (km)	Failure (1/year)	Repair time (hour)
4071 To bus 4012	300	1.53	15.15
4044 To bus 4043	70	0.357	3.54
4043 To bus 4046	60	0.306	3.03
4046 To bus 4047	100	0.51	5.05
4043 To bus 4047	130	0.663	6.57
4044 To bus 4045	170	0.867	8.59
4044 To bus 4045	170	0.867	8.59
4045 To bus 4051	280	1.428	14.14
4045 To bus 4051	280	1.428	14.14
4062 To bus 4045	760	3.876	38.39
4041 To bus 4044	260	1.326	13.13
4041 To bus 4061	395	2.0145	19.95
4061 To bus 4062	130	0.663	6.57
4062 To bus 4063	230	1.173	11.62
4062 To bus 4063	269	1.173	11.62
4031 To bus 4032	93.7	0.408	4.04

Table 18.3 Total system indices of the conventional AC system (Base Case).

Index	Unit	Value	Description
N	–	22	Total number of customers served
SAIFI	1/year	0.537	System average interruption duration index
SAIDI	Minute/year	530.339	Customer average interruption duration index
CAIDI	Hour	16.475	Average service availability index
ASAI	%	99.899	System load interruption frequency
F	1/year	1.609	System load interruption mean duration
T	Hour	16.718	System load interruption probability
Q	Minute/year	1614.476	Total interrupted load power
P	MW/year	6949.589	Total load energy not supplied
W	MWh/year	114 597.339	Total load energy not supplied

Table 18.4 Total system indices of Case 1.

Index	Unit	Value	Description
N	–	22	Total number of customers served
SAIFI	1/year	0.732	System average interruption frequency index
SAIDI	Minute/year	42.7169	System average interruption duration index
CAIDI	Hour	9.7314	Customer average interruption duration index
ASAI	%	99.9919	Average service availability index
F	1/year	0.1888	System load interruption frequency
T	Hour	9.2892	System load interruption mean duration
Q	Minute/year	105.2046	System load interruption probability
P	MW/year	949.8795	Total interrupted load power
W	MWh/year	9236.336	Total load energy not supplied

Table 18.5 Reliability indices of all cases.

No	HVDC cases	*SII* (%)	*SAIFI*	*SAIDI*	*CAIDI*	*ASAI*	*P* (MW/year)	*T* (hour)	*F* (1/year)
0	Without HVDC (Base case)	94.60	0.537	530.399	16.475	99.899	6949.589	16.718	1.609
1	HVDC btn 4032–4043 (added)	98.94	0.0732	42.7169	9.7314	99.9919	949.8795	9.2892	0.1888
2	HVDC btn 4021–4042 (added)	98.96	0.0874	50.1334	9.5619	99.9905	1146.6004	8.9944	0.2372
3	HVDC btn 4021–4042 (replaced)	98.33	0.1098	96.8032	14.6913	99.9816	1446.8168	14.415	0.3129
4	HVDC btn 4031–4042 (added)	99.27	0.039	22.979	9.825	99.996	492.866	9.175	0.145
5	HVDC btn 4032–4042 (added)	98.78	0.0731	42.7516	9.7436	99.9919	949.4004	9.2285	0.1905
6	HVDC btn 4032–4042 (replaced)	96.20	0.0871	77.736	14.8693	99.9852	1139.1299	15.388	0.2323
7	HVDC btn 4032–4044 (added)	98.90	0.0729	42.6608	9.7549	99.9919	946.7034	9.3304	0.1877
8	HVDC btn 4032–4044 (replaced)	98.71	0.1045	115.0563	18.3534	99.9781	1375.2667	18.97	0.279
9	HVDC btn 4032–4044 (replaced)HVDC btn 4021–4042 (replaced)	99.19	0.075	52.834	11.725	99.99	979.713	11.566	0.199

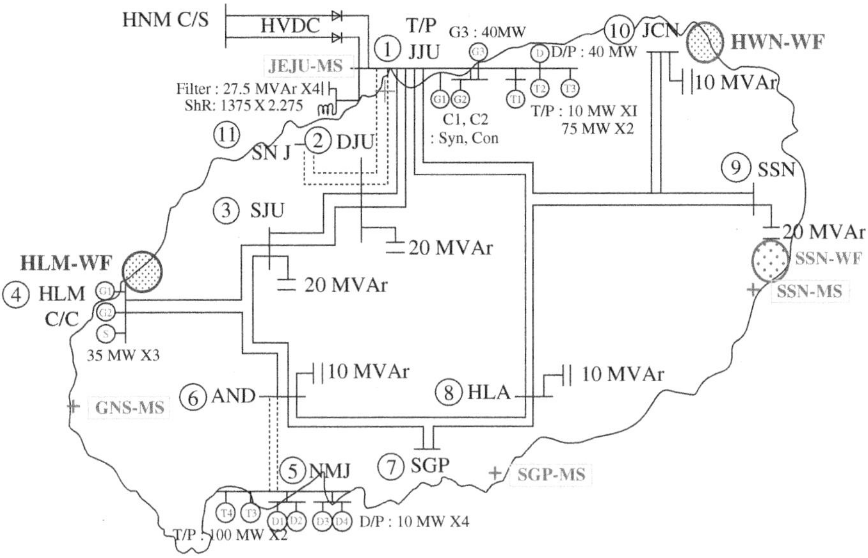

Figure 18.12 Jeju Island-sized model power system diagram.

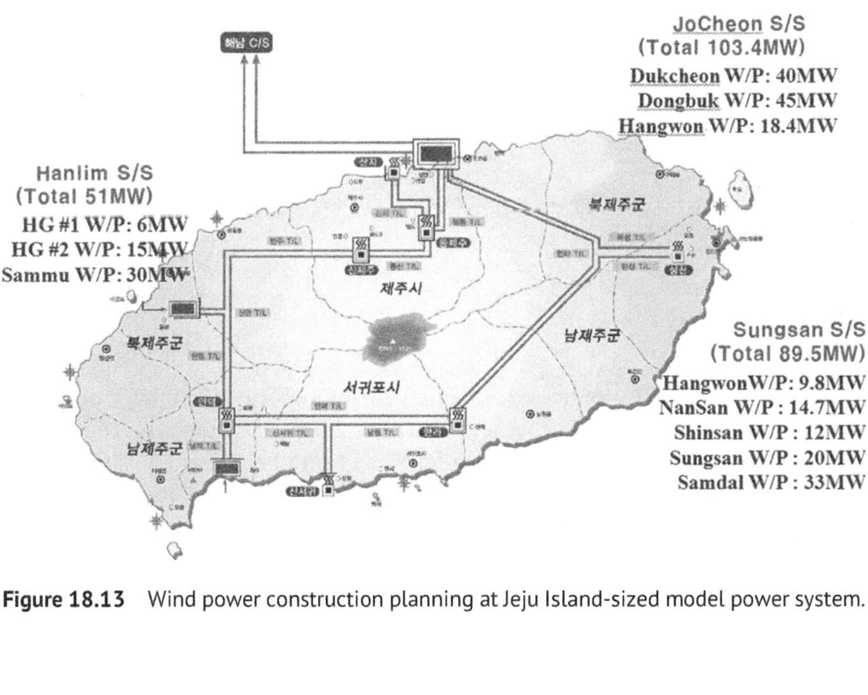

Figure 18.13 Wind power construction planning at Jeju Island-sized model power system.

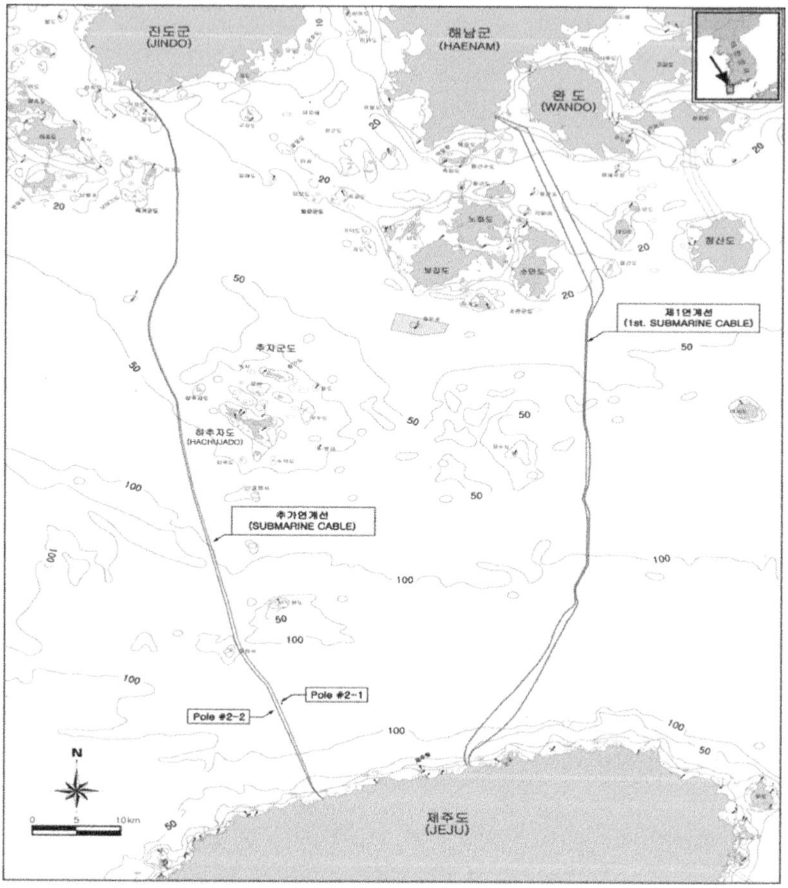

Figure 18.14 New HVDC in Jeju Island-sized model power system diagram application for new HVDC construction.

The system reliability indices are computed for two cases: connect the new HVDC to Bus 1, and connect the new HVDC to Bus 11 as shown in Table 18.6 where both with and without transmission constraints. From the simulation results, the new HVDC should be suggested to be connected with Bus # 11 rather than Bus # 1 as the Bus #11-connected scenario reliability indices, LOLE and EENS, are lower.

Table 18.6 System reliability indices of Jeju system.

	Grid (Transmission) not constrained	Grid (Transmission) constrained	New HVDC+	
			Bus #11	Bus #1
LOLE (hours/day)	1.26	6.893 07	0.285 807	3.552 56
EENS (MWh/day)	44.52	3587.35	142.011	1851.91
EIR (pu)	0.999 99	0.999 222	0.999 969	0.999 598

References

1 Kim, K.J., Park, Y.M., and Lee, K.Y. (1988). Optimal long-term transmission planning based on maximum principle. *IEEE Trans. Power Syst.* 3 (4): 1494–1501.

2 Electric generation expansion analysis system (EGAS)(2014). *EPRI Project RP 1529*, Final Report EL 2561.

3 Park, Y.M., Lee, K.Y., and Youn, L.T.O. (1985). New analytical approach for long-term generation expansion planning based on maximum principle and gaussian distribution function. *IEEE Trans. Power Appr. Syst.* PAS-104 (2): 390–397.

4 Youn, L.T.O., Lee, K.Y., and Park, Y.M. (1986). Optimal long-range generation expansion planning for hydro thermal system based on analytical production costing model. *IEEE/PES Winter Meeting*, #86 WM 080-6.

5 Lee, B.Y., Park, Y.M., and Lee, K.Y. (1986). Optimal generation planning for a thermal system with pumped-storage based on analytical production costing model. *IEEE/PES Summer Meeting*, #86 SM 301-6, July 1986.

6 Lee, K.Y., Park, Y.M., and Ortiz, J.L. (1985). A united approach to optimal real and reactive power dispatch. *IEEE Trans. Power Appr. Syst.* PAS-104 (5): 1147–1153.

7 Lee, K.Y., Oritz, J.L., Park, Y.M., and Pond, L.G. (1986). An optimization technique for reactive power planning of subtransmission network under normal operation. *IEEE Trans. Power Syst.* PWRS-1 (2): 153–159.

8 Fischl, R. (1975). Optimal system expansion: a critical review. *Systems Engineering for Power: Status and Prospects*, DOE, Henniker, New Hempshire, 17–22 August 1975.

9 Monticelli, A., Santos, A., Periera, M.V.F. et al. (1982). Iterative transmission network planning using a least-effort criterion. *IEEE Trans. Power Appr. Syst.* PAS-101: 3919–3925.

10 Billington, R. and Bhavaraju, M.P. (1970). Transmission planning using reliability criterion Part I: a reliability criterion. *IEEE Trans. Power Appr. Syst.* PAS-89 (1): 28–34.

Probabilistic Power System Expansion Planning with Renewable Energy Resources and Energy Storage Systems, First Edition. Jaeseok Choi and Kwang Y. Lee.

Published 2022 by John Wiley & Sons, Inc.

11 Kaltenbach, J., Peschon, J., and Behrig, E.H. (1970). A mathemaatical optimization technique for the expansion of electric power transmission systems. *IEEE Trans. Power Appr. Syst.* PAS -89: 113–119.

12 Carver, L.L. (1970). Transmission network expansion using linear programming. *IEEE Trans. Power Appr. Syst.* PAS-89: 1688–1697.

13 Lee, S.Y., Hicks, K.L., and Hnylicza, E. (1974). Transmission expansion by branch-and-bound integer programming with optimal cost-capacity curves. *IEEE Trans. Power Appr. Syst.* PAS-93: 1390–1400.

14 Fischl, R., and Puntel, W.R. (1972). Computer-aided design of electric power transmission networks. *IEEE PES Winter Power Meeting*, #C72-168-8, January 1972.

15 Puntel, W.R., Reppen, N.D., Ringlee, J.R., Platts, J.E., Ryan, W.A., and Sullivan, P.J. (1973). An automated method for long-range of transmission networks. *PICA Conference Proceedings*, June 1973, pp. 38–46.

16 Adams, R.N. and Laugton, M.A. (1974). Optimal planning of power networks using mixed-integer programming Part-I – static and time-phased network synthesis. *Proc. IEEE* 121 (2): 139–147.

17 Dusondhet, Y.P. and El-Abaid, A. (1973). Transmission planning using discrete dynamic optimization. *IEEE Trans. Power Appr. Syst.* PAS-92 (4): 1358–1371.

18 Henault, R.H., Eastyedt, R.B., Peschon, J., and Hajdu, L.P. (1970). Power system long-term planning in the presence of uncertainty. *IEEE Trans. Power Appr. Syst.* PAS-89: 156–164.

19 Choi, J., Trungtinh, T., El-Keib, A. et al. (2005). A method for transmission system expansion planning considering probabilistic reliability criteria. *IEEE Trans. PS* 20 (3): 1406–1415.

20 Alguacil, N., Motto, A.L., and Conejo, A.J. (2003). Transmission expansion planning: a mixed-integer LP approach. *IEEE Trans. Power Syst.* 18 (3): 1070–1077.

21 Romero, R., Rochay, C., Mantovani, M., and Mantovani, J.R.S. (2007). Evaluation of hybrid models for multistage transmission system planning. *Revista Controle Automação* 18 (1): 106–114, Janeiro, Fevereiro e Março.

22 Kaltenbach, J.-C., Peschon, J., and Gehrig, E.H. (1970). A mathematical optimization technique for the expansion of electric power transmission systems. *IEEE Trans. Power App. Syst.* PAS-89 (1): 113–119.

23 Das, P.K., Chanda, R.S., and Bhattacharjee, P.K. (2005). Combined generation and transmission system expansion planning sing implicit enumeration and linear programming technique. *IE(I) J. EL* L (86): 110–116.

24 Dusonchet, Y.P. and El-Abiad, A. (1973). Transmission planning using discrete dynamic optimizing. *IEEE Trans. Power Appa. Syst.* PAS-92 (4): 1358–1371.

25 Shayeghi, H., Mahdavi, M., and Haddadian, H. (2008). DCGA based-transmission network expansion planning considering network adequacy. *Int. J. Inf. Technol.* 4 (4): 299–304.

26 Tran, T.T., Choi, J., Moon, S., and Billinton, R. (2003). Transmission system expansion planning using a branch and bound approach. *IASTED2003*, EuroPES, Marbella, Spain, 3–5 September 2003.

27 Choi, J., Billinton, R., and Futuhi-Firuzabed, M. (2005). Development of a nodal effective load model considering transmission system element unavailabilities. *IEE Proceedings on GTD* 152 (1): 79–89.

28 Choi, J., Tran, T.T., Kwon, J. et al. (2008). Nodal probabilistic production cost simulation considering transmission system unavailability. *IET Generation, Transmission and Distribution* 2 (1): 34–42.

29 Choi, J., Mount, T., and Thomas, R. (2007). Transmission expansion planning using contingency criteria. *IEEE Trans. PS* 22 (4): 2249–2261.

30 Choi, J., Cha, J., Watada, J., Mount, T.D., and Thomas, R. (2007). Transmission system expansion planning considering outage cost. *ICICIC 2007*, Kumamoto, Kusyu, Japan, 5–7 September 2007.

31 Huiling (Michelle) Bao (2007). Application of the jint deterministic-probabilistic (D-P) creiteria to bulk electric system planning. A thesis for mater of science in the Dept. Electrical Engineering University of Saskatchewan Saskatoon.

32 Choi, J., El-Keib, A.A., and Tran, T. (2005). A fuzzy branch and bound-based transmission system expansion planning for the highest satisfaction level of the decision maker. *IEEE Trans. PS* 20 (1): 479–484.

33 Tran, T., Kwon, J., Choi, J., El-Keib, A.A., and Watada, J. (2007). Application of fuzzy set theory to transmission system expansion planning. *ISME2007*, Waseda University, Kitakyushu, Japan, 10–12 March 2007.

34 Choi, J., Thomas, R., Wang, Z., Rahim, A., El-Keib, A., and Billinton, R. (2005). A study on probabilistic optimal reliability criterion determination in composite power system expansion planning. *IEEE PES GM2005*, San Francisco, CA, 12–16 June 2005.

35 Park, J., Choi, J., Jung, Y., Yun, Y., and Cha, J. (2009). Determination of a deterministic reliability criterion for composite power system expansion planning. *IEEE PES GM2009*, Calgary, AB, 26–30 July 2009.

36 Cho, K., Park, J., and Choi, J. (2012). Probabilistic reliability based grid expansion planning of power system including wind turbine generators. *J. Electr. Eng. Technol.* 7 (5): 698–704.

37 Wang, H. and Redfern, M.A. (2010). The advantages and disadvantages of using HVDC to interconnect AC networks. *UPEC2010*.

38 Tjernberg, L.B. (2008). Introduction to HVDC technology for reliable electrical power systems. *IEEE Xplore*.

39 Li, W. (2006). Expected energy not served (EENS) study for vancouver island transmission reinforcement project. *Report-BCTC*, 9 January 2006.

40 Wang, T. and Guan, L. (2014). Improvements on the hvdc reliability assessment index system considering preventive maintenance. *Eksploatacja I Niezawodnosc – Mainten. Reliab.* 16 (4): 629–636.

41 HVDC: technology for energy efficiency and grid reliability. *ABB*. https://studylib.net/doc/18268252/hvdc--technology-for-energy-efficiency-and-grid-reliability.

42 Brand, L., de Silva, R., Bebbington, E., and Chilukuri, K. (2014). Grid west project HVDC technology review. *PSC*, 17 December 2014.

43 Waterworth, D.A., Arnold, C.P., and Watson, N.R. (1998). Reliability assessment technique for HVdc systems. *IPENZ Trans.* 25 (1): 12–19 EMCh.

44 Swaitti, O. (2007). *Assessing the Impacts of Increasing Penetration of HVDC Lines on Power System Reliability. KTH Royal Institute of Technology.*

45 Baleriaux, H. (1967). Simulation de 1'exploitation d'unpare de machines thermiques de production d'electricite couple a des stations de pompage. *Revue E(edition S.R.B.E.), S.A. EBES, Brussels, Belgium* 5 (7): 1–24.

46 Booth, R.R. (1972). Optimal generation planning considering uncertainty. *IEEE Trans.* PAS-91: 70–77.

47 Stremel, J.P., Jenkins, R.T., Babb, R.A., and Bayless, W.D. (1980). Production costing using the cumulant method of representing the equivalent load curve. *IEEE* PAS-99 (5): 1947–1956.

48 Gros, G., Garapic, N.V., and McNutt, B. (1988). The mixture of normals approximation technique for equivalent load curves. *IEEE Trans. Power Syst.* 3 (2): 368–374.

49 Bellman, R.E. and Zadeh, L.A. (1970). Decision-making in a fuzzy environment. *Manag. Sci.* 17 (4): B141–B164.

50 Lee, K.Y. and El-Sharkawi, M.A. (eds.) (2008). *Modern Heuristic Optimization Techniques: Theory amd Applications to Power Systems*. Wiley-IEEE Press.

Index

Probabilistic Power System Expansion Planning with Renewable Energy Resources and Energy Storage Systems, First Edition. Jaeseok Choi and Kwang Y. Lee.

Published 2022 by John Wiley & Sons, Inc.

h

i

j

k

l

u

v

w

Books in the IEEE Press Series on Power Engineering

Series Editor: ME El-Hawary, Dalhousie University, Halifax, Nova Scotia, Canada

The mission of IEEE Press Series on Power Engineering is to publish leading-edge books that cover the broad spectrum of current and forward-looking technologies in this fast-moving area. The series attracts highly acclaimed authors from industry/academia to provide accessible coverage of current and emerging topics in power engineering and allied fields. Our target audience includes the power engineering professional who is interested in enhancing their knowledge and perspective in their areas of interest.

1. *Electric Power Systems: Design and Analysis, Revised Printing*
Mohamed E. El-Hawary

2. *Power System Stability*
Edward W. Kimbark

3. *Analysis of Faulted Power Systems*
Paul M. Anderson

4. *Inspection of Large Synchronous Machines: Checklists, Failure Identification, and Troubleshooting*
Isidor Kerszenbaum

5. *Electric Power Applications of Fuzzy Systems*
Mohamed E. El-Hawary

6. *Power System Protection*
Paul M. Anderson

7. *Subsynchronous Resonance in Power Systems*
Paul M. Anderson, B.L. Agrawal, J.E. Van Ness

8. *Understanding Power Quality Problems: Voltage Sags and Interruptions*
Math H. Bollen

9. *Analysis of Electric Machinery*
Paul C. Krause, Oleg Wasynczuk, and S.D. Sudhoff

10. *Power System Control and Stability, Revised Printing*
Paul M. Anderson, A.A. Fouad

11. *Principles of Electric Machines with Power Electronic Applications*, Second Edition

Mohamed E. El-Hawary

12. *Pulse Width Modulation for Power Converters: Principles and Practice*
D. Grahame Holmes and Thomas Lipo

13. *Analysis of Electric Machinery and Drive Systems*, Second Edition

Paul C. Krause, Oleg Wasynczuk, and S.D. Sudhoff

14. *Risk Assessment for Power Systems: Models, Methods, and Applications*
Wenyuan Li

Probabilistic Power System Expansion Planning with Renewable Energy Resources and Energy Storage Systems, First Edition. Jaeseok Choi and Kwang Y. Lee.

Published 2022 by John Wiley & Sons, Inc.

15. *Optimization Principles: Practical Applications to the Operations of Markets of the Electric Power Industry*
Narayan S. Rau

16. *Electric Economics: Regulation and Deregulation*
Geoffrey Rothwell and Tomas Gomez

17. *Electric Power Systems: Analysis and Control*
Fabio Saccomanno

18. *Electrical Insulation for Rotating Machines: Design, Evaluation, Aging, Testing, and Repair*
Greg C. Stone, Edward A. Boulter, Ian Culbert, and Hussein Dhirani

19. *Signal Processing of Power Quality Disturbances*
Math H. J. Bollen and Irene Y. H. Gu

20. *Instantaneous Power Theory and Applications to Power Conditioning*
Hirofumi Akagi, Edson H. Watanabe and Mauricio Aredes

21. *Maintaining Mission Critical Systems in a 24/7 Environment*
Peter M. Curtis

22. *Elements of Tidal-Electric Engineering*
Robert H. Clark

23. *Handbook of Large Turbo-Generator Operation and Maintenance,* Second Edition

Geoff Klempner and Isidor Kerszenbaum

24. *Introduction to Electrical Power Systems*
Mohamed E. El-Hawary

25. *Modeling and Control of Fuel Cells: Distributed Generation Applications*
M. Hashem Nehrir and Caisheng Wang

26. *Power Distribution System Reliability: Practical Methods and Applications*
Ali A. Chowdhury and Don O. Koval

27. *Introduction to FACTS Controllers: Theory, Modeling, and Applications*
Kalyan K. Sen, Mey Ling Sen

28. *Economic Market Design and Planning for Electric Power Systems*
James Momoh and Lamine Mili

29. *Operation and Control of Electric Energy Processing Systems*
James Momoh and Lamine Mili

30. *Restructured Electric Power Systems: Analysis of Electricity Markets with Equilibrium Models*
Xiao-Ping Zhang

31. *An Introduction to Wavelet Modulated Inverters*
S.A. Saleh and M.A. Rahman

32. *Control of Electric Machine Drive Systems*
Seung-Ki Sul

33. *Probabilistic Transmission System Planning*
Wenyuan Li

34. *Electricity Power Generation: The Changing Dimensions*
Digambar M. Tagare

35. *Electric Distribution Systems*
Abdelhay A. Sallam and Om P. Malik

36. *Practical Lighting Design with LEDs*
Ron Lenk, Carol Lenk

37. *High Voltage and Electrical Insulation Engineering*
Ravindra Arora and Wolfgang Mosch

38. *Maintaining Mission Critical Systems in a 24/7 Environment, Second Edition*
Peter Curtis

39. *Power Conversion and Control of Wind Energy Systems*
Bin Wu, Yongqiang Lang, Navid Zargari, Samir Kouro

40. *Integration of Distributed Generation in the Power System*
Math H. Bollen, Fainan Hassan

41. *Doubly Fed Induction Machine: Modeling and Control for Wind Energy Generation Applications*
Gonzalo Abad, Jesús López, Miguel Rodrigues, Luis Marroyo, and Grzegorz Iwanski

42. *High Voltage Protection for Telecommunications*
Steven W. Blume

43. *Smart Grid: Fundamentals of Design and Analysis*
James Momoh

44. *Electromechanical Motion Devices, Second Edition*
Paul Krause, Oleg Wasynczuk, Steven Pekarek

45. *Electrical Energy Conversion and Transport: An Interactive Computer-Based Approach, Second Edition*
George G. Karady, Keith E. Holbert

46. *ARC Flash Hazard and Analysis and Mitigation*
J.C. Das

47. *Handbook of Electrical Power System Dynamics: Modeling, Stability, and Control*
Mircea Eremia, Mohammad Shahidehpour

48. *Analysis of Electric Machinery and Drive Systems, Third Edition*
Paul C. Krause, Oleg Wasynczuk, S.D. Sudhoff, Steven D. Pekarek

49. *Extruded Cables for High-Voltage Direct Current Transmission: Advances in Research and Development*
Giovanni Mazzanti, Massimo Marzinotto

50. *Power Magnetic Devices: A Multi-Objective Design Approach*
S.D. Sudhoff

51. *Risk Assessment of Power Systems: Models, Methods, and Applications, Second Edition*
Wenyuan Li

52. *Practical Power System Operation*
Ebrahim Vaahedi

53. *The Selection Process of Biomass Materials for the Production of Bio-Fuels and Co-Firing*
Najib Altawell

54. *Electrical Insulation for Rotating Machines: Design, Evaluation, Aging, Testing, and Repair, Second Edition*
Greg C. Stone, Ian Culbert, Edward A. Boulter, and Hussein Dhirani

55. *Principles of Electrical Safety*
Peter E. Sutherland

56. *Advanced Power Electronics Converters: PWM Converters Processing AC Voltages*
Euzeli Cipriano dos Santos Jr., Edison Roberto Cabral da Silva

57. *Optimization of Power System Operation, Second Edition*
Jizhong Zhu

58. *Power System Harmonics and Passive Filter Designs*
J.C. Das

59. *Digital Control of High-Frequency Switched-Mode Power Converters*
Luca Corradini, Dragan Maksimoviæ, Paolo Mattavelli, Regan Zane

60. *Industrial Power Distribution, Second Edition*
Ralph E. Fehr, III

61. *HVDC Grids: For Offshore and Supergrid of the Future*
Dirk Van Hertem, Oriol Gomis-Bellmunt, Jun Liang

62. *Advanced Solutions in Power Systems: HVDC, FACTS, and Artificial Intelligence*
Mircea Eremia, Chen-Ching Liu, Abdel-Aty Edris

63. *Operation and Maintenance of Large Turbo-Generators*
Geoff Klempner, Isidor Kerszenbaum

64. *Electrical Energy Conversion and Transport: An Interactive Computer-Based Approach*
George G. Karady, Keith E. Holbert

65. *Modeling and High-Performance Control of Electric Machines*
John Chiasson

66. *Rating of Electric Power Cables in Unfavorable Thermal Environment*
George J. Anders

67. *Electric Power System Basics for the Nonelectrical Professional*
Steven W. Blume

68. *Modern Heuristic Optimization Techniques: Theory and Applications to Power Systems*
Kwang Y. Lee, Mohamed A. El-Sharkawi

69. *Real-Time Stability Assessment in Modern Power System Control Centers*
Savu C. Savulescu

70. *Optimization of Power System Operation*
Jizhong Zhu

71. *Insulators for Icing and Polluted Environments*
Masoud Farzaneh, William A. Chisholm

72. *PID and Predictive Control of Electric Devices and Power Converters Using MATLAB®/ Simulink®*
Liuping Wang, Shan Chai, Dae Yoo, Lu Gan, Ki Ng

73. *Power Grid Operation in a Market Environment: Economic Efficiency and Risk Mitigation*
Hong Chen

74. *Electric Power System Basics for Nonelectrical Professional, Second Edition*
Steven W. Blume

75. *Energy Production Systems Engineering*
Thomas Howard Blair

76. *Model Predictive Control of Wind Energy Conversion Systems*
Venkata Yaramasu, Bin Wu

77. *Understanding Symmetrical Components for Power System Modeling*
J.C. Das

78. *High-Power Converters and AC Drives, Second Edition*
Bin Wu, Mehdi Narimani

79. *Current Signature Analysis for Condition Monitoring of Cage Induction Motors: Industrial Application and Case Histories*
William T. Thomson, Ian Culbert

80. *Introduction to Electric Power and Drive Systems*
Paul Krause, Oleg Wasynczuk, Timothy O'Connell, Maher Hasan

81. *Instantaneous Power Theory and Applications to Power Conditioning, Second Edition*
Hirofumi, Edson Hirokazu Watanabe, Mauricio Aredes

82. *Practical Lighting Design with LEDs, Second Edition*
Ron Lenk, Carol Lenk

83. *Introduction to AC Machine Design*
Thomas A. Lipo

84. *Advances in Electric Power and Energy Systems: Load and Price Forecasting*
Mohamed E. El-Hawary

85. *Electricity Markets: Theories and Applications*
Jeremy Lin, Jernando H. Magnago

86. *Multiphysics Simulation by Design for Electrical Machines, Power Electronics and Drives*
Marius Rosu, Ping Zhou, Dingsheng Lin, Dan M. Ionel, Mircea Popescu, Frede Blaabjerg, Vandana Rallabandi, David Staton

87. *Modular Multilevel Converters: Analysis, Control, and Applications*
Sixing Du, Apparao Dekka, Bin Wu, Navid Zargari

88. *Electrical Railway Transportation Systems*
Morris Brenna, Federica Foiadelli, Dario Zaninelli

89. *Energy Processing and Smart Grid*
James A. Momoh

90. *Handbook of Large Turbo-Generator Operation and Maintenance, 3rd Edition*
Geoff Klempner, Isidor Kerszenbaum

91. *Advanced Control of Doubly Fed Induction Generator for Wind Power Systems*
Dehong Xu, Dr. Frede Blaabjerg, Wenjie Chen, Nan Zhu

92. *Electric Distribution Systems, 2nd Edition*
Abdelhay A. Sallam, Om P. Malik

93. *Power Electronics in Renewable Energy Systems and Smart Grid: Technology and Applications*
Bimal K. Bose

94. *Distributed Fiber Optic Sensing and Dynamic Rating of Power Cables*
Sudhakar Cherukupalli, and George J Anders

95. *Power System and Control and Stability, Third Edition*
Vijay Vittal, James D. McCalley, Paul M. Anderson, and A. A. Fouad

96. *Electromechanical Motion Devices: Rotating Magnetic Field-Based Analysis and Online Animations, Third Edition*
Paul Krause, Oleg Wasynczuk, Steven D. Pekarek, and Timothy O'Connell

97. *Applications of Modern Heuristic Optimization Methods in Power and Energy Systems*
Kwang Y. Lee and Zita A. Vale

98. *Handbook of Large Hydro Generators: Operation and Maintenance*
Glenn Mottershead, Stefano Bomben, Isidor Kerszenbaum, and Geoff Klempner

99. *Advances in Electric Power and Energy: Static State Estimation*
Mohamed E. El-hawary

100. *Arc Flash Hazard Analysis and Mitigation, Second Edition*
J.C. Das

101. *Maintaining Mission Critical Systems in a 24/7 Environment, Third Edition*
Peter M. Curtis

102. *Probabilistic Power System Expansion Planning with Renewable Energy Resources and Energy Storage Systems*
Jaeseok Choi and Kwang Y. Lee

Printed and bound by CPI Group (UK) Ltd, Croydon, CR0 4YY